Numerical Methods for Chemical Engineering

Suitable for a first-year graduate course, this textbook unites the applications of numerical mathematics and scientific computing to the practice of chemical engineering. Written in a pedagogic style, the book describes basic linear and nonlinear algebraic systems all the way through to stochastic methods, Bayesian statistics, and parameter estimation. These subjects are developed at a nominal level of theoretical mathematics suitable for graduate engineers. The implementation of numerical methods in MATLAB® is integrated within each chapter and numerous examples in chemical engineering are provided, together with a library of corresponding MATLAB programs. Although the applications focus on chemical engineering, the treatment of the topics should also be of interest to non-chemical engineers and other applied scientists that work with scientific computing. This book will provide the graduate student with the essential tools required by industry and research alike.

Supplementary material includes solutions to homework problems set in the text, MATLAB programs and tutorial, lecture slides, and complicated derivations for the more advanced reader. These are available online at www.cambridge.org/9780521859714.

KENNETH J BEERS has been Assistant Professor at MIT since 2000. He has taught extensively across the engineering discipline at both the undergraduate and graduate level. This book is a result of the successful course the author devised at MIT for numerical methods applied to chemical engineering.

Numerical Methods for Chemical Engineering

Applications in MATLAB®

KENNETH J. BEERS
Massachusetts Institute of Technology

CAMBRIDGE
UNIVERSITY PRESS

CAMBRIDGE UNIVERSITY PRESS
Cambridge, New York, Melbourne, Madrid, Cape Town, Singapore, São Paulo

Cambridge University Press
The Edinburgh Building, Cambridge CD2 2RU, UK

Published in the United States of America by Cambridge University Press, New York

www.cambridge.org
Information on this title: www.cambridge.org/9780521859714

First published 2007

Printed in the United Kingdom at the University Press, Cambridge

A catalog record for this publication is available from the British Library

ISBN-13 978-0-521-85971-4 hardback
ISBN-10 0-521-85971-9 hardback

Contents

Preface

This text focuses on the application of quantitative analysis to the field of chemical engineering. Modern engineering practice is becoming increasingly more quantitative, as the use of scientific computing becomes ever more closely integrated into the daily activities of all engineers. It is no longer the domain of a small community of specialist practitioners. Whereas in the past, one had to hand-craft a program to solve a particular problem, carefully husbanding the limited memory and CPU cycles available, now we can very quickly solve far more complex problems using powerful, widely-available software. This has introduced the need for research engineers and scientists to become computationally literate – to know the possibilities that exist for applying computation to their problems, to understand the basic ideas behind the most important algorithms so as to make wise choices when selecting and tuning them, and to have the foundational knowledge necessary to navigate independently through the literature.

This text meets this need, and is written at the level of a first-year graduate student in chemical engineering, a consequence of its development for use at MIT for the course 10.34, "Numerical methods applied to chemical engineering." This course was added in 2001 to the graduate core curriculum to provide all first-year Masters and Ph.D. students with an overview of quantitative methods to augment the existing core courses in transport phenomena, thermodynamics, and chemical reaction engineering. Care has been taken to develop any necessary material specific to chemical engineering, so this text will prove useful to other engineering and scientific fields as well. The reader is assumed to have taken the traditional undergraduate classes in calculus and differential equations, and to have some experience in computer programming, although not necessarily in MATLAB$^{\circledR}$.

Even a cursory search of the holdings of most university libraries shows there to be a great number of texts with titles that are variations of "Advanced Engineering Mathematics" or "Numerical Methods." *So why add yet another?*

I find that there are two broad classes of texts in this area. The first focuses on introducing numerical methods, applied to science and engineering, at the level of a junior or senior undergraduate elective course. The scope is necessarily limited to rather simple techniques and applications. The second class is targeted to research-level workers, either higher graduate-level applied mathematicians or computationally-focused researchers in science and engineering. These may be either advanced treatments of numerical methods for mathematicians, or detailed discussions of scientific computing as applied to a specific subject such as fluid mechanics.

Neither of these classes of text is appropriate for teaching the fundamentals of scientific computing to beginning chemical engineering graduate students. Examples should be typical of those encountered in graduate-level chemical engineering research, and while the students should gain an understanding of the basis of each method and an appreciation of its limitations, they do not need exhaustive theory-proof treatments of convergence, error analysis, etc. It is a challenge for beginning students to identify how their own problems may be mapped into ones amenable to quantitative analysis; therefore, any appropriate text should have an extensive library of worked examples, with code available to serve later as templates. Finally, the text should address the important topics of model development and parameter estimation. This book has been developed with these needs in mind.

This text first presents a fundamental discussion of linear algebra, to provide the necessary foundation to read the applied mathematical literature and progress further on one's own. Next, a broad array of simulation techniques is presented to solve problems involving systems of nonlinear algebraic equations, initial value problems of ordinary differential and differential-algebraic (DAE) systems, optimizations, and boundary value problems of ordinary and partial differential equations. A treatment of matrix eigenvalue analysis is included, as it is fundamental to analyzing these simulation techniques.

Next follows a detailed discussion of probability theory, stochastic simulation, statistics, and parameter estimation. As engineering becomes more focused upon the molecular level, stochastic simulation techniques gain in importance. Particular attention is paid to Brownian dynamics, stochastic calculus, and Monte Carlo simulation. Statistics and parameter estimation are addressed from a Bayesian viewpoint, in which Monte Carlo simulation proves a powerful and general tool for making inferences and testing hypotheses from experimental data.

In each of these areas, topically relevant examples are given, along with MATLAB (www.mathworks.com) programs that serve the students as templates when later writing their own code. An accompanying website includes a MATLAB tutorial, code listings of all examples, and a supplemental material section containing further detailed proofs and optional topics. Of course, while significant effort has gone into testing and validating these programs, no guarantee is provided and the reader should use them with caution.

The problems are graded by difficulty and length in each chapter. Those of grade A are simple and can be done easily by hand or with minimal programming. Those of grade B require more programming but are still rather straightforward extensions or implementations of the ideas discussed in the text. Those of grade C either involve significant thinking beyond the content presented in the text or programming effort at a level beyond that typical of the examples and grade B problems.

The subjects covered are broad in scope, leading to the considerable (though hopefully not excessive) length of this text. The focus is upon providing a fundamental understanding of the underlying numerical algorithms without necessarily exhaustively treating all of their details, variations, and complexities of use. Mastery of the material in this text should enable first-year graduate students to perform original work in applying scientific computation to their research, and to read the literature to progress independently to the use of more sophisticated techniques.

Writing a book is a lengthy task, and one for which I have enjoyed much help and support. Professor William Green of MIT, with whom I taught this course for one semester, generously shared his opinions of an early draft. The teaching assistants who have worked on the course have also been a great source of feedback and help in problem-development, as have, of course, the students who have wrestled with intermediate drafts and my evolving approach to teaching the subject. My Ph.D. students Jungmee Kang, Kirill Titievskiy, Erik Allen, and Brian Stephenson have shown amazing forbearance and patience as the text became an additional, and sometimes demanding, member of the group. Above all, I must thank my family, and especially my supportive wife Jen, who have been tracking my progress and eagerly awaiting the completion of the book.

1 Linear algebra

This chapter discusses the solution of sets of linear algebraic equations and defines basic vector/matrix operations. The focus is upon elimination methods such as Gaussian elimination, and the related LU and Cholesky factorizations. Following a discussion of these methods, the existence and uniqueness of solutions are considered. Example applications include the modeling of a separation system and the solution of a fluid mechanics boundary value problem. The latter example introduces the need for sparse-matrix methods and the computational advantages of banded matrices. Because linear algebraic systems have, under well-defined conditions, a unique solution, they serve as fundamental building blocks in more-complex algorithms. Thus, linear systems are treated here at a high level of detail, as they will be used often throughout the remainder of the text.

Linear systems of algebraic equations

We wish to solve a system of N simultaneous linear algebraic equations for the N unknowns $x_1, x_2, \ldots, x_N$, that are expressed in the general form

$$
\begin{aligned}
a_{11}x_1 + a_{12}x_2 + \cdots + a_{1N}x_N &= b_1 \\
a_{21}x_1 + a_{22}x_2 + \cdots + a_{2N}x_N &= b_2 \\
&\vdots \\
a_{N1}x_1 + a_{N2}x_2 + \cdots + a_{NN}x_N &= b_N
\end{aligned}
\tag{1.1}
$$

a_{ij} is the constant coefficient (assumed real) that multiplies the unknown x_j in equation i. b_i is the constant "right-hand-side" coefficient for equation i, also assumed real. As a particular example, consider the system

$$
\begin{aligned}
x_1 + x_2 + x_3 &= 4 \\
2x_1 + x_2 + 3x_3 &= 7 \\
3x_1 + x_2 + 6x_3 &= 2
\end{aligned}
\tag{1.2}
$$

for which

$$
\begin{array}{llll}
a_{11} = 1 & a_{12} = 1 & a_{13} = 1 & b_1 = 4 \\
a_{21} = 2 & a_{22} = 1 & a_{23} = 3 & b_2 = 7 \\
a_{31} = 3 & a_{32} = 1 & a_{33} = 6 & b_3 = 2
\end{array}
\tag{1.3}
$$

It is common to write linear systems in *matrix/vector form* as

$$A\boldsymbol{x} = \boldsymbol{b} \tag{1.4}$$

where

$$A = \begin{bmatrix} a_{11} & a_{12} & a_{13} & \cdots & a_{1N} \\ a_{21} & a_{22} & a_{23} & \cdots & a_{2N} \\ \vdots & \vdots & \vdots & & \vdots \\ a_{N1} & a_{N2} & a_{N3} & \cdots & a_{NN} \end{bmatrix} \qquad \boldsymbol{x} = \begin{bmatrix} x_1 \\ x_2 \\ \vdots \\ x_N \end{bmatrix} \qquad \boldsymbol{b} = \begin{bmatrix} b_1 \\ b_2 \\ \vdots \\ b_N \end{bmatrix} \tag{1.5}$$

Row i of A contains the values $a_{i1}, a_{i2}, \ldots, a_{iN}$ that are the coefficients multiplying each unknown $x_1, x_2, \ldots, x_N$ in equation i. Column j contains the coefficients $a_{1j}, a_{2j}, \ldots, a_{Nj}$ that multiply x_j in each equation $i = 1, 2, \ldots, N$. Thus, we have the following associations,

$$\text{rows} \Leftrightarrow \text{equations} \qquad \text{columns} \Leftrightarrow \begin{array}{c} \text{coefficients multiplying} \\ \text{a specific unknown} \\ \text{in each equation} \end{array}$$

We often write $A\boldsymbol{x} = \boldsymbol{b}$ explicitly as

$$\begin{bmatrix} a_{11} & a_{12} & \cdots & a_{1N} \\ a_{21} & a_{22} & \cdots & a_{2N} \\ \vdots & \vdots & & \vdots \\ a_{N1} & a_{N2} & \cdots & a_{NN} \end{bmatrix} \begin{bmatrix} x_1 \\ x_2 \\ \vdots \\ x_N \end{bmatrix} = \begin{bmatrix} b_1 \\ b_2 \\ \vdots \\ b_N \end{bmatrix} \tag{1.6}$$

For the example system (1.2),

$$A = \begin{bmatrix} 1 & 1 & 1 \\ 2 & 1 & 3 \\ 3 & 1 & 6 \end{bmatrix} \qquad \boldsymbol{b} = \begin{bmatrix} 4 \\ 7 \\ 2 \end{bmatrix} \tag{1.7}$$

In MATLAB we solve $A\boldsymbol{x} = \boldsymbol{b}$ with the single command, **x = A\b**. For the example (1.2), we compute the solution with the code

```
A = [1 1 1; 2 1 3; 3 1 6];
b = [4; 7; 2];
x = A\b,
x =
    19.0000
    -7.0000
    -8.0000
```

Thus, we are tempted to assume that, as a practical matter, we need to know little about how to solve a linear system, as someone else has figured it out and provided us with this handy linear solver. Actually, we shall need to understand the fundamental properties of linear systems in depth to be able to master methods for solving more complex problems, such as sets of nonlinear algebraic equations, ordinary and partial differential equations, etc. Also, as we shall see, this solver fails for certain common classes of very large systems of equations, and we need to know enough about linear algebra to diagnose such situations and to propose other methods that do work in such instances. This chapter therefore contains not only an explanation of how the MATLAB solver is implemented, but also a detailed, fundamental discussion of the properties of linear systems.

Our discussion is intended only to provide a foundation in linear algebra for the practice of numerical computing, and is continued in Chapter 3 with a discussion of matrix eigenvalue analysis. For a broader, more detailed, study of linear algebra, consult Strang (2003) or Golub & van Loan (1996).

Review of scalar, vector, and matrix operations

As we use vector notation in our discussion of linear systems, a basic review of the concepts of vectors and matrices is necessary.

Scalars, real and complex

Most often in basic mathematics, we work with *scalars*, i.e., single-valued numbers. These may be real, such as $3, 1.4, 5/7, 3.14159\ldots.$, or they may be complex, $1 + 2i, 1/2\,i$, where $i = \sqrt{-1}$. The *set of all real scalars* is denoted $\Re$. The *set of all complex scalars* we call C. For a complex number $z \in C$, we write $z = a + ib$, where $a, b \in \Re$ and

$$a = Re\{z\} = \text{real part of } z$$
$$b = \text{Im}\{z\} = \text{imaginary part of } z \tag{1.8}$$

The *complex conjugate*, $\bar{z} = z^*$, of $z = a + ib$ is

$$\bar{z} = z^* = a - ib \tag{1.9}$$

Note that the product $\bar{z}z$ is always real and nonnegative,

$$\bar{z}z = (a - ib)(a + ib) = a^2 - iab + iab - i^2 b^2 = a^2 + b^2 \geq 0 \tag{1.10}$$

so that we may define the real-valued, nonnegative *modulus* of z, $|z|$, as

$$|z| = \sqrt{\bar{z}z} = \sqrt{a^2 + b^2} \geq 0 \tag{1.11}$$

Often, we write complex numbers in polar notation,

$$z = a + ib = |z|(\cos\theta + i\sin\theta) \qquad \theta = \tan^{-1}(b/a) \tag{1.12}$$

Using the important *Euler formula*, a proof of which is found in the supplemental material found at the website that accompanies this book,

$$e^{i\theta} = \cos\theta + i\sin\theta \tag{1.13}$$

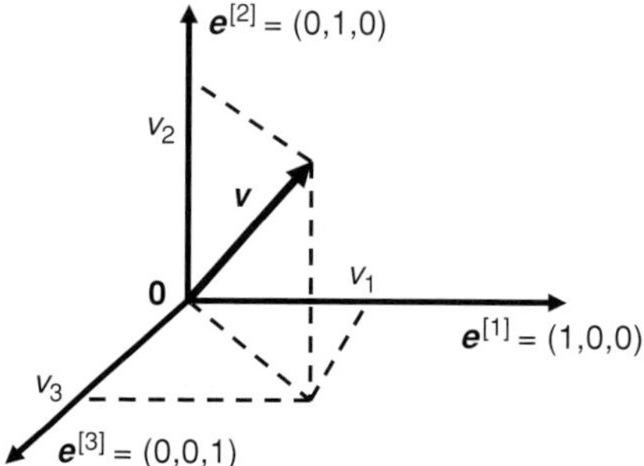

Figure 1.1 Physical interpretation of a 3-D vector.

we can write z as

$$z = |z|e^{i\theta} \tag{1.14}$$

Vector notation and operations

We write a three-dimensional (3-D) vector v (Figure 1.1) as

$$v = \begin{bmatrix} v_1 \\ v_2 \\ v_3 \end{bmatrix} \tag{1.15}$$

v is *real* if $v_1, v_2, v_3 \in \Re$; we then say $v \in \Re^3$. We can easily visualize this vector in 3-D space, defining the three coordinate basis vectors in the $1(x)$, $2(y)$, and $3(z)$ directions as

$$e^{[1]} = \begin{bmatrix} 1 \\ 0 \\ 0 \end{bmatrix} \qquad e^{[2]} = \begin{bmatrix} 0 \\ 1 \\ 0 \end{bmatrix} \qquad e^{[3]} = \begin{bmatrix} 0 \\ 0 \\ 1 \end{bmatrix} \tag{1.16}$$

to write $v \in \Re^3$ as

$$v = v_1 e^{[1]} + v_2 e^{[2]} + v_3 e^{[3]} \tag{1.17}$$

We extend this notation to define $\Re^N$, the *set of N-dimensional real vectors*,

$$v = \begin{bmatrix} v_1 \\ v_2 \\ \vdots \\ v_N \end{bmatrix} \tag{1.18}$$

where $v_j \in \Re$ for $j = 1, 2, \ldots, N$. By writing v in this manner, we define a *column vector*; however, v can also be written as a *row vector*,

$$v = [v_1 \quad v_2 \ldots \quad v_N] \tag{1.19}$$

The difference between column and row vectors only becomes significant when we start combining them in equations with matrices.

We write $v \in \Re^N$ as an expansion in coordinate basis vectors as

$$v = v_1 e^{[1]} + v_2 e^{[2]} + \cdots + v_N e^{[N]} \tag{1.20}$$

where the components of $e^{[j]}$ are *Kroenecker deltas* δ_{jk},

$$e^{[j]} = \begin{bmatrix} e_1^{[j]} \\ e_2^{[j]} \\ \vdots \\ e_N^{[j]} \end{bmatrix} = \begin{bmatrix} \delta_{j1} \\ \delta_{j2} \\ \vdots \\ \delta_{jN} \end{bmatrix} \qquad \delta_{jk} = \begin{cases} 1, & \text{if } j = k \\ 0, & \text{if } j \neq k \end{cases} \tag{1.21}$$

Addition of two real vectors $v \in \Re^N$, $w \in \Re^N$ is straightforward,

$$v + w = \begin{bmatrix} v_1 \\ v_2 \\ \vdots \\ v_N \end{bmatrix} + \begin{bmatrix} w_1 \\ w_2 \\ \vdots \\ w_N \end{bmatrix} = \begin{bmatrix} v_1 + w_1 \\ v_2 + w_2 \\ \vdots \\ v_N + w_N \end{bmatrix} \tag{1.22}$$

as is multiplication of a vector $v \in \Re^N$ by a real scalar $c \in \Re$,

$$cv = c \begin{bmatrix} v_1 \\ v_2 \\ \vdots \\ v_N \end{bmatrix} = \begin{bmatrix} cv_1 \\ cv_2 \\ \vdots \\ cv_N \end{bmatrix} \tag{1.23}$$

For all $u, v, w \in \Re^N$ and all $c_1, c_2 \in \Re$,

$$\begin{aligned} u + (v + w) = (u + v) + w \qquad & c(v + u) = cv + cu \\ u + v = v + u \qquad & (c_1 + c_2)v = c_1 v + c_2 v \\ v + 0 = v \qquad & (c_1 c_2)v = c_1(c_2 v) \\ v + (-v) = 0 \qquad & 1v = v \end{aligned} \tag{1.24}$$

where the *null vector* $\mathbf{0} \in \Re^N$ is

$$\mathbf{0} = \begin{bmatrix} 0 \\ 0 \\ \vdots \\ 0 \end{bmatrix} \tag{1.25}$$

We further add to the list of operations associated with the vectors $v, w \in \Re^N$ the *dot (inner, scalar) product*,

$$v \cdot w = v_1 w_1 + v_2 w_2 + \cdots + v_N w_N = \sum_{k=1}^{N} v_k w_k \tag{1.26}$$

For example, for the two vectors

$$v = \begin{bmatrix} 1 \\ 2 \\ 3 \end{bmatrix} \qquad w = \begin{bmatrix} 4 \\ 5 \\ 6 \end{bmatrix} \tag{1.27}$$

$$\begin{aligned} v \cdot w = v_1 w_1 + v_2 w_2 + v_3 w_3 &= (1)(4) + (2)(5) + (3)(6) \\ &= 4 + 10 + 18 = 32 \end{aligned} \tag{1.28}$$

For 3-D vectors, the dot product is proportional to the product of the lengths and the cosine of the angle between the two vectors,

$$v \cdot w = |v||w| \cos \theta \tag{1.29}$$

where the *length* of v is

$$|v| = \sqrt{v \cdot v} \geq 0 \tag{1.30}$$

Therefore, when two vectors are parallel, the magnitude of their dot product is maximal and equals the product of their lengths, and when two vectors are perpendicular, their dot product is zero. These ideas carry completely into N- dimensions. The *length* of a vector $v \in \Re^N$ is

$$|v| = \sqrt{v \cdot v} = \sqrt{\sum_{k=1}^{N} v_k^2} \geq 0 \tag{1.31}$$

If $v \cdot w = 0$, v and w are said to be *orthogonal*, the extension of the adjective "perpendicular" from $\Re^3$ to $\Re^N$. If $v \cdot w = 0$ and $|v| = |w| = 1$, i.e., both vectors are *normalized* to unit length, v and w are said to be *orthonormal*.

The formula for the length $|v|$ of a vector $v \in \Re^N$ satisfies the more general properties of a *norm* $\|v\|$ of $v \in \Re^N$. A norm $\|v\|$ is a rule that assigns a real scalar, $\|v\| \in \Re$, to each vector $v \in \Re^N$ such that for every $v, w \in \Re^N$, and for every $c \in \Re$, we have

$$\begin{aligned} \|v\| \geq 0 \qquad \|\mathbf{0}\| &= 0 \\ \|v\| = 0 \qquad \text{if and only if (iff)} \qquad v &= \mathbf{0} \\ \|cv\| &= |c| \|v\| \\ \|v + w\| &\leq \|v\| + \|w\| \end{aligned} \tag{1.32}$$

Each norm also provides an accompanying *metric*, a measure of how different two vectors are

$$d(v, w) = \|v - w\| \tag{1.33}$$

In addition to the length, many other possible definitions of norm exist. The *p-norm*, $\|v\|_p$, of $v \in \Re^N$ is

$$\|v\|_p = \left[\sum_{k=1}^{N} |v_k|^p \right]^{1/p} \tag{1.34}$$

Table 1.1 p-norm values for the 3-D vector (1, −2, 3)

p	$\|v\|_p$
1	6
2	$\sqrt{14} = 3.742$
10	3.005
50	3.00000000009

The *length* of a vector is thus also the *2-norm*. For $v = [1\ {-}2\ 3]$, the values of the p-norm, computed from (1.35), are presented in Table 1.1.

$$\|v\|_p = [|1|^p + |-2|^p + |3|^p]^{1/p} = [(1)^p + (2)^p + (3)^p]^{1/p} \tag{1.35}$$

We define the *infinity norm* as the limit of $\|v\|_p$ as $p \to \infty$, which merely extracts from v the largest magnitude of any component,

$$\|v\|_\infty \equiv \lim_{p\to\infty} \|v\|_p = \max_{j\in[1,N]}\{|v_j|\} \tag{1.36}$$

For $v = [1\ {-}2\ 3]$, $\|v\|_\infty = 3$.

Like scalars, vectors can be complex. We define the *set of complex N-dimensional vectors* as C^N, and write each component of $v \in C^N$ as

$$v_j = a_j + ib_j \qquad a_j, b_j \in \Re \qquad i = \sqrt{-1} \tag{1.37}$$

The *complex conjugate* of $v \in C^N$, written as $\bar{v}$ or v^*, is

$$v^* = \begin{bmatrix} a_1 + ib_1 \\ a_2 + ib_2 \\ \vdots \\ a_N + ib_N \end{bmatrix}^* = \begin{bmatrix} a_1 - ib_1 \\ a_2 - ib_2 \\ \vdots \\ a_N - ib_N \end{bmatrix} \tag{1.38}$$

For complex vectors $v, w \in C^N$, to form the dot product $v \cdot w$, we take the complex conjugates of the first vector's components,

$$v \cdot w = \sum_{k=1}^{N} v_k^* w_k \tag{1.39}$$

This ensures that the length of any $v \in C$ is always real and nonnegative,

$$\|v\|_2^2 = \sum_{k=1}^{N} v_k^* v_k = \sum_{k=1}^{N}(a_k - ib_k)(a_k + ib_k) = \sum_{k=1}^{N} (a_k^2 + b_k^2) \geq 0 \tag{1.40}$$

For $v, w \in C^N$, the order of the arguments is significant,

$$v \cdot w = (w \cdot v)^* \neq w \cdot v \tag{1.41}$$

Matrix dimension

For a linear system $A\boldsymbol{x} = \boldsymbol{b}$,

$$
A = \begin{bmatrix} a_{11} & a_{12} & \cdots & a_{1N} \\ a_{21} & a_{22} & \cdots & a_{2N} \\ \vdots & \vdots & & \vdots \\ a_{N1} & a_{N2} & \cdots & a_{NN} \end{bmatrix} \qquad \boldsymbol{x} = \begin{bmatrix} x_1 \\ x_2 \\ \vdots \\ x_N \end{bmatrix} \qquad \boldsymbol{b} = \begin{bmatrix} b_1 \\ b_2 \\ \vdots \\ b_N \end{bmatrix} \tag{1.42}
$$

to have a unique solution, there must be as many equations as unknowns, and so typically A will have an equal number N of columns and rows and thus be a *square matrix*. A matrix is said to be of *dimension* $M \times N$ if it has M rows and N columns. We now consider some simple matrix operations.

Multiplication of an M × N matrix A by a scalar c

$$
cA = c \begin{bmatrix} a_{11} & a_{12} & \cdots & a_{1N} \\ a_{21} & a_{22} & \cdots & a_{2N} \\ \vdots & \vdots & & \vdots \\ a_{M1} & a_{M2} & \cdots & a_{MN} \end{bmatrix} = \begin{bmatrix} ca_{11} & ca_{12} & \cdots & ca_{1N} \\ ca_{21} & ca_{22} & \cdots & ca_{2N} \\ \vdots & \vdots & & \vdots \\ ca_{M1} & ca_{M2} & \cdots & ca_{MN} \end{bmatrix} \tag{1.43}
$$

Addition of an M × N matrix A with an equal-sized M × N matrix B

$$
\begin{bmatrix} a_{11} & \cdots & a_{1N} \\ a_{21} & \cdots & a_{2N} \\ \vdots & & \vdots \\ a_{M1} & \cdots & a_{MN} \end{bmatrix} + \begin{bmatrix} b_{11} & \cdots & b_{1N} \\ b_{21} & \cdots & b_{2N} \\ \vdots & & \vdots \\ b_{M1} & \cdots & b_{MN} \end{bmatrix}
$$

$$
= \begin{bmatrix} a_{11} + b_{11} & \cdots & a_{1N} + b_{1N} \\ a_{21} + b_{21} & \cdots & a_{2N} + b_{2N} \\ \vdots & & \vdots \\ a_{M1} + b_{M1} & \cdots & a_{MN} + b_{MN} \end{bmatrix} \tag{1.44}
$$

Note that $A + B = B + A$ and that two matrices can be added *only* if both the number of rows and the number of columns are equal for each matrix. Also, $c(A + B) = cA + cB$.

Multiplication of a square N × N matrix A with an N-dimensional vector v

This operation must be defined as follows if we are to have equivalence between the coefficient and matrix/vector representations of a linear system:

$$
A\boldsymbol{v} = \begin{bmatrix} a_{11} & a_{12} & \cdots & a_{1N} \\ a_{21} & a_{22} & \cdots & a_{2N} \\ \vdots & \vdots & & \vdots \\ a_{N1} & a_{N2} & \cdots & a_{NN} \end{bmatrix} \begin{bmatrix} v_1 \\ v_2 \\ \vdots \\ v_N \end{bmatrix} = \begin{bmatrix} a_{11}v_1 + a_{12}v_2 + \cdots + a_{1N}v_N \\ a_{21}v_1 + a_{22}v_2 + \cdots + a_{2N}v_N \\ \vdots \\ a_{N1}v_1 + a_{N2}v_2 + \cdots + a_{NN}v_N \end{bmatrix}
$$

$$
\tag{1.45}
$$

Av is also an N-dimensional vector, whose jth component is

$$(Av)_j = a_{j1}v_1 + a_{j2}v_2 + \cdots + a_{jN}v_N = \sum_{k=1}^{N} a_{jk}v_k \tag{1.46}$$

We compute $(Av)_j$ by summing $a_{jk}v_k$ along rows of A and down the vector,

$$\begin{bmatrix} \Rightarrow & \Rightarrow a_{jk} & \Rightarrow & \Rightarrow \end{bmatrix} \begin{bmatrix} \Downarrow \\ v_k \\ \Downarrow \end{bmatrix}$$

Multiplication of an M × N matrix A with an N-dimensional vector v

From the rule for forming Av, we see that the number of columns of A must equal the dimension of v; however, we also can define Av when $M \neq N$,

$$Av = \begin{bmatrix} a_{11} & a_{12} & \cdots & a_{1N} \\ a_{21} & a_{22} & \cdots & a_{2N} \\ \vdots & \vdots & & \vdots \\ a_{M1} & a_{M2} & \cdots & a_{MN} \end{bmatrix} \begin{bmatrix} v_1 \\ v_2 \\ \vdots \\ v_N \end{bmatrix} = \begin{bmatrix} a_{11}v_1 + a_{12}v_2 + \cdots + a_{1N}v_N \\ a_{21}v_1 + a_{22}v_2 + \cdots + a_{2N}v_N \\ \vdots \\ a_{M1}v_1 + a_{M2}v_2 + \cdots + a_{MN}v_N \end{bmatrix} \tag{1.47}$$

If $v \in \Re^N$, for an $M \times N$ matrix A, $Av \in \Re^M$. Consider the following examples:

$$\begin{bmatrix} 1 & 2 & 3 & 4 \\ 4 & 3 & 2 & 1 \\ 11 & 12 & 13 & 14 \end{bmatrix} \begin{bmatrix} 1 \\ 2 \\ 3 \\ 4 \end{bmatrix} = \begin{bmatrix} 30 \\ 20 \\ 130 \end{bmatrix} \qquad \begin{bmatrix} 1 & 2 & 3 \\ 3 & 1 & 2 \\ 4 & 5 & 6 \\ 5 & 6 & 4 \end{bmatrix} \begin{bmatrix} 1 \\ 2 \\ 3 \end{bmatrix} = \begin{bmatrix} 14 \\ 11 \\ 32 \\ 29 \end{bmatrix} \tag{1.48}$$

Note also that $A(cv) = cAv$ and $A(v + w) = Av + Aw$.

Matrix transposition

We define for an $M \times N$ matrix A the *transpose* A^T to be the $N \times M$ matrix

$$A^T = \begin{bmatrix} a_{11} & a_{12} & \cdots & a_{1N} \\ a_{21} & a_{22} & \cdots & a_{2N} \\ \vdots & \vdots & & \vdots \\ a_{M1} & a_{M2} & \cdots & a_{MN} \end{bmatrix}^T = \begin{bmatrix} a_{11} & a_{21} & \cdots & a_{M1} \\ a_{12} & a_{22} & \cdots & a_{M2} \\ \vdots & \vdots & & \vdots \\ a_{1N} & a_{2N} & \cdots & a_{NM} \end{bmatrix} \tag{1.49}$$

The transpose operation is essentially a mirror reflection across the *principal diagonal* $a_{11}, a_{22}, a_{33}, \ldots$. Consider the following examples:

$$\begin{bmatrix} 1 & 2 & 3 \\ 4 & 5 & 6 \end{bmatrix}^{\mathrm{T}} = \begin{bmatrix} 1 & 4 \\ 2 & 5 \\ 3 & 6 \end{bmatrix} \qquad \begin{bmatrix} 1 & 2 & 3 \\ 4 & 5 & 6 \\ 7 & 8 & 9 \end{bmatrix}^{\mathrm{T}} = \begin{bmatrix} 1 & 4 & 7 \\ 2 & 5 & 8 \\ 3 & 6 & 9 \end{bmatrix} \qquad (1.50)$$

If a matrix is equal to its transpose, $A = A^{\mathrm{T}}$, it is said to be *symmetric*. Then,

$$a_{ij} = \left(A^{\mathrm{T}}\right)_{ij} = a_{ji} \qquad \forall i, j \in \{1, 2, \ldots, N\} \qquad (1.51)$$

Complex-valued matrices

Here we have defined operations for real matrices, however, matrices may also be complex-valued,

$$C = \begin{bmatrix} c_{11} & \cdots & c_{1N} \\ c_{21} & \cdots & c_{2N} \\ \vdots & & \vdots \\ c_{M1} & \cdots & c_{MN} \end{bmatrix} = \begin{bmatrix} (a_{11} + ib_{21}) & \cdots & (a_{1N} + ib_{1N}) \\ (a_{21} + ib_{21}) & \cdots & (a_{2N} + ib_{2N}) \\ \vdots & & \vdots \\ (a_{M1} + ib_{M1}) & \cdots & (a_{MN} + ib_{MN}) \end{bmatrix} \qquad (1.52)$$

For the moment, we are concerned with the properties of real matrices, as applied to solving linear systems in which the coefficients are real.

Vectors as matrices

Finally, we note that the matrix operations above can be extended to vectors by considering a vector $v \in \Re^N$ to be an $N \times 1$ matrix if in column form and to be a $1 \times N$ matrix if in row form. Thus, for $v, w \in \Re^N$, expressing vectors by default as column vectors, we write the dot product as

$$v \cdot w = v^{\mathrm{T}} w = [v_1 \quad \cdots \quad v_N] \begin{bmatrix} w_1 \\ \vdots \\ w_N \end{bmatrix} = v_1 w_1 + \cdots + v_N w_N \qquad (1.53)$$

The notation $v^{\mathrm{T}} w$ for the dot product $v \cdot w$ is used extensively in this text.

Elimination methods for solving linear systems

With these basic definitions in hand, we now begin to consider the solution of the linear system $Ax = b$, in which $x, b \in \Re^N$ and A is an $N \times N$ real matrix. We consider here *elimination* methods in which we convert the linear system into an equivalent one that is easier to solve. These methods are straightforward to implement and work generally for any linear system that has a unique solution; however, they can be quite costly (perhaps prohibitively so) for large systems. Later, we consider iterative methods that are more effective for certain classes of large systems.

Gaussian elimination

We wish to develop an algorithm for solving the set of N linear equations

$$
\begin{aligned}
a_{11}x_1 + a_{12}x_2 + \cdots + a_{1N}x_N &= b_1 \\
a_{21}x_1 + a_{22}x_2 + \cdots + a_{2N}x_N &= b_2 \\
&\ \ \vdots \\
a_{N1}x_1 + a_{N2}x_2 + \cdots + a_{NN}x_N &= b_N
\end{aligned}
\tag{1.54}
$$

The basic strategy is to define a sequence of operations that converts the original system into a simpler, but equivalent, one that may be solved easily.

Elementary row operations

We first note that we can select any two equations, say j and k, and add them to obtain another one that is equally valid,

$$
\frac{(a_{j1}x_1 + a_{j2}x_2 + \cdots + a_{jN}x_N = b_j) + (a_{k1}x_1 + a_{k2}x_2 + \cdots + a_{kN}x_N = b_k)}{(a_{j1} + a_{k1})x_1 + (a_{j2} + a_{k2})x_2 + \cdots + (a_{jN} + a_{kN})x_N = (b_j + b_k)}
\tag{1.55}
$$

If equation j is satisfied, and the equation obtained by summing j and k is satisfied, it follows that equation k must be satisfied as well. We are thus free to replace in our system the equation

$$
a_{k1}x_1 + a_{k2}x_2 + \cdots + a_{kN}x_N = b_k
\tag{1.56}
$$

with

$$
(a_{j1} + a_{k1})x_1 + (a_{j2} + a_{k2})x_2 + \cdots + (a_{jN} + a_{kN})x_N = (b_j + b_k)
\tag{1.57}
$$

with no effect upon the solution x. Similarly, we can take any equation, say j, multiply it by a nonzero scalar c, to obtain

$$
ca_{j1}x_1 + ca_{j2}x_2 + \cdots + ca_{jN}x_N = cb_j
\tag{1.58}
$$

which we then can substitute for equation j without affecting the solution. In general, in the linear system

$$
\begin{aligned}
a_{11}x_1 + a_{12}x_2 + \cdots + a_{1N}x_N &= b_1 \\
&\ \ \vdots \\
a_{j1}x_1 + a_{j2}x_2 + \cdots + a_{jN}x_N &= b_j \\
&\ \ \vdots \\
a_{k1}x_1 + a_{k2}x_2 + \cdots + a_{kN}x_N &= b_k \\
&\ \ \vdots \\
a_{N1}x_1 + a_{N2}x_2 + \cdots + a_{NN}x_N &= b_N
\end{aligned}
\tag{1.59}
$$

we can choose any $j \in [1, N]$, $k \in [1, N]$ and any scalar $c \neq 0$ to form the following equivalent system that has exactly the same solution(s):

$$a_{11}x_1 + \cdots + a_{1N}x_N = b_1$$
$$\vdots$$
$$a_{j1}x_1 + \cdots + a_{jN}x_N = b_j$$
$$\vdots \tag{1.60}$$
$$(ca_{j1} + a_{k1})x_1 + \cdots + (ca_{jN} + a_{kN})x_N = (cb_j + b_k)$$
$$\vdots$$
$$a_{N1}x_1 + \cdots + a_{NN}x_N = b_N$$

This procedure is known as an *elementary row operation*. The particular one shown here we denote by $k \leftarrow c \times j + k$.

We use matrix-vector notation, $Ax = b$, and write the system (1.59) as

$$
\begin{bmatrix}
a_{11} & a_{12} & \cdots & a_{1N} \\
\vdots & \vdots & & \vdots \\
a_{j1} & a_{j2} & \cdots & a_{jN} \\
\vdots & \vdots & & \vdots \\
a_{k1} & a_{k2} & \cdots & a_{kN} \\
\vdots & \vdots & & \vdots \\
a_{N1} & a_{N2} & \cdots & a_{NN}
\end{bmatrix}
\begin{bmatrix}
x_1 \\ \vdots \\ x_j \\ \vdots \\ x_k \\ \vdots \\ x_N
\end{bmatrix}
=
\begin{bmatrix}
b_1 \\ \vdots \\ b_j \\ \vdots \\ b_k \\ \vdots \\ b_N
\end{bmatrix}
\tag{1.61}
$$

After the row operation $k \leftarrow c \times j + k$, we have an equivalent system $A'x = b'$,

$$
\begin{bmatrix}
a_{11} & a_{12} & \cdots & a_{1N} \\
\vdots & \vdots & & \vdots \\
a_{j1} & a_{j2} & \cdots & a_{jN} \\
\vdots & \vdots & & \vdots \\
(ca_{j1} + a_{k1}) & (ca_{j2} + a_{k2}) & \cdots & (ca_{jN} + a_{kN}) \\
\vdots & \vdots & & \vdots \\
a_{N1} & a_{N2} & \cdots & a_{NN}
\end{bmatrix}
\begin{bmatrix}
x_1 \\ \vdots \\ x_j \\ \vdots \\ x_k \\ \vdots \\ x_N
\end{bmatrix}
=
\begin{bmatrix}
b_1 \\ \vdots \\ b_j \\ \vdots \\ (cb_j + b_k) \\ \vdots \\ b_N
\end{bmatrix}
\tag{1.62}
$$

As we must change both A and b, it is common to perform row operations on the *augmented matrix* (A, b) of dimension $N \times (N+1)$,

$$(A, b) = \begin{bmatrix} a_{11} & \cdots & a_{1N} & b_1 \\ \vdots & & \vdots & \vdots \\ a_{j1} & \cdots & a_{jN} & b_j \\ \vdots & & \vdots & \vdots \\ a_{k1} & \cdots & a_{kN} & b_k \\ \vdots & & \vdots & \vdots \\ a_{N1} & \cdots & a_{NN} & b_N \end{bmatrix} \tag{1.63}$$

After the operation $k \leftarrow c \times j + k$, the augmented matrix is

$$(A, b)' = \begin{bmatrix} a_{11} & \cdots & a_{1N} & b_1 \\ \vdots & & \vdots & \vdots \\ a_{j1} & \cdots & a_{jN} & b_j \\ \vdots & & \vdots & \vdots \\ (ca_{j1} + a_{k1}) & \cdots & (ca_{jN} + a_{kN}) & (cb_j + b_k) \\ \vdots & & \vdots & \vdots \\ a_{N1} & \cdots & a_{NN} & b_N \end{bmatrix} \tag{1.64}$$

Gaussian elimination to place A**x** = **b** in upper triangular form

We now build a systematic approach to solving $Ax = b$ based upon a sequence of elementary row operations, known as *Gaussian elimination*. First, we start with the original augmented matrix

$$(A, b) = \begin{bmatrix} a_{11} & a_{12} & a_{13} & \cdots & a_{1N} & b_1 \\ a_{21} & a_{22} & a_{23} & \cdots & a_{2N} & b_2 \\ a_{31} & a_{32} & a_{33} & \cdots & a_{3N} & b_3 \\ \vdots & \vdots & \vdots & \vdots & \vdots & \vdots \\ a_{N1} & a_{N2} & a_{N3} & \cdots & a_{NN} & b_N \end{bmatrix} \tag{1.65}$$

As long as $a_{11} \neq 0$ (later we consider what to do if $a_{11} = 0$), we can define the finite scalar

$$\lambda_{21} = a_{21}/a_{11} \tag{1.66}$$

and perform the row operation $2 \leftarrow 2 - \lambda_{21} \times 1$,

$$\frac{-\lambda_{21}(a_{11}x_1 + a_{12}x_2 + \cdots + a_{1N}x_N = b_1) + (a_{21}x_1 + a_{22}x_2 + \cdots + a_{2N}x_N = b_2)}{(a_{21} - \lambda_{21}a_{11})x_1 + (a_{22} - \lambda_{21}a_{12})x_2 + \cdots + (a_{2N} - \lambda_{21}a_{1N})x_N = (b_2 - \lambda_{21}b_1)} \tag{1.67}$$

The coefficient multiplying x_1 in this new equation is zero:

$$a_{21} - \lambda_{21}a_{11} = a_{21} - \left(\frac{a_{21}}{a_{11}}\right)a_{11} = a_{21} - a_{21} = 0 \tag{1.68}$$

We use the notation $(A, b)^{(2,1)}$ for the augmented matrix obtained after placing a zero at the $(2,1)$ position through the operation $2 \leftarrow 2 - \lambda_{21} \times 1$,

$$(A, b)^{(2,1)} = \begin{bmatrix} a_{11} & a_{12} & \cdots & a_{1N} & b_1 \\ 0 & (a_{22} - \lambda_{21} a_{12}) & \cdots & (a_{2N} - \lambda_{21} a_{1N}) & (b_2 - \lambda_{21} b_1) \\ a_{31} & a_{32} & \cdots & a_{3N} & b_3 \\ \vdots & \vdots & & \vdots & \vdots \\ a_{N1} & a_{N2} & \cdots & a_{NN} & b_N \end{bmatrix} \tag{1.69}$$

Again, it is important to note that the linear system $A^{(2,1)}x = b^{(2,1)}$ has the *same* solution(s) x as the original system $Ax = b$.

As we develop this method, let us consider the example (1.2),

$$\begin{aligned} x_1 + x_2 + x_3 &= 4 \\ 2x_1 + x_2 + 3x_3 &= 7 \\ 3x_1 + x_2 + 6x_3 &= 2 \end{aligned} \qquad (A, b) = \begin{bmatrix} 1 & 1 & 1 & 4 \\ 2 & 1 & 3 & 7 \\ 3 & 1 & 6 & 2 \end{bmatrix} \tag{1.70}$$

Since $a_{11} \neq 0$,

$$\lambda_{21} = \frac{a_{21}}{a_{11}} = \frac{2}{1} = 2 \tag{1.71}$$

and the row operation $2 \leftarrow 2 - \lambda_{21} \times 1$ yields

$$\begin{aligned} (A, b)^{(2,1)} &= \begin{bmatrix} 1 & 1 & 1 & 4 \\ [2 - (2)(1)] & [1 - (2)(1)] & [3 - (2)(1)] & [7 - (2)(4)] \\ 3 & 1 & 6 & 2 \end{bmatrix} \\ &= \begin{bmatrix} 1 & 1 & 1 & 4 \\ 0 & -1 & 1 & -1 \\ 3 & 1 & 6 & 2 \end{bmatrix} \end{aligned} \tag{1.72}$$

We now define

$$a_{2j}^{(2,1)} \equiv a_{2j} - \lambda_{21} a_{1j} \qquad b_2^{(2,1)} \equiv b_2 - \lambda_{21} b_1 \tag{1.73}$$

and write $(A, b)^{(2,1)}$ as

$$(A, b)^{(2,1)} = \begin{bmatrix} a_{11} & a_{12} & a_{13} & \cdots & a_{1N} & b_1 \\ 0 & a_{22}^{(2,1)} & a_{23}^{(2,1)} & \cdots & a_{2N}^{(2,1)} & b_2^{(2,1)} \\ a_{31} & a_{32} & a_{33} & \cdots & a_{3N} & b_3 \\ \vdots & \vdots & \vdots & \vdots & \vdots & \vdots \\ a_{N1} & a_{N2} & a_{N3} & \cdots & a_{NN} & b_N \end{bmatrix} \tag{1.74}$$

We continue this process and place zeros in all elements in the first column below the diagonal, considering next the element in row 3. If $a_{11} \neq 0$,

$$\lambda_{31} = \frac{a_{31}}{a_{11}} \tag{1.75}$$

and the row operation $3 \leftarrow 3 - \lambda_{31} \times 1$ yields the new $(3, 1)$ element

$$a_{31} - \lambda_{31} a_{11} = a_{31} - \left(\frac{a_{31}}{a_{11}}\right) a_{11} = a_{31} - a_{31} = 0 \tag{1.76}$$

The augmented matrix after this operation is

$$(A, b)^{(3,\,1)} = \begin{bmatrix} a_{11} & a_{12} & a_{13} & \cdots & a_{1N} & b_1 \\ 0 & a_{22}^{(2,\,1)} & a_{23}^{(2,\,1)} & \cdots & a_{2N}^{(2,\,1)} & b_2^{(2,\,1)} \\ 0 & a_{32}^{(3,\,1)} & a_{33}^{(3,\,1)} & \cdots & a_{3N}^{(3,\,1)} & b_3^{(3,1)} \\ \vdots & \vdots & \vdots & \vdots & \vdots & \vdots \\ a_{N1} & a_{N2} & a_{N3} & \cdots & a_{NN} & b_N \end{bmatrix} \tag{1.77}$$

where

$$a_{3j}^{(3,\,1)} \equiv a_{3j} - \lambda_{31} a_{1j} \qquad b_3^{(3,\,1)} \equiv b_3 - \lambda_{31} b_1 \tag{1.78}$$

For the example system (1.72),

$$\lambda_{31} = \frac{a_{31}}{a_{11}} = \frac{3}{1} = 3 \tag{1.79}$$

and

$$\begin{aligned} (A, b)^{(3,1)} &= \begin{bmatrix} 1 & 1 & 1 & 4 \\ 0 & -1 & 1 & -1 \\ [3 - 3(1)] & [1 - 3(1)] & [6 - 3(1)] & [2 - 3(4)] \end{bmatrix} \\[2mm] &= \begin{bmatrix} 1 & 1 & 1 & 4 \\ 0 & -1 & 1 & -1 \\ 0 & -2 & 3 & -10 \end{bmatrix} \end{aligned} \tag{1.80}$$

In general, for $N > 3$, we continue this procedure until all elements of the first column are zero except a_{11}. The augmented system matrix after this sequence of row operations is

$$(A, b)^{(N,1)} = \begin{bmatrix} a_{11} & a_{12} & a_{13} & \cdots & a_{1N} & b_1 \\ 0 & a_{22}^{(2,\,1)} & a_{23}^{(2,\,1)} & \cdots & a_{2N}^{(2,\,1)} & b_2^{(2,\,1)} \\ 0 & a_{32}^{(3,\,1)} & a_{33}^{(3,\,1)} & \cdots & a_{3N}^{(3,\,1)} & b_3^{(3,\,1)} \\ \vdots & \vdots & \vdots & \vdots & \vdots & \vdots \\ 0 & a_{N2}^{(N,\,1)} & a_{N3}^{(N,\,1)} & \cdots & a_{NN}^{(N,\,1)} & b_N^{(N,\,1)} \end{bmatrix} \tag{1.81}$$

We now move to the second column and perform row operations to place zeros everywhere below the diagonal $(2, 2)$ position. If $a_{22}^{(2,1)} \neq 0$, we define

$$\lambda_{32} = a_{32}^{(3,\,1)} \big/ a_{22}^{(2,\,1)} \tag{1.82}$$

and perform the row operation $3 \leftarrow 3 - \lambda_{32} \times 2$, to obtain

$$(A, b)^{(3,\,2)} = \begin{bmatrix} a_{11} & a_{12} & a_{13} & \cdots & a_{1N} & b_1 \\ 0 & a_{22}^{(2,\,1)} & a_{23}^{(2,\,1)} & \cdots & a_{2N}^{(2,\,1)} & b_2^{(2,\,1)} \\ 0 & 0 & a_{33}^{(3,\,2)} & \cdots & a_{3N}^{(3,\,2)} & b_3^{(3,\,2)} \\ \vdots & \vdots & \vdots & \vdots & \vdots & \vdots \\ 0 & a_{N2}^{(N,\,1)} & a_{N3}^{(N,\,1)} & \cdots & a_{NN}^{(N,\,1)} & b_N^{(N,\,1)} \end{bmatrix} \tag{1.83}$$

where

$$a_{3j}^{(3,2)} \equiv a_{3j}^{(3,1)} - \lambda_{32}a_{2j}^{(2,1)} \qquad b_3^{(3,2)} \equiv b_3^{(3,1)} - \lambda_{32}b_2^{(2,1)} \qquad (1.84)$$

For our example $(A, \boldsymbol{b})^{(3,1)}$ (1.80) we obtain

$$\lambda_{32} = a_{32}^{(3,1)}/a_{22}^{(2,1)} = (-2)/(-1) = 2 \qquad (1.85)$$

so that

$$(A, \boldsymbol{b})^{(3,2)} = \begin{bmatrix} 1 & 1 & 1 & 4 \\ 0 & -1 & 1 & -1 \\ 0 & [(-2)-(2)(-1)] & [3-(2)(1)] & [(-10)-(2)(-1)] \end{bmatrix}$$

$$= \begin{bmatrix} 1 & 1 & 1 & 4 \\ 0 & -1 & 1 & -1 \\ 0 & 0 & 1 & -8 \end{bmatrix} \qquad (1.86)$$

For $N > 3$, we perform another $N - 3$ row operations to place all zeros below the principal diagonal in the second column, yielding

$$(A, \boldsymbol{b})^{(N,2)} = \begin{bmatrix} a_{11} & a_{12} & a_{13} & \dots & a_{1N} & b_1 \\ 0 & a_{22}^{(2,1)} & a_{23}^{(2,1)} & \dots & a_{2N}^{(2,1)} & b_2^{(2,1)} \\ 0 & 0 & a_{33}^{(3,2)} & \dots & a_{3N}^{(3,2)} & b_3^{(3,2)} \\ \vdots & \vdots & \vdots & \vdots & \vdots & \vdots \\ 0 & 0 & a_{N3}^{(N,2)} & \dots & a_{NN}^{(N,2)} & b_N^{(N,2)} \end{bmatrix} \qquad (1.87)$$

We then perform a sequence of row operations to place all zeros in the third column below the $(3, 3)$ position, then in column 4, we put all zeros below $(4, 4)$, etc. When finished, the augmented matrix is *upper triangular*, containing only zeros below the *principal diagonal* $a_{11}, a_{22}, \dots, a_{NN}$:

$$(A, \boldsymbol{b})^{(N, N-1)} = \begin{bmatrix} a_{11} & a_{12} & a_{13} & a_{14} & \dots & a_{1N} & b_1 \\ 0 & a_{22}^{(2,1)} & a_{23}^{(2,1)} & a_{24}^{(2,1)} & \dots & a_{2N}^{(2,1)} & b_2^{(2,1)} \\ 0 & 0 & a_{33}^{(3,2)} & a_{34}^{(3,2)} & \dots & a_{3N}^{(3,2)} & b_3^{(3,2)} \\ 0 & 0 & 0 & a_{44}^{(4,3)} & \dots & a_{4N}^{(4,3)} & b_4^{(4,3)} \\ \vdots & \vdots & \vdots & \vdots & & \vdots & \vdots \\ 0 & 0 & 0 & 0 & \dots & a_{NN}^{(N,N-1)} & b_N^{(N,N-1)} \end{bmatrix} \qquad (1.88)$$

For our example, the final augmented matrix is

$$(A, \boldsymbol{b})^{(3,2)} = \begin{bmatrix} 1 & 1 & 1 & 4 \\ 0 & -1 & 1 & -1 \\ 0 & 0 & 1 & -8 \end{bmatrix} \qquad (1.89)$$

Solving triangular systems by substitution

The equations defined by $A^{(N,N-1)}x = b^{(N,N-1)}$ (1.88) are

$$
\begin{aligned}
a_{11}x_1 + a_{12}x_2 + a_{13}x_3 + \cdots + a_{1N}x_N &= b_1 \\
a_{22}^{(2,\,1)}x_2 + a_{23}^{(2,\,1)}x_3 + \cdots + a_{2N}^{(2,\,1)}x_N &= b_2^{(2,\,1)} \\
&\vdots \\
a_{N-1,N-1}^{(N-1,N-2)}x_{N-1} + a_{N-1,N}^{(N-1,N-2)}x_N &= b_{N-1}^{(N-1,N-2)} \\
a_{N,N}^{(N,N-1)}x_N &= b_N^{(N,N-1)}
\end{aligned}
\tag{1.90}
$$

For (1.89), these equations are

$$
\begin{aligned}
x_1 + x_2 + x_3 &= 4 \\
-x_2 + x_3 &= -1 \\
x_3 &= -8
\end{aligned}
\tag{1.91}
$$

From the last equation, we obtain immediately the last unknown,

$$
x_N = b_N^{(N,N-1)} \big/ a_{NN}^{(N,N-1)}
\tag{1.92}
$$

For the example, $x_3 = -8/1 = -8$.

We then work backwards, next solving for x_{N-1},

$$
x_{N-1} = \left[b_{N-1}^{(N-1,N-2)} - a_{N-1,N}^{(N-1,N-2)}x_N \right] \big/ a_{N-1,N-1}^{(N-1,N-2)}
\tag{1.93}
$$

then for x_{N-2},

$$
x_{N-2} = \left[b_{N-2}^{(N-2,N-3)} - a_{N-2,N-1}^{(N-2,N-3)}x_{N-1} - a_{N-2,N}^{(N-2,N-3)}x_N \right] \big/ a_{N-2,N-2}^{(N-2,N-3)}
\tag{1.94}
$$

until we obtain the values of all unknowns through this procedure of *backward substitution*. For our example, we have

$$
\begin{aligned}
x_2 &= [(-1) - (1)(-8)]/(-1) = -7 \\
x_1 &= [4 - (1)(-7) - (1)(-8)]/1 = 19
\end{aligned}
\tag{1.95}
$$

so that a solution to the example system (1.70) is

$$
x = \begin{bmatrix} x_1 \\ x_2 \\ x_3 \end{bmatrix} = \begin{bmatrix} 19 \\ -7 \\ -8 \end{bmatrix}
\tag{1.96}
$$

While we have found a solution, it remains to be established that this is the *only* solution.

Basic algorithm for solving Ax = b by Gaussian elimination

The following algorithm transforms a linear system to upper triangular form by *Gaussian elimination*:

```
allocate N² memory locations to store A, and N for b
for i = 1, 2, ..., N − 1; iterate over columns from left to right
    if aᵢᵢ = 0, STOP to avoid division by zero
    for j = i + 1, i + 2, ..., N; iterate over rows below diagonal
        λ ← aⱼᵢ/aᵢᵢ
        for k = i, i + 1, ..., N; iterate in row j from column # i to right
            aⱼₖ ← aⱼₖ − λaᵢₖ
        end for k = i, i + 1, ..., N
        bⱼ ← bⱼ − λbᵢ
    end for j = i + 1, i + 2, ..., N
end for i = 1, 2, ..., N − 1
```

The basic computational unit of this algorithm is the scalar operation

$$d \leftarrow a + b \times c \tag{1.97}$$

that comprises two *floating point operations (FLOPs)*, one a scalar multiplication and the other a scalar addition. The term "floating point operation" originates from the binary system for representing real numbers in digital memory that is described in the supplemental material found at the accompanying website. It is common to describe the computational effort required by an algorithm in terms of the number of FLOPs that it performs during execution.

The algorithm for Gaussian elimination contains three nested loops that each run over all or part of the set of indices $\{1, 2, \ldots, N\}$. Therefore, the total number of FLOPs is proportional to N^3 (the exact number, $2N^3/3$ is computed in the supplemental material). If we increase N by a factor of 10, the amount of work required increases by a factor of $10^3 = 1000$. We see here the major problem with Gaussian elimination; it can become very costly for large systems!

Backward substitution follows the algorithm:

```
allocate N memory locations for the components of x
for i = N, N − 1, ..., 1; iterate over rows from bottom to top
    sum = 0
    for j = i + 1, i + 2, ..., N; iterate over lower columns
        sum ← sum + aᵢⱼ × xⱼ
    end for j = i + 1, i + 2, ..., N
    xᵢ ← [bᵢ − sum]/aᵢᵢ
end i = N, N − 1, ..., 1
```

As there are only two nested loops, the number of FLOPs scales only as N^2. Therefore, for large systems, it takes far (N times) longer to transform the system into triangular form by elimination ($2N^3/3$ FLOPs) than it does to solve the triangular system by substitution (N^2 FLOPs).

Gauss–Jordan elimination

In Gaussian elimination, we convert the system $Ax = b$ into an upper triangular form $Ux = c$ after $2N^3/3$ FLOPs,

$$
\begin{bmatrix}
u_{11} & u_{12} & u_{13} & \cdots & u_{1N} \\
 & u_{22} & u_{23} & \cdots & u_{2N} \\
 & & u_{33} & \cdots & u_{3N} \\
 & & & \ddots & \vdots \\
 & & & & u_{NN}
\end{bmatrix}
\begin{bmatrix}
x_1 \\ x_2 \\ x_3 \\ \vdots \\ x_N
\end{bmatrix}
=
\begin{bmatrix}
c_1 \\ c_2 \\ c_3 \\ \vdots \\ c_N
\end{bmatrix}
\tag{1.98}
$$

that we then solve by backward substitution in another N^2 FLOPs. Instead of performing backward substitution, we could continue the row operations to place zeros above the diagonal, to obtain a diagonal system $Dx = f$,

$$
\begin{bmatrix}
d_{11} \\
 & d_{22} \\
 & & \ddots \\
 & & & d_{NN}
\end{bmatrix}
\begin{bmatrix}
x_1 \\ x_2 \\ \vdots \\ x_N
\end{bmatrix}
=
\begin{bmatrix}
f_1 \\ f_2 \\ \vdots \\ f_N
\end{bmatrix}
\tag{1.99}
$$

The solution is now computed very easily, $x_j = f_j/d_{jj}$. This approach, *Gauss–Jordan elimination*, is rarely used in computational practice.

Partial pivoting

Let us consider again the first row operation in Gaussian elimination. We start with the original augmented matrix of the system,

$$
(A, \, b) =
\begin{bmatrix}
a_{11} & a_{12} & a_{13} & \cdots & a_{1N} & b_1 \\
a_{21} & a_{22} & a_{23} & \cdots & a_{2N} & b_2 \\
a_{31} & a_{32} & a_{33} & \cdots & a_{3N} & b_3 \\
\vdots & \vdots & \vdots & & \vdots & \vdots \\
a_{N1} & a_{N2} & a_{N3} & \cdots & a_{NN} & b_N
\end{bmatrix}
\tag{1.100}
$$

and perform the row operation $2 \leftarrow 2 + \lambda_{21} \times 1$ to obtain

$$
(A, \, b)^{(2, \, 1)} =
\begin{bmatrix}
a_{11} & a_{12} & \cdots & a_{1N} & b_1 \\
(a_{21} - \lambda_{21}a_{11}) & (a_{22} - \lambda_{21}a_{12}) & \cdots & (a_{2N} - \lambda_{21}a_{1N}) & (b_2 - \lambda_{21}b_1) \\
a_{31} & a_{32} & \cdots & a_{3N} & b_3 \\
\vdots & \vdots & & \vdots & \vdots \\
a_{N1} & a_{N2} & \cdots & a_{NN} & b_N
\end{bmatrix}
\tag{1.101}
$$

To place a zero at the (2,1) position, we define λ_{21} as

$$
\lambda_{21} = a_{21}/a_{11}
\tag{1.102}
$$

but if $a_{11} = 0$, λ_{21} blows up to $\pm\infty$. What do we do then?

We avoid such divisions by zero through the technique of *partial pivoting*. Before beginning any row operations, let us examine the first column of A,

$$A(:, 1) = \begin{bmatrix} a_{11} \\ a_{21} \\ a_{31} \\ \vdots \\ a_{N1} \end{bmatrix} \tag{1.103}$$

and search all elements in this column to find the row j that contains the element with the largest magnitude,

$$|a_{j1}| = \max_{k\in[1, N]}\{|a_{k1}|\} \tag{1.104}$$

Since the order in which the equations appear is irrelevant, we are perfectly free to exchange rows 1 and j to form the equivalent system

$$(\bar{A}, \bar{b}) = \begin{bmatrix} a_{j1} & a_{j2} & a_{j3} & \dots & a_{jN} & b_j \\ a_{21} & a_{22} & a_{23} & \dots & a_{2N} & b_2 \\ \vdots & \vdots & \vdots & & \vdots & \vdots \\ a_{11} & a_{12} & a_{13} & \dots & a_{1N} & b_1 \\ \vdots & \vdots & \vdots & & \vdots & \vdots \\ a_{N1} & a_{N2} & a_{N3} & \dots & a_{NN} & b_N \end{bmatrix} \tag{1.105}$$

As long as any of the elements of the first column of A are nonzero, $a_{j1} = \bar{a}_{11}$ is nonzero and the scalar $\lambda_{21} = \bar{a}_{21}/\bar{a}_{11}$ is finite. We may then perform without difficulty the row operations on $(\bar{A}, \bar{b})$ that place zeros in the first column below the diagonal. By contrast, if all of the elements of the first column are zero, as they must be if $\bar{a}_{11} = a_{j1} = 0$, there can be no unique solution. We thus stop the elimination procedure at this point.

In *Gaussian elimination with partial pivoting*, when moving to column i, we first examine all elements in this column at or below the diagonal, and select the row $j \geq i$ with the largest magnitude element,

$$|a_{ji}| \geq |a_{ki}| \qquad \forall k = i, i+1, \dots, N \tag{1.106}$$

If $j \neq i$, we swap rows j and i, and thus, unless all elements at or below the diagonal in column i are zero, $|a_{ji}| \neq 0$. We then can compute the scalars $\lambda_{i+1,i,\dots,}\lambda_{N,i}$ without having to divide by zero. If $|a_{ji}|$ is zero, then all of the elements $a_{i,i}, a_{i+1,i,\dots,} a_{N,i}$ must be zero.

To see what happens in this case, consider the following system, obtained after placing zeros successfully below the diagonal in the first three columns. When preparing to eliminate the values in the fourth column, we find that all values at or below the diagonal are already zero,

$$(A, b)^{(N, 3)} = \begin{bmatrix} a_{11} & a_{12} & a_{13} & a_{14} & \cdots & a_{1N} & b_1 \\ 0 & a_{22}^{(2, 1)} & a_{23}^{(2, 1)} & a_{24}^{(2, 1)} & \cdots & a_{2N}^{(2, 1)} & b_2^{(2, 1)} \\ 0 & 0 & a_{33}^{(3, 2)} & a_{34}^{(3, 2)} & \cdots & a_{3N}^{(3, 2)} & b_3^{(3, 2)} \\ 0 & 0 & 0 & 0 & \cdots & a_{4N}^{(4, 3)} & b_4^{(4, 3)} \\ 0 & 0 & 0 & 0 & \cdots & a_{5N}^{(5, 3)} & b_5^{(5, 3)} \\ \vdots & \vdots & \vdots & \vdots & & \vdots & \vdots \\ 0 & 0 & 0 & 0 & \cdots & a_{NN}^{(N, 3)} & b_N^{(N, 3)} \end{bmatrix} \tag{1.107}$$

Thus, we can write the fourth column as a linear combination of the first three columns, for some c_1, c_2, c_3,

$$a_{j4} = c_1 a_{j1} + c_2 a_{j2} + c_3 a_{j3} \qquad \forall j = 1, 2, \ldots, N \tag{1.108}$$

so that any equation j in this system can be written as

$$a_{j1}x_1 + a_{j2}x_2 + a_{j3}x_3 + (c_1 a_{j1} + c_2 a_{j2} + c_3 a_{j3})x_4 + \cdots + a_{jN}x_N = b_j$$
$$a_{j1}(x_1 + c_1 x_4) + a_{j2}(x_2 + c_2 x_4) + a_{j3}(x_3 + c_3 x_4) + \cdots + a_{jN}x_N = b_j \tag{1.109}$$

This means that we can make any change to x of the form

$$\begin{aligned} x_1 &\to x_1 - c_1 \Delta x_4 & x_2 &\to x_2 - c_2 \Delta x_4 \\ x_3 &\to x_3 - c_3 \Delta x_4 & x_4 &\to x_4 + \Delta x_4 \end{aligned} \tag{1.110}$$

without changing the value of Ax. Obviously, there can be no unique solution of $Ax = b$, and we thus stop the elimination procedure at this point.

If a unique solution exists, Gaussian elimination with partial pivoting will find it, even if there are zeros along the principal diagonal of the original augmented matrix. It is called partial pivoting because we only swap rows; if we swap columns as well (which necessitates more complex book-keeping, but results in better propagation of round-off error), we perform *Gaussian elimination with full pivoting*.

The algorithm for *Gaussian elimination with partial pivoting* is

for $i = 1, 2, \ldots, N - 1$; iterate over columns
    select row $j \geq i$ such that $|a_{ji}| = max_{j \geq i}\{|a_{ii}|, |a_{i+1, i}|, \ldots, |a_{N1}|\}$
    if $a_{ji} = 0$, no unique solution exists, STOP
    if $j \neq i$, interchange rows i and j of augmented matrix
    for $j = i + 1, i + 2, \ldots, N$; iterate over rows $j > i$
        $\lambda \leftarrow a_{ji}/a_{ii}$
        for $k = i, i + 1, \ldots, N$; iterate over elements in row j from left
            $a_{jk} \leftarrow a_{jk} - \lambda a_{ik}$
        end for $k = i, i + 1, \ldots, N$
        $b_j \leftarrow b_j - \lambda b_i$
    end for $j = i + 1, i + 2, \ldots, N$
end for $i = 1, 2, \ldots, N - 1$

Backward substitution proceeds in the same manner as before. Even if rows must be swapped for each column, the computational overhead of partial pivoting is relatively small.

To demonstrate Gaussian elimination with partial pivoting, consider the system of equations (1.70) with the augmented matrix

$$(A, b) = \begin{bmatrix} 1 & 1 & 1 & 4 \\ 2 & 1 & 3 & 7 \\ 3 & 1 & 6 & 2 \end{bmatrix} \tag{1.111}$$

First, we examine the first column to see that the element of largest magnitude is in row 3. We thus swap rows 1 and 3,

$$(\bar{A}, \bar{b}) = \begin{bmatrix} 3 & 1 & 6 & 2 \\ 2 & 1 & 3 & 7 \\ 1 & 1 & 1 & 4 \end{bmatrix} \tag{1.112}$$

Note that we swap the rows even though $a_{11} \neq 0$. As is shown in the supplemental material in the accompanying website, we do this to improve the numerical stability with respect to round-off error. We next do a row operation to zero the $(2, 1)$ element.

$$\lambda_{21} = \bar{a}_{21}/\bar{a}_{11} = 2/3 \tag{1.113}$$

$$(A, b)^{(2, 1)} = \begin{bmatrix} 3 & 1 & 6 & 2 \\ \left[2 - \left(\frac{2}{3}\right)3\right] & \left[1 - \left(\frac{2}{3}\right)1\right] & \left[3 - \left(\frac{2}{3}\right)6\right] & \left[7 - \left(\frac{2}{3}\right)2\right] \\ 1 & 1 & 1 & 4 \end{bmatrix}$$

$$= \begin{bmatrix} 3 & 1 & 6 & 2 \\ 0 & \frac{1}{3} & -1 & 5\frac{2}{3} \\ 1 & 1 & 1 & 4 \end{bmatrix} \tag{1.114}$$

We then perform another row operation to zero the $(3, 1)$ element.

$$\lambda_{31} = a_{31}^{(2, 1)}/a_{11}^{(2, 1)} = 1/3 \tag{1.115}$$

$$(A, b)^{(3, 1)} = \begin{bmatrix} 3 & 1 & 6 & 2 \\ 0 & \frac{1}{3} & -1 & 5\frac{2}{3} \\ \left[1 - \left(\frac{1}{3}\right)3\right] & \left[1 - \left(\frac{1}{3}\right)1\right] & \left[1 - \left(\frac{1}{3}\right)6\right] & \left[4 - \left(\frac{1}{3}\right)2\right] \end{bmatrix}$$

$$= \begin{bmatrix} 3 & 1 & 6 & 2 \\ 0 & \frac{1}{3} & -1 & 5\frac{2}{3} \\ 0 & \frac{2}{3} & -1 & 3\frac{1}{3} \end{bmatrix} \tag{1.116}$$

We now move to the second column, and note that the element of largest magnitude appears in the third row. We thus swap rows 2 and 3,

$$(\bar{A}, \bar{b})^{(3,1)} = \begin{bmatrix} 3 & 1 & 6 & 2 \\ 0 & \frac{2}{3} & -1 & 3\frac{1}{3} \\ 0 & \frac{1}{3} & -1 & 5\frac{2}{3} \end{bmatrix} \tag{1.117}$$

and perform a row operation to zero the $(3, 2)$ element.

$$\lambda_{32} = \left(\frac{1}{3}\right) \Big/ \left(\frac{2}{3}\right) = \frac{1}{2} \tag{1.118}$$

$$(A, b)^{(3,2)} = \begin{bmatrix} 3 & 1 & 6 & 2 \\ 0 & \frac{2}{3} & -1 & 3\frac{1}{3} \\ 0 & [\frac{1}{3} - \frac{1}{2}\left(\frac{2}{3}\right)] & [-1 - \frac{1}{2}(-1)] & [5\frac{2}{3} - \left(\frac{1}{2}\right)\left(3\frac{1}{3}\right)] \end{bmatrix}$$

$$= \begin{bmatrix} 3 & 1 & 6 & 2 \\ 0 & \frac{2}{3} & -1 & 3\frac{1}{3} \\ 0 & 0 & -\frac{1}{2} & 4 \end{bmatrix} \tag{1.119}$$

We now have an upper triangular system to solve by backward substitution,

$$\begin{aligned} 3x_1 + x_2 + 6x_3 &= 2 \\ \tfrac{2}{3}x_2 - x_3 &= 3\tfrac{1}{3} \\ -\tfrac{1}{2}x_3 &= 4 \end{aligned} \tag{1.120}$$

First, $x_3 = -8$ from the last equation. Then, from the second equation,

$$x_2 = \left[3\tfrac{1}{3} + x_3\right] \Big/ \left(\tfrac{2}{3}\right) = -7 \tag{1.121}$$

Finally, from the first equation,

$$x_1 = (2 - 6x_3 - x_2)/3 = 19 \tag{1.122}$$

The solution to (1.70) is thus $(x_1, x_2, x_3) = (19, -7, -8)$.

Existence and uniqueness of solutions

With Gaussian elimination and partial pivoting, we have a method for solving linear systems that either finds a solution or fails under conditions in which no unique solution exists. In this section, we consider at more depth the question of when a linear system possesses a real solution (*existence*) and if so, whether there is exactly one (*uniqueness*). These questions are vitally important, for linear algebra is the basis upon which we build algorithms for solving nonlinear equations, ordinary and partial differential equations, and many other tasks.

Interpreting Ax = b as a linear transformation

As a first step, we consider the equation $Ax = b$ from a somewhat more abstract viewpoint. We note that A is an $N \times N$ real matrix and x and b are both N-dimensional real vectors.

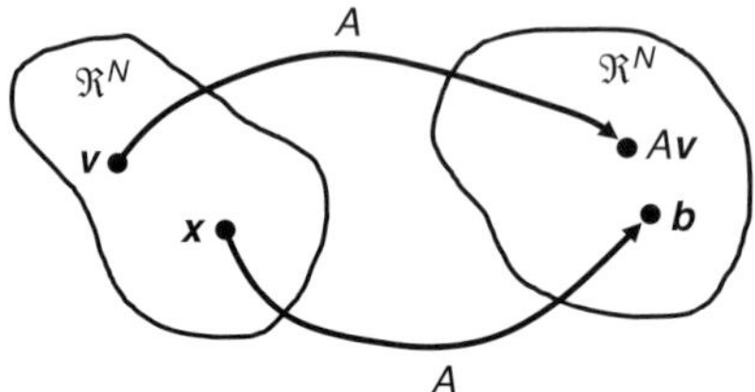

Figure 1.2 Interpretation of a real $N \times N$ matrix A as a linear transformation from the *domain* $\mathfrak{R}^N$ into the *codomain* $\mathfrak{R}^N$.

We have introduced the notation $\mathfrak{R}^N$ for the set of all real N-dimensional vectors. The term *set* merely refers to a collection of objects; however, $\mathfrak{R}^N$ possesses many other properties, including

closure under addition

$$\text{if } v \in \mathfrak{R}^N \text{ and } w \in \mathfrak{R}^N, \text{ then the vector } v + w \text{ is also in } \mathfrak{R}^N;$$

closure under multiplication by a real scalar

$$\text{if } v \in \mathfrak{R}^N \text{ and } c \in \mathfrak{R}, \text{ then } cv \in \mathfrak{R}^N.$$

Also, for any $u, v, w \in \mathfrak{R}^N$, any $c_1, c_2 \in \mathfrak{R}$, a null vector $\mathbf{0} \in \mathfrak{R}^N$, and for every $v \in \mathfrak{R}^N$ defining an additive inverse, $-v \in \mathfrak{R}^N$, we have the identities

$$\begin{aligned}
u + (v + w) &= (u + v) + w & c_1(v + u) &= c_1 v + c_1 u \\
u + v &= v + u & (c_1 + c_2)v &= c_1 v + c_2 v \\
v + 0 &= v & (c_1 c_2)v &= c_1(c_2 v) \\
v + (-v) &= 0 & 1v &= v
\end{aligned} \tag{1.123}$$

As these properties hold for $\mathfrak{R}^N$, we say that $\mathfrak{R}^N$ not only constitutes a set, but that it is also a *vector space*.

Using the concept of the vector space $\mathfrak{R}^N$, we now interpret the $N \times N$ real matrix A in a new fashion. We note that for any $v \in \mathfrak{R}^N$, the matrix-vector product with A is also in $\mathfrak{R}^N$, $Av \in \mathfrak{R}^N$. This product is formed by the rule

$$Av = \begin{bmatrix} a_{11} & a_{12} & \cdots & a_{1N} \\ a_{21} & a_{22} & \cdots & a_{2N} \\ \vdots & \vdots & & \vdots \\ a_{N1} & a_{N2} & \cdots & a_{NN} \end{bmatrix} \begin{bmatrix} v_1 \\ v_2 \\ \vdots \\ v_N \end{bmatrix} = \begin{bmatrix} a_{11}v_1 + a_{12}v_2 + \cdots + a_{1N}v_N \\ a_{21}v_1 + a_{22}v_2 + \cdots + a_{2N}v_N \\ \vdots \\ a_{N1}v_1 + a_{N2}v_2 + \cdots + a_{NN}v_N \end{bmatrix} \tag{1.124}$$

Thus, A maps any vector $v \in \mathfrak{R}^N$ into another vector $Av \in \mathfrak{R}^N$.

Also, since for any $v, w \in \mathfrak{R}^N$, $c \in \mathfrak{R}$, we have the linearity properties

$$A(v + w) = Av + Aw \qquad A(cv) = cAv \tag{1.125}$$

we say that A is a *linear transformation* mapping $\mathfrak{R}^N$ into itself, $A : \mathfrak{R}^N \to \mathfrak{R}^N$. The action of A upon vectors $v, w \in \mathfrak{R}^N$ is sketched in Figure 1.2. From this interpretation of A as

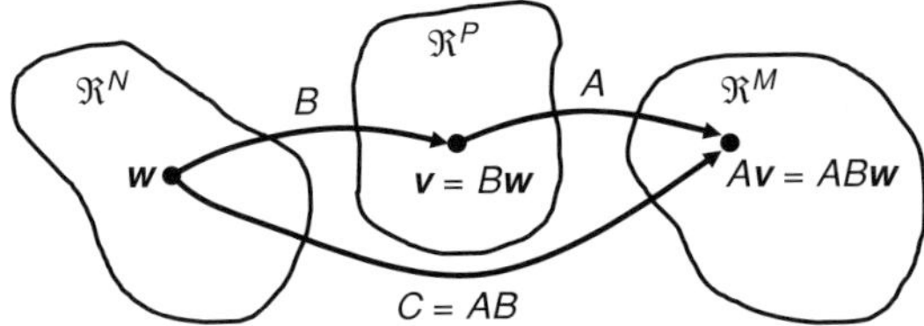

Figure 1.3 Interpreting matrix multiplication $C = AB$ as sequence of linear transformations.

a linear transformation, we can understand better the existence and uniqueness properties of $Ax = b$. We see that a solution exists if there is some vector $x \in \mathfrak{R}^N$ that is mapped into $b \in \mathfrak{R}^N$ by A, and that this solution is unique if there exists no other $y \neq x$ that is also mapped into b by A.

Multiplication of matrices

Before continuing with our discussion of existence and uniqueness, let us see how the interpretation of a matrix as a linear transformation immediately dictates a rule for multiplying two matrices. Let us say that we have an $M \times P$ real matrix A that maps every $v \in \mathfrak{R}^P$ into some $Av \in \mathfrak{R}^M$. In addition, we also have some $P \times N$ real matrix B that maps every $w \in \mathfrak{R}^N$ into some $Bw \in \mathfrak{R}^P$. We therefore make the association $v = Bw$ and define a composite mapping from $\mathfrak{R}^N$ into $\mathfrak{R}^M$, C, such that for $w \in \mathfrak{R}^N$, we obtain $Cw \in \mathfrak{R}^M$ by first multiplying w by B and then by A (Figure 1.3). That is,

$$v = Bw \qquad Cw = Av = A(Bw) = (AB)w \tag{1.126}$$

This defines the $M \times N$ matrix C, obtained by *matrix multiplication*,

$$C = AB \tag{1.127}$$

To construct C, we compute for $w \in \mathfrak{R}^N$ the vector $Bw \in \mathfrak{R}^P$,

$$Bw = \begin{bmatrix} b_{11} & b_{12} & \ldots & b_{1N} \\ b_{21} & b_{22} & \ldots & b_{2N} \\ \vdots & \vdots & & \vdots \\ b_{P1} & b_{P2} & \ldots & b_{PN} \end{bmatrix} \begin{bmatrix} w_1 \\ w_2 \\ \vdots \\ w_N \end{bmatrix} = \begin{bmatrix} \sum_{j=1}^{N} b_{1j} w_j \\ \vdots \\ \sum_{j=1}^{N} b_{Pj} w_j \end{bmatrix} \tag{1.128}$$

We next apply A to $Bw \in \mathfrak{R}^P$,

$$A(Bw) = \begin{bmatrix} a_{11} & a_{12} & \ldots & a_{1P} \\ a_{21} & a_{22} & \ldots & a_{2P} \\ \vdots & \vdots & & \vdots \\ a_{M1} & a_{M2} & \ldots & a_{MP} \end{bmatrix} \begin{bmatrix} \sum_{j=1}^{N} b_{1j} w_j \\ \vdots \\ \sum_{j=1}^{N} b_{Pj} w_j \end{bmatrix} = \begin{bmatrix} \sum_{k=1}^{P} a_{1k} \sum_{j=1}^{N} b_{kj} w_j \\ \vdots \\ \sum_{k=1}^{P} a_{Mk} \sum_{j=1}^{N} b_{kj} w_j \end{bmatrix} \tag{1.129}$$

Comparing this to the product of an $M \times N$ matrix C and $\boldsymbol{w} \in \mathfrak{R}^N$ yields

$$
C\boldsymbol{w} =
\begin{bmatrix}
c_{11} & c_{12} & \cdots & c_{1N} \\
c_{21} & c_{22} & \cdots & c_{2N} \\
\vdots & \vdots & & \vdots \\
c_{M1} & c_{M2} & \cdots & c_{MN}
\end{bmatrix}
\begin{bmatrix}
w_1 \\ w_2 \\ \vdots \\ w_N
\end{bmatrix}
=
\begin{bmatrix}
\sum\limits_{j=1}^{N} c_{1j} w_j \\
\vdots \\
\sum\limits_{j=1}^{N} c_{Mj} w_j
\end{bmatrix}
$$

$$
=
\begin{bmatrix}
\sum\limits_{j=1}^{N} \left(\sum\limits_{k=1}^{P} a_{1k} b_{kj} \right) w_j \\
\vdots \\
\sum\limits_{j=1}^{N} \left(\sum\limits_{k=1}^{P} a_{Mk} b_{kj} \right) w_j
\end{bmatrix}
\tag{1.130}
$$

We therefore have the following rule for *matrix multiplication* to form the $M \times N$ matrix product $C = AB$ of an $M \times P$ matrix A and a $P \times N$ matrix B:

$$
c_{ij} = \sum_{k=1}^{P} a_{ik} b_{kj}
\tag{1.131}
$$

To obtain c_{ij}, we sum the products of the elements of A in row i with those of B in column j. The matrix product AB is defined only if the number of columns of A equals the number of rows of B.

We also note that, in general, matrix multiplication is not commutative,

$$
AB \neq BA
\tag{1.132}
$$

From this rule for matrix multiplication, we obtain the following relation for the transpose of a product of two equal-sized square matrices,

$$
(AB)^{\mathrm{T}} = B^{\mathrm{T}} A^{\mathrm{T}}
\tag{1.133}
$$

Vector spaces and basis sets

We must discuss one more topic before considering the existence and uniqueness of solutions to linear systems: the use of basis sets. The set of vectors $\{ \boldsymbol{b}^{[1]}, \boldsymbol{b}^{[2]}, \ldots, \boldsymbol{b}^{[P]} \}$ in $\mathfrak{R}^N$ is said to be *linearly independent* if there exists no set of real scalars $\{ c_1, c_2, \ldots, c_P \}$ except that of all zeros, $c_j = 0$, $j = 1, 2, \ldots, P$, such that

$$
c_1 \boldsymbol{b}^{[1]} + c_2 \boldsymbol{b}^{[2]} + \cdots + c_P \boldsymbol{b}^{[P]} = \boldsymbol{0}
\tag{1.134}
$$

In other words, if a set is linearly independent, no member of the set can be expressed as a linear combination of the others. For vectors in $\mathfrak{R}^N$, a set $\{ \boldsymbol{b}^{[1]}, \boldsymbol{b}^{[2]}, \ldots, \boldsymbol{b}^{[P]} \}$ can be linearly independent only if $P \leq N$.

If $P = N$, we say that the set of N linearly independent vectors $\{ \boldsymbol{b}^{[1]}, \boldsymbol{b}^{[2]}, \ldots, \boldsymbol{b}^{[N]} \}$ forms a *basis set* for $\mathfrak{R}^N$. Then any vector $\boldsymbol{v} \in \mathfrak{R}^N$ may be expressed as a linear combination

$$
\boldsymbol{v} = c_1 \boldsymbol{b}^{[1]} + c_2 \boldsymbol{b}^{[2]} + \cdots + c_N \boldsymbol{b}^{[N]} \qquad c_j \in \mathfrak{R}
\tag{1.135}
$$

A common question is

Given $v \in \Re^N$ and a linearly independent basis $\{b^{[1]}, b^{[2]}, \ldots, b^{[N]}\}$, what is the set of scalar coefficients $\{c_1, c_2, \ldots, c_N\}$ that represent v as the *basis set expansion* $v = c_1 b^{[1]} + c_2 b^{[2]} + \cdots + c_N b^{[N]}$?

To answer this question, we note that we obtain a set of N linear equations by forming the dot products of v with each $b^{[k]}$,

$$
\begin{aligned}
c_1\left(b^{[1]} \cdot b^{[1]}\right) + c_2\left(b^{[1]} \cdot b^{[2]}\right) + \cdots + c_N\left(b^{[1]} \cdot b^{[N]}\right) &= \left(b^{[1]} \cdot v\right) \\
c_1\left(b^{[2]} \cdot b^{[1]}\right) + c_2\left(b^{[2]} \cdot b^{[2]}\right) + \cdots + c_N\left(b^{[2]} \cdot b^{[N]}\right) &= \left(b^{[2]} \cdot v\right) \\
&\ \ \vdots \\
c_1\left(b^{[N]} \cdot b^{[1]}\right) + c_2\left(b^{[N]} \cdot b^{[2]}\right) + \cdots + c_N\left(b^{[N]} \cdot b^{[N]}\right) &= \left(b^{[N]} \cdot v\right)
\end{aligned}
\tag{1.136}
$$

In matrix-vector form, this system is written as

$$
\begin{bmatrix}
\left(b^{[1]} \cdot b^{[1]}\right) & \left(b^{[1]} \cdot b^{[2]}\right) & \cdots & \left(b^{[1]} \cdot b^{[N]}\right) \\
\left(b^{[2]} \cdot b^{[1]}\right) & \left(b^{[2]} \cdot b^{[2]}\right) & \cdots & \left(b^{[2]} \cdot b^{[N]}\right) \\
\vdots & \vdots & & \vdots \\
\left(b^{[N]} \cdot b^{[1]}\right) & \left(b^{[N]} \cdot b^{[2]}\right) & \cdots & \left(b^{[N]} \cdot b^{[N]}\right)
\end{bmatrix}
\begin{bmatrix} c_1 \\ c_2 \\ \vdots \\ c_N \end{bmatrix}
=
\begin{bmatrix}
\left(b^{[1]} \cdot v\right) \\
\left(b^{[2]} \cdot v\right) \\
\vdots \\
\left(b^{[N]} \cdot v\right)
\end{bmatrix}
\tag{1.137}
$$

To compute the scalar coefficients in this manner, we must solve a system of N linear equations, requiring $\sim N^3$ FLOPs. However, if we were to use a basis set $\{w^{[1]}, w^{[2]}, \ldots, w^{[N]}\}$ that is *orthogonal*, i.e., the dot product between any two unlike members is zero,

$$
w^{[i]} \cdot w^{[j]} = \left|w^{[i]}\right|^2 \delta_{ij} =
\begin{cases}
\left|w^{[i]}\right|^2, & i = j \\
0, & i \neq j
\end{cases}
\tag{1.138}
$$

the coefficients are obtained far more easily,

$$
\begin{bmatrix}
\left|w^{[1]}\right|^2 & & & \\
& \left|w^{[2]}\right|^2 & & \\
& & \ddots & \\
& & & \left|w^{[3]}\right|^2
\end{bmatrix}
\begin{bmatrix} c_1 \\ c_2 \\ \vdots \\ c_N \end{bmatrix}
=
\begin{bmatrix}
\left(w^{[1]} \cdot v\right) \\
\left(w^{[2]} \cdot v\right) \\
\vdots \\
\left(w^{[N]} \cdot v\right)
\end{bmatrix}
\qquad
\begin{aligned}
c_j &= \frac{\left(w^{[j]} \cdot v\right)}{\left|w^{[j]}\right|^2} \\
j &= 1, 2, \ldots, N
\end{aligned}
\tag{1.139}
$$

For this reason, the use of orthogonal basis sets, and of *orthonormal basis sets* $\{u^{[1]}, u^{[2]}, \ldots, u^{[N]}\}$ that in addition have all $|u^{[k]}| = 1$, is common. For an orthonormal basis, we have simply

$$
c_j = u^{[j]} \cdot v \qquad v = \sum_{j=1}^{N} \left(u^{[j]} \cdot v\right) u^{[j]}
\tag{1.140}
$$

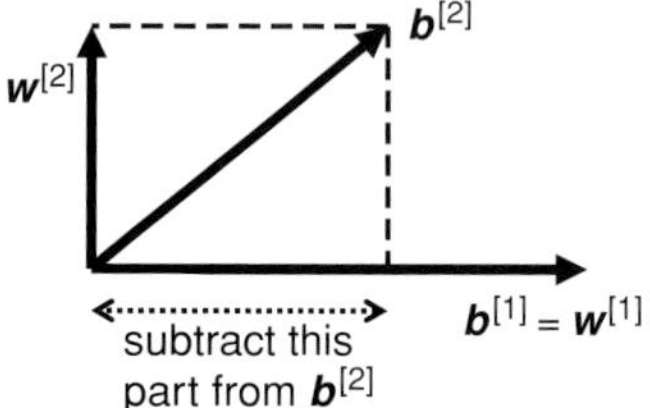

Figure 1.4 Gram–Schmidt method makes vectors orthogonal through projection operations.

Gram–Schmidt orthogonalization

We describe here a simple method for forming an orthogonal basis set $\{w^{[1]}, w^{[2]}, \ldots, w^{[N]}\}$ for $\Re^N$ from one, $\{b^{[1]}, b^{[2]}, \ldots, b^{[N]}\}$, that is merely linearly independent. As a first step, we simply assign $w^{[1]}$ to be

$$w^{[1]} = b^{[1]} \tag{1.141}$$

Next, we write $w^{[2]}$ as a linear combination of $b^{[2]}$ and $w^{[1]}$,

$$w^{[2]} = b^{[2]} + s_{21} w^{[1]} \qquad s_{21} \in \Re \tag{1.142}$$

and enforce that $w^{[2]}$ be orthogonal to $w^{[1]}$ through our choice of s_{21},

$$w^{[1]} \cdot w^{[2]} = 0 = w^{[1]} \cdot b^{[2]} + s_{21}\left(w^{[1]} \cdot w^{[1]}\right) \qquad s_{21} = -\frac{w^{[1]} \cdot b^{[2]}}{\left|w^{[1]}\right|^2} \tag{1.143}$$

so that

$$w^{[2]} = b^{[2]} - \left[\frac{w^{[1]} \cdot b^{[2]}}{\left|w^{[1]}\right|^2}\right] w^{[1]} \tag{1.144}$$

Essentially, we "project out" the component of $b^{[2]}$ in the direction of $w^{[1]}$, as shown in Figure 1.4.

We next write

$$w^{[3]} = b^{[3]} + s_{31} w^{[1]} + s_{32} w^{[2]} \tag{1.145}$$

and choose s_{31} and s_{32} such that $w^{[3]} \cdot w^{[1]} = w^{[3]} \cdot w^{[2]} = 0$, to obtain

$$w^{[3]} = b^{[3]} - \left[\frac{w^{[1]} \cdot b^{[3]}}{\left|w^{[1]}\right|^2}\right] w^{[1]} - \left[\frac{w^{[2]} \cdot b^{[3]}}{\left|w^{[2]}\right|^2}\right] w^{[2]} \tag{1.146}$$

This process may be continued to fill out the orthogonal basis set $\{w^{[1]}, w^{[2]}, \ldots, w^{[N]}\}$, where

$$w^{[k]} = b^{[k]} - \sum_{j=1}^{k-1} \left[\frac{w^{[j]} \cdot b^{[k]}}{\left|w^{[j]}\right|^2}\right] w^{[j]} \tag{1.147}$$

From this orthogonal basis set $\{w^{[1]}, w^{[2]}, \ldots, w^{[N]}\}$, an orthonormal one $\{u^{[1]}, u^{[2]}, \ldots, u^{[N]}\}$ may be formed by

$$u^{[j]} = w^{[j]} / \left| w^{[j]} \right| \tag{1.148}$$

or we can normalize the vectors as we generate them,

$$w^{[k]} = b^{[k]} - \sum_{j=1}^{k-1} \left[u^{[j]} \cdot b^{[k]} \right] u^{[j]} \qquad u^{[k]} = \frac{w^{[k]}}{\left| w^{[k]} \right|} \qquad k = 1, 2, \ldots, N \tag{1.149}$$

While this method is straightforward, for very large N, the propagation of round-off errors is a problem. We discuss in Chapter 3 the use of eigenvalue analysis to generate an orthonormal basis set with better error properties.

Subspaces and the span of a set of vectors

Let us say that we have some set of vectors $\{b^{[1]}, b^{[2]}, \ldots, b^{[P]}\}$ in $\Re^N$ with $P \leq N$. We then define the *span* of $\{b^{[1]}, b^{[2]}, \ldots, b^{[P]}\}$ as the set of all vectors $v \in \Re^N$ that can be written as a linear combination of members of the set,

$$\mathrm{span}\{b^{[1]}, b^{[2]}, \ldots, b^{[P]}\} \equiv \{v \in \Re^N | v = c_1 b^{[1]} + c_2 b^{[2]} + \cdots + c_P b^{[P]}\} \tag{1.150}$$

We see that span $\{b^{[1]}, b^{[2]}, \ldots, b^{[P]}\}$ possesses the properties of closure under addition, closure under multiplication by a real scalar, and all of the other properties (1.123) required to define a vector space. We thus say that span$\{b^{[1]}, b^{[2]}, \ldots, b^{[P]}\}$ is a *subspace* of $\Re^N$.

Let S be some subspace in $\Re^N$. The *dimension* of S is P, $\dim(S) = P$, if there exists some spanning set $\{b^{[1]}, b^{[2]}, \ldots, b^{[P]}\}$ that is linearly independent. If $\dim(S) = P$, then any spanning set of S with more than P members must be dependent. Since any linearly independent spanning set of $\Re^N$ must contain N vectors, $\dim(\Re^N) = N$.

The null space and the existence/uniqueness of solutions

We are now in a position to consider the existence and uniqueness properties of the linear system $Ax = b$, where $x, b \in \Re^N$ and A is an $N \times N$ real matrix. Viewing A as a linear transformation, the problem of solving $Ax = b$ may be envisioned as finding some $x \in \Re^N$ that is mapped by A into a specific $b \in \Re^N$. We now ask the following questions:

For particular A and b, when does there exist some $x \in \Re^N$ such that $Ax = b$? *(existence of solutions)*

For particular A and b, if a solution $x \in \Re^N$ exists, when is it the only solution? *(uniqueness of solution)*

The answers to these questions depend upon the nature of the *null space*, or *kernel*, of A, K_A, that is defined as the subspace of all vectors $w \in \Re^N$ that are mapped by A into the *null vector*, $\mathbf{0}$ (Figure 1.5). We know that the null space must contain at least the null vector itself, as for any A,

$$A\mathbf{0} = \mathbf{0} \tag{1.151}$$

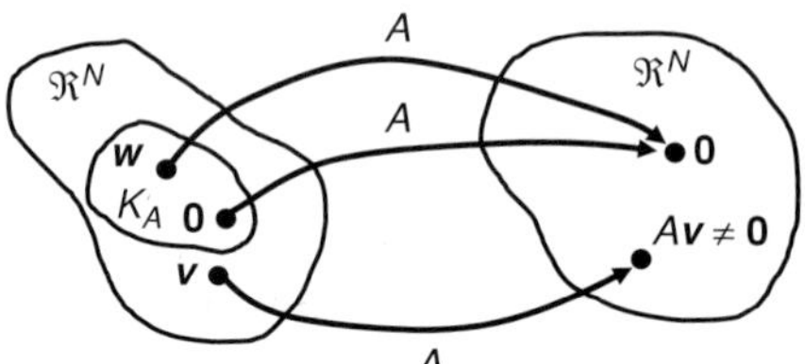

Figure 1.5 The null space (kernel) of A, K_A.

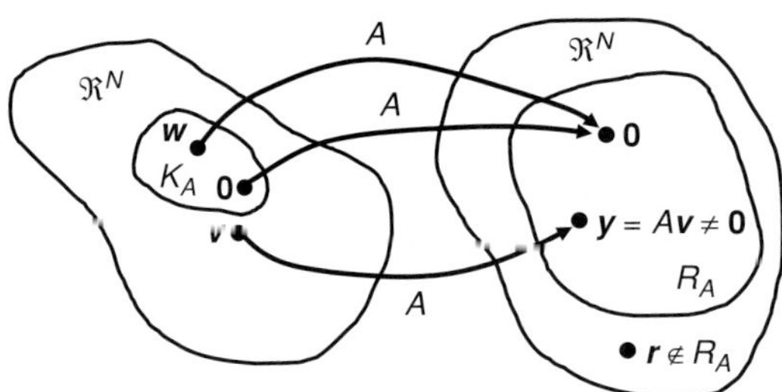

Figure 1.6 Venn diagram of linear transformation by A from domain $\Re^N$ into codomain $\Re^N$ showing the kernel and the range subspaces.

If the null space contains only the null vector, K_A is said to be *empty*. As we now show, if this is the case, then $Ax = b$ must have a unique solution $x \in \Re^N$ for any possible $b \in \Re^N$. However, if the null space contains any other non zero vectors (*i.e.*, $Aw = 0$ with $w \neq 0$), there is no unique solution. Then, depending upon the particular $b \in \Re^N$, there may be either no solution at all or an infinite number of them.

Theorem Uniqueness of solution for $Ax = b$ *Let $x \in \Re^N$ be a solution to the linear system $Ax = b$, where $b \in \Re^N$ and A is an $N \times N$ real matrix. If the null space (kernel) of A contains only the null vector, $K_A = 0$, this solution is unique.*

Proof Let $y \in \Re^N$ be some vector, not necessarily x, that satisfies the system of equations, $Ay = b$. We then define $v = y - x$, so that

$$Ay = A(x + v) = Ax + Av = b + Av \tag{1.152}$$

If $Ay = b$, then $v \in K_A$, as $Av = 0$. If the null space is empty, $K_A = 0$, we must have $v = 0$ and the solution x is unique. *QED*

Now that we have a theorem for uniqueness, let us consider existence. To do so, we define the *range* of A, R_A, as the subspace of all vectors $y \in \Re^N$ for which there exists some $v \in \Re^N$ such that $Av = y$. Formally, we write

$$R_A \equiv \left\{ y \in \Re^N \middle| \exists v \in \Re^N, \, Av = y \right\} \tag{1.153}$$

Figure 1.6 shows the relationship between the range and the kernel of A.

Theorem Existence of solutions for $Ax = b$ *Let A be a real $N \times N$ matrix with a null space (kernel) K_A and range R_A, and let $b \in \Re^N$. Then,*

(I) *the dimensions of K_A and R_A satisfy the* dimension theorem,

$$\dim(K_A) + \dim(R_A) = N \tag{1.154}$$

(II) *If K_A contains only the null vector $\mathbf{0}$, $\dim(K_A) = 0$, and so $\dim(R_A) = N$. The range therefore completely fills $\mathfrak{R}^N$ so that every $\mathbf{b} \in \mathfrak{R}^N$ is in the range, $\mathbf{b} \in R_A$, and for any $\mathbf{b} \in \mathfrak{R}^N$, the system $A\mathbf{x} = \mathbf{b}$ has a solution.*

Proof (I) Let us define an orthonormal basis $\{\mathbf{u}^{[1]}, \mathbf{u}^{[2]}, \ldots, \mathbf{u}^{[P]}, \mathbf{u}^{[P+1]}, \ldots, \mathbf{u}^{[N]}\}$ for $\mathfrak{R}^N$ such that the first P members of the basis span the kernel of A,

$$K_A = \operatorname{span}\{\mathbf{u}^{[1]}, \mathbf{u}^{[2]}, \ldots, \mathbf{u}^{[P]}\} \tag{1.155}$$

Since the kernel is a subspace and thus satisfies all of the properties of a vector space, we can always form such a basis. Therefore, we can write any $\mathbf{w} \in K_A$ as

$$\mathbf{w} = c_1 \mathbf{u}^{[1]} + c_2 \mathbf{u}^{[2]} + \cdots + c_P \mathbf{u}^{[P]} \tag{1.156}$$

and

$$\dim(K_A) = P \tag{1.157}$$

We now write any arbitrary vector $\mathbf{v} \in \mathfrak{R}^N$ as an expansion in this orthonormal basis with the scalar coefficients $v_j = \mathbf{v} \cdot \mathbf{u}^{[j]}$,

$$\mathbf{v} = v_1 \mathbf{u}^{[1]} + \cdots + v_P \mathbf{u}^{[P]} + v_{P+1} \mathbf{u}^{[P+1]} + \cdots + v_N \mathbf{u}^{[N]} \tag{1.158}$$

We now operate on this vector by A,

$$\begin{aligned} A\mathbf{v} &= \left[v_1 A\mathbf{u}^{[1]} + \cdots + v_P A\mathbf{u}^{[P]} \right] + \left[v_{P+1} A\mathbf{u}^{[P+1]} + \cdots + v_N A\mathbf{u}^{[N]} \right] \\ A\mathbf{v} &= A\left[v_1 \mathbf{u}^{[1]} + \cdots + v_P \mathbf{u}^{[P]} \right] + \left[v_{P+1} A\mathbf{u}^{[P+1]} + \cdots + v_N A\mathbf{u}^{[N]} \right] \end{aligned} \tag{1.159}$$

As $v_1 \mathbf{u}^{[1]} + \cdots + v_P \mathbf{u}^{[P]} \in K_A$, this equation becomes

$$A\mathbf{v} = v_{P+1} A\mathbf{u}^{[P+1]} + \cdots + v_N A\mathbf{u}^{[N]} \tag{1.160}$$

Since the range of A is the set of $A\mathbf{v}$ for all $\mathbf{v} \in \mathfrak{R}^N$, we see that any vector in the range can be written as a linear combination of the $N - P$ basis vectors $\{A\mathbf{u}^{[P+1]}, \ldots, A\mathbf{u}^{[N]}\}$, and so $\dim(R_A) = N - P$. Thus, $\dim(K_A) + \dim(R_A) = N$.

(II) follows directly from (I). $\qquad QED$

What happens to the existence and uniqueness of solutions to $A\mathbf{x} = \mathbf{b}$ if K_A is not empty? Let us say that $\dim(K_A) = P > 0$, and that we form the orthonormal basis for A as in the proof above, such that we can write any $\mathbf{w} \in K_A$ as

$$\mathbf{w} = c_1 \mathbf{u}^{[1]} + c_2 \mathbf{u}^{[2]} + \cdots + c_P \mathbf{u}^{[P]} \tag{1.161}$$

We now consider the arbitrary vector $\mathbf{v} \in \mathfrak{R}^N$, written as

$$\mathbf{v} = v_1 \mathbf{u}^{[1]} + \cdots + v_P \mathbf{u}^{[P]} + v_{P+1} \mathbf{u}^{[P+1]} + \cdots + v_N \mathbf{u}^{[N]} \tag{1.162}$$

and use a similar expansion for our specific $\boldsymbol{b} \in \mathfrak{R}^N$ in $A\boldsymbol{x} = \boldsymbol{b}$,

$$\boldsymbol{b} = b_1\boldsymbol{u}^{[1]} + \cdots + b_P\boldsymbol{u}^{[P]} + b_{P+1}\boldsymbol{u}^{[P+1]} + \cdots + b_N\boldsymbol{u}^{[N]} \tag{1.163}$$

We now write $A\boldsymbol{v} = \boldsymbol{b}$ to consider when a solution $\boldsymbol{v} = \boldsymbol{x}$ exists,

$$A\big[v_1\boldsymbol{u}^{[1]} + \cdots + v_P\boldsymbol{u}^{[P]}\big] + A\big[v_{P+1}\boldsymbol{u}^{[P+1]} + \cdots + v_N\boldsymbol{u}^{[N]}\big]$$
$$= b_1\boldsymbol{u}^{[1]} + \cdots + b_P\boldsymbol{u}^{[P]} + b_{P+1}\boldsymbol{u}^{[P+1]} + \cdots + b_N\boldsymbol{u}^{[N]} \tag{1.164}$$

Noting that $v_1\boldsymbol{u}^{[1]} + \cdots + v_P\boldsymbol{u}^{[P]} \in K_A$, we have

$$A\big[v_{P+1}\boldsymbol{u}^{[P+1]} + \cdots + v_N\boldsymbol{u}^{[N]}\big]$$
$$= b_1\boldsymbol{u}^{[1]} + \cdots + b_P\boldsymbol{u}^{[P]} + b_{P+1}\boldsymbol{u}^{[P+1]} + \cdots + b_N\boldsymbol{u}^{[N]} \tag{1.165}$$

The vector $A\boldsymbol{v}$ on the left-hand side must be in R_A, and thus must have no nonzero component in K_A; i.e., $A[v_{P+1}\boldsymbol{u}^{[P+1]} + \cdots + v_N\boldsymbol{u}^{[N]}] \cdot \boldsymbol{w} = 0$ for any $\boldsymbol{w} \in K_A$. If any of the coefficients $b_1, \ldots, b_P$ on the right-hand side are nonzero, the two sides of the equation cannot agree. Therefore, a solution to $A\boldsymbol{x} = \boldsymbol{b}$ exists if for every $\boldsymbol{w} \in K_A$, $\boldsymbol{b} \cdot \boldsymbol{w} = 0$. However, if the null space is not empty, this solution cannot be unique, because for any $\boldsymbol{w} \in K_A$, we also have

$$A(\boldsymbol{x} + \boldsymbol{w}) = A\boldsymbol{x} + A\boldsymbol{w} = \boldsymbol{b} + \boldsymbol{0} = \boldsymbol{b} \tag{1.166}$$

A system $A\boldsymbol{x} = \boldsymbol{b}$ whose matrix has a nonempty kernel either has no solution, or an infinite number of them.

We have now identified when the linear system $A\boldsymbol{x} = \boldsymbol{b}$ will have exactly one solution, no solution, or an infinite number of solutions; however, these conditions are rather abstract. Later, in our discussion of eigenvalue analysis, we see how to implement these conditions for specific matrices A and vectors $\boldsymbol{b}$.

Here, we have introduced some rather abstract concepts (vector spaces, linear transformations) to analyze the properties of linear algebraic systems. For a fuller theoretical treatment of these concepts, and their extension to include systems of differential equations, consult Naylor & Sell (1982).

The determinant

In the previous section we have found that the null space of A is very important in determining whether $A\boldsymbol{x} = \boldsymbol{b}$ has a unique solution. For a single equation, $ax = b$, it is easy to find whether the null space is empty. If $a \neq 0$, the equation has a unique solution $x = b/a$. If $a = 0$, there is no solution if $b \neq 0$ and an infinite number if $b = 0$.

We would like to determine similarly whether $A\boldsymbol{x} = \boldsymbol{b}$ has a unique solution for $N \geq 1$. We thus define the *determinant* of A as

$$\det(A) = |A| = \begin{cases} c \neq 0, & \text{if } K_A = \boldsymbol{0} \\ 0, & \text{if } \exists \boldsymbol{w} \in K_A, \boldsymbol{w} \neq \boldsymbol{0} \end{cases} \tag{1.167}$$

If $\det(A) \neq 0$; i.e., if A is *nonsingular*, then $A\mathbf{x} = \mathbf{b}$ has a unique solution for all $\mathbf{b} \in \Re^N$. Otherwise, if $\det(A) = 0$; i.e., if A is *singular*, then no unique solution to $A\mathbf{x} = \mathbf{b}$ exists (either there are no solutions or an infinite number of them).

A fuller discussion of matrix determinants is provided in the supplemental material in the accompanying website, but here we summarize the main results. The determinant of a matrix is computed from its matrix elements by the formula

$$\det(A) = \sum_{i_1=1}^{N} \sum_{i_2=1}^{N} \cdots \sum_{i_N=1}^{N} \varepsilon_{i_1,i_2,\dots,i_N} a_{i_1,1} a_{i_2,2} \cdots a_{i_N,N} \tag{1.168}$$

$a_{i_k,k}$ is the element of A in row i_k, column k and $\varepsilon_{i_1,i_2,\dots,i_N}$ takes the values

$$\varepsilon_{i_1,i_2,\dots,i_N} = \begin{cases} 0, & \text{if any two of } \{i_1, i_2, \dots, i_N\} \text{ are equal} \\ 1, & \text{if } (i_1, i_2, \dots, i_N) \text{ is an even permuation of } (1, 2, \dots, N) \\ -1, & \text{if } (i_1, i_2, \dots, i_N) \text{ is an odd permuation of } (1, 2, \dots, N) \end{cases} \tag{1.169}$$

By an even or odd permutation of $(1, 2, \dots, N)$ we mean the following: if no two of $\{i_1, i_2, \dots, i_N\}$ are equal, then we can rearrange the ordered set $(i_1, i_2, \dots, i_N)$ into the ordered set $(1, 2, \dots, N)$ by some sequence of exchanges. Consider the set $(i_1, i_2, \dots, i_N) = (3, 2, 4, 1)$ which we can reorder into $(1, 2, 3, 4)$ by any of the following sequences, where at each step we underline the two members that are exchanged,

$$(3, 2, \underline{4}, \underline{1}) \to (3, \underline{2}, \underline{1}, 4) \to (\underline{3}, \underline{1}, 2, 4) \to (1, \underline{3}, \underline{2}, 4) \to (1, 2, 3, 4)$$
$$(\underline{3}, 2, 4, \underline{1}) \to (1, 2, \underline{4}, \underline{3}) \to (1, 2, 3, 4)$$
$$(\underline{3}, 2, \underline{4}, 1) \to (\underline{4}, 2, 3, \underline{1}) \to (\underline{2}, 4, 3, \underline{1}) \to (1, 4, \underline{3}, \underline{2}) \to (1, 4, \underline{2}, \underline{3})$$
$$\to (1, \underline{4}, 3, \underline{2}) \to (1, 2, 3, 4)$$

Some sequences reorder $(3, 2, 4, 1)$ to $(1, 2, 3, 4)$ in fewer steps than others, but for $(3, 2, 4, 1)$, any possible sequence of simple exchanges that yield $(1, 2, 3, 4)$ must comprise an even number of steps; $(3, 2, 4, 1)$ is therefore an *even permutation*. Similarly, because $(3, 2, 1, 4)$ can be reordered by the three exchanges

$$(3, \underline{2}, \underline{1}, 4) \to (\underline{3}, \underline{1}, 2, 4) \to (1, \underline{3}, \underline{2}, 4) \to (1, 2, 3, 4)$$

it is an *odd permutation*. The even or odd nature of a permutation is called its *parity*.

We can determine whether an ordered set $(i_1, i_2, \dots, i_N)$ has even or odd parity by the following procedure: for each $m = 1, 2, \dots, N$, let α_m be the number of integers in the set $\{i_{m+1}, i_{m+2}, \dots, i_N\}$ that are smaller than i_m. We sum these α_m to obtain

$$v = \sum_{m=1}^{N} \alpha_m \tag{1.170}$$

If $v = 0, 2, 4, 6, \dots, (i_1, i_2, \dots, i_N)$ is even. If $v = 1, 3, 5, 7, \dots$, it is odd.

Expansion by minors

From (1.168), it may be shown that $\det(A)$ may be written as an *expansion in minors* along any row j or column j as

$$\det(A) = \sum_{k=1}^{N} a_{jk}\, C_{jk} = \sum_{k=1}^{N} a_{kj}\, C_{kj} \qquad C_{jk} = (-1)^{j+k}\, M_{jk} \tag{1.171}$$

M_{jk} (the *minor* of a_{jk}) is the determinant of the $(N-1) \times (N-1)$ matrix obtained by deleting row j and column k of A. The quantity C_{jk} is the *cofactor* of a_{jk}.

The determinant of 2×2 and 3×3 matrices

Expansion of minors can be useful when computing the determinant of a matrix, as the minors are determinants of smaller matrices that are easier to compute. For example, the determinant of a 3×3 matrix can be written as

$$\det(A) = a_{11} \begin{vmatrix} a_{22} & a_{23} \\ a_{32} & a_{33} \end{vmatrix} - a_{12} \begin{vmatrix} a_{21} & a_{23} \\ a_{31} & a_{33} \end{vmatrix} + a_{13} \begin{vmatrix} a_{21} & a_{22} \\ a_{31} & a_{32} \end{vmatrix} \tag{1.172}$$

As the determinant of a 2×2 matrix is

$$\det(A) = \begin{vmatrix} a_{11} & a_{12} \\ a_{21} & a_{22} \end{vmatrix} = \varepsilon_{12} a_{11} a_{22} + \varepsilon_{21} a_{12} a_{21} = a_{11} a_{22} - a_{12} a_{21} \tag{1.173}$$

expansion by minors provides an easy means to compute the determinants of small matrices. A general numerical approach is discussed below.

General properties of the determinant function

We now consider some general properties of the determinant. The proofs for some are given in the supplemental material in the accompanying website.

Property I

The determinant of an $N \times N$ real matrix A equals that of its transpose, A^{T}.

Property II

If every element in a row (column) of A is zero, $\det(A) = 0$.

Property III

If every element in a row (column) of a matrix A is multiplied by a scalar c to form a matrix B, then $\det(B) = c \times det(A)$.

Property IV

If two rows (columns) of A are swapped to form B, $\det(B) = -\det(A)$.

Property V

If two rows (columns) of A are the same, then $\det(A) = 0$.

Property VI

If we decompose $\boldsymbol{a}^{(m)}$, the row of a matrix A, as

$$\boldsymbol{a}^{(m)} = \boldsymbol{b}^{(m)} + \boldsymbol{d}^{(m)} \tag{1.174}$$

to form the matrices

$$A = \begin{bmatrix} - & \boldsymbol{a}^{(1)} & - \\ & \vdots & \\ - & \boldsymbol{a}^{(m)} & - \\ & \vdots & \\ - & \boldsymbol{a}^{(N)} & - \end{bmatrix} \qquad B = \begin{bmatrix} - & \boldsymbol{a}^{(1)} & - \\ & \vdots & \\ - & \boldsymbol{b}^{(m)} & - \\ & \vdots & \\ - & \boldsymbol{a}^{(N)} & - \end{bmatrix} \qquad D = \begin{bmatrix} - & \boldsymbol{a}^{(1)} & - \\ & \vdots & \\ - & \boldsymbol{d}^{(m)} & - \\ & \vdots & \\ - & \boldsymbol{a}^{(N)} & - \end{bmatrix} \tag{1.175}$$

then

$$\det(A) = \det(B) + \det(D) \tag{1.176}$$

Property VII

If a matrix B is obtained from A by adding c times one row (column) of A to another row (column) of A, then $\det(B) = \det(A)$. That is, elementary row operations do not change the value of the determinant.

Property VIII

$$\det(AB) = \det(A) \times \det(B) \tag{1.177}$$

Property IX

If A is upper triangular or lower triangular, $\det(A)$ equals the product of the elements on the principal diagonal, $\det(A) = a_{11} \times a_{22} \times \cdots \times a_{NN}$.

Computing the determinant value

Properties VI–IX give us the fastest method to compute the determinant. Note that the general formula for $\det(A)$ is a sum of $N!$ nonzero terms, each requiring N scalar multiplications, and is therefore very costly to evaluate. Since Gaussian elimination merely consists of a sequence of elementary row operations that by property VII do not change the determinant

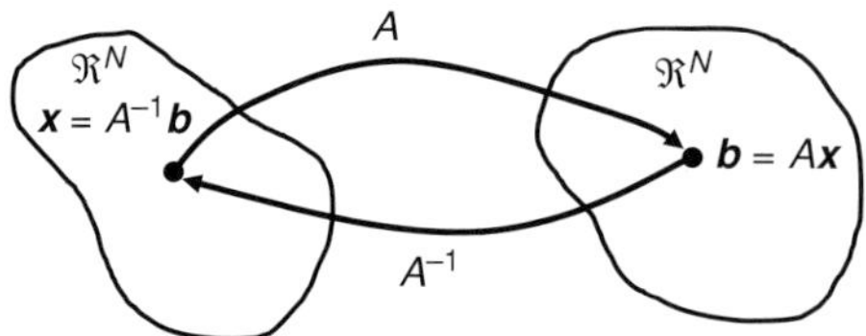

Figure 1.7 Defining A^{-1} as the inverse transformation of A.

value and some row exchanges (partial pivots), that by property IV only change the sign of the determinant, we obtain, after $\sim N^3$ FLOPs an upper triangular system U such that

$$\det(A) = \pm u_{11} \times u_{22} \times \cdots \times u_{NN} \tag{1.178}$$

Note that it takes nearly as long to obtain the value of the determinant (within a sign change) as it does to solve the system. Therefore, when faced with a new system, we attempt to solve it using Gaussian elimination without first checking that the determinant is nonzero.

Matrix inversion

Let us consider an $N \times N$ real matrix A with $\det(A) \neq 0$ so that for every $\boldsymbol{b} \in \Re^N$, there exists exactly one vector $\boldsymbol{x} \in \Re^N$ such that $A\boldsymbol{x} = \boldsymbol{b}$. We have interpreted A as a linear transformation because it maps every $\boldsymbol{v} \in \Re^N$ into a vector $A\boldsymbol{v} \in \Re^N$, and the properties of linearity hold:

$$A(\boldsymbol{v} + \boldsymbol{w}) = A\boldsymbol{v} + A\boldsymbol{w} \qquad A(c\boldsymbol{v}) = cA\boldsymbol{v} \tag{1.179}$$

Similarly, we define the inverse transformation, A^{-1}, such that if $A\boldsymbol{x} = \boldsymbol{b}$, then $\boldsymbol{x} = A^{-1}\boldsymbol{b}$. This mapping assigns a unique $\boldsymbol{x}$ to every $\boldsymbol{b}$ as long as $\det(A) \neq 0$. The relationship between A and A^{-1} is shown in Figure 1.7. The matrix A^{-1} that accomplishes this inverse transformation is called the *inverse* of A.

If A is singular, $\det(A) = 0$, $\dim(K_A) > 0$ and by the dimension theorem, the range of A cannot fill completely $\Re^N$. It is therefore possible to find some $\boldsymbol{r} \in \Re^N$ that is *not* in the range of A, such that there exist no $\boldsymbol{z} \in \Re^N$ for which $A\boldsymbol{z} = \boldsymbol{r}$ (Figure 1.8). If $\det(A) = 0$ it is therefore impossible to define an inverse A^{-1} that assigns to every $\boldsymbol{v} \in \Re^N$ a vector $A^{-1}\boldsymbol{v} \in \Re^N$ such that $A(A^{-1}\boldsymbol{v}) = \boldsymbol{v}$. If $\det(A) = 0$, A^{-1} does not exist (is not defined).

We now have a definition of A^{-1} as a linear transformation, but given a particular matrix A that is nonsingular, how do we compute A^{-1}? The (j, k) element of A^{-1} may be written in terms of the cofactor of a_{jk} as Cramer's rule

$$(A^{-1})_{jk} = \frac{C_{jk}}{\det(A)} \tag{1.180}$$

Numerical use of this equation is not very practical.

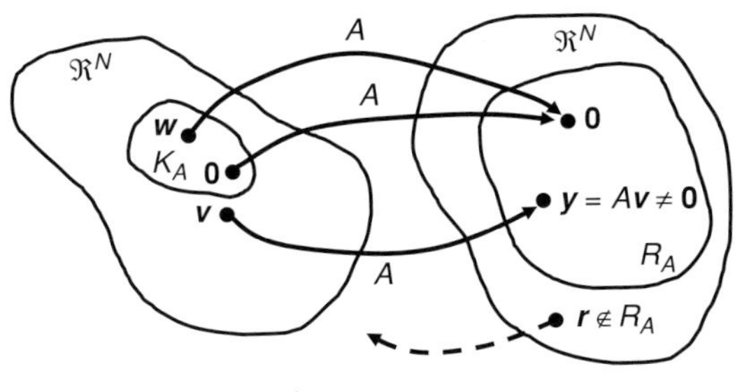

Figure 1.8 Defining A^{-1} is impossible when $\det(A) = 0$.

We obtain a more efficient method of computing A^{-1} by first noting that if A^{-1} exists, then for every $v \in \mathfrak{R}^N$, $A(A^{-1}v) = v$. We next define the identity matrix, I, for which

$$Iv = v \qquad \forall v \in \mathfrak{R}^N \tag{1.181}$$

By the rule of matrix multiplication, I must take the form

$$I = \begin{bmatrix} 1 & & & & \\ & 1 & & & \\ & & 1 & & \\ & & & \ddots & \\ & & & & 1 \end{bmatrix} \tag{1.182}$$

Since $A(A^{-1}v) = v = A^{-1}(Av)$, the inverse matrix A^{-1} must be related to A by

$$A^{-1}A = AA^{-1} = I \tag{1.183}$$

We compute A^{-1} using the rule for matrix multiplication

$$AB = A \begin{bmatrix} | & | & & | \\ b^{(1)} & b^{(2)} & \cdots & b^{(N)} \\ | & | & & | \end{bmatrix} = \begin{bmatrix} | & | & & | \\ Ab^{(1)} & Ab^{(2)} & \cdots & Ab^{(N)} \\ | & | & & | \end{bmatrix} \tag{1.184}$$

Writing A^{-1} and I in terms of their column vectors, $AA^{-1} = I$ becomes

$$A \begin{bmatrix} | & | & & | \\ \tilde{a}^{(1)} & \tilde{a}^{(2)} & \cdots & \tilde{a}^{(N)} \\ | & | & & | \end{bmatrix} = \begin{bmatrix} | & | & & | \\ A\tilde{a}^{(1)} & A\tilde{a}^{(2)} & \cdots & A\tilde{a}^{(N)} \\ | & | & & | \end{bmatrix}$$

$$= \begin{bmatrix} | & | & & | \\ e^{[1]} & e^{[2]} & \cdots & e^{[N]} \\ | & | & & | \end{bmatrix} \tag{1.185}$$

The column vectors of A^{-1} are obtained by solving the N linear systems

$$A\tilde{a}^{(k)} = e^{[k]} \qquad k = 1, 2, \ldots, N \tag{1.186}$$

At first glance, it would appear that obtaining A^{-1} requires a factor N more effort than solving a single linear system $Ax = b$; however, this overlooks the very important fact that every linear system $A\tilde{a}^{(k)} = e^{[k]}$ to be solved has the same matrix A. Therefore, during Gaussian elimination, the sequence of row operations and pivoting is the same for each system. It would be very helpful if we could do this elimination only once, and store all of the details of how Gaussian elimination is performed for A, so that for subsequent solution of any system $Ax = b$, we need only perform the backward substitution step for the new b. Then, the work of computing the inverse scales only as N^3 rather than N^4.

Matrix factorization

We have seen that the column vectors of A^{-1} may be computed one-by-one by solving the N linear systems

$$A\tilde{a}^{(k)} = e^{[k]} \qquad k = 1, 2, \ldots, N \tag{1.187}$$

More generally, let us say that we wish to solve some set of M linear systems that have the same matrix A but different vectors $b^{[1]}, b^{[2]}, \ldots, b^{[M]}$. In this section, we show that it is possible in this case to do the work of Gaussian elimination only once, and to factor A into a product of two triangular matrices. Thus, each system with a new vector $b^{[k]}$ can be solved with two successive substitutions. As these substitutions require only N^2 FLOPs each, a much smaller number than the $2N^3/3$ FLOPs required for Gaussian elimination, this factorization can save much effort.

LU decomposition

Let us say that we want to solve M linear systems that have the same matrix,

$$Ax^{[k]} = b^{[k]} \qquad k = 1, 2, \ldots, M \tag{1.188}$$

Let us assume that it is possible to decompose A into the product of a lower triangular matrix L and an upper triangular matrix U, $A = LU$. Then,

$$Ax^{[k]} = LUx^{[k]} = b^{[k]} \tag{1.189}$$

We obtain $x^{[k]}$ quickly by solving in succession two triangular problems, the first by forward substitution and the second by backward substitution,

$$\begin{aligned} Lc^{[k]} &= b^{[k]} \\ Ux^{[k]} &= c^{[k]} \end{aligned} \tag{1.190}$$

We now show that with some additional book-keeping, Gaussian elimination without partial pivoting returns just such an LU factorization, $A = LU$. We now perform Gaussian elimination on matrix A alone, without augmenting it with b,

$$A = \begin{bmatrix} a_{11} & a_{12} & a_{13} & \cdots & a_{1N} \\ a_{21} & a_{22} & a_{23} & \cdots & a_{2N} \\ a_{31} & a_{32} & a_{33} & \cdots & a_{3N} \\ \vdots & \vdots & \vdots & \vdots & \vdots \\ a_{N1} & a_{N2} & a_{N3} & \cdots & a_{NN} \end{bmatrix} \tag{1.191}$$

We perform the row operation $2 \leftarrow 2 - \lambda_{21} \times 1$ with $\lambda_{21} = a_{21}/a_{11}$ to obtain

$$A^{(2,1)} = \begin{bmatrix} a_{11} & a_{12} & a_{13} & \cdots & a_{1N} \\ 0 & a_{22}^{(2,1)} & a_{23}^{(2,1)} & \cdots & a_{2N}^{(2,1)} \\ a_{31} & a_{32} & a_{33} & \cdots & a_{3N} \\ \vdots & \vdots & \vdots & \vdots & \vdots \\ a_{N1} & a_{N2} & a_{N3} & \cdots & a_{NN} \end{bmatrix} \tag{1.192}$$

We know from our choice of $\lambda_{21} = a_{21}/a_{11}$ that $a_{21}^{(2,1)} = 0$; therefore, we are free to use this location in memory to store something else. Let us take the advantage of this free location to store the value of λ_{21},

$$A^{(2,1)} = \begin{bmatrix} a_{11} & a_{12} & a_{13} & \cdots & a_{1N} \\ \lambda_{21} & a_{22}^{(2,1)} & a_{23}^{(2,1)} & \cdots & a_{2N}^{(2,1)} \\ a_{31} & a_{32} & a_{33} & \cdots & a_{3N} \\ \vdots & \vdots & \vdots & \vdots & \vdots \\ a_{N1} & a_{N2} & a_{N3} & \cdots & a_{NN} \end{bmatrix} \tag{1.193}$$

Next, we perform the row operation $3 \leftarrow 3 - \lambda_{31} \times 1$ with $\lambda_{31} = a_{31}/a_{11}$, and use the location freed by the fact that $a_{31}^{(3,1)} = 0$ to store λ_{31},

$$A^{(3,1)} = \begin{bmatrix} a_{11} & a_{12} & a_{13} & \cdots & a_{1N} \\ \lambda_{21} & a_{22}^{(2,1)} & a_{23}^{(2,1)} & \cdots & a_{2N}^{(2,1)} \\ \lambda_{31} & a_{32}^{(3,1)} & a_{33}^{(3,1)} & \cdots & a_{3N}^{(3,1)} \\ \vdots & \vdots & \vdots & \vdots & \vdots \\ a_{N1} & a_{N2} & a_{N3} & \cdots & a_{NN} \end{bmatrix} \tag{1.194}$$

After completing Gaussian elimination, storing after each row operation $k \leftarrow k - \lambda_{kj} \times j$ the value of λ_{kj} in the position freed by $a_{jk} = 0$, we have in memory

$$A^{(N,N-1)} = \begin{bmatrix} a_{11} & a_{12} & a_{13} & a_{14} & \cdots & a_{1N} \\ \lambda_{21} & a_{22}^{(2,1)} & a_{23}^{(2,1)} & a_{24}^{(2,1)} & \cdots & a_{2N}^{(2,1)} \\ \lambda_{31} & \lambda_{32} & a_{33}^{(3,2)} & a_{34}^{(3,2)} & \cdots & a_{3N}^{(3,2)} \\ \lambda_{41} & \lambda_{42} & \lambda_{43} & a_{44}^{(4,3)} & \cdots & a_{4N}^{(4,3)} \\ \vdots & \vdots & \vdots & \vdots & \vdots & \vdots \\ \lambda_{N1} & \lambda_{N2} & \lambda_{N3} & \lambda_{N4} & \cdots & a_{NN}^{(N,N-1)} \end{bmatrix} \tag{1.195}$$

We now extract from this matrix the lower and upper triangular matrices

$$
L = \begin{bmatrix} 1 \\ \lambda_{21} & 1 \\ \lambda_{31} & \lambda_{32} & 1 \\ \vdots & \vdots & \vdots & \ddots \\ \lambda_{N1} & \lambda_{N2} & \lambda_{N3} & \cdots & 1 \end{bmatrix} \qquad U = \begin{bmatrix} a_{11} & a_{12} & a_{13} & \cdots & a_{1N} \\ & a_{22}^{(2,1)} & a_{23}^{(2,1)} & \cdots & a_{2N}^{(2,1)} \\ & & a_{33}^{(3,2)} & \cdots & a_{3N}^{(3,2)} \\ & & & \ddots & \vdots \\ & & & & a_{NN}^{(N,N-1)} \end{bmatrix}
$$

$$(1.196)$$

We demonstrate that $A = LU$ by forming the product

$$
LU = \begin{bmatrix} 1 \\ \lambda_{21} & 1 \\ \lambda_{31} & \lambda_{32} & 1 \\ \vdots & \vdots & \vdots & \ddots \\ \lambda_{N1} & \lambda_{N2} & \lambda_{N3} & \cdots & 1 \end{bmatrix} \begin{bmatrix} a_{11} & a_{12} & a_{13} & \cdots & a_{1N} \\ & a_{22}^{(2,1)} & a_{23}^{(2,1)} & \cdots & a_{2N}^{(2,1)} \\ & & a_{33}^{(3,2)} & \cdots & a_{3N}^{(3,2)} \\ & & & \ddots & \vdots \\ & & & & a_{NN}^{(N,N-1)} \end{bmatrix} \qquad (1.197)
$$

For the elements in the first row of LU, matrix multiplication yields

$$(LU)_{1j} = (1)(a_{1j}) = a_{1j} \qquad (1.198)$$

Next, in the second row, we have

$$(LU)_{2j} = (\lambda_{21})(a_{1j}) + a_{2j}^{(2,1)} \qquad (1.199)$$

But, since in Gaussian elimination, $a_{2j}^{(2,1)} = a_{2j} - \lambda_{21} \times a_{1j}$,

$$(LU)_{2j} = (\lambda_{21})(a_{1j}) + [a_{2j} - \lambda_{21} \times a_{1j}] = a_{2j} \qquad (1.200)$$

We can continue this process to find that each row of LU equals the corresponding row of A, and thus $A = LU$.

As an example, consider the system (1.2),

$$
\begin{bmatrix} 1 & 1 & 1 \\ 2 & 1 & 3 \\ 3 & 1 & 6 \end{bmatrix} \begin{bmatrix} x_1 \\ x_2 \\ x_3 \end{bmatrix} = \begin{bmatrix} 4 \\ 7 \\ 2 \end{bmatrix} \qquad (1.201)
$$

Gaussian elimination (without partial pivoting) for this system was demonstrated earlier, and from (1.70)–(1.89) we obtain the factorization

$$
L = \begin{bmatrix} 1 \\ 2 & 1 \\ 3 & 2 & 1 \end{bmatrix} \qquad U = \begin{bmatrix} 1 & 1 & 1 \\ & -1 & 1 \\ & & 1 \end{bmatrix} \qquad (1.202)
$$

Multiplying these two matrices shows that indeed they satisfy $A = LU$.

We have seen that to make Gaussian elimination robust, we must include partial pivoting so that all λ_{kj} are finite. When the factorization is performed using Gaussian elimination with partial pivoting, the book-keeping is a bit more complex, but the result is similar. We obtain lower and upper triangular matrices L and U, and a permutation matrix P, such that

$$PA = LU \qquad (1.203)$$

A *permutation matrix* is a matrix that can be obtained from the identity matrix by performing some sequence of row or column interchanges. Since $\det(I) = 1$, $\det(P) = \pm 1$. For $PA = LU$, P records the cumulative effect of the pivot operations conducted during Gaussian elimination. An example of a permutation matrix is

$$P = \begin{bmatrix} 1 & 0 & 0 \\ 0 & 0 & 1 \\ 0 & 1 & 0 \end{bmatrix} \qquad Pv = \begin{bmatrix} 1 & 0 & 0 \\ 0 & 0 & 1 \\ 0 & 1 & 0 \end{bmatrix} \begin{bmatrix} v_1 \\ v_2 \\ v_3 \end{bmatrix} = \begin{bmatrix} v_1 \\ v_3 \\ v_2 \end{bmatrix} \tag{1.204}$$

To solve $Ax^{[k]} = b^{[k]}$, we premultiply the system by P and substitute for PA,

$$\begin{aligned} PAx^{[k]} &= Pb^{[k]} \\ LUx^{[k]} &= Pb^{[k]} \\ Lc^{[k]} = Pb^{[k]} \qquad Ux^{[k]} &= c^{[k]} \end{aligned} \tag{1.205}$$

In MATLAB, LU factorization is invoked through the command **lu**. The following code computes the LU factorization for the example of (1.70),

```
A = [1 1 1; 2 1 3; 3 1 6];
[L, U, P] = lu(A),
L =
    1.0000        0        0
    0.3333   1.0000        0
    0.6667   0.5000   1.0000
U =
    3.0000   1.0000   6.0000
    0        0.6667  -1.0000
    0        0       -0.5000
P =
    0   0   1
    1   0   0
    0   1   0
```

Thus, we have

$$L = \begin{bmatrix} 1 & & \\ 0.3333 & 1 & \\ 0.6667 & 0.500 & 1 \end{bmatrix} \quad U = \begin{bmatrix} 3 & 1 & 6 \\ & 0.6667 & -1 \\ & & -0.5 \end{bmatrix} \quad P = \begin{bmatrix} 0 & 0 & 1 \\ 1 & 0 & 0 \\ 0 & 1 & 0 \end{bmatrix} \tag{1.206}$$

The following code uses this LU decomposition to solve (1.70),

```
b = [4; 7; 2];
x = U\(L\(P* b)),
x =
   19.0000
   -7.0000
   -8.0000
```

For further discussion of LU factorization with pivoting, consult Quateroni et al. (2000).

Cholesky decomposition

Since the transpose of a lower triangular matrix L is an upper triangular matrix, we might wonder if it is possible for some matrix A, that in the LU decomposition, we can set $U = L^\mathrm{T}$ to obtain

$$A = LL^\mathrm{T} \tag{1.207}$$

First, by taking the transpose of this equation, we see that any A that can be written in this manner must be symmetric, $A = A^\mathrm{T}$,

$$A^\mathrm{T} = (LL^\mathrm{T})^\mathrm{T} = LL^\mathrm{T} = A \tag{1.208}$$

Also, for A to be nonsingular, L and L^T must be nonsingular, so that for any $v \in \Re^N$, $v \neq 0$,

$$\left(L^\mathrm{T} v\right) \cdot \left(L^\mathrm{T} v\right) > 0 \tag{1.209}$$

From

$$v^\mathrm{T} A v = v^\mathrm{T}(LL^\mathrm{T})v = \left(L^\mathrm{T} v\right)^\mathrm{T}\left(L^\mathrm{T} v\right) = \left(L^\mathrm{T} v\right) \cdot \left(L^\mathrm{T} v\right) \tag{1.210}$$

we see that a matrix A could be written as $A = LL^\mathrm{T}$ only if it were symmetric and *positive-definite*; i.e.,

$$v^\mathrm{T} A v > 0 \qquad \forall v \in \Re^N, v \neq 0 \tag{1.211}$$

While only a subset of all matrices are symmetric, positive-definite, they in fact form an important subset, especially in numerical optimization. Thus $A = LL^\mathrm{T}$, known as a *Cholesky decomposition*, is used frequently.

The special structure of a positive-definite matrix allows us to perform Cholesky factorization more quickly than LU decomposition. We start by writing $A = LL^\mathrm{T}$ explicitly,

$$\begin{bmatrix} a_{11} & a_{21} & a_{31} & \cdots & a_{N1} \\ a_{21} & a_{22} & a_{32} & \cdots & a_{N2} \\ a_{31} & a_{32} & a_{33} & \cdots & a_{N3} \\ \vdots & \vdots & \vdots & \vdots & \vdots \\ a_{N1} & a_{N2} & a_{N3} & \cdots & a_{NN} \end{bmatrix} = \begin{bmatrix} L_{11} & & & & \\ L_{21} & L_{22} & & & \\ L_{31} & L_{32} & L_{33} & & \\ \vdots & \vdots & \vdots & \ddots & \\ L_{N1} & L_{N2} & L_{N3} & \cdots & L_{NN} \end{bmatrix} \times \begin{bmatrix} L_{11} & L_{21} & L_{31} & \cdots & L_{N1} \\ & L_{22} & L_{32} & \cdots & L_{N2} \\ & & L_{33} & \cdots & L_{N3} \\ & & & \ddots & \vdots \\ & & & & L_{NN} \end{bmatrix} \tag{1.212}$$

From the $(1, 1)$ element we obtain

$$a_{11} = L_{11} \times L_{11} \qquad L_{11} = (a_{11})^{1/2} \tag{1.213}$$

From the $(1, 2)$ element,

$$a_{12} = L_{11} \times L_{21} \qquad L_{21} = a_{12}/L_{11} \qquad (1.214)$$

Similarly, we compute the first column of L for $j = 3, 4, \ldots, N$,

$$L_{j1} = a_{1j}/L_{11} \qquad (1.215)$$

Next, we move to the second column, and from the $(2, 2)$ element obtain

$$a_{22} = L_{21}L_{21} + L_{22}L_{22} \qquad L_{22} = \left(a_{22} - L_{21}^2\right)^{1/2} \qquad (1.216)$$

and for $j = 3, 4, \ldots, N$,

$$a_{2j} = L_{21}L_{j1} + L_{22}L_{j2} \qquad L_{j2} = (a_{2j} - L_{21}L_{j1})/L_{22} \qquad (1.217)$$

Continuing this process for columns 3, 4, etc. yields the algorithm:

for $i = 1, 2, \ldots, N$; iterate over each column of L

$$L_{ii} \leftarrow \left[a_{ii} - \sum_{k=1}^{i-1} L_{ik}^2 \right]^{1/2}$$

for $j = i + 1, i + 2, \ldots, N$; iterate over elements below the diagonal

$$L_{ji} \leftarrow \frac{1}{L_{ii}} \left[a_{ij} - \sum_{k=1}^{i-1} L_{ik} L_{jk} \right]$$

$$\text{end } j = i + 1, i + 2, \ldots, N$$
$$\text{end } i = 1, 2, \ldots, N$$

As there are only two nested loops, the FLOPs required scale only as N^2.

In MATLAB, Cholesky decomposition is invoked by **chol**; however, this function returns not a lower-triangular matrix L with $A = LL^T$ but an upper-triangular matrix $R = L^T$ with $A = R^T R$. For the positive-definite matrix

$$A = \begin{bmatrix} 2 & 1 & 1 \\ 1 & 4 & 2 \\ 1 & 2 & 6 \end{bmatrix} \qquad (1.218)$$

the Cholesky factorization $A = R^T R$ is performed by the code,

```
A = [2 1 1; 1 4 2; 1 2 6];
R = chol(A),
R =
    1.4142    0.7071    0.7071
    0         1.8708    0.8018
    0         0         2.2039
```

We see that $A = R^T R$ by

```
R'* R,
ans =
    2.0000    1.0000    1.0000
    1.0000    4.0000    2.0000
    1.0000    2.0000    6.0000
```

We use this Cholesky decomposition to solve quickly the system

$$\begin{bmatrix} 2 & 1 & 1 \\ 1 & 4 & 2 \\ 1 & 2 & 6 \end{bmatrix} \begin{bmatrix} x_1 \\ x_2 \\ x_3 \end{bmatrix} = \begin{bmatrix} 7 \\ 15 \\ 23 \end{bmatrix} \tag{1.219}$$

through the command

```
x = R\(R'\b),
x =
   1.0000
   2.0000
   3.0000
```

Matrix norm and rank

We now introduce two important definitions for matrices. We describe how "large" a vector $v \in \mathfrak{R}^N$ is through the use of a norm $\|v\|$. For any particular choice of a vector norm $\|v\|$, we can generate a corresponding *matrix norm*

$$\|A\| = \max_{v \neq 0} \frac{\|Av\|}{\|v\|} = \max_{\|v\|=1} \|Av\| \tag{1.220}$$

that measures how "large" the matrix is.

The determinant, $\det(A)$ remains the proper measure of singularity. However, we might want some more information on just how singular a particular matrix A is if $\det(A) = 0$. The *rank* of a matrix A is the number of the linearly independent rows (or columns) of the matrix. Therefore, a nonsingular $N \times N$ matrix must be of rank N, and is said to be of *full rank*. The rank also may be defined as the dimension of the range of A. Matrix rank is discussed later in Chapter 3 within the context of singular value decomposition (SVD).

Submatrices and matrix partitions

The matrix operations that we have defined previously also extend to matrices that are block-partitioned into submatrices. For example, consider the 4×4 matrices

$$A = \begin{bmatrix} 1 & 2 & 3 & 4 \\ 2 & 1 & 3 & 4 \\ 3 & 2 & 1 & 4 \\ 4 & 2 & 3 & 1 \end{bmatrix} \qquad B = \begin{bmatrix} 8 & 7 & 6 & 5 \\ 7 & 8 & 6 & 5 \\ 6 & 7 & 8 & 5 \\ 5 & 7 & 6 & 8 \end{bmatrix} \tag{1.221}$$

We define from A the *submatrices*

$$A_{11} = \begin{bmatrix} 1 & 2 \\ 2 & 1 \end{bmatrix} \quad A_{12} = \begin{bmatrix} 3 & 4 \\ 3 & 4 \end{bmatrix} \quad A_{21} = \begin{bmatrix} 3 & 2 \\ 4 & 2 \end{bmatrix} \quad A_{22} = \begin{bmatrix} 1 & 4 \\ 3 & 1 \end{bmatrix} \tag{1.222}$$

to write A in *block-partitioned* form as

$$A = \begin{bmatrix} 1 & 2 & 3 & 4 \\ 2 & 1 & 3 & 4 \\ 3 & 2 & 1 & 4 \\ 4 & 2 & 3 & 1 \end{bmatrix} = \begin{bmatrix} \begin{bmatrix} 1 & 2 \\ 2 & 1 \end{bmatrix} & \begin{bmatrix} 3 & 4 \\ 3 & 4 \end{bmatrix} \\ \begin{bmatrix} 3 & 2 \\ 4 & 2 \end{bmatrix} & \begin{bmatrix} 1 & 4 \\ 3 & 1 \end{bmatrix} \end{bmatrix} = \begin{bmatrix} A_{11} & A_{12} \\ A_{21} & A_{22} \end{bmatrix} \tag{1.223}$$

Similarly,

$$B = \begin{bmatrix} 8 & 7 & 6 & 5 \\ 7 & 8 & 6 & 5 \\ 6 & 7 & 8 & 5 \\ 5 & 7 & 6 & 8 \end{bmatrix} = \begin{bmatrix} \begin{bmatrix} 8 & 7 \\ 7 & 8 \end{bmatrix} & \begin{bmatrix} 6 & 5 \\ 6 & 5 \end{bmatrix} \\ \begin{bmatrix} 6 & 7 \\ 5 & 7 \end{bmatrix} & \begin{bmatrix} 8 & 5 \\ 6 & 8 \end{bmatrix} \end{bmatrix} = \begin{bmatrix} B_{11} & B_{12} \\ B_{21} & B_{22} \end{bmatrix} \tag{1.224}$$

where

$$B_{11} = \begin{bmatrix} 8 & 7 \\ 7 & 8 \end{bmatrix} \quad B_{12} = \begin{bmatrix} 6 & 5 \\ 6 & 5 \end{bmatrix} \quad B_{21} = \begin{bmatrix} 6 & 7 \\ 5 & 7 \end{bmatrix} \quad B_{22} = \begin{bmatrix} 8 & 5 \\ 6 & 8 \end{bmatrix} \tag{1.225}$$

$A + B$ can be obtained by summing the corresponding submatrices,

$$A + B = \begin{bmatrix} A_{11} & A_{12} \\ A_{21} & A_{22} \end{bmatrix} + \begin{bmatrix} B_{11} & B_{12} \\ B_{21} & B_{22} \end{bmatrix} = \begin{bmatrix} (A_{11} + B_{11}) & (A_{12} + B_{12}) \\ (A_{21} + B_{21}) & (A_{22} + B_{22}) \end{bmatrix} \tag{1.226}$$

More surprisingly, the rules for matrix multiplication and transposition also can be applied to block-partitioned matrices, if the matrices are *conformally partitioned*; i.e., sized such that all necessary products of submatrices are defined. Thus,

$$AB = \begin{bmatrix} A_{11} & A_{12} \\ A_{21} & A_{22} \end{bmatrix} \begin{bmatrix} B_{11} & B_{12} \\ B_{21} & B_{22} \end{bmatrix} = \begin{bmatrix} (A_{11}B_{11} + A_{12}B_{21}) & (A_{11}B_{12} + A_{12}B_{22}) \\ (A_{21}B_{11} + A_{22}B_{21}) & (A_{21}B_{12} + A_{22}B_{22}) \end{bmatrix}$$

$$\tag{1.227}$$

$$A^{\mathrm{T}} = \begin{bmatrix} A_{11} & A_{12} \\ A_{21} & A_{22} \end{bmatrix}^{\mathrm{T}} = \begin{bmatrix} A_{11}^{T} & A_{21}^{T} \\ A_{12}^{T} & A_{22}^{T} \end{bmatrix} \tag{1.228}$$

Example. Modeling a separation system

We consider a simple mass balance problem to demonstrate the use of MATLAB to solve a system of linear equations. For the separation system of Figure 1.9, we know the inlet mass flow rate (in kilograms per hour) and the mass fractions of each species in the inlet (stream 1) and each outlet (streams 2, 4, and 5). We wish to compute the mass flow rates of each outlet stream.

Here we use the notation that ^{i}F is the mass flow rate of stream i, and $^{i}w_{j}$ is the mass fraction of species j in stream # i. We define the unknowns

$$x_1 = {}^2F \qquad x_2 = {}^4F \qquad x_3 = {}^5F \tag{1.229}$$

and set up balances for

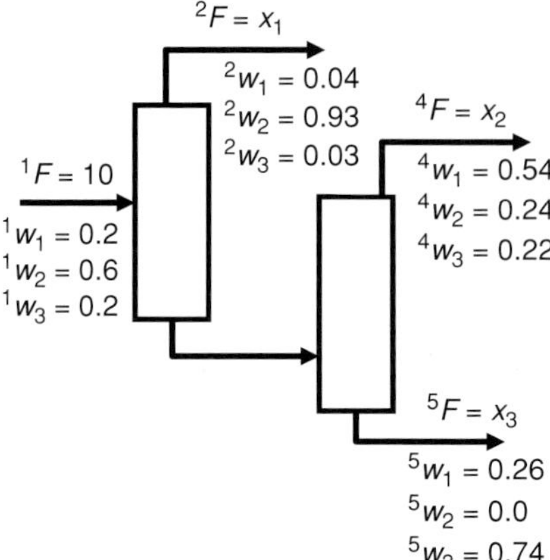

Figure 1.9 Process diagram for separation system.

1.   the total mass flow rate

$$^2F + {}^4F + {}^5F = {}^1F \tag{1.230}$$

2.   the mass flow rate of species 1

$$({}^2w_1)({}^2F) + ({}^4w_1)({}^4F) + ({}^5w_1)({}^5F) = ({}^1w_1)({}^1F) \tag{1.231}$$

3.   the mass flow rate of species 2

$$({}^2w_2)({}^2F) + ({}^4w_2)({}^4F) + ({}^5w_2)({}^5F) = ({}^1w_2)({}^1F) \tag{1.232}$$

This yields the set of three algebraic equations

$$\begin{aligned}
x_1 + x_2 + x_3 &= 10 \\
(0.04)x_1 + (0.54)x_2 + (0.26)x_3 &= 2 \\
(0.93)x_1 + (0.24)x_2 + (0.0)x_3 &= 6
\end{aligned} \tag{1.233}$$

Gaussian elimination yields

$$x_1 = 5.8238 \qquad x_2 = 2.4330 \qquad x_3 = 1.7433 \tag{1.234}$$

sep_system_example.m performs this calculation. For further discussion of the formulation of material and energy balances, and algorithms for their solution, consult Reklaitis (1983).

Sparse and banded matrices

In the example above, the mathematical formulation of the problem was indeed a linear system, and we could apply Gaussian elimination directly. Most mathematical problems, however, are not expressed naturally as linear systems. Still, the availability for linear systems of rigorous existence and uniqueness conditions and an automated solution procedure

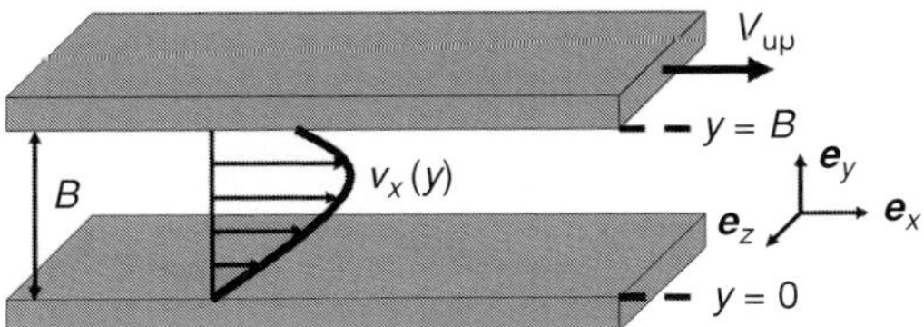

Figure 1.10 Pressure-driven flow between two infinite, parallel, flat plates.

makes them ideal building blocks upon which to construct algorithms for more complex problems.

Here, we solve a boundary value problem from fluid mechanics numerically by converting it into a linear algebraic system. As this example makes clear, it is sometimes possible to reduce greatly the computational burden of elimination when the matrix is *banded*; i.e., all nonzero elements are found near the principal diagonal.

Example. Solving a boundary value problem from fluid mechanics

Consider the case of a Newtonian fluid undergoing laminar, pressure-driven flow between two parallel, infinite flat plates separated by a distance B (Figure 1.10). The bottom plate is stationary and the top plate moves at a constant velocity V_{up}. For a constant dynamic pressure gradient, $\Delta P / \Delta x$, $P = p - \boldsymbol{g} \cdot \boldsymbol{r}$, we wish to calculate the resulting velocity profile.

If we assume a velocity profile of the form

$$\boldsymbol{v}(\boldsymbol{r}, t) = v_x(y)\boldsymbol{e}_x \tag{1.235}$$

the equation of continuity for an incompressible fluid, $\nabla \cdot \boldsymbol{v} = 0$, is satisfied automatically and the Navier–Stokes equation of motion

$$\rho \frac{D\boldsymbol{v}}{Dt} = \rho \frac{\partial}{\partial t}\boldsymbol{v} + \rho \boldsymbol{v} \cdot \nabla \boldsymbol{v} = -\nabla P + \mu \nabla^2 \boldsymbol{v} \tag{1.236}$$

reduces to

$$0 = -\left(\frac{\Delta P}{\Delta x}\right) + \mu \frac{d^2 v_x}{dy^2} \tag{1.237}$$

A brief discussion of these equations is provided in the supplemental material in the accompanying website. For a more detailed treatment, see Bird *et al.* (2002) and Deen (1998).

We wish to solve this differential equation subject to the no-slip boundary conditions

$$v_x(y = 0) = 0 \qquad v_x(y = B) = V_{up} \tag{1.238}$$

This is a classic problem from fluid mechanics that is solved easily by integrating the differential equation twice and using the boundary conditions to obtain the constants of integration. The resulting solution is

$$v_x(y) = V_{up}\left(\frac{y}{B}\right) + \frac{1}{2\mu}\left(\frac{\Delta P}{\Delta x}\right)(y^2 - yB) \tag{1.239}$$

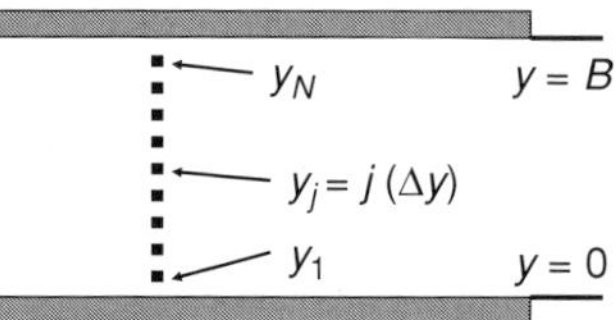

Figure 1.11 Placement of grid points for finite difference computation.

We now employ a numerical method to "solve" this problem by converting it into a set of algebraic equations. For this particular example, there is little actual need to do so since an analytical solution is available; however, the technique that we develop here can be used to obtain numerical approximations to the solution even when no analytical solution exists.

For this example, we use the conceptually-simple *method of finite differences* that is based on the following definition of the derivative of $f(x)$:

$$\frac{df}{dx} = \lim_{\Delta x \to 0} \frac{f(x + \Delta x) - f(x - \Delta x)}{2\Delta x}$$
$$= \lim_{\Delta x \to 0} \frac{f(x + \Delta x) - f(x)}{\Delta x} = \lim_{\Delta x \to 0} \frac{f(x) - f(x - \Delta x)}{\Delta x} \tag{1.240}$$

In the limit as $\Delta x \to 0$, all three formulas agree if the derivative indeed exists. In the method of finite differences, we use finite, but "small," values of Δx in one of (1.240) to approximate the derivative by an algebraic form. We study this method in further detail in Chapter 6; however, for now we merely note that the first approximation formula given above, the *central-difference approximation*, is the most accurate.

Our differential equation in this example involves the second derivative of the velocity; therefore, we need to construct an algebraic approximation to this higher-order derivative. We place a grid of N points along the computational domain $y \in [0, B]$ (Figure 1.11) at the locations

$$y_j = j(\Delta y) \qquad \Delta y = \frac{B}{N + 1} \qquad j = 1, 2, \ldots, N \tag{1.241}$$

At grid point j, we use a central-difference formula to approximate the local value of the second derivative of the velocity,

$$\left.\frac{d^2 v_x}{dy^2}\right|_{y_j} \approx \frac{\left(\dfrac{dv_x}{dy}\right)\bigg|_{y_j+(\Delta y)/2} - \left(\dfrac{dv_x}{dy}\right)\bigg|_{y_j-(\Delta y)/2}}{\Delta y} \tag{1.242}$$

Here, the values of the first derivatives are evaluated at the mid-points between the grid locations. We then use yet other central-difference formulas for these mid-point values,

$$\left.\left(\frac{dv_x}{dy}\right)\right|_{y_j+(\Delta y)/2} \approx \frac{v_x(y_{j+1}) - v_x(y_j)}{\Delta y} \qquad \left.\left(\frac{dv_x}{dy}\right)\right|_{y_j-(\Delta y)/2} \approx \frac{v_x(y_j) - v_x(y_{j-1})}{\Delta y}$$
$$\tag{1.243}$$

to obtain the approximation of the second derivative at y_j

$$\left.\frac{d^2 v_x}{dy^2}\right|_{y_j} \approx \frac{v_x(y_{j+1}) - 2v_x(y_j) + v_x(y_{j-1})}{(\Delta y)^2} \tag{1.244}$$

In general, this approximation is not exact, and we must reduce the value of Δy by increasing N until the magnitude of the approximation error is below some acceptable value. For this particular problem, as the true solution is a quadratic function, we are lucky and this algebraic approximation is exact.

To "solve" a boundary value problem using the method of finite differences, we formulate a set of N algebraic equations for the set of N unknowns $\{v_x(y_1), v_x(y_2), \ldots, v_x(y_N)\}$. For each grid point, we obtain an algebraic equation by requiring the differential equation to be satisfied locally

$$0 = -\left(\frac{\Delta P}{\Delta x}\right) + \mu \left.\frac{d^2 v_x}{dy^2}\right|_{y_j} \tag{1.245}$$

If we insert the central-difference approximation for the second derivative, the algebraic equation for grid point j is

$$0 = -\left(\frac{\Delta P}{\Delta x}\right) + \mu \frac{v_x(y_{j+1}) - 2v_x(y_j) + v_x(y_{j-1})}{(\Delta y)^2} \tag{1.246}$$

We write this in a more compact form by defining the column vector

$$v = \begin{bmatrix} v_1 \\ v_2 \\ \vdots \\ v_N \end{bmatrix} = \begin{bmatrix} v_x(y_1) \\ v_x(y_2) \\ \vdots \\ v_x(y_N) \end{bmatrix} \tag{1.247}$$

so that the algebraic equation for grid point j becomes

$$v_{j+1} - 2v_j + v_{j-1} = \frac{(\Delta y)^2}{\mu}\left(\frac{\Delta P}{\Delta x}\right) \tag{1.248}$$

It is standard practice to make the diagonal elements positive,

$$-v_{j+1} + 2v_j - v_{j-1} = -\frac{(\Delta y)^2}{\mu}\left(\frac{\Delta P}{\Delta x}\right) \tag{1.249}$$

If we assemble these equations in matrix form, we obtain the system

$$\begin{bmatrix} 2 & -1 & & & & \\ -1 & 2 & -1 & & & \\ & -1 & 2 & -1 & & \\ & & \ddots & \ddots & \ddots & \\ & & & -1 & 2 & -1 \\ & & & & -1 & 2 \end{bmatrix} \begin{bmatrix} v_1 \\ v_2 \\ v_3 \\ \vdots \\ v_{N-1} \\ v_N \end{bmatrix} = \begin{bmatrix} G + v_0 \\ G \\ G \\ \vdots \\ G \\ G + v_{N+1} \end{bmatrix} \tag{1.250}$$

where

$$G = -\frac{(\Delta y)^2}{\mu}\left(\frac{\Delta P}{\Delta x}\right) \tag{1.251}$$

and

$$v_0 = v_x(y = 0) \qquad v_{N+1} = v_x(y = B) \tag{1.252}$$

To enforce the no-slip boundary conditions, we merely set in the right-hand side vector of (1.250)

$$v_0 = 0 \qquad v_{N+1} = V_{\text{up}} \tag{1.253}$$

By this technique, we have converted the original differential equation into a set of algebraic equations for the values of the velocity at each grid point.

The matrix in (1.250) has a special structure: the only nonzero elements are located along the main diagonal and on the diagonals immediately above and below. Such a matrix is said to be *tridiagonal*. This special structure allows us to solve this system in a fraction of the time required by brute-force Gaussian elimination.

Remember that the number of FLOPs required to perform Gaussian elimination for a system of N equations is $2N^3/3$. If we want, in general, to obtain an accurate solution of a differential equation, we may need to use a grid of 100 or more points, so that the number of FLOPs required is on the order of one million. In addition to CPU time, the memory required to store the matrix is significant. A matrix for a system with N unknowns contains N^2 elements, each requiring its own location in memory. As N increases, these numbers become much larger. The number of FLOPs required to perform full Gaussian elimination on a system of 1000 unknowns is on the order of one billion, and storing the matrix requires one million locations in memory.

Because here nearly all of the matrix elements are zero, much of the effort of brute-force Gaussian elimination is a waste of time. As the matrix is tridiagonal, we only need to perform one row operation per column to zero the element immediately below the diagonal. Also, since each row only contains three non zero values, the number of FLOPs required for each row operation is a small number, not on the order of N as is the case generally. Thus, we can remove two of the nested for loops of the Gaussian elimination algorithm so that the total number of FLOPs scales only linearly with N. This is a very important point, as this trick of taking advantage of the tridiagonal structure makes feasible the numerical solution of this system even when the number of grid points is large.

Banded and sparse matrices

More generally, a matrix is said to be *sparse* if most of its elements are zero. A sparse matrix can be stored in memory efficiently by recording only the positions and values of its nonzero elements. Let us say that we have an $N \times N$ matrix A with at most only three non-zero elements per row. While the total number of elements is N^2, at most only $3N$ are non-zero, and the matrix is sparse for large N. If $N_{\text{nz}} \ll N^2$ is the number of nonzero elements in A, or at least an upper bound to the number of nonzero elements, we can store all the information that we need about A in two integer vectors of length N_{nz} (for the row and column positions of the nonzero elements) and a single real vector of length N_{nz} for their values. For example, the matrix

$$
A = \begin{bmatrix}
2 & -1 & & \\
-1 & 2 & -1 & \\
& -1 & 2 & -1 \\
& & -1 & 2
\end{bmatrix}
\tag{1.254}
$$

can be stored as

$$
i_{\text{row}} = \begin{bmatrix} 1 \\ 1 \\ 2 \\ 2 \\ 2 \\ 3 \\ 3 \\ 3 \\ 4 \\ 4 \end{bmatrix}
\qquad
j_{\text{col}} = \begin{bmatrix} 1 \\ 2 \\ 1 \\ 2 \\ 3 \\ 2 \\ 3 \\ 4 \\ 3 \\ 4 \end{bmatrix}
\qquad
a = \begin{bmatrix} 2 \\ -1 \\ -1 \\ 2 \\ -1 \\ -1 \\ 2 \\ -1 \\ -1 \\ 2 \end{bmatrix}
\qquad \Leftrightarrow \qquad
\begin{array}{l}
a_{11} = 2 \\
a_{12} = -1 \\
a_{21} = -1 \\
a_{22} = 2 \\
a_{23} = -1 \\
a_{32} = -1 \\
a_{33} = 2 \\
a_{34} = -1 \\
a_{43} = -1 \\
a_{44} = 2
\end{array}
\tag{1.255}
$$

Obviously, this format requires much less memory for sparse matrices than does allocating a separate location to store a real value for each element, whether it is zero or not.

Even though it is possible to store a sparse matrix efficiently, can we efficiently solve a system by Gaussian elimination using this notation? We can if the matrix is *banded*; i.e., if the nonzero values are clustered in the vicinity of the principal diagonal.

Definition A matrix A is said to be *banded* with *bandwidth m* if all nonzero elements are located along diagonals removed at most by m positions from the principal diagonal. The following matrices demonstrate this terminology, where asterisks denote the allowed locations of nonzero elements and the principal diagonal is denoted by Ds,

$$
\begin{bmatrix}
D & * & * & & & & & & & \\
* & D & * & * & & & & & & \\
* & * & D & * & * & & & & & \\
& * & * & D & * & * & & & & \\
& & * & * & D & * & * & & & \\
& & & * & * & D & * & * & & \\
& & & & * & * & D & * & * & \\
& & & & & * & * & D & * & * \\
& & & & & & * & * & D & * \\
& & & & & & & * & * & D
\end{bmatrix}
$$

$$\text{bandwidth} = 2$$

$$\begin{bmatrix} D & * & * & * & & & & & & & \\ * & D & * & * & * & & & & & & \\ * & * & D & * & * & * & & & & & \\ * & * & * & D & * & * & * & & & & \\ & * & * & * & D & * & * & * & & & \\ & & * & * & * & D & * & * & * & & \\ & & & * & * & * & D & * & * & * & \\ & & & & * & * & * & D & * & * \\ & & & & & * & * & * & D & * \\ & & & & & & * & * & * & D \end{bmatrix}$$

bandwidth $= 3$

$$\begin{bmatrix} D & * & * & * & * & & & & & \\ * & D & * & * & * & * & & & & \\ * & * & D & * & * & * & * & & & \\ * & * & * & D & * & * & * & * & & \\ * & * & * & * & D & * & * & * & * & \\ & * & * & * & * & D & * & * & * & * \\ & & * & * & * & * & D & * & * & * \\ & & & * & * & * & * & D & * & * \\ & & & & * & * & * & * & D & * \\ & & & & & * & * & * & * & D \end{bmatrix}$$

bandwidth $= 4$

Gaussian elimination for tightly banded matrices is particularly efficient, because for each row there are at most m elements below the principal diagonal that must be eliminated. Likewise, for each row operation, one need perform only about $m + 1$ eliminations. Therefore, the number of FLOPs required to perform Gaussian elimination on an $N \times N$ matrix of bandwidth m scales only as $m^2 N$. For $m \ll N$, $m^2 N \ll N^3$, and taking advantage of the banded nature of A speeds up the calculation significantly. In particular, for (1.250), taking advantage of the tridiagonal nature of A means that the computational effort scales linearly with N, rather than as N^3 with full Gaussian elimination.

Treatment of sparse, banded matrices in MATLAB

MATLAB is structured to employ sparse-matrix notation very easily, often with no further complication to the user beyond initially declaring the matrix to be sparse. The command

A = spalloc (M,N,Nnz);

allocates space in memory to store an $M \times N$ matrix in sparse format, for which an upper bound on the number of nonzero elements is **Nnz**. This matrix is initialized to contain all zeros; however, the nonzero elements are declared similarly to the full matrix format. For example, the following code sets the matrix for the 1-D flow system,

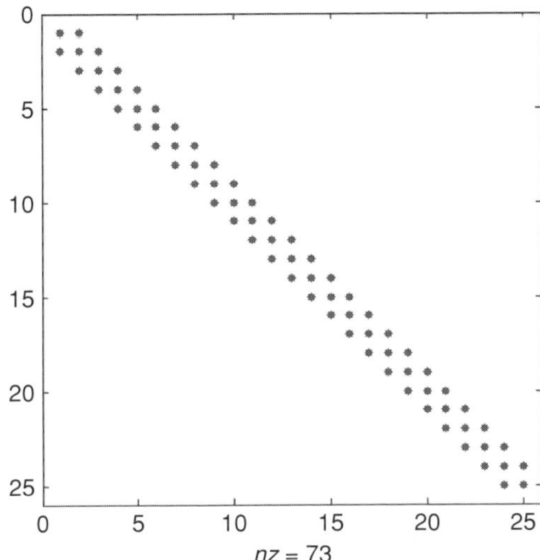

Figure 1.12 Spy plot of tridiagonal matrix ($N = 25$) showing location of nonzero elements.

```
A = spalloc (N,N,3*N);
A(1,1) = 2; A(1,2)= -1;
for k = 2:(N-1)
    A(k,k-1) = -1; A(k,k) = 2; A(k,k+1) = -1;
end
A(N,N-1) = - 1; A(N,N) = 2;
```

The locations of the nonzero elements in a sparse matrix are shown using the command **spy**. The spy plot for the matrix produced by the code above is shown in Figure 1.12.

A listing of MATLAB functions for manipulating sparse matrices is returned by the command **sparfun**. With these sparse functions, the matrix above can be generated more efficiently by the code,

```
A = spalloc(N,N,3* N);
v = ones(N,1);
A = spdiags([- v 2*v - v],- 1:1, N,N);
```

The MATLAB elimination solver "\," also known as the **mldivide** function, can handle matrices stored in sparse-matrix format. If the matrix is banded, the bandwidth is determined and the elimination algorithm modified accordingly. If the matrix is not banded, the solver attempts to reduce the bandwidth as much as possible by applying a heuristic algorithm that interchanges rows and columns.

Let us solve the problem $Ax = b$, where A has been declared as above and b is a vector containing all ones. Then, we obtain x quickly, taking advantage of the sparse, banded structure of A, by typing the code

```
b = ones(N,1);
x = A\b;
```

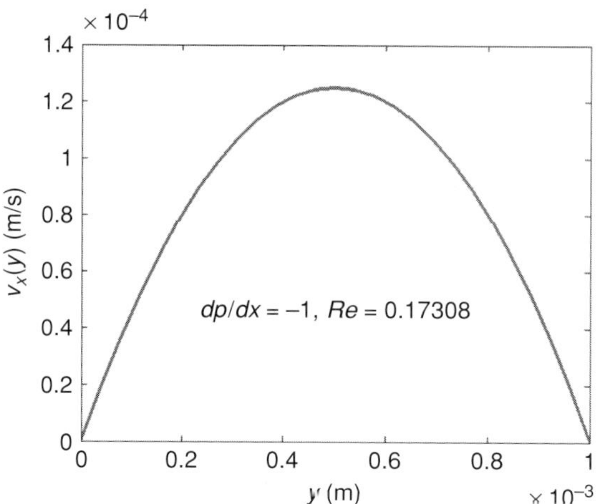

Figure 1.13 Velocity profile for 1-D laminar flow with a moving upper plate.

Solving the 1-D fluid flow problem in MATLAB

simple_flow_1D.m solves the 1-D flow example above where the fluid is water ($\rho = 10^3 \text{kg/m}^3$, $\mu = 10^{-3} \text{Pa s}$), the upper plate is stationary, and the separation between plates is 1 mm. The dynamic pressure gradient is selected to give a Reynolds' number near 1. The computed velocity profile is shown in Figure 1.13.

Fill-in (why Gaussian elimination is sometimes impractical)

It is important to note that many sparse systems cannot be placed in a banded form, and elimination remains costly. For such systems, the iterative techniques discussed later in our discussion of boundary value problems are preferred. Even if a matrix is banded; however, elimination may be too costly due to *fill-in*.

The flow example (1.237) and (1.238) was of the form of a 1-D boundary value problem,

$$-\frac{d^2\varphi}{dy^2} = f(y) \qquad \varphi(0) = \varphi_0 \qquad \varphi(B) = \varphi_B \tag{1.256}$$

Each row of the linear system resulting from finite differences on a uniform grid of spacing Δy has only three nonzero elements

$$A_{k,k-1} = \frac{-1}{(\Delta y)^2} \quad A_{k,k} = \frac{2}{(\Delta y)^2} \quad A_{k,k+1} = \frac{-1}{(\Delta y)^2} \quad b_k = f(y_k) \tag{1.257}$$

As is shown in Chapter 6, for the analogous problem on a 2-D domain,

$$-\nabla^2\varphi = -\frac{\partial^2\varphi}{\partial x^2} - \frac{\partial^2\varphi}{\partial y^2} = f(x,y) \qquad 0 \le x \le L \qquad 0 \le y \le H$$

$$\begin{aligned} \text{BC1} \quad & \varphi(0,y) = 0 \quad 0 \le y \le H \\ \text{BC2} \quad & \varphi(L,y) = 0 \quad 0 \le y \le H \\ \text{BC3} \quad & \varphi(x,0) = 0 \quad 0 \le x \le L \\ \text{BC4} \quad & \varphi(x,H) = 0 \quad 0 \le x \le L \end{aligned} \tag{1.258}$$

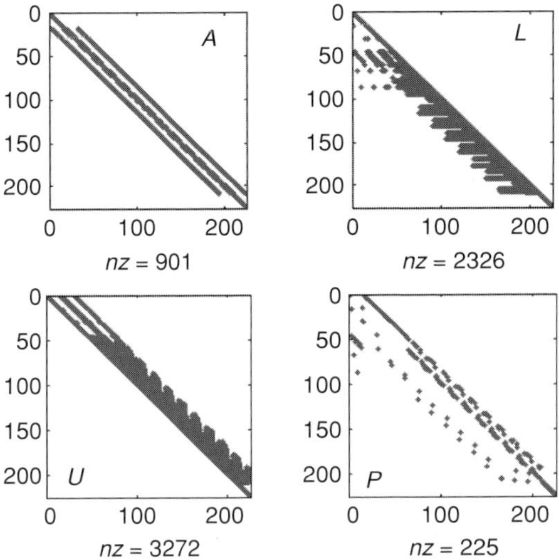

Figure 1.14 Sparsity patterns of A and its LU factors, $PA = LU$, showing fill-in.

the method of finite differences on a uniform grid of $N_x \times N_y$ points, with the labeling scheme

$$\varphi(x_i, y_j) = \varphi_n \qquad n = (i - 1) \times N_y + j \qquad (1.259)$$

yields a linear system $A\varphi = b$ of the form

$$A_{n,n-N_y}\varphi_{n-N_y} + A_{n,n-1}\varphi_{n-1} + A_{nn}\varphi_n + A_{n,n+1}\varphi_{n+1} + A_{n,n+N_y}\varphi_{n+N_y} = b_n$$

$$A_{n,n-N_y} = A_{n,n+N_y} = \left[\frac{-1}{(\Delta x)^2} \right]$$

$$A_{n,n-1} = A_{n,n+1} = \left[\frac{-1}{(\Delta y)^2} \right] \qquad (1.260)$$

$$A_{nn} = \left[\frac{2}{(\Delta x)^2} + \frac{2}{(\Delta y)^2} \right] \qquad b_n = f(x_i, y_j)$$

The nonzero elements in this system are located now on five diagonals: the principal diagonal, the diagonals immediately above and below the principal, and the two diagonals offset by N_y diagonals from the principal. The sparsity pattern of this matrix for a grid of 15×15 points is shown in the upper left of Figure 1.14. While A is banded, the banded region itself is quite sparse, containing mostly zeros except on five diagonals. The LU factors from $PA = LU$ are shown in the upper right and lower left of Figure 1.14.

Even though A contains only 901 nonzero elements (two-fifths of which are below the diagonal), U contains 3272 nonzero elements. Partially, this is due to the fact that if A has a bandwidth m, U generally has a bandwidth of $2m + 1$; however, we also see that the banded region of U is significantly more dense than that of A. That is, Gaussian elimination fills-in zero positions of the original matrix with nonzero values. For the relatively small system here, the increased memory and CPU-time demands due to fill-in are manageable; however, in many problems, elimination cannot be applied because of severe fill-in.

Boundary value problems in three dimensions, for example, cannot be solved with Gaussian elimination due to fill-in. For a boundary value problem involving a single field on a regular grid in d dimensions with N grid points in each direction, the following scaling laws hold for the number of nonzero elements before and after fill-in from elimination:

$$
\begin{aligned}
\text{total number of grid points} &= N^d \\
\text{matrix dimension} &= N^d \times N^d \\
\text{total number of elements in } A &= (N^d)^2 = N^{2d} \\
\text{number of nonzero elements per row} &= 1 + 2d \\
\text{total number of nonzero elements in } A &= N^d(1 + 2d) \\
\text{fraction of } A \text{ elements that are nonzero} &\propto N^{-d} \\
\text{bandwidth of } A &= N^{(d-1)} \\
\text{bandwidth of } U \text{ following elimination} &= 2N^{(d-1)} + 1 \\
\text{number of nonzero elements in } U \text{ if band is dense} &= N^d[2N^{(d-1)} + 1] \propto N^{(2d-1)}
\end{aligned}
\tag{1.261}
$$

As the dimension increases, the fill-in problem becomes more acute, and we may not have enough memory to store even the contents of A. For these reasons, in our later discussion of boundary value problems (Chapter 6), we consider iterative methods for solving linear systems that are not susceptible to fill-in and that do not require us even to store in memory the components of A.

MATLAB summary

The primary tool to solve systems of linear algebraic equations is Gaussian elimination. In MATLAB, this calculation is performed using the "backslash" operator "\." The following code demonstrates its use:

```
A = [1 -2 2; 3 1 -4; 3 -2 1];
b = [7; -3; 8];
x = A\b,
```

A matrix can be stored either in full format, allocating a memory location for each element, as above, or in sparse format. To allocate memory for a sparse $M \times N$ matrix containing no more than NZ nonzero elements, use

```
A = spalloc(M, N, NZ);
```

The matrix then may be treated essentially as one stored in full format,

```
N = 100;
A = spalloc(N, N, 3*N);
for k = 1:N
    A(k,k) = 2;
    if(k > 1)
        A(k,k-1) = -1
```

```
   end
   if(k < N)
      A(k,k+1) = - 1;
   end
end
b = ones(N,1);
x = A\b;
```

Such sparse systems may be generated more efficiently through use of specialty functions such as **spdiags**. A list of sparse matrix function is generated by the command **sparfun**.

Gaussian elimination is used to form the decomposition $PA = LU$, where L is lower triangular, U is upper triangular, and P is a permutation matrix, by

```
[L, U, P] = lu(A);
```

When A is positive-definite, the decomposition $A = R^T R$, with R upper triangular, is performed by Cholesky factorization,

```
R = chol(A);
```

The determinant of a matrix is computed, through LU factorization, by

```
det_A = det(A);
```

A list of functions available for matrix manipulation is returned by

```
matfun
```

Problems

1.A.1. Solve the following linear system by hand, using Gaussian elimination with partial pivoting, followed by backward substitution.

$$\begin{aligned}
-2x_1 + 3x_2 + x_3 &= -6 \\
-x_1 + 3x_2 + 3x_3 &= -8 \\
2x_1 - x_2 + x_3 &= 2
\end{aligned} \tag{1.262}$$

1.A.2. For the matrix of the system in problem 1.A.1, compute by hand the LU decomposition (you do not need to use pivoting).

1.A.3. Consider the problem of fitting a polynomial to the values of $f(x)$ at points $x_0 < x_1 < x_2 < \cdots < x_N$. We wish to find the coefficients of a polynomial

$$p(x) = a_0 + a_1 x + a_2 x^2 + \cdots + a_N x^N \tag{1.263}$$

such that $p(x_k) = f(x_k)$ for all $k = 0, 1, \ldots, N$. Using the "\" linear solver, write a routine that accomplishes this task, of the form

```
a = calc_poly_coeff(x,f);
```

x, f, and a are vectors of x_k, $f(x_k)$, and a_k respectively.

1.A.4. Using Gram–Schmidt orthogonalization, write a routine that takes as input a vector $v \in \Re^N$ and returns an $N \times N$ matrix V,

$$V = \begin{bmatrix} | & | & & | \\ v & v^{[2]} & \cdots & v^{[N]} \\ | & | & & | \end{bmatrix} \tag{1.264}$$

such that $\{v, v^{[2]}, \ldots, v^{[N]}\}$ forms an orthonormal basis for $\Re^N$.

1.B.1. Compare the number of FLOPs necessary to solve a system of N linear algebraic equations by Gaussian and Gauss–Jordan elimination. Which one requires less work?

1.B.2. You are studying the kinetics of an enzyme-catalyzed reaction converting a substrate S to a product P. The reaction occurs through a two-stage mechanism of reversible substrate S binding to enzyme E to form a complex ES, followed by irreversible conversion to the product P,

$$E + S \Leftrightarrow ES \rightarrow E + P \tag{1.265}$$

The rate of conversion of substrate to product is

$$r = k_2[ES] \tag{1.266}$$

A mole balance on ES,

$$\frac{d}{dt}[ES] = k_1[E][S] - k_{-1}[ES] - k_2[ES] \tag{1.267}$$

yields, under a quasi-steady state assumption (QSSA), $d[ES]/dt = 0$,

$$[ES] = \frac{k_1[E][S]}{(k_{-1} + k_2)} \tag{1.268}$$

If $[E]_0$ is the total concentration of enzyme, bound and unbound, added to the system, $[E] = [E]_0 - [ES]$, yielding

$$[ES] = \frac{[E]_0[S]}{K_m + [S]} \qquad K_m = \frac{k_{-1} + k_2}{k_1} \tag{1.269}$$

Thus, the reaction rate follows Michaelis–Menten kinetics,

$$r = \frac{k_2[E]_0[S]}{K_m + [S]} \tag{1.270}$$

$[E]_0$ is the original enzyme concentration added to start the reaction in units of grams of E per liter. For r in units of grams of S converted per liter per minute, the units of $[S]$ and K_m are grams of S per liter, and the units of k_2 are grams of S converted per minute per gram of E. You have conducted a series of experiments to measure the rate of substrate conversion at various substrate and enzyme concentrations, and have obtained the data in Table 1.2. The rate law can be written in the form

$$\frac{[E]_0}{r} = \frac{1}{k_2} + \frac{K_m}{k_2}\frac{1}{[S]} \tag{1.271}$$

Table 1.2 Rate data for grams of S converted
per liter per minute

[S] g s/l	$[E]_0 = 0.005$ g$_E$/l	$[E]_0 = 0.01$ g$_E$/l
1.0	0.055	0.108
2.0	0.099	0.196
5.0	0.193	0.383
7.5	0.244	0.488
10.0	0.280	0.560
15.0	0.333	0.665
20.0	0.365	0.733
30.0	0.407	0.815

to obtain the linear model

$$y = b_0 + b_1 x \quad y = \frac{[E]_0}{r} \quad x = \frac{1}{[S]} \quad b_0 = \frac{1}{k_2} \quad b_1 = \frac{K_m}{k_2} \tag{1.272}$$

Let us number each experiment in the table as $k = 1, 2, \ldots, N$ and let values of x and y for experiment k be $x^{[k]}$ and $y^{[k]}$ respectively. Then, using the laws of matrix multiplication, we can write

$$X b = \begin{bmatrix} 1 & x^{[1]} \\ 1 & x^{[2]} \\ \vdots & \vdots \\ 1 & x^{[N]} \end{bmatrix} \begin{bmatrix} b_0 \\ b_1 \end{bmatrix} = \begin{bmatrix} y^{[1]} \\ y^{[2]} \\ \vdots \\ y^{[N]} \end{bmatrix} = y \tag{1.273}$$

By multiplying each side by X^T we obtain a system of equations for the coefficients b_0, b_1 that fit the linear model to the data,

$$X^T X b = X^T y \tag{1.274}$$

Using this linear regression technique, compute the values of k_2 and K_m that fit the rate data. Plot for each $[E]_0$ the data of r vs. $[S]$ along with the curve from the fitted model.

1.B.3. Consider reaction and diffusion of a species A in a thin catalyst slab of thickness B. Inside the slab, the concentration field of A is governed at steady state by a diffusion equation with a source term from a first-order chemical reaction,

$$0 = D_A \frac{d^2 c_A}{dx^2} - k c_A \tag{1.275}$$

The catalyst slab is in contact with a gas phase, with a partial pressure of A of p_A. At the solid–gas interfaces at $x = \pm B/2$, the local concentration of A in the slab is in equilibrium with the gas phase, providing the boundary conditions

$$c_A(B/2) = c_A(-B/2) = H_A p_A \tag{1.276}$$

Write the boundary value problem in dimensionless form to reduce the number of independent parameters, and solve using the finite difference method. Plot the dimensionless concentration profiles for various values of the dimensionless parameter(s). Make sure to increase the number of grid points until you no longer see a significant effect upon the solution.

1.B.4. When solving by finite differences the Poisson equation, $-\nabla^2 \varphi = f$, in d dimensions on a regular hypercube grid of N points in each direction, the A matrix has the following nonzero elements in each row (for $\Delta x_j = 1$),

$$A(k, k) = 2d \quad A(k, k \pm N^m) = -1 \quad m = 0, 1, \ldots, d-1 \tag{1.277}$$

For 2-D, 3-D, and 4-D grids generate the resulting A matrix for various N (for higher d, you might only be able to use very small N). For each d, plot as functions of N the numbers of nonzero elements in A and in the Cholesky factor R. Show also the spy plots of A and R for the largest N value considered for each d. Using **cputime**, plot as well the computational time required to solve $A\varphi = b$, b being a vector of all ones, both by the "\" operator and by Cholesky factorization. NOTE High dimension boundary value problems do indeed arise in practice, especially when one is computing the positional and orientational distribution of objects.

1.C.1. Consider the following boundary value problem involving two coupled fields governed by linear differential equations,

$$-\frac{d^2\varphi}{dx^2} = 1 + \psi(x) \qquad -\frac{d^2\psi}{dx^2} = \varphi(x) \qquad \text{on} \qquad 0 \le x \le 1 \tag{1.278}$$

$$\begin{array}{lll} \text{BC1} & \varphi(0) = 0 & \varphi(1) = 0 \\ \text{BC2} & \psi(0) = 0 & \psi(1) = 0 \end{array}$$

We wish to solve this problem numerically using the method of finite differences for a uniform grid of points x_j, $j = 1, 2, \ldots, N$, in the interior of the domain, as was done in the flow example for a single field. We compute numerically the field values at each point, $\varphi(x_j) = \varphi_j$ and $\psi(x_j) = \psi_j$, by solving through elimination the resulting linear algebraic system.

Which of the following ways of stacking the unknowns into a single vector do you recommend using? Explain your reasoning.

$$x = \begin{bmatrix} \varphi_1 \\ \varphi_2 \\ \vdots \\ \varphi_N \\ \psi_1 \\ \psi_2 \\ \vdots \\ \psi_N \end{bmatrix} \qquad \text{or} \qquad x = \begin{bmatrix} \varphi_1 \\ \psi_1 \\ \varphi_2 \\ \psi_2 \\ \vdots \\ \varphi_N \\ \psi_N \end{bmatrix} \tag{1.279}$$

Write a MATLAB program to solve the boundary value problem and plot the solution.

2 Nonlinear algebraic systems

When a set of algebraic equations is nonlinear, there are no general uniqueness and existence criteria, and solution can be quite difficult, even for sets of equations that appear simple. This chapter discusses iterative techniques, in which we make an initial guess of the solution that is refined by solving successive sets of linear equations. Hopefully, this sequence of estimates converges to a solution. These methods are first introduced for a single nonlinear algebraic equation, and then extended to systems of multiple nonlinear equations. The use of MATLAB nonlinear algebraic solvers is demonstrated.

Existence and uniqueness of solutions to a nonlinear algebraic equation

A single linear algebraic equation, $ax = b$, is easily solved, and the condition for existence and uniqueness of the solution $x = b/a$, $a \neq 0$, is trivial. For a single nonlinear algebraic equation

$$f(x) = 0 \tag{2.1}$$

there is, in general, no way to tell a priori whether a solution exists, and if so, whether it is unique. It is easy to identify nonlinear algebraic equations with multiple real roots,

$$f(x) = (x - 3)(x - 2)(x - 1) = x^3 - 6x^2 + 11x - 6 \tag{2.2}$$

with only a single real root,

$$f(x) = (x - 3)(x - i)(x + i) = x^3 - 3x^2 + x - 3 \tag{2.3}$$

or with no real roots at all,

$$f(x) = 3x^4 + 2x^2 + 1 \tag{2.4}$$

Typically, we are presented with a nonlinear function that is not simple to factorize, and so we know nothing about the number of real solutions. The methods described in this chapter are designed to search for a real solution starting from an initial guess and will be demonstrated on systems with varying numbers of solutions.

Iterative methods and the use of Taylor series

The techniques that we will use are iterative. We start with some initial guess $x^{[0]}$ of the solution, and apply an algorithm to refine this guess to generate a sequence $x^{[1]}, x^{[2]}, \ldots$ that hopefully converges to a solution x_s with $f(x_s) = 0$; i.e.,

$$\lim_{k \to \infty} \left| x^{[k]} - x_s \right| = 0 \tag{2.5}$$

We first consider Newton's method, an iterative technique that is based on the use of Taylor series expansions. As Taylor series are used extensively in numerical mathematics, we briefly review their use.

Let us say that we have some function $f(x)$ that we wish to represent as a polynomial in the vicinity of some point x_0,

$$f(x) = \sum_{m=0}^{\infty} a_m (x - x_0)^m \qquad \forall x \text{ in } |x - x_0| < \Delta \tag{2.6}$$

Given only information of the function at x_0 – the value of the function itself plus the values of all of its derivatives (we assume derivatives of $f(x)$ to all orders exist at x_0), what coefficients $a_0, a_1, \ldots$ should we use to match the polynomial to $f(x)$? First, we see that

$$f(x_0) = \sum_{m=0}^{\infty} a_m (x_0 - x_0)^m = \sum_{m=0}^{\infty} a_m (0)^m = a_0 \tag{2.7}$$

so that

$$f(x) = f(x_0) + \sum_{m=1}^{\infty} a_m (x - x_0)^m \tag{2.8}$$

We differentiate to obtain

$$\left. \frac{df}{dx} \right|_{x_0} = \sum_{m=1}^{\infty} m a_m (x - x_0)^{m-1} \bigg|_{x_0} = a_1 + \sum_{m=2}^{\infty} m a_m (0)^{m-1} = a_1 \tag{2.9}$$

Taking yet another derivative yields

$$\left. \frac{d^2 f}{dx^2} \right|_{x_0} = \sum_{m=2}^{\infty} m(m-1) a_m (x - x_0)^{m-2} \bigg|_{x_0} = 2a_2 + \sum_{m=3}^{\infty} m(m-1) a_m (0)^{m-2} \tag{2.10}$$

Continuing this process, we find that if all derivatives to infinite order of $f(x)$ exist at x_0, we may represent the function as the infinite series

$$f(x) = f(x_0) + \left. \frac{df}{dx} \right|_{x_0} (x - x_0) + \left. \frac{1}{2!} \frac{d^2 f}{dx^2} \right|_{x_0} (x - x_0)^2 + \left. \frac{1}{3!} \frac{d^3 f}{dx^3} \right|_{x_0} (x - x_0)^3 + \cdots \tag{2.11}$$

If we only wish to approximate the function in the vicinity of x_0, we can truncate the series at order n:

$$f(x) = \sum_{m=0}^{n} \left. \frac{1}{m!} \frac{d^m f}{dx^m} \right|_{x_0} (x - x_0)^m + R_n(x) \tag{2.12}$$

where the truncation error is

$$R_n(x) = \frac{1}{n!}\frac{d^n f}{dx^n}\bigg|_{\zeta}(x - x_0)^n \qquad \text{for some } \zeta \in [x_0, x] \tag{2.13}$$

When $|x - x_0|$ is very small,

$$|x - x_0| \gg |x - x_0|^2 \gg |x - x_0|^3 \gg \cdots \tag{2.14}$$

and we expect that, unless the higher-order derivatives become very large at x_0, the truncated expansion should be a reasonable description of the local behavior of $f(x)$ near x_0. The smaller the truncation order, in general, the nearer that we will have to be to x_0 for the approximation to be accurate.

Newton's method for a single equation

We now use this truncated Taylor series to develop an iterative technique for solving a non-linear algebraic equation $f(x) = 0$ known as *Newton's method*. To start Newton's method, we make an initial guess $x^{[0]}$ of the solution that we hope is close to the true value x_s where $f(x_s) = 0$. We use a Taylor series to approximate $f(x)$ in the vicinity of $x^{[0]}$,

$$f(x) = f\left(x^{[0]}\right) + \frac{df}{dx}\bigg|_{x^{[0]}}\left(x - x^{[0]}\right) + \frac{1}{2!}\frac{d^2 f}{dx^2}\bigg|_{x^{[0]}}\left(x - x^{[0]}\right)^2 + \cdots \tag{2.15}$$

At the solution, $f(x_s) = 0$, and the Taylor series yields

$$0 = f\left(x^{[0]}\right) + \frac{df}{dx}\bigg|_{x^{[0]}}\left(x_s - x^{[0]}\right) + \frac{1}{2!}\frac{d^2 f}{dx^2}\bigg|_{x^{[0]}}\left(x_s - x^{[0]}\right)^2 + \cdots \tag{2.16}$$

Now, if $x^{[0]}$ is sufficiently close to x_s, then

$$\left|x_s - x^{[0]}\right| \gg \left|x_s - x^{[0]}\right|^2 \gg \left|x_s - x^{[0]}\right|^3 \gg \cdots \tag{2.17}$$

In this case, as long as the first derivative is nonzero at $x^{[0]}$, we obtain a reasonable approximation of the solution, $x^{[1]}$, from the rule

$$0 = f\left(x^{[0]}\right) + \frac{df}{dx}\bigg|_{x^{[0]}}\left(x^{[1]} - x^{[0]}\right) \tag{2.18}$$

Successive application of this rule yields *Newton's method* for solving a single nonlinear algebraic equation,

$$x^{[k+1]} = x^{[k]} - \frac{f\left(x^{[k]}\right)}{f^{(1)}\left(x^{[k]}\right)} \tag{2.19}$$

where we have used the notation $f^{(m)}(x)$ for the mth derivative of $f(x)$. The iterations are stopped when the function value satisfies

$$\left|f\left(x^{[k]}\right)\right| \le \delta_{\text{abs}} \qquad \text{and/or} \qquad \left|f\left(x^{[k]}\right)\right| \le \delta_{\text{rel}}\left|f\left(x^{[0]}\right)\right| \tag{2.20}$$

Table 2.1 *Performance of Newton's method*
for $f(x) = (x - 3)(x - i)(x + i)$

$x^{[0]}$	1	2	4	10
$x^{[1]}$	-1	7	3.3200	7.0064
$x^{[2]}$	-0.2	5.1132	3.0013	5.1558
$x^{[3]}$	1.2345	3.9367	3.0000	3.9621
$x^{[4]}$	-1.1938	3.2894	3.0000	3.3016
$x^{[5]}$	-0.3761	3.0401	3	3.0432
$x^{[6]}$	0.6707	3.0009		3.0011
$x^{[7]}$	-1.3458	3.0000		3.0000
$x^{[8]}$	-0.5037	3.0000		3.0000
$x^{[9]}$	0.4146	3.0000		3.0000
$x^{[10]}$	-2.7029	3		3

Performance of Newton's method for a single equation

We demonstrate the performance of Newton's method for various cubic polynomials, starting with one that possesses only a single real root,

$$f(x) = (x - 3)(x - i)(x + i) = x^3 - 3x^2 + x - 3 \qquad (2.21)$$

In Table 2.1, the trajectories of Newton's method are presented starting from various values of the initial guess, $x^{[0]}$. We see that the number of iterations required to converge to the solution at $x = 3$ depends strongly upon the initial guess. To see why this is so, we will examine graphically the progress of Newton's method. In Figure 2.1, we plot the function and its first derivative in the vicinity of the solution. Both of these functions seem rather uncomplicated, so why does Newton's method converge so erratically for this example? The reason lies in the fact that the update function

$$u\left(x^{[k]}\right) = x^{[k+1]} - x^{[k]} \qquad (2.22)$$

involves the ratio of two functions,

$$u(x) = -\frac{f(x)}{f^{(1)}(x)} \qquad (2.23)$$

Because the derivative of $f(x)$ is zero at two locations, $u(x)$ blows up to $\pm\infty$ at these points, leading to large, erratic steps when the estimates are in the vicinity of such "flat" points in $f(x)$. Figure 2.2 plots the update function in the vicinity of the solution and the number of iterations required for convergence to a specified accuracy, as a function of the initial guess. In the top plot, the solid line is $u(x)$ vs. x and the dots are the converged solutions x_s vs. the initial guess $x^{[0]}$. The bottom graph shows the number of iterations required for convergence vs. $x^{[0]}$. Even for this simple cubic polynomial, Newton's method may take large steps far away from the true solution, making convergence extremely slow for some choices of the initial guess. This is a general characteristic of Newton's method – the convergence properties depend strongly upon the choice of the initial guess.

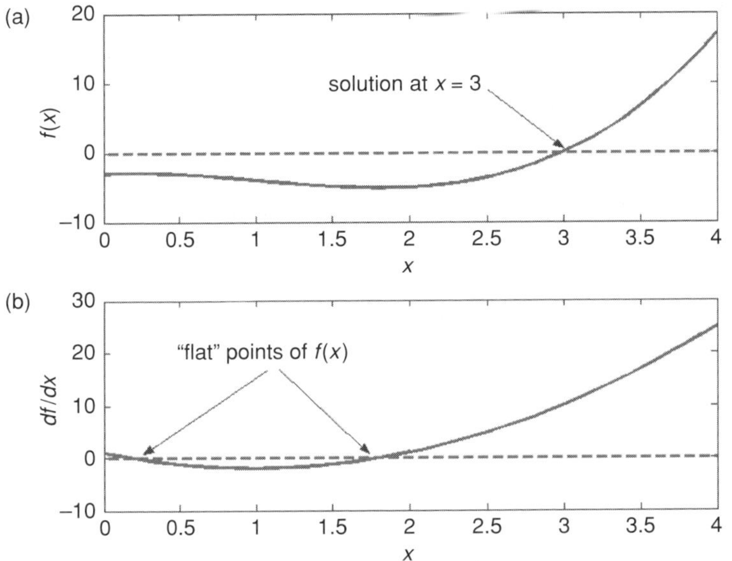

Figure 2.1 Plot of (a) the function and (b) the first derivative with a single real root, $f(x) = (x - 3)(x - i)(x + i)$.

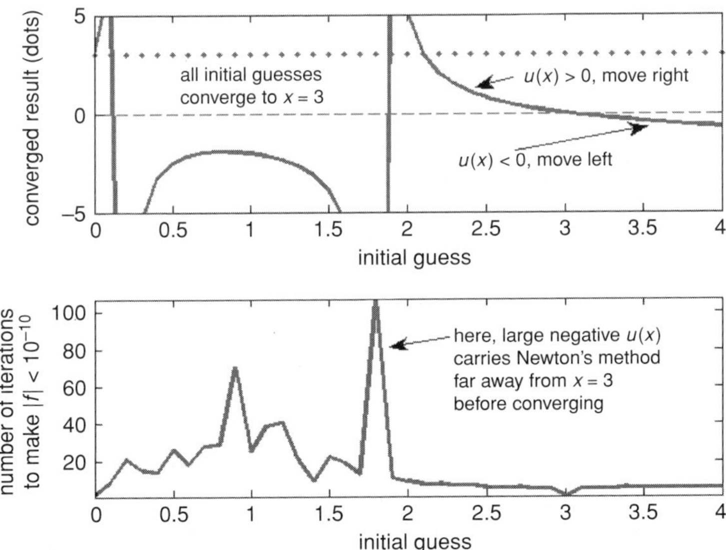

Figure 2.2 Update function and effect of initial guess on convergence, $f(x) = (x - 3)(x - i)(x + i)$. In the upper plot, the line is $u(x)$ vs. x and the dots indicate the converged result of Newton's method vs. $x^{[0]}$.

We next consider the performance of Newton's method for a system with multiple real roots to ask the question: is there any clear relation between the choice of initial guess and the identity of the root that is found? Consider the cubic polynomial

$$f(x) = (x - 3)(x - 2)(x - 1) = x^3 - 6x^2 + 11x - 6 \qquad (2.24)$$

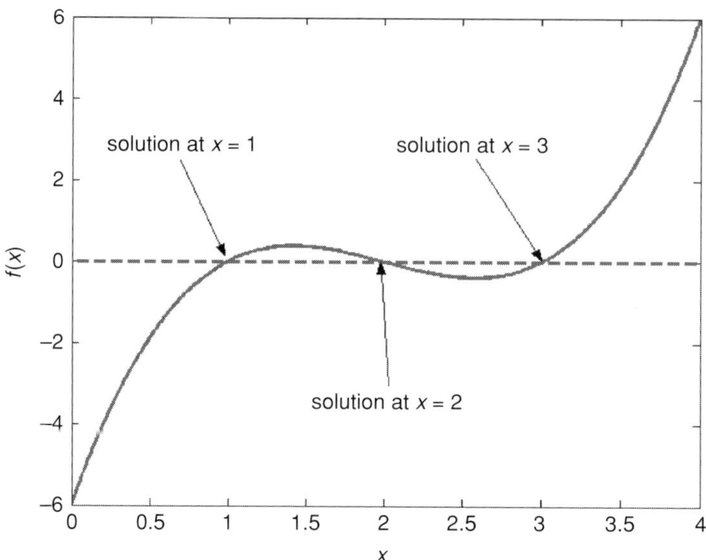

Figure 2.3 Cubic polynomial $f(x) = (x - 3)\,(x - 2)\,(x - 1)$.

The function is plotted in Figure 2.3, showing the positions of three real roots. Note that this function has "flat points" where $f^{(1)}(x) = 0$ near 1.4 and 2.6, where we expect Newton's method to behave erratically.

The performance of Newton's method for this example can be understood by considering the plot of the update function in Figure 2.4 (plotted as a solid line). In addition to this plot, for various values of the initial guess along the x-axis, the values of the corresponding solution found are plotted along the y-axis as dots. For initial guesses that are less than 1, the update function is positive and Newton's method moves towards the right to find the solution $x = 1$. Similarly, above 3, Newton's method moves to the left to find the solution at $x = 3$.

In $1 \leq x \leq 3$, however, the behavior is more complex. Near an initial guess of 1.5, Newton's method finds the root at $x = 3$ before entering a window in which it finds the root at $x = 2$ (and at least once more returns to $x = 1$ briefly). This unusual behavior of the convergence of Newton's method can be understood from the locations of divergence of the update function. For an initial guess of 1.5, a large positive value of the update function generates a new estimate $x^{[1]}$ that is far to the right, and so the trajectory enters the region where the solution at $x = 3$ is obtained. For slightly larger values of the initial guess, there is a brief window in which the positive, but only moderately large, value of the update function generates a new estimate $x^{[1]}$ in the vicinity of 2.5, where the large negative $u(x^{[1]})$ carries Newton's method far to the left for $x^{[2]}$ so that the solution $x = 1$ is found. Even for such a simple cubic polynomial, we see that Newton's method can yield erratic behavior and return different roots depending sensitively upon the choice of initial guess.

We also see from this example that there are many initial guesses that will identify the roots at $x = 1$ and $x = 3$, but that there exists only a small window of initial guesses that

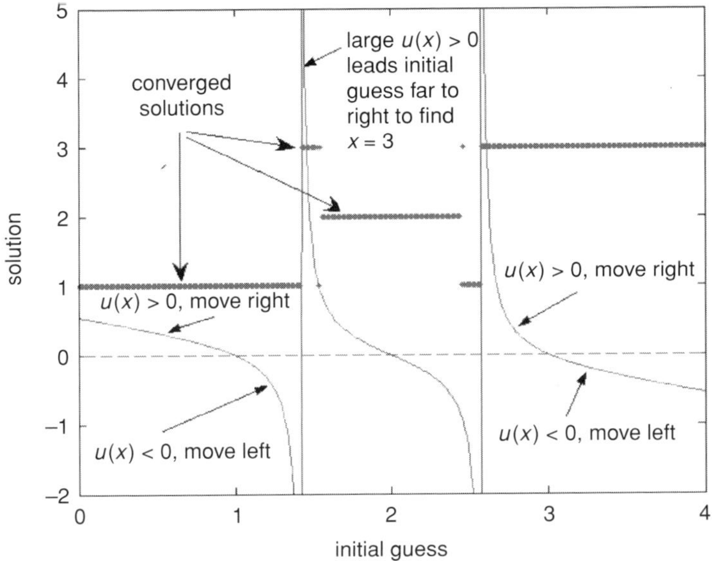

Figure 2.4 Update function (line) of Newton's method for various initial guesses for $f(x) = (x - 3)$ $(x - 2)(x - 1)$. Converged solutions are plotted as dots vs. the initial guess.

return the root at $x = 2$. This demonstrates another feature of nonlinear equations: it is usually very difficult to identify all solutions, and while guess after guess may yield one of some particular set of solutions, this does *not* guarantee that there are no other solutions waiting to be found. We describe here a simple technique to search for additional solutions. Let us say that we have identified the roots at $x = 1$ and $x = 3$, but want to look for additional solutions. We can factor these roots out formally,

$$f(x) = (x - 1)(x - 3)g(x) \tag{2.25}$$

and then apply Newton's method to solve

$$g(x) = \frac{f(x)}{(x - 1)(x - 3)} = 0 \tag{2.26}$$

Here, because $f(x)$ is a polynomial, we can obtain $g(x)$ analytically. In general, this is not possible, and $g(x)$ must be defined as above, in a form that behaves poorly ($\sim 0/0$) at $x = 1$ and $x = 3$. Care must be taken to avoid these regions, or to provide an upper limit on the allowable magnitude of $g(x)$ to avoid numerical overflow. As Figure 2.5 shows, this technique finds the third root at $x = 2$ from any initial guess except close to $x = 1$ and $x = 3$ where the algorithm terminates without convergence.

Formal convergence properties of Newton's method for a single equation

We now consider the convergence properties of Newton's method more formally. We have seen that when the initial guess is not very close to a solution, Newton's method behaves

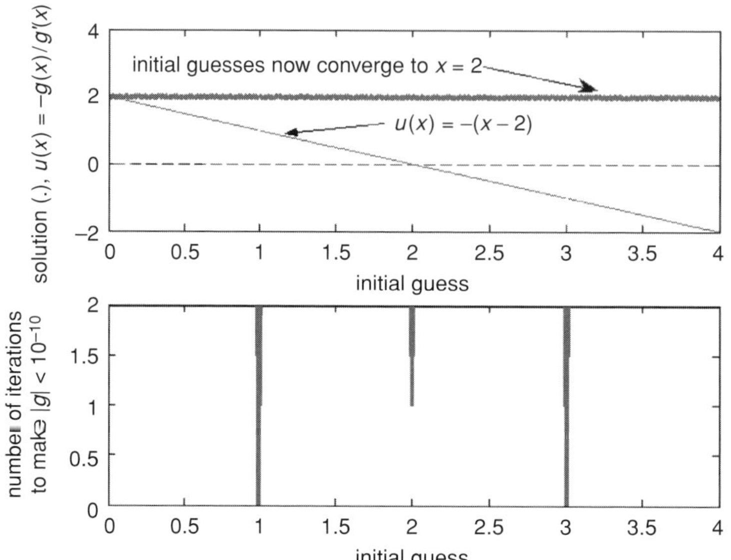

Figure 2.5 Convergence of a simple factoring method to find remaining root at $x = 2$ for $f(x) = (x - 3)(x - 2)(x - 1)$.

erratically. Let us assume now that the current estimate of the solution is indeed near a solution. Then, can we say anything about the rate at which successive Newton iterations converge upon the true solution value? First, we write a Taylor series of $f(x)$ about the current estimate $x^{[k]}$,

$$f\left(x^{[k]} + \varepsilon\right) = f\left(x^{[k]}\right) + \varepsilon f^{(1)}\left(x^{[k]}\right) + \frac{1}{2!}\varepsilon^2 f^{(2)}\left(x^{[k]}\right) + \frac{1}{3!}\varepsilon^3 f^{(3)}\left(x^{[k]}\right) + \cdots \qquad (2.27)$$

We define the error at iteration k as

$$\varepsilon_k \equiv x_s - x^{[k]} \qquad (2.28)$$

If the current estimate is very close to the true solution, ε_k is very small and

$$|\varepsilon_k| \gg \left|\varepsilon_k^2\right| \gg \left|\varepsilon_k^3\right| \gg \cdots \qquad (2.29)$$

Therefore, if we can neglect terms of third order and higher in ε_k, we can approximate

$$f\left(x^{[k]} + \varepsilon_k\right) = f(x_s) = 0 \approx f\left(x^{[k]}\right) + \varepsilon_k f^{(1)}\left(x^{[k]}\right) + \tfrac{1}{2}\varepsilon_k^2 f^{(2)}\left(x^{[k]}\right) \qquad (2.30)$$

Now, from Newton's method (2.19), we have

$$\left(x^{[k+1]} - x_s\right) = \left(x^{[k]} - x_s\right) - \frac{f\left(x^{[k]}\right)}{f^{(1)}\left(x^{[k]}\right)} \qquad (2.31)$$

so that the errors at successive iterations are related by

$$\varepsilon_{k+1} = \varepsilon_k + \frac{f\left(x^{[k]}\right)}{f^{(1)}\left(x^{[k]}\right)} \tag{2.32}$$

If we divide the Taylor series approximation (2.30) by $f^{(1)}\left(x^{[k]}\right)$,

$$0 \approx \frac{f\left(x^{[k]}\right)}{f^{(1)}\left(x^{[k]}\right)} + \varepsilon_k + \varepsilon_k^2 \frac{f^{(2)}\left(x^{[k]}\right)}{2 f^{(1)}\left(x^{[k]}\right)} \tag{2.33}$$

and use (2.32), we obtain

$$\varepsilon_{k+1} = -\varepsilon_k^2 \frac{f^{(2)}\left(x^{[k]}\right)}{2 f^{(1)}\left(x^{[k]}\right)} \tag{2.34}$$

This means that if we are very close to the solution, Newton's method converges quadratically. For example, assume that we are sufficiently close to a solution for this quadratic convergence to hold and that $|\varepsilon_k| = 10^{-1}$. Then, the sequence of errors in the next few iterations is approximately

$$|\varepsilon_{k+1}| = 10^{-2} \quad |\varepsilon_{k+2}| = 10^{-4} \quad |\varepsilon_{k+3}| = 10^{-8} \quad |\varepsilon_{k+4}| = 10^{-16} \tag{2.35}$$

Once Newton's method is close enough to the real solution for the second-order Taylor series approximation to be accurate, the sequence of estimates converges very rapidly (quadratically) to the solution.

The secant method

Each iteration of Newton's method requires not only an evaluation of the function, but also an evaluation of the first derivative. In some cases, the algebraic function may be of such complexity that it is inconvenient to derive the analytical form of the derivative. One alternative would be to use a finite difference approximation,

$$\left.\frac{df}{dx}\right|_{x^{[k]}} \approx \frac{f\left(x^{[k]} + \varepsilon\right) - f\left(x^{[k]}\right)}{\varepsilon} \qquad \varepsilon \approx \sqrt{eps} \tag{2.36}$$

where *eps* is the machine precision. This approach, however, requires two function evaluations per iteration, so that in practice it is more common to use the *secant method*, in which the value of the derivative is approximated by the two most recent function evaluations,

$$\left.\frac{df}{dx}\right|_{x^{[k]}} \approx \frac{f\left(x^{[k]}\right) - f\left(x^{[k-1]}\right)}{x^{[k]} - x^{[k-1]}} \tag{2.37}$$

Thus, the new estimate is

$$x^{[k+1]} = x^{[k]} - \frac{f\left(x^{[k]}\right)\left[x^{[k]} - x^{[k-1]}\right]}{f\left(x^{[k]}\right) - f\left(x^{[k-1]}\right)} \tag{2.38}$$

In this method, only one new function evaluation per iteration is necessary. In comparison to Newton's method, there is no need to evaluate analytically the value of the first derivative; however, convergence is slower than with Newton's method,

$$|\varepsilon_{k+1}| \approx C|\varepsilon_k|^{1.618} \tag{2.39}$$

When an analytical expression for the first derivative is available, Newton's method is preferred due to its faster, quadratic rate of convergence; otherwise, the secant method is suggested. In practice, the loss of quadratic convergence is not as bad as one might expect, for, in general, it is found only very near the solution

Bracketing and bisection methods

It is far easier to find a suitable initial guess for a single equation than it is for multiple equations. For a single equation, we can locate the region of a solution by scanning only a single variable x; however, with even only two equations, for each trial value of x_1, there are infinitely many possible guesses of x_2. Only for a single equation does it become really practical to try various initial guesses in some planned, rigorous manner to search for a solution.

Let us say that we have two values of x, $x^{[k]}$ and $x^{[k+1]}$, and their function values, $f(x^{[k]})$ and $f(x^{[k+1]})$. If the signs of the two function values differ and $f(x)$ is continuous, we then know that $f(x)$ must cross $f = 0$ at least once between $x^{[k]}$ and $x^{[k+1]}$. Therefore, $f(x)$ must have at least one solution in $[x^{[k]}, x^{[k+1]}]$. Once we have found such a segment that brackets a solution, we can narrow the bracketing range by setting the bisecting point $x^{[k+2]} = (x^{[k]} + x^{[k+1]})/2$ and computing $f(x^{[k+2]})$. We then select two consecutive members of $\{x^{[k]}, x^{[k+1]}, x^{[k+2]}\}$ whose signs of $f(x)$ differ, and repeat the bisection. This approach is robust, but rather slow.

Once the bracket becomes sufficiently small that we feel that Newton's method or the secant method should be able to find the solution, we switch to one of those more efficient procedures. If this fails, we continue with bisection until the initial guess is sufficiently close for the iterative method to succeed. In MATLAB, the routine **fzero** takes such an approach. For further discussion of iterative methods to solve a single equation $f(x) = 0$, consult Press *et al.* (*1992*) and Quateroni *et al.* (*2000*).

Finding complex solutions

The previous discussion focused upon methods for finding solutions to $f(x) = 0$, where both x and $f(x)$ are real. While this is generally the case in engineering and scientific

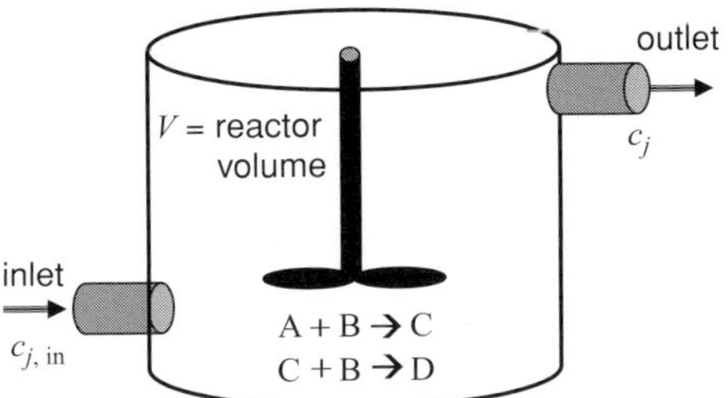

Figure 2.6 CSTR with two chemical reactions.

practice, we may wish to identify complex solutions to complex-valued functions. To do so, we write x as

$$x = a + ib \qquad a = \text{Re}\{x\} \in \Re \qquad b = \text{Im}\{x\} \in \Re \tag{2.40}$$

and compute a and b by solving the two coupled real-valued equations

$$\text{Re}\{f(a + ib)\} = 0 \qquad \text{Im}\{f(a + ib)\} = 0 \tag{2.41}$$

Thus, techniques for treating multiple nonlinear algebraic equations are required in this instance.

Systems of multiple nonlinear algebraic equations

We now extend Newton's method to solve a set of N simultaneous nonlinear algebraic equations for N unknowns

$$\begin{aligned}
f_1(x_1, x_2, \ldots, x_N) &= 0 \\
f_2(x_1, x_2, \ldots, x_N) &= 0 \\
&\vdots \\
f_N(x_1, x_2, \ldots, x_N) &= 0
\end{aligned} \tag{2.42}$$

More compactly, we define the state vector of unknowns

$$\boldsymbol{x} = [x_1 \, x_2 \ldots x_N]^{\text{T}} \tag{2.43}$$

and write the system of equations as

$$\boldsymbol{f}(\boldsymbol{x}) = \boldsymbol{0} \tag{2.44}$$

As an example, consider a ***continuous stirred-tank reactor*** (CSTR), operated isothermally, with negligible volume change due to reaction, in overflow mode with a constant fluid volume V, and with the two chemical reactions (assumed elementary) (Figure 2.6)

$$\begin{aligned}
\text{A} + \text{B} &\rightarrow \text{C} \qquad r_{\text{R1}} = k_1 c_{\text{A}} c_{\text{B}} \\
\text{C} + \text{B} &\rightarrow \text{D} \qquad r_{\text{R2}} = k_2 c_{\text{C}} c_{\text{B}}
\end{aligned} \tag{2.45}$$

In a CSTR, we assume that the reactor is so perfectly mixed that the concentration field of each species is spatially uniform. That is, every point in the reactor has the same concentration of each species, governed by the set of mass balances

$$\frac{d}{dt}(Vc_A) = v(c_{A,\,in} - c_A) + V(-k_1 c_A c_B)$$
$$\frac{d}{dt}(Vc_B) = v(c_{B,\,in} - c_B) + V(-k_1 c_A c_B - k_2 c_C c_B)$$
$$\frac{d}{dt}(Vc_C) = v(c_{C,\,in} - c_C) + V(k_1 c_A c_B - k_2 c_C c_B)$$
$$\frac{d}{dt}(Vc_D) = v(c_{D,\,in} - c_D) + V(k_2 c_C c_B)$$

(2.46)

v is the volumetric flow rate of the feed and outlet streams, V is the fixed reactor volume, c_j is the concentration of species j in the reactor (and in the output stream), $c_{j,in}$ is the concentration of species j in the inlet stream, and k_1, k_2 are the rate constants of each chemical reaction.

At steady state, the time derivatives on the left are zero, and the concentrations of each species within the reactor satisfy a set of four nonlinear algebraic equations. To put these equations in standard form, we define the unknowns

$$x_1 = c_A \quad x_2 = c_B \quad x_3 = c_C \quad x_4 = c_D$$

(2.47)

to obtain the algebraic system

$$v(c_{A,\,in} - x_1) + V(-k_1 x_1 x_2) = 0$$
$$v(c_{B,\,in} - x_2) + V(-k_1 x_1 x_2 - k_2 x_3 x_2) = 0$$
$$v(c_{C,\,in} - x_3) + V(k_1 x_1 x_2 - k_2 x_3 x_2) = 0$$
$$v(c_{D,\,in} - x_4) + V(k_2 x_3 x_2) = 0$$

(2.48)

We see that in addition to the four unknowns, there are a number of other model parameters whose values we must know before attempting to solve the equations. We collect these quantities into a parameter vector Θ:

$$\Theta = [v \; V \; k_1 \; k_2 \; c_{A,\,in} \; c_{B,\,in} \; c_{C,\,in} \; c_{D,\,in}]^T$$

(2.49)

and write the set of nonlinear algebraic equations as

$$f(x; \Theta) = 0$$

(2.50)

Newton's method for multiple nonlinear equations

Again we use a Taylor series expansion to obtain Newton's method, representing the ith function in the vicinity of the current estimate $x^{[k]}$ as

$$f_i(x) = f_i(x^{[k]}) + \sum_{m=1}^{N} \left.\frac{\partial f_i}{\partial x_m}\right|_{x^{[k]}} (x_m - x_m^{[k]})$$

$$+ \frac{1}{2} \sum_{m=1}^{N} \sum_{n=1}^{N} (x_m - x_m^{[k]}) \left.\frac{\partial^2 f_i}{\partial x_m \partial x_n}\right|_{x^{[k]}} (x_n - x_n^{[k]}) + \cdots$$

(2.51)

Assuming that $x^{[k]}$ is sufficiently close to the true solution x_s that we introduce little error by dropping the terms of quadratic and higher order,

$$f_i(x_s) = 0 \approx f_i\left(x^{[k]}\right) + \sum_{m=1}^{N} \left.\frac{\partial f_i}{\partial x_m}\right|_{x^{[k]}} \left(x_{s,m} - x_m^{[k]}\right) \tag{2.52}$$

For convenience, we collect the first partial derivatives into the $N \times N$ *Jacobian matrix* $J^{[k]} = J(x^{[k]})$, with elements

$$J_{im}^{[k]} = \left.\frac{\partial f_i}{\partial x_m}\right|_{x^{[k]}} \tag{2.53}$$

The truncated Taylor series expansion then becomes

$$0 \approx f_i\left(x^{[k]}\right) + \sum_{m=1}^{N} J_{im}^{[k]}\left(x_{s,m} - x_m^{[k]}\right) \tag{2.54}$$

We thus generate the new estimate of the solution, $x^{[k+1]} \approx x_s$ by solving the set of linear algebraic equations

$$0 = f_i\left(x^{[k]}\right) + \sum_{m=1}^{N} J_{im}^{[k]}\left(x_m^{[k+1]} - x_m^{[k]}\right) \tag{2.55}$$

Defining the update vector

$$\Delta x^{[k]} \equiv x^{[k+1]} - x^{[k]} \tag{2.56}$$

the set of Newton update linear equations is

$$\sum_{m=1}^{N} J_{im}^{[k]} \Delta x_m^{[k]} = -f_i\left(x^{[k]}\right) \tag{2.57}$$

We recognize the term on the left-hand side as the ith component of a matrix–vector product and write the linear system as

$$J^{[k]} \Delta x^{[k]} = -f\left(x^{[k]}\right) \tag{2.58}$$

At each Newton iteration, we must solve a linear system of N equations, e.g. through Gaussian elimination. This procedure is repeated until the norm of the function vector becomes smaller than some tolerance value,

$$\left\| f\left(x^{[k]}\right) \right\| \leq \delta_{abs} \qquad \text{and/or} \qquad \left\| f\left(x^{[k]}\right) \right\| \leq \delta_{rel} \left\| f\left(x^{[0]}\right) \right\| \tag{2.59}$$

The choice of norm is somewhat subjective, but a common selection is

$$\left\| f\left(x^{[k]}\right) \right\|_{\infty} = \max\{|f_1|, |f_2|, \ldots, |f_N|\} \tag{2.60}$$

As was the case for a single equation, convergence in the close vicinity of the solution is quadratic. The proof is somewhat more involved with multiple equations, and is presented in the supplemental material in the accompanying website.

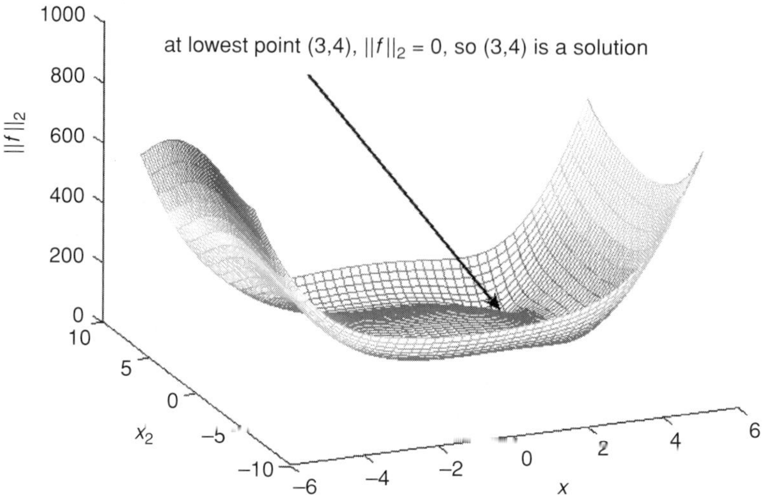

Figure 2.7 3-D surface plot of $\|f\|_2$ for a system with a solution at (3,4).

Performance of Newton's method for an example system of two equations

Let us next consider the performance of Newton's method for the following system of two equations with a real solution at $x_s = [3\ 4]^T$:

$$\begin{aligned}
f_1(x_1, x_2) &= 3x_1^3 + 4x_2^2 - 145 = 0 \\
f_2(x_1, x_2) &= 4x_1^2 - x_2^3 + 28 = 0
\end{aligned} \tag{2.61}$$

The Jacobian matrix for this system is

$$J = \begin{bmatrix} \dfrac{\partial f_1}{\partial x_1} & \dfrac{\partial f_1}{\partial x_2} \\ \dfrac{\partial f_2}{\partial x_1} & \dfrac{\partial f_2}{\partial x_2} \end{bmatrix} = \begin{bmatrix} 9x_1^2 & 8x_2 \\ 8x_1 & -3x_2^2 \end{bmatrix} \tag{2.62}$$

Figure 2.7 shows a surface plot of the 2-norm of the function vector, $\|f(x)\|_2$, vs. (x_1, x_2). The solution is a global minimum of the 2-norm, and we would like Newton's method to march steadily "downhill" on this surface until we reach a minimum elevation that we hope is a solution with $\|f\|_2 = 0$. To understand the performance of Newton's method, we need to keep the shape of this 2-norm surface in mind. Figure 2.8 presents a contour plot of $\|f(x)\|_2$ with lines drawn at constant values of the 2-norm and arrows pointing in the local direction of increasing 2-norm. "Steep" regions of rapidly varying 2-norm are identified by larger arrows and contour lines that are close together.

Figure 2.9 overlays upon this contour plot of the 2-norm, a trajectory of solution estimates obtained from Newton's method with an initial guess of (2,2). Newton's method converges in seven iterations to the desired accuracy; however, the first step carries the estimate too far into a region of increasing 2-norm. Such a step is not very helpful for finding a solution.

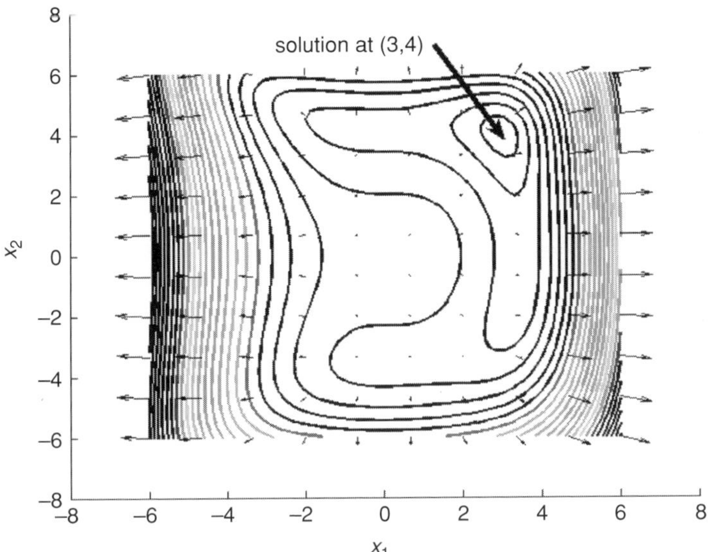

Figure 2.8 Contour plot of $\|f\|_2$ for a system with a solution at (3,4).

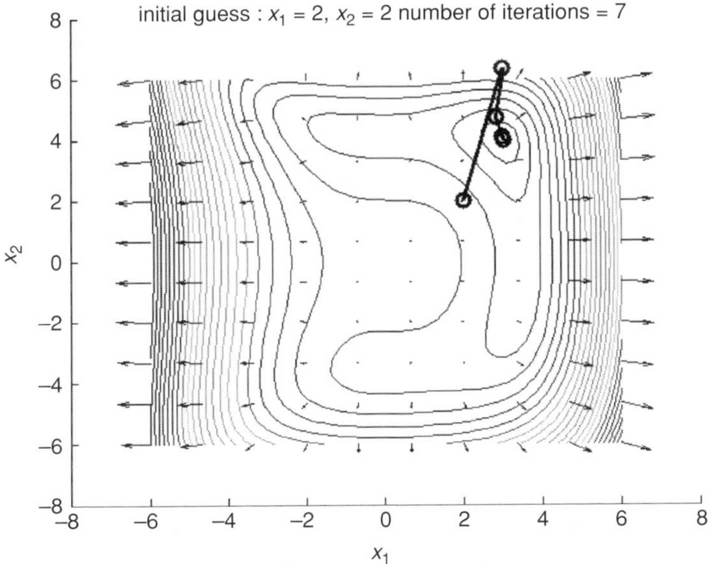

Figure 2.9 Trajectory of Newton's method with the initial guess (2,2).

Figure 2.10 repeats this plot for an initial guess of (1,1). While Newton's method again converges (now in eleven iterations to the desired accuracy), a much larger step into a region of high 2-norm is made. With an initial guess of (2,–1), such a departure is even more pronounced (Figure 2.11); however, convergence to the desired accuracy still occurs after 21 iterations. While Newton's method does converge (at least for these initial guesses), it

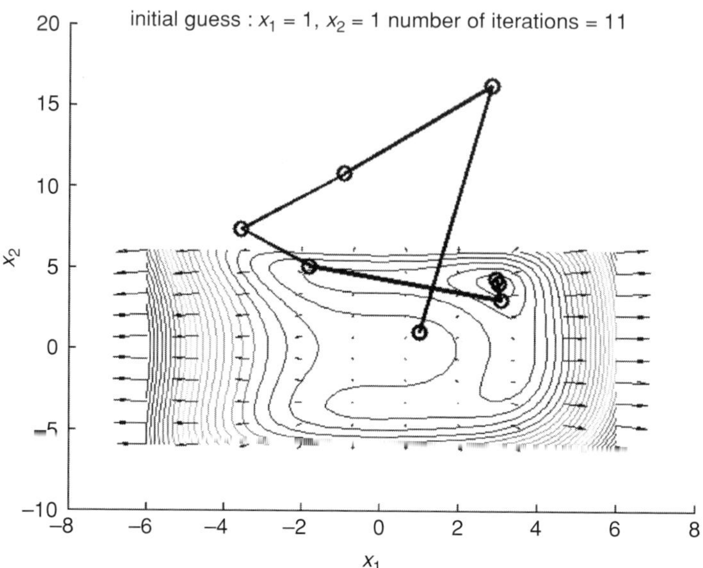

Figure 2.10 Trajectory of Newton's method with the initial guess (1,1).

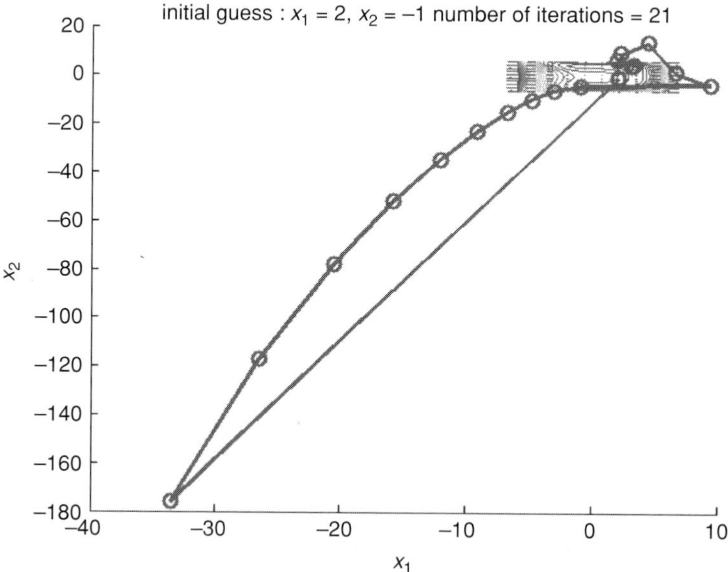

Figure 2.11 Trajectory of Newton's method with the initial guess (2,−1).

certainly appears as if the choice of updates could be improved, by moderating the tendency of the algorithm to make large, erratic steps. A reduced-step algorithm is proposed later that avoids generating these erratic steps and makes Newton's method more robust. These plots are generated by Newton_2D_test.m.

Estimating the Jacobian and quasi-Newton methods

With Newton's method, to update our estimate of the solution we need to evaluate at $x^{[k]}$ the function vector $f(x^{[k]})$ and the Jacobian matrix $J^{[k]}$. In many instances, we do not wish to take the time to derive an analytical form for the elements of the Jacobian matrix and code a subroutine to evaluate them. It is often easier merely to provide a subroutine that evaluates the function vector, so we consider now how the human effort of computing the Jacobian can be eliminated. One approach gaining use is to employ an automatic differentiation program, a routine that reads the code that we write to evaluate the function vector and automatically generates additional code that evaluates the Jacobian matrix. Here, we describe two more traditional approaches: finite difference approximations and the use of approximate Jacobian matrices generated by the past history of function evaluations.

In these methods, we use not the exact value of the Jacobian matrix, but rather some approximation

$$B^{[k]} \approx J\left(x^{[k]}\right) \tag{2.63}$$

We then compute the update by solving the linear system

$$B^{[k]} \Delta x^{[k]} = -f\left(x^{[k]}\right) \tag{2.64}$$

The use of such an approximation does not change the value of the solution that we obtain, as at a solution x_s, $f(x_s) = 0$ and $\Delta x = (B^{[k]})^{-1} f(x_s) = 0$ no matter the value of $B^{[k]}$ (provided that it is nonsingular). This allows some freedom in the choice of $B^{[k]}$ to balance accuracy in the approximation of the true Jacobian against computational efficiency.

The most straight forward approach to approximate the Jacobian is to use a finite difference approximation

$$\left.\frac{\partial f_m}{\partial x_n}\right|_{x^{[k]}} = J^{[k]}_{mn} \approx B^{[k]}_{mn} = \frac{f_m\left(x^{[k]} + \varepsilon e^{[n]}\right) - f_m\left(x^{[k]}\right)}{\varepsilon} \tag{2.65}$$

$e^{[n]}$ is the unit vector comprising all zeros except for a value of 1 for the nth component, $e^{[n]}_j = \delta_{jn}$. Here ε is some small number, typically chosen for reasons of numerical accuracy to be the square root of the machine precision, $\varepsilon \approx \sqrt{eps}$. This finite difference approximation can yield quite accurate approximations of the Jacobian, but at the cost of N additional function evaluations per Newton iteration. When the Jacobian has a known sparsity structure, the number of function evaluations may be reduced, but, in general, the overhead associated with these additional function evaluations can be quite significant for large systems.

To avoid the numerous and costly function evaluations required by the finite difference method, various *quasi-Newton* methods have been developed in which the approximation $B^{[k]}$ of the Jacobian is constructed from the recent values of the function vector $f(x^{[k]})$, $f(x^{[k-1]})$, $f(x^{[k-2]})$, …. We describe here the popular *Broyden's method* (Broyden, 1965) for updating the estimate of the Jacobian, which can be considered an extension of the secant method to systems of multiple equations.

Let $B^{[k]}$ be the current estimate of the Jacobian. We update the solution estimate by the rule

$$B^{[k]}\Delta x^{[k]} = -f(x^{[k]}) \tag{2.66}$$
$$x^{[k+1]} \leftarrow x^{[k]} + \Delta x^{[k]}$$

At $x^{[k+1]}$, we evaluate the function value $f(x^{[k+1]})$, but to obtain the new update by solving the linear system

$$B^{[k+1]}\Delta x^{[k+1]} = -f(x^{[k+1]}) \tag{2.67}$$

we need a new Jacobian estimate $B^{[k+1]}$.

Broyden's method generates $B^{[k+1]}$ from $B^{[k]}$, $f(x^{[k]})$, and $f(x^{[k+1]})$ using the path integral formula

$$f(x^{[k+1]}) = f(x^{[k]}) + \int_0^1 J(x^{[k]} + s\Delta x^{[k]})\Delta x^{[k]}ds \tag{2.68}$$

and its approximation for very small $\Delta x^{[k]}$, obtained by neglecting the variation in the Jacobian,

$$f(x^{[k+1]}) \approx f(x^{[k]}) + J(x^{[k+1]})\Delta x^{[k]} \tag{2.69}$$

In Broyden's method, we require that $B^{[k+1]}$ exactly satisfies this approximation:

$$f(x^{[k+1]}) = f(x^{[k]}) + B^{[k+1]}\Delta x^{[k]} \tag{2.70}$$

Using (2.64), this becomes

$$f(x^{[k+1]}) + B^{[k]}\Delta x^{[k]} = B^{[k+1]}\Delta x^{[k]} \tag{2.71}$$

We postmultiply by $(\Delta x^{[k]})^{\mathrm{T}}$ and use the identity $Bvv^{\mathrm{T}} = |v|^2 B$ to write

$$f(x^{[k+1]})(\Delta x^{[k]})^{\mathrm{T}} + |\Delta x^{[k]}|^2 B^{[k]} = |\Delta x^{[k]}|^2 B^{[k+1]} \tag{2.72}$$

Division by the scalar $|\Delta x^{[k]}|^2$ yields the *Broyden rank-one update* rule

$$B^{[k+1]} = B^{[k]} + \frac{f(x^{[k+1]})(\Delta x^{[k]})^{\mathrm{T}}}{|\Delta x^{[k]}|^2} \tag{2.73}$$

To start, we typically use a crude approximation of the Jacobian such as $B^{[0]} = I$. In general, convergence is slower with Broyden's method than with Newton's method. Note, however, that the quadratic convergence of Newton's method is only obtained near the solution, and that this update formula does not require the N additional function evaluations of the finite difference approximation. This reduction in workload per Newton iteration makes the Broyden method a popular choice when we do not supply a routine to evaluate the Jacobian matrix analytically. For more information on quasi-Newton methods and Jacobian estimation, consult Nocedal & Wright (1999).

Robust reduced-step Newton method

In the study of the performance of Newton's method for a simple 2-D system, we have seen that far away from the solution, the update steps are large and lie in erratic directions. The efficiency and robustness of Newton's method (and quasi-Newton variations such as that of Broyden) are improved dramatically through use of a reduced-step algorithm, in which only a fraction of the update vector is accepted. The full update vector is generated by solving the linear system

$$B^{[k]}\Delta x^{[k]} = -f(x^{[k]}) \tag{2.74}$$

but now only a fraction $\alpha^{[k]} \in [0, 1]$ of the full step is accepted,

$$x^{[k+1]} = x^{[k]} + \alpha^{[k]}\Delta x^{[k]} \tag{2.75}$$

This fractional step is chosen such that the norm of the function vector at the new estimate is smaller than at the old one, yielding the *descent criterion*

$$\left\| f(x^{[k+1]}) \right\| = \left\| f(x^{[k]} + \alpha^{[k]}\Delta x^{[k]}) \right\| < \left\| f(x^{[k]}) \right\| \tag{2.76}$$

Since a solution x_s has $f(x_s) = 0$, x_s is a global minimum of $\| f(x)\|$. So as long as we are decreasing the value of the norm, we feel that we are making good progress towards finding a solution.

We now ask two questions:

Is there always some $\alpha^{[k]} > 0$ such that the descent criterion using the 2-norm, $\|v\|_2^2 = v \cdot v$, is satisfied?

Under what conditions will this method find a limiting point that is *not* a solution?

To answer the first question, we use the path integral relation

$$f(x^{[k]} + \varepsilon\Delta x^{[k]}) = f(x^{[k]}) + \int_0^\varepsilon J(x^{[k]} + s\Delta x^{[k]})\Delta x^{[k]}ds \tag{2.77}$$

which we may approximate for very small ε by neglecting the variation in the Jacobian as

$$f(x^{[k]} + \varepsilon\Delta x^{[k]}) \approx f(x^{[k]}) + \varepsilon J(x^{[k]})\Delta x^{[k]} \tag{2.78}$$

Taking the dot product of this equation with itself yields

$$\left\| f(x^{[k]} + \varepsilon\Delta x^{[k]}) \right\|_2^2 = \left[f(x^{[k]}) + \varepsilon J(x^{[k]})\Delta x^{[k]} \right] \cdot \left[f(x^{[k]}) + \varepsilon J(x^{[k]})\Delta x^{[k]} \right] \tag{2.79}$$

Retaining the terms up to first order in ε yields

$$\left\| f(x^{[k]} + \varepsilon\Delta x^{[k]}) \right\|_2^2 \approx \left\| f(x^{[k]}) \right\|_2^2 + 2\varepsilon f(x^{[k]}) \cdot J(x^{[k]})\Delta x^{[k]} \tag{2.80}$$

Assuming $B^{[k]}$ is nonsingular, $\Delta x^{[k]} = -(B^{[k]})^{-1} f(x^{[k]})$, and

$$\left\| f(x^{[k]} + \varepsilon\Delta x^{[k]}) \right\|_2^2 \approx \left\| f(x^{[k]}) \right\|_2^2 - 2\varepsilon f(x^{[k]}) \cdot \left[J(x^{[k]})(B^{[k]})^{-1} \right] f(x^{[k]}) \tag{2.81}$$

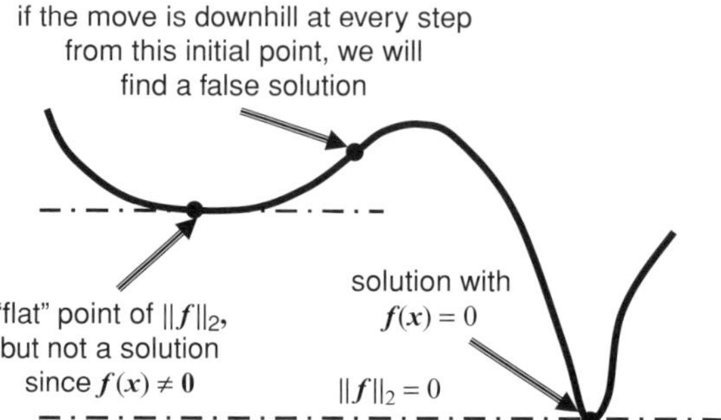

Figure 2.12 The reduced-step norms reduction method may also find a local minimum of the norm that is not a solution.

Now, if the exact Jacobian is used, $B^{[k]} = J^{[k]}$, then $J\left(x^{[k]}\right)\left(B^{[k]}\right)^{-1} = I$ and

$$\left\| f\left(x^{[k]} + \varepsilon \Delta x^{[k]}\right)\right\|_2^2 \approx \left\| f\left(x^{[k]}\right)\right\|_2^2 - 2\varepsilon f\left(x^{[k]}\right) \cdot f\left(x^{[k]}\right) = (1 - 2\varepsilon)\left\| f\left(x^{[k]}\right)\right\|_2^2 \tag{2.82}$$

For $0 < \varepsilon \ll 1$, $\| f(x^{[k]} + \varepsilon \Delta x^{[k]})\|_2^2 < \| f(x^{[k]})\|_2^2$; therefore, when using the exact Jacobian, there always exists some very small, positive value of the fractional step length that results in a decrease of the norm, and so it will be possible always to find some $\alpha^{[k]} \in [0, 1]$ that satisfies the descent criterion. Even if we do not use the exact Jacobian, but only an approximation of it, as long as the matrix $J(x^{[k]})(B^{[k]})^{-1}$ exists and is positive-definite, it is possible to find an $\alpha^{[k]} \in [0, 1]$ that satisfies the descent criterion.

We now consider the second question. As Figure 2.12 demonstrates, merely reducing the norm of the cost function is not sufficient to ensure that we end up at a solution, since we could find instead a local minimum in the cost function that is not a solution (the norm value is nonzero).

We can identify the condition necessary for such a "false solution" through the relation

$$\nabla \| f \|_2^2 = 2 J^{\mathrm{T}} f \tag{2.83}$$

At a flat point in the 2-norm, we must have $J^{\mathrm{T}} f = 0$, and the only way that this can occur if we are not at a solution is if $\det(J^{\mathrm{T}}) = \det(J) = 0$.

With reduced-step Newton algorithms, one of two results generally occurs. Either we converge to a solution (eventually), or the method locates a "false solution" where the Jacobian is singular. If the latter occurs, the iterations stop since we cannot solve the linear update equations and we must start again using a different initial guess.

The backtracking weak line search method

A common method to generate the fractional step length in a reduced-step algorithm is the *backtracking weak line search*. First, we attempt to take the full Newton step. If this step satisfies the descent criterion (the 2-norm is reduced), it is accepted. Otherwise, the fractional step length is reduced by one-half iteratively until the descent criterion is satisfied. The algorithm of the reduced-step Newton method with a weak line search is

Guess $x^{[0]}$; compute $f(x^{[0]})$; initialize $B^{[0]}$
for $k = 0, 1, 2, \ldots, k_{\max}$
    if $\|f(x^{[k]})\|_2 \leq \delta_{\text{tol}}$, STOP and ACCEPT solution $x_s \approx x^{[k]}$
    solve linear system $B^{[k]} \Delta x^{[k]} = -f(x^{[k]})$
    for $m = 0, 1, 2, \ldots, m_{\max}$
      $\alpha^{[k]} = 2^{-m}$
      $x^{[k+1]} = x^{[k]} + \alpha^{[k]} \Delta x^{[k]}$
      calculate $f(x^{[k+1]})$
      if $\|f(x^{[k+1]})\|_2^2 \| f(x^{[k]})\|_2^2$, STOP m iterations and accept $\alpha^{[k]}$
    end of $m = 0, 1, 2, \ldots, m_{\max}$ for loop
end of $k = 0, 1, 2, \ldots, k_{\max}$ for loop

Performance of the reduced-step Newton method for an example system of two equations

We now revisit the example system

$$f_1(x_1, x_2) = 3x_1^3 + 4x_2^2 - 145 = 0$$
$$f_2(x_1, x_2) = 4x_1^2 + x_2^3 + 28 = 0$$

(2.84)

that has a real solution at $x_s = [3\ 4]^T$, using now the reduced-step Newton method. Once again, the trajectories of the solution estimates are plotted as connected circles superimposed on the contour plot of the 2-norm. Figure 2.13 shows the trajectory for the guess $(2,1)$ that converges in six iterations without the erratic first step of the full-step Newton trajectory in Figure 2.9. With the guess $(1,1)$, the reduced-step trajectory of Figure 2.14 likewise shows better performance than the full-step trajectory of Figure 2.10. But, with a guess of $(2, -1)$, Figure 2.15 demonstrates that the reduced-step method can "hang up" by taking only very small steps when the direction of the Newton step lies nearly perpendicular to the local gradient of the 2-norm. Then, the trust-region Newton method, described below, does better as it varies concurrently with both the step length and direction.

The trust-region Newton method

As noted in the example above, the reduced-step Newton method can fail when the search direction is nearly perpendicular to the steepest descent direction so that only very short steps are taken. This problem originates from the fact that the direction of the full Newton step is always accepted; we reduce only the magnitude of the step to attain reduction of

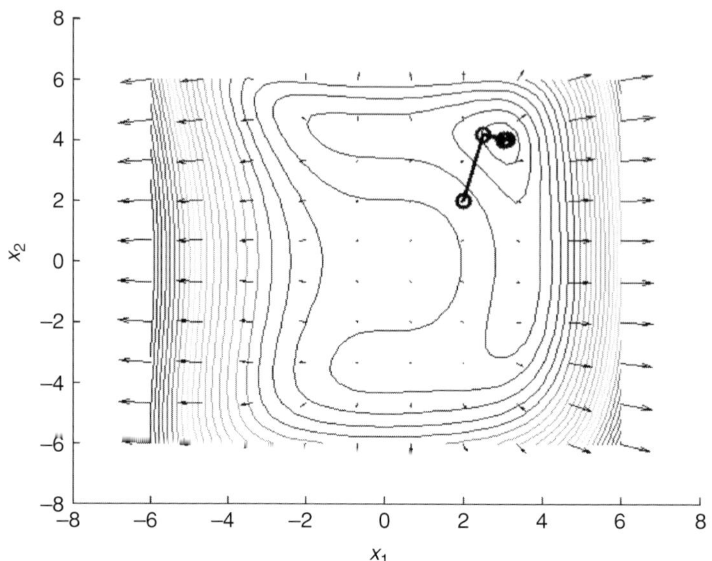

Figure 2.13 Trajectory of reduced-step Newton method with the guess (2,2) (number of iterations = 6).

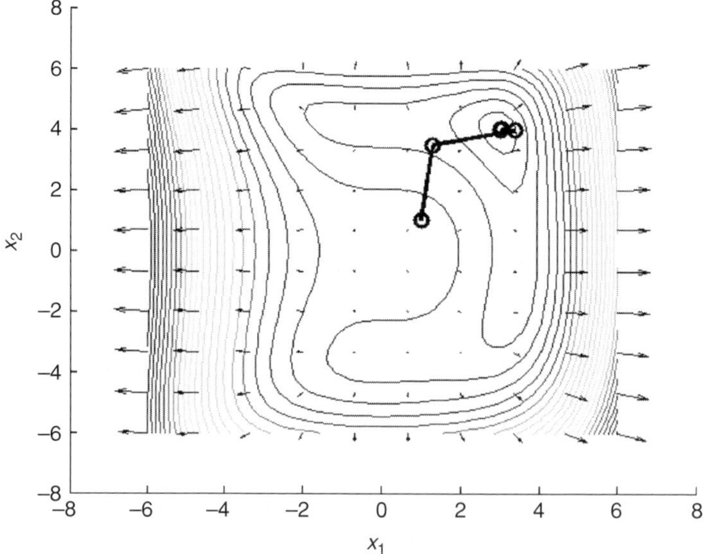

Figure 2.14 Trajectory of reduced-step Newton method with the guess (1,1) (number of iterations = 6).

the function norm. The *trust-region Newton method* is more robust, especially when the Jacobian is nearly singular, because it simultaneously varies both the step direction and the step magnitude to find an acceptable reduction of the function norm. Being a form of numerical optimization, a detailed discussion of this algorithm is postponed until Chapter 5; however, we present here the basic concepts.

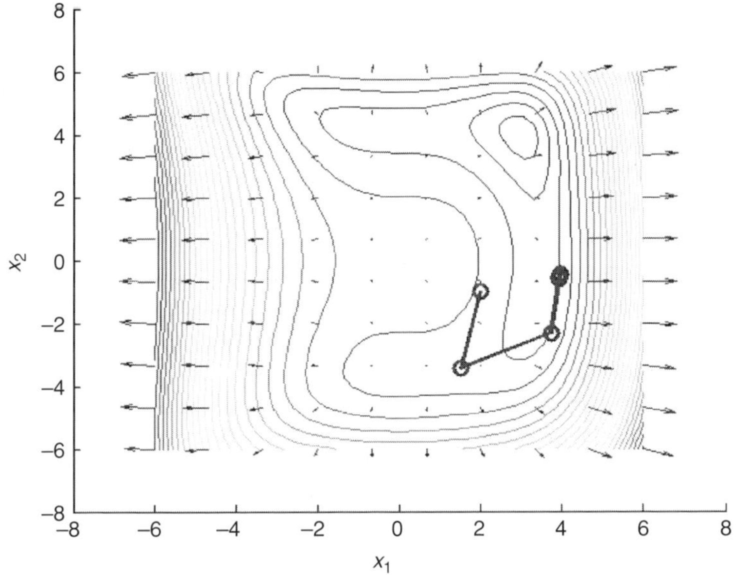

Figure 2.15 Trajectory of reduced-step Newton method with the guess $(2,-1)$.

The full Newton step at $x^{[k]}$ satisfies the linear system

$$B^{[k]} p^{(n)} = -f(x^{[k]})$$

(2.85)

Therefore, it is also a global minimum of the quadratic cost function

$$m^{[k]}(p) = \tfrac{1}{2} \left\| B^{[k]} p + f(x^{[k]}) \right\|_2^2$$

(2.86)

$$= \tfrac{1}{2} \left\| f(x^{[k]}) \right\|_2^2 + f(x^{[k]}) \cdot B^{[k]} p + \tfrac{1}{2} p^{\mathrm{T}} (B^{[k]})^{\mathrm{T}} B^{[k]} p$$

If we were to apply a minimization algorithm to $m^{[k]}(p)$, we would find the full Newton step $p^{(n)}$. We note, however, that $p^{(n)}$ becomes erratic when we are far from the solution. We therefore expect there to be some trust radius $\Delta^{[k]}$ (that we modify at each iteration) such that we expect the quadratic model $m^{[k]}(p)$ to be a valid measure of locating the solution only if $\|p\|_2 < \Delta^{[k]}$. We obtain the update vector by solving the constrained problem

$$\min m^{[k]}(p) \quad \text{subject to } |p| < \Delta^{[k]}$$

(2.87)

This technique is the basis of the MATLAB routine **fsolve**, whose use is demonstrated below. A further discussion of the trust-region method, and the efficient dogleg method of implementing it, is provided in Chapter 5.

Solving nonlinear algebraic systems in MATLAB

Although MATLAB contains a built-in routine, **fsolve**, for solving systems of multiple nonlinear algebraic equations (using the trust-region Newton method discussed above), it is part of the Optimization Toolkit and so is not available in every installation of MATLAB. reduced_Newton.m implements the reduced-step Newton algorithm, and can be used if

fsolve is unavailable. For further information on the use of reduced_Newton.m, consult the help section at the beginning of that file.

The syntax for using **fsolve** to find an approximate solution x and its function value f (near zero) is

[x, f] = fsolve(fun_name, x0, Options, P1, P2, ...);

fun_name is the name of the function that returns the function vector,

f = fun_name(x, P1, P2, ...);

or, if optionally it returns the Jacobian matrix as well,

[f, Jac] = fun_name(x, P1, P2, ...);

P1, P2, ... are optional parameters passed through **fsolve**. x0 is the initial guess. Options is a data structure managed by the command **optimset** that allows us to modify the performance of **fsolve**. Unless the number of equations is very large, it is best to start with

Options = optimset('LargeScale', 'off');

We then can add additional fields to override the default options (type help optimset for further details). For example, if your routine returns as a second argument the Jacobian matrix evaluated at x, let **fsolve** know this by

Options = optimset(Options, 'Jacobian', 'on');

For the example system (2.61), with an easily-computed Jacobian,

$$f_1(x_1, x_2) = 3x_1^3 + 4x_2^2 - 145 = 0 \qquad J = \begin{bmatrix} 9x_1^2 & 8x_2 \\ 8x_1 & -3x_2^2 \end{bmatrix} \qquad (2.88)$$
$$f_2(x_1, x_2) = 4x_1^2 + x_2^3 + 28 = 0$$

the following code computes the solution, starting from $x^{[0]} = [1\ 1]^T$,

Options = optimset('LargeScale', 'off', 'Jacobian', 'on');
x = fsolve('calc_f_ex1',x0,Options),

calc_f_ex1.m returns the value of the function vector and the Jacobian,

```
% calc_f_ex1.m
function [f, Jac] = calc_f_ex1(x);
f = zeros(2,1);
f(1) = 3*x(1)^3 + 4*x(2)^2 - 145;
f(2) = 4*x(1)^2-x(2)^3 + 28;

Jac = zeros(2,2);
Jac(1,1) = 9*x(1)^2; Jac(1,2) = 8*x(2);
Jac(2,1) = 8*x(1); Jac(2,2) = -3*x(2)^2;

return ;
```

We could also use @calc_f_ex1 as the first **fsolve** argument rather than 'calc_f_ex1'. For the CSTR example, with the two chemical reactions

$$A + B \rightarrow C \quad r_{R1} = k_1 c_A c_B$$
$$C + B \rightarrow D \quad r_{R2} = k_2 c_C c_B \tag{2.89}$$

the steady-state concentrations

$$x_1 = c_A \quad x_2 = c_B \quad x_3 = c_C \quad x_4 = c_D \tag{2.90}$$

are obtained by solving the nonlinear algebraic system

$$\begin{aligned}
&v(c_{A,\,in} - x_1) + V(-k_1 x_1 x_2) = 0 \\
&v(c_{B,\,in} - x_2) + V(-k_1 x_1 x_2 - k_2 x_3 x_2) = 0 \\
&v(c_{C,\,in} - x_3) + V(k_1 x_1 x_2 - k_2 x_3 x_2) = 0 \\
&v(c_{D,\,in} - x_4) + V(k_2 x_3 x_2) = 0
\end{aligned} \tag{2.91}$$

CSTR_SS_ex1.m prompts the user for the values of the parameters and computes the concentrations, using the inlet concentrations as the initial guesses. For the parameter values

$$v = 1 \quad V = 100 \quad k_1 = 1 \quad k_2 = 1$$
$$c_{A,\,in} = 1 \quad c_{B,\,in} = 2 \quad c_{C,\,in} = 0 \quad c_{D,\,in} = 0 \tag{2.92}$$

running this program yields the output

Steady state concentrations:
[A] = 0.056614
[B] = 0.16664
[C] = 0.053409
[D] = 0.88998
infinity norm of f = 2.0326e-009

Example. 1-D laminar flow of a shear-thinning polymer melt

Again, let us consider laminar flow between two parallel plates separated by a distance B, but now for simplicity we assume both plates to be stationary. Previously in Chapter 1, we employed finite differences to compute the velocity profile $v_x(y)$ for a Newtonian fluid. We now consider the same problem, but for a nonNewtonian fluid whose viscosity decreases with increasing shear-rate, common behavior for many polymer solutions and melts. In this problem, the (y-dependent) shear-rate is

$$\dot{\gamma}(y) = \left| \frac{dv_x}{dy} \right|_y \tag{2.93}$$

A common model of shear-thinning behavior is that of Carreau and Yasuda (Yasuda *et al.*, 1981), for which the shear-rate dependent viscosity $\eta(\dot{\gamma})$ is

$$\frac{\eta(\dot{\gamma}) - \eta_\infty}{\eta_0 - \eta_\infty} = [1 + (\lambda\dot{\gamma})^a]^{(n-1)/a} \tag{2.94}$$

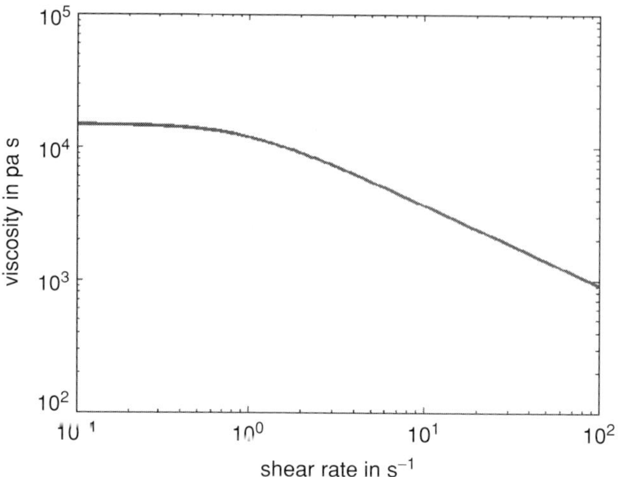

Figure 2.16 Shear-dependent viscosity of poly(styrene) melt at 453 K, estimated from the Carreau–Yasuda model.

Abdel-Khalik *et al.* (1974) fit the poly(styrene) melt data at 453 K of Ballenger *et al.* (1971) to obtain

$$\eta_\infty = 0 \quad \eta_0 = 1.48 \times 10^4 \, Pa \, s \quad \lambda = 1.04 \, s \quad n = 0.398 \quad a = 2 \tag{2.95}$$

The predicted shear-thinning behavior is shown in Figure 2.16.

With this nonlinear dependence of the viscosity upon the local shear-rate, numerical solution of the equation of motion is now required. For 1-D laminar flow, the equation of motion

$$\rho \frac{Dv}{Dt} = \rho \frac{\partial}{\partial t} v + \rho(v \cdot \nabla v) = \nabla \cdot \tau - \nabla P \tag{2.96}$$

reduces to

$$0 = \frac{d\tau_{yx}}{dy} - \frac{dP}{dx} \quad \tau_{yx} = \eta(\dot{\gamma})\frac{dv_x}{dy} \tag{2.97}$$

τ_{yx} is the shear-stress and P is the dynamic pressure. For a constant dynamic pressure gradient, the equation of motion

$$\frac{d\tau_{yx}}{dy} = \left(\frac{\Delta P}{\Delta x}\right) \tag{2.98}$$

yields a shear-stress that varies linearly with y,

$$\tau_{yx}(y) = \tau_{yx}(0) + \left(\frac{\Delta P}{\Delta x}\right) y \tag{2.99}$$

By symmetry, the velocity gradient mid-way between the plates is zero, and thus the shear-stress at $y = B/2$ is zero. Hence,

$$\tau_{yx}\left(\frac{B}{2}\right) = 0 = \tau_{yx}(0) + \left(\frac{\Delta P}{\Delta x}\right)\left(\frac{B}{2}\right) \tag{2.100}$$

and the linear shear-stress profile is

$$\tau_{yx}(y) = \left(\frac{\Delta P}{\Delta x}\right)\left(y - \frac{B}{2}\right) \tag{2.101}$$

Using the Carreau–Yasuda constitutive equation, this yields a nonlinear first-order differential equation for the velocity field

$$\eta(\dot{\gamma})\frac{dv_x}{dy} = \left(\frac{\Delta P}{\Delta x}\right)\left(y - \frac{B}{2}\right) \tag{2.102}$$

To solve this system, we again use finite differences, and place evenly spaced points between the plates,

$$y_k = k(\Delta y) \quad \Delta y = \frac{B}{(N+1)} \tag{2.103}$$

The local velocity and shear-rate at y_k are

$$v_k = v_x(y_k) \quad \dot{\gamma}_k = \left|\frac{dv_x}{dy}\right|_{y_k} \tag{2.104}$$

At each grid point, we require (2.102) to be satisfied locally,

$$\eta(\dot{\gamma}_k)\frac{dv_x}{dy}\bigg|_{y_k} = \left(\frac{\Delta P}{\Delta x}\right)\left(y_k - \frac{B}{2}\right) \tag{2.105}$$

The finite difference approximations

$$\frac{dv_x}{dy}\bigg|_{y_k} \approx \frac{v_k - v_{k-1}}{(\Delta y)} \tag{2.106}$$

yield for each point except the first a nonlinear algebraic equation

$$f_k = \eta(\dot{\gamma}_k)\left[\frac{v_k - v_{k-1}}{(\Delta y)}\right] - \left(\frac{\Delta P}{\Delta x}\right)\left(y_k - \frac{B}{2}\right) = 0 \quad k = 2, 3, \ldots, N \tag{2.107}$$

where

$$\dot{\gamma}_k = \left|\frac{v_k - v_{k-1}}{\Delta y}\right| \tag{2.108}$$

At the first grid point, we use the boundary condition $v_x(0) = 0$ to obtain

$$f_1 = \eta(\dot{\gamma}_1)\left[\frac{v_1}{\Delta y}\right] - \left(\frac{\Delta P}{\Delta x}\right)\left(y_1 - \frac{B}{2}\right) = 0 \quad \dot{\gamma}_1 = \left|\frac{v_1}{\Delta y}\right| \tag{2.109}$$

The Jacobian is sparse as each row has at most two nonzero elements,

$$\begin{aligned} J_{k,k-1} &= \frac{\partial f_k}{\partial v_{k-1}} = \eta(\dot{\gamma}_k)\left[\frac{-1}{\Delta y}\right] + \left[\frac{v_k - v_{k-1}}{(\Delta y)}\right]\frac{d\eta}{d\dot{\gamma}}\bigg|_{\dot{\gamma}_k}\frac{d\dot{\gamma}_k}{dv_{k-1}}\bigg|_{v_{k-1},v_k} \\ J_{k,k} &= \frac{\partial f_k}{\partial v_k} = \eta(\dot{\gamma}_k)\left[\frac{1}{\Delta y}\right] + \left[\frac{v_k - v_{k-1}}{(\Delta y)}\right]\frac{d\eta}{d\dot{\gamma}}\bigg|_{\dot{\gamma}_k}\frac{d\dot{\gamma}_k}{dv_k}\bigg|_{v_{k-1},v_k} \end{aligned} \tag{2.110}$$

We could reduce the effort required to solve the problem by providing the Jacobian as a second output argument of our function routine. In polymer_flow_1D.m, **fsolve** is provided with a sparse matrix that is nonzero only at those elements that are nonzero in the Jacobian, through the JacobPattern option in **optimset**. This sparsity information helps **fsolve** to

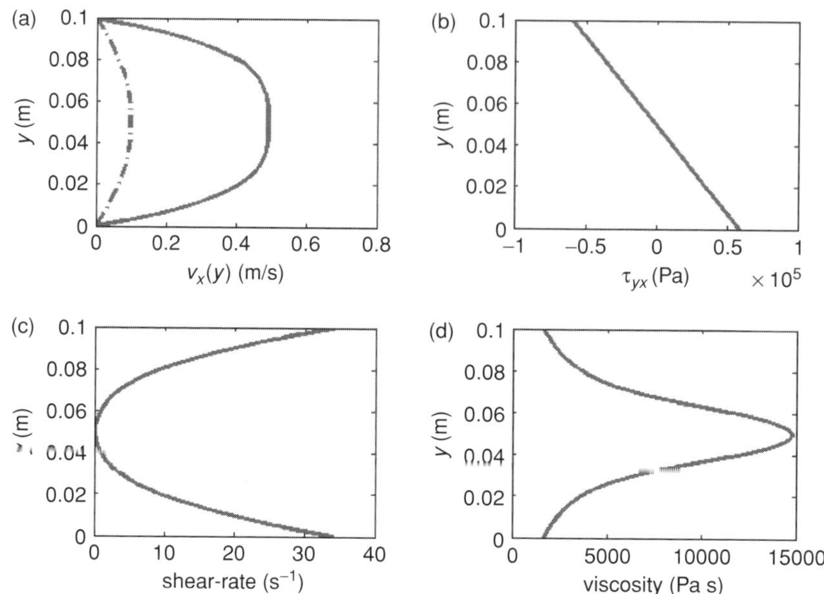

Figure 2.17 Profiles for laminar flow of poly(styrene) melt at 453 K between two parallel plates separated by a distance of 10 cm. The imposed pressure gradient is $-1\,184\,000$ Pa/m, which in the absence of shear-thinning yields a centerline velocity of 10 cm/s: (a) velocity profile of Newtonian (dash-dot) and shear-thinning (solid) fluids; (b) linear shear-stress profile; (c) shear-rate profile; (d) viscosity profile.

estimate the Jacobian more effectively, while avoiding the analytical work necessary to evaluate the second terms on the right-hand side in (2.110).

For an initial guess, we use the analytical profile for a Newtonian fluid with the zero shear-rate viscosity

$$v_x(y) = \frac{y(y-B)}{2\eta_0}\left(\frac{\Delta P}{\Delta x}\right) \tag{2.111}$$

The computed velocity, shear-stress, shear-rate, and viscosity profiles are shown in Figure 2.17 for the case of plates separated by 10 cm, with a pressure gradient that yields a centerline velocity of 10 cm/s in the absence of shear-thinning. Here, shear-thinning results in a five-fold increase in centerline velocity. Rather than a parabolic profile, the shear-thinning fluid has large velocity gradients near the walls, yet a relatively flat center "plug-flow" region.

Homotopy

Above, we have used **fsolve** to solve the steady-state CSTR model

$$
\begin{aligned}
v(c_{A,\,in} - x_1) + V(-k_1 x_1 x_2) &= 0 \\
v(c_{B,\,in} - x_2) + V(-k_1 x_1 x_2 - k_2 x_3 x_2) &= 0 \\
v(c_{C,\,in} - x_3) + V(k_1 x_1 x_2 - k_2 x_3 x_2) &= 0 \\
v(c_{D,\,in} - x_4) + V(k_2 x_3 x_2) &= 0
\end{aligned}
\tag{2.112}
$$

In addition to coding a routine to return the function vector, we also had to provide an initial guess of the solution. Above, we used simply the inlet concentration values, but ideally, we want our initial guess to be as close as possible to the true solution for Newton's method to be robust and efficient. Here, we were able to find a solution with this initial guess, but this does not always occur. We thus may have to propose many different guesses, perhaps at random, before we find a solution.

We can take a more systematic approach to the generation of initial guesses by using some insight into the nature of the equations through a technique known as *homotopy*. If we study the structure of the equations for the CSTR model, we note that the nonlinearity is associated with the reaction terms. In the limit that the residence time, V/υ, goes to zero (i.e., if we make the flow rate very large), the convective terms will be much larger in magnitude than the reaction terms, and the system of equations becomes

$$\begin{aligned}
\upsilon(c_{A,in} - x_1) &\approx 0 \\
\upsilon(c_{B,in} - x_2) &\approx 0 \\
\upsilon(c_{C,in} - x_3) &\approx 0 \\
\upsilon(c_{D,in} - x_4) &\approx 0
\end{aligned} \tag{2.113}$$

As $\upsilon/V \rightarrow \infty$, the reactor concentrations are very near those of the inlet, providing very natural and accurate initial guesses,

$$x_1 \approx c_{A,in} \quad x_2 \approx c_{B,in} \quad x_3 \approx c_{C,in} \quad x_4 \approx c_{D,in} \tag{2.114}$$

If we were to find it difficult to generate an initial guess that converges to a solution for some smaller value of υ/V, a good strategy would be to solve the system first with a very large value of the flow rate, for which Newton's method converges quickly. Then, to compute the concentrations at the flow rate of interest, we decrease the flow rate value incrementally, and for each new value, use as an initial guess the solution from the previous step.

This approach, known as *homotopy*, allows us to move gradually from a region of parameter space in which it is easy to solve the set of equations to a region where solution is difficult, but always to operate Newton's method in the vicinity of a solution where convergence is robust and rapid. With a bit of insight into the structure of the equations, this approach is very powerful. An efficient implementation of homotopy, *arc-length continuation*, is described in Chapter 4. CSTR_2D_NAE.m demonstrates the use of the simple homotopy algorithm described above to solve the steady-state CSTR system.

Example. Steady-state modeling of a condensation polymerization reactor

The most common form of nylon, nylon-6,6, is made by polycondensation of the two monomers hexamethylene diamine (HMD) and adipic acid (ADA). The first step in the reaction sequence is the condensation of two monomers to form a dimer with an amide linkage, $-CONH-$, and a water molecule (the condensate).

$$\begin{aligned}
HOOC(CH_2)_4&COOH + H_2N(CH_2)_6NH_2 \\
&\Leftrightarrow HOOC(CH_2)_4CONH(CH_2)_6NH_2 + H_2O
\end{aligned}$$

The double-headed arrow denotes that this reaction is reversible, with an equilibrium constant on the order of 100. The dimer still has functional groups on each end, and so may continue to react to produce even larger molecules. If the water is removed by evaporation to drive the equilibrium to the right, polymer chains with molecular weights of $\sim$20–30 kg/mol may be produced. To achieve very high molecular weights, we use equimolar amounts of each monomer so that the stoichiometry between the acid ends ($-COOH$) and the base ends ($-NH_2$) is balanced.

We describe the hierarchy of reactions in polycondensation using the following notation. Let $[P_x]$ be the total concentration of chains in the system that contain x monomer units, i.e., are x-mers. We then have the following hierarchy of reactions:

$$\begin{aligned} P_1 + P_1 &\Leftrightarrow P_2 + W \\ P_1 + P_2 &\Leftrightarrow P_3 + W \\ P_2 + P_2 &\Leftrightarrow P_4 + W \\ P_1 + P_3 &\Leftrightarrow P_4 + W \\ P_2 + P_3 &\Leftrightarrow P_5 + W \end{aligned}$$
(2.115)

The number of reactions and species in the system quickly grows very large. We can, however, derive simple equations that describe the average chain length and breadth of the chain length distribution with only a few state variables. We define the kth moment of the chain length distribution, λ_k, as

$$\lambda_k = \sum_{m=1}^{\infty} m^k [P_m]$$
(2.116)

Of particular interest are the three leading moments

$$\lambda_0 = \sum_{m=1}^{\infty} [P_m] \quad \lambda_1 = \sum_{m=1}^{\infty} m [P_m] \quad \lambda_2 = \sum_{m=1}^{\infty} m^2 [P_m]$$
(2.117)

From these three values, we can calculate the number-averaged chain length, $\bar{x}_n = \lambda_1/\lambda_0$, and the weight-averaged chain length, $\bar{x}_w = \lambda_2/\lambda_1$. Since the weight-averaged chain length biases more the contributions of the larger chains, $\bar{x}_w \geq \bar{x}_n$, with the equality holding only if all chains are of the same length. The ratio of these two averages, $Z = \bar{x}_w/\bar{x}_n$, the polydispersity, provides a simple measure of the breadth of the chain length distribution. The larger the polydispersity, the greater the disparity in the lengths (molecular weights) of individual chains. We now need only calculate the rates of change of these three moments to predict, for given polymerization conditions, the characteristics of the polymer produced. For further discussion of the use of population balances and moment equations in polymer reaction engineering, consult Ray (1972) and Dotson et al. (1996).

Rate equations for polycondensation

We obtain rate equations for these moments by counting how each x-mer species is produced or consumed in each individual reaction in the hierarchy above. First, we represent the sequence of reactions (2.115) as

$$P_m + P_n \Leftrightarrow P_{m+n} + W$$
(2.118)

that should be read as "A chain with m monomer units reacts reversibly with a chain containing n monomer units, to produce a combined chain containing $m+n$ monomer units and a single condensate (water) molecule." We then combine the contributions from each individual reaction in the forward and reverse directions to calculate the net rate of change of each species' concentration and those of the leading moments.

Forward condensation reaction

$$P_m + P_n \rightarrow P_{m+n} + W \tag{2.119}$$

The net rate of change of m-mer from forward condensation is

$$r_{P_m(\text{fc})} = -2k_{\text{fc}}[P_m]\sum_{n=1}^{\infty}[P_n] + k_{\text{fc}}\sum_{n=1}^{m-1}[P_n][P_{m-n}] \tag{2.120}$$

The first term is the rate of disappearance of m-mer from reaction with all other species, and the second term is the rate of m-mer creation through the reaction of two smaller species. The rate of change of the kth moment is

$$r_{\lambda_k(\text{fc})} = \sum_{m=1}^{\infty} m^k r_{P_m(\text{fc})} \tag{2.121}$$

After some mathematics, we get for the three leading moments of interest,

$$r_{\lambda_0(\text{fc})} = -k_{\text{fc}}\lambda_0^2 \quad r_{\lambda_1(\text{fc})} = 0 \quad r_{\lambda_2(\text{fc})} = 2k_{\text{fc}}\lambda_1^2 \tag{2.122}$$

Reverse condensation reaction

The reverse reaction

$$P_{m+n} + W \rightarrow P_m + P_n \tag{2.123}$$

has a rate constant $k_{\text{fc}}/K_{\text{eq}}$, where K_{eq} is the equilibrium constant. The net rate of change of m-mer concentration is

$$r_{P_m(\text{rc})} = k_{\text{fc}}K_{\text{eq}}^{-1}[W]\left\{2\sum_{n=m+1}^{\infty}[P_n] - (m-1)[P_m]\right\} \tag{2.124}$$

The first term is the rate of production of m-mer through the scission of larger molecules, and the second term is the rate at which m-mer disappears when one of its linkages is attacked by water. The rates of change of the three leading moments from reverse condensation are

$$r_{\lambda_0(\text{rc})} = k_{\text{fc}}K_{\text{eq}}^{-1}[W](\lambda_1 - \lambda_0) \qquad r_{\lambda_1(\text{rc})} = 0$$

$$r_{\lambda_2(\text{rc})} = k_{\text{fc}}K_{\text{eq}}^{-1}[W]\left(\tfrac{1}{3}\lambda_1 - \lambda_3\right) \tag{2.125}$$

From these results, we have the following rates of change of each leading moment due to forward and reverse condensation:

$$r_{\lambda_0} = -k_{\text{fc}}\lambda_0^2 + k_{\text{fc}}K_{\text{eq}}^{-1}[W](\lambda_1 - \lambda_0) \qquad r_{\lambda_1} = 0$$

$$r_{\lambda_2} = 2k_{\text{fc}}\lambda_1^2 + k_{\text{fc}}K_{\text{eq}}^{-1}[W]\left(\tfrac{1}{3}\lambda_1 - \lambda_3\right) \tag{2.126}$$

The first moment is unchanged by the reaction, as is expected since the total number of monomer units is conserved. The equation for the second moment requires that we know the value of the third moment. To obtain a closed set of equations, it is common to postulate a mathematical form of the chain length distribution (see Chapter 7) to relate the unknown λ_3 to the known λ_0, λ_1, λ_2. This approach yields the closure approximation

$$\lambda_3 \approx \frac{\lambda_2\left(2\lambda_2\lambda_0 - \lambda_1^2\right)}{\lambda_1\lambda_0} \tag{2.127}$$

Steady-state model of a stirred-tank polycondensation reactor

We now use these results to model the steady-state behavior of a polycondensation CSTR. The mole balance on m-mer is

$$\frac{d}{dt}\{M[\mathrm{P}_m]\} = F^{(\mathrm{in})}[\mathrm{P}_m]^{(\mathrm{in})} - F[\mathrm{P}_m] + Mr_{\mathrm{P}_m} \tag{2.128}$$

Here, we represent concentrations on a per-mass basis, due to volume changes during reaction. M is the total mass of the reaction medium in the reactor. $F^{(\mathrm{in})}$ and F are the inlet and outlet mass flow rates.

The first term on the right-hand side of (2.128) is the flux of m-mer into the reactor from the inlet stream, the second term is the flux of m-mer out of the reactor, and the last term is the rate of change of m-mer concentration due to chemical reaction. Multiplying this equation by m^k and summing over all m yields

$$\frac{d}{dt}\left\{M\sum_{m=1}^{\infty}m^k[\mathrm{P}_m]\right\} = F^{(\mathrm{in})}\sum_{m=1}^{\infty}m^k[\mathrm{P}_m]^{(\mathrm{in})} - F\sum_{m=1}^{\infty}m^k[\mathrm{P}_m] + M\sum_{m=1}^{\infty}m^k r_{\mathrm{P}_m} \tag{2.129}$$

This is simply the balance for the kth moment, and at steady state yields

$$\frac{d}{dt}\{M\lambda_k\} = 0 = F^{(\mathrm{in})}\lambda_k^{(\mathrm{in})} - F\lambda_k + Mr_{\lambda_k} \tag{2.130}$$

With λ_1 held constant (as it is unchanged by the reaction), we have

$$\frac{d}{dt}\{M\lambda_0\} = 0 = F^{(\mathrm{in})}\lambda_0^{(\mathrm{in})} - F\lambda_0 - Mk_{\mathrm{fc}}\lambda_0^2 + Mk_{\mathrm{fc}}K_{\mathrm{eq}}^{-1}[\mathrm{W}](\lambda_1 - \lambda_0)$$

$$\frac{d}{dt}\{M\lambda_2\} = 0 = F^{(\mathrm{in})}\lambda_2^{(\mathrm{in})} - F\lambda_2 + 2Mk_{\mathrm{fc}}\lambda_1^2 + Mk_{\mathrm{fc}}K_{\mathrm{eq}}^{-1}[\mathrm{W}]\left(\tfrac{1}{3}\lambda_1 - \lambda_3\right) \tag{2.131}$$

and for λ_3 use the auxiliary moment closure equation

$$\lambda_3 \approx \frac{\lambda_2\left(2\lambda_2\lambda_0 - \lambda_1^2\right)}{\lambda_1\lambda_0} \tag{2.132}$$

We contact the reaction medium with a purge gas stream to remove the water, such that the total mass balance on the reaction medium is

$$F = F^{(\mathrm{in})} - (k_m a)M\overline{M}_{\mathrm{W}}[\mathrm{W}] \tag{2.133}$$

$\overline{M}_W$ is the molecular weight of water, a is the interfacial mass transfer area per unit mass in the reactor and k_m is a mass transfer coefficient. The mole balance on water is

$$\frac{d}{dt}\{M[W]\} = 0 = F^{(in)}[W]^{(in)} - F[W] - (k_m a)M[W] + Mr_W \tag{2.134}$$

$r_W = -r_{\lambda_0}$ is the rate of generation of condensate due to reaction, yielding

$$0 = F^{(in)}[W]^{(in)} - F[W] - (k_m a)M[W] + Mk_{fc}\lambda_0^2 - Mk_{fc}K_{eq}^{-1}[W](\lambda_1 - \lambda_0) \tag{2.135}$$

We next define the dimensionless quantities

$$\mu_k^{(in)} = \frac{\lambda_k^{(in)}}{\lambda_1} \qquad \mu_k = \frac{\lambda_k}{\lambda_1} \qquad \omega = \frac{[W]}{\lambda_1} \qquad \phi = \frac{F}{F^{(in)}} \tag{2.136}$$

the dimensionless Damköhler number

$$Da = \frac{k_{fc}M\lambda_1}{F^{(in)}} \tag{2.137}$$

and the dimensionless strength of mass transfer

$$\gamma = \frac{(k_m a)M}{F^{(in)}} \tag{2.138}$$

such that the three dimensionless balances for $\{\mu_0,\ \mu_2,\ \omega\}$ are

$$\begin{aligned}
f_1 &= \mu_0^{(in)} - \phi\mu_0 - (Da)\mu_0^2 + (Da)K_{eq}^{-1}\omega(1 - \mu_0) = 0 \\
f_2 &= \mu_2^{(in)} - \phi\mu_2 + 2(Da) + (Da)K_{eq}^{-1}\omega\left(\tfrac{1}{3} - \mu_3\right) = 0 \\
f_3 &= \omega^{(in)} - \phi\omega - \gamma\omega + (Da)\mu_0^2 - (Da)K_{eq}^{-1}\omega(1 - \mu_0) = 0
\end{aligned} \tag{2.139}$$

In addition to these three coupled nonlinear equations, we have two auxiliary equations from the moment closure approximation and the overall mass balance on the reaction medium,

$$\mu_3 \approx \frac{\mu_2(2\mu_2\mu_0 - 1)}{\mu_0} \qquad \phi = 1 - \gamma\zeta\omega \tag{2.140}$$

ζ is the molecular weight of a condensate relative to that of a monomer,

$$\zeta = \lambda_1\overline{M}_W \tag{2.141}$$

Effect of Da and mass transfer upon polymer chain length

polycond_CSTR.m plots the number-averaged chain length $\bar{x}_n = \mu_1/\mu_0$, the polydispersity, $Z = \bar{x}_w/\bar{x}_n = \mu_2\mu_0$, the dimensionless condensate concentration ω, and the relative outlet flow rate ϕ as functions of Da and γ, for input values of the fixed parameters $\{\zeta,\ K_{eq},\ \omega^{(in)},\ \mu_0^{(in)},\ \mu_2^{(in)}\}$. The calculations are performed for a monomer-fed reactor with $\zeta = 0.2$, $K_{eq} = 10^2$, $\omega^{(in)} = 0$, and $\mu_0^{(in)} = \mu_2^{(in)} = 1$.

For this system, convergence of Newton's method can be difficult, especially at high Da. Homotopy is used; for each γ the sequence of simulations is started with the lowest Da, for which the inlet values are reasonable guesses. Convergence can still be difficult at high Da, thus we use here an additional convergence trick: we simulate an approximate set of

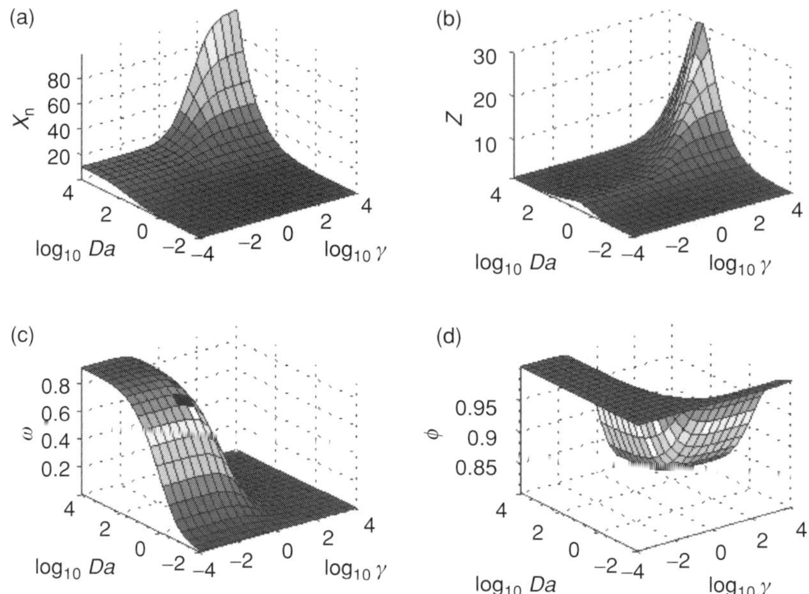

Figure 2.18 Effect of Da and mass transfer efficiency upon operation of a monomer-fed CSTR for polycondensation: (a) number average chain length; (b) polydispersity; (c) dimensionless condensate concentration; (d) output mass flow rate relative to input value.

dynamics forward in time, to approach robustly the vicinity of a stable steady state before beginning Newton's method. Defining a dimensionless time, $\tau = (F^{(\mathrm{in})}t)/M$, we have the dynamics

$$\frac{d\mu_0}{d\tau} = f_1 \quad \frac{d\mu_2}{d\tau} = f_2 \quad \frac{d\omega}{d\tau} = f_3 \tag{2.142}$$

Dynamic simulation of ordinary differential equation (ODE) systems is discussed in Chapter 4. For now, we merely note that we use **ode23s** to propagate the dynamics until the function norm is decreased to a value deemed sufficiently small for the robust start of Newton's method.

Figure 2.18 plots $\bar{x}_{\mathrm{n}}$, Z, ω, ϕ as functions of Da and γ. High-molecular-weight polymer is achieved only for high values of both Da and γ. At low Da, there is insufficient residence time in the reactor to achieve much conversion, and at low γ the water is not removed, limiting the achievable conversion at equilibrium.

Bifurcation analysis

We have stated that we do not in general know the number or even the existence of solutions to a nonlinear algebraic system. This is true; however, it is possible to identify points at which the existence properties of the system change through locating *bifurcation points*; i.e., choices of parameters at which the Jacobian, evaluated at the solution, is singular.

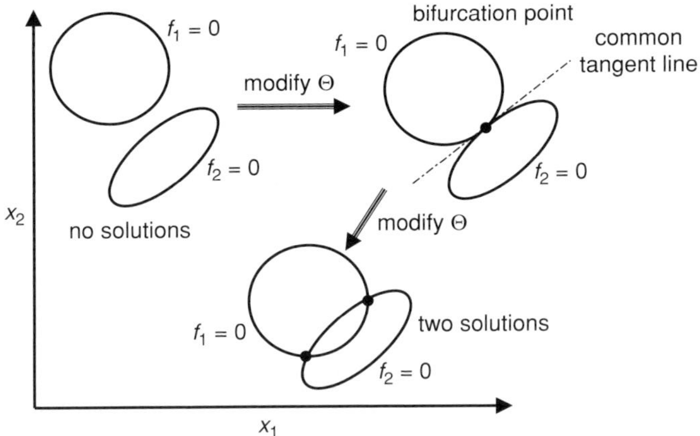

Figure 2.19 Bifurcation point separating the region in parameter space that has two solutions from that which has none. At the bifurcation point, both curves share a common tangent at contact.

Consider a simple case of two nonlinear equations whose solution(s) depend upon some parameter vector Θ.

$$f_1(x_1, x_2; \Theta) = 0$$
$$f_2(x_1, x_2; \Theta) = 0 \tag{2.143}$$

As a solution x_s must satisfy both equations, it appears as a point of intersection between the curves $f_1 = 0$ and $f_2 = 0$. Let us modify the parameter vector Θ to move the locations of these curves in x space as shown in Figure 2.19. Initially, there are no solutions, but as the curves approach each other, they overlap to generate two solutions. Consider the particular parameter vector Θ_c at which the curves first touch. Clearly this is an important choice of parameter(s), as it separates the region of parameter space in which there are solutions from that in which there are none. If the curves just touch at Θ_c and do not cross, they must be parallel to each other at the point of contact. That is, the slopes of the tangent lines in (x_1, x_2) space of the two curves $f_1 = 0$ and $f_2 = 0$ must be equal at the point of contact, which we note is a solution. At Θ_c, with solution $x_s(\Theta_c)$, if $S = dx_2/dx_1$ is the common tangent slope,

$$O = \left(\frac{\partial f_1}{\partial x_1}\right) + \left(\frac{\partial f_1}{\partial x_2}\right) S = \left(\frac{\partial f_2}{\partial x_1}\right) + \left(\frac{\partial f_2}{\partial x_2}\right) S \tag{2.144}$$

Therefore, the Jacobian evaluated at the solution,

$$J(x_s(\Theta_c)) = \begin{bmatrix} \dfrac{\partial f_1}{\partial x_1}\Big|_{x_s(\Theta_c)} & \dfrac{\partial f_1}{\partial x_2}\Big|_{x_s(\Theta_c)} \\[2ex] \dfrac{\partial f_2}{\partial x_1}\Big|_{x_s(\Theta_c)} & \dfrac{\partial f_2}{\partial x_2}\Big|_{x_s(\Theta_c)} \end{bmatrix} \tag{2.145}$$

is singular (i.e. Θ_c is a bifurcation point).

We can search for a bifurcation point along some path $\Theta(\lambda)$ in parameter space by solving the augmented set of $N+1$ equations for $x_s(\Theta_c)$ and the value of λ_c, $\Theta(\lambda_c) = \Theta_c$, at which the bifurcation occurs,

$$f(x_s; \Theta(\lambda)) = 0$$
$$|J(x_s; \Theta(\lambda))| = 0 \tag{2.146}$$

Computing the locations of bifurcation points allows us to carve up parameter space into different regions, to find one in which one or more real solutions do indeed exist.

Example. Bifurcation points of a simple quadratic equation

As an example, consider the quadratic equation

$$f(x) = x^2 + \theta x + 1 \tag{2.147}$$

that has solutions at

$$x = \frac{-\theta \pm \sqrt{\theta^2 - 4}}{2} \tag{2.148}$$

Obviously, this equation has two real solutions for $\theta^2 > 4$ and no real solutions within $-2 < \theta < 2$. There exist two bifurcation points at $\theta = \pm 2$ at which the two real solutions are degenerate. Let us consider the bifurcation point at $\theta = 2$ for which the solution is $x_s = -1$. The Jacobian of this system (here a 1×1 matrix, a scalar) is

$$J(x, \theta) = \frac{\partial f}{\partial x} = 2x + \theta \tag{2.149}$$

At each bifurcation point $\theta = \pm 2$, the Jacobian is singular, as

$$J(x, \theta) = 2 \left[\frac{-\theta \pm \sqrt{\theta^2 - 4}}{2} \right] + \theta = \pm\sqrt{\theta^2 - 4} \tag{2.150}$$

Here, we can compute the bifurcation point analytically, but let us see how we might do so numerically (as we would have to do for more complex systems). Let us say that we wish to look for bifurcation points along the path $\theta(\lambda) = 4\lambda$, where we encompass in the range $0 \le \lambda \le 1$ the points $0 \le \theta(\lambda) \le 4$. Obviously, there will be a bifurcation point at $\lambda = 0.5$, but if we did not know this, we could try different values of λ as the initial guess and use Newton's method in (x, λ) space to solve the augmented system

$$f_1(x, \lambda) = f(x) = x^2 + \theta(\lambda)x + 1 = 0$$
$$f_2(x, \lambda) = det[J(x, \lambda)] = 2x + \theta(\lambda) = 0 \tag{2.151}$$

The Newton update rule is

$$\begin{bmatrix} x^{[k+1]} \\ \lambda^{[k+1]} \end{bmatrix} = \begin{bmatrix} x^{[k]} \\ \lambda^{[k]} \end{bmatrix} + \begin{bmatrix} \Delta x^{[k]} \\ \Delta \lambda^{[k]} \end{bmatrix} \tag{2.152}$$

where

$$
\begin{bmatrix}
\left(\dfrac{\partial f_1}{\partial x}\Big|_{[k]}\right) & \left(\dfrac{\partial f_1}{\partial \lambda}\Big|_{[k]}\right) \\[2ex]
\left(\dfrac{\partial f_2}{\partial x}\Big|_{[k]}\right) & \left(\dfrac{\partial f_2}{\partial \lambda}\Big|_{[k]}\right)
\end{bmatrix}
\begin{bmatrix}
\Delta x^{[k]} \\[1ex]
\Delta \lambda^{[k]}
\end{bmatrix}
=
\begin{bmatrix}
-f_1(x,\lambda) \\[1ex]
-f_2(x,\lambda)
\end{bmatrix}
\tag{2.153}
$$

The elements of the Jacobian of the augmented system are

$$
\frac{\partial f_1}{\partial x} = \frac{\partial}{\partial x}\{x^2 + \theta(\lambda)x + 1\} = 2x + \theta(\lambda) = 2x + 4\lambda
$$

$$
\frac{\partial f_1}{\partial \lambda} = \frac{\partial}{\partial \lambda}\{x^2 + \theta(\lambda)x + 1\} = x\frac{d\theta}{d\lambda} = 4x
$$

$$
\frac{\partial f_2}{\partial x} = \frac{\partial}{\partial x}\{2x + \theta(\lambda)\} = 2 \qquad \frac{\partial f_2}{\partial \lambda} = \frac{\partial}{\partial \lambda}\{2x + \theta(\lambda)\} = \frac{d\theta}{d\lambda} = 4
\tag{2.154}
$$

Therefore, the augmented Jacobian is

$$
J^{(a)}(x,\lambda) =
\begin{bmatrix}
(2x + 4\lambda) & (4x) \\
2 & 4
\end{bmatrix}
\tag{2.155}
$$

At the bifurcation point, the augmented Jacobian and its determinant are

$$
J^{(a)}(-1, 0.5) =
\begin{bmatrix}
0 & -4 \\
2 & 4
\end{bmatrix}
\qquad
\left| J^{(a)} \right| = (0)(4) - (2)(-4) = 8
\tag{2.156}
$$

Thus, the augmented Jacobian is not singular at the bifurcation point. Newton's method should be able to find it, with a suitable initial guess.

Numerical calculation of bifurcation points

As a more general formulation of the bifurcation point problem, let us search for a bifurcation point along the linear path in parameter space

$$
\Theta(\lambda) = (1 - \lambda)\Theta_0 + \lambda\Theta_1
\tag{2.157}
$$

We apply Newton's method to the augmented system for $x_s,\ \lambda$

$$
f(x_s; \Theta(\lambda)) = 0
$$
$$
|J(x_s; \Theta(\lambda))| = 0
\tag{2.158}
$$

Clearly, as we must compute the determinant of the Jacobian at each Newton iteration, and in general must obtain the Jacobian by finite differences, finding a bifurcation point is more costly than merely computing the solution for a fixed parameter vector. But, for systems in which we cannot find a solution to the system at the parameter vector of interest, and for which we wonder if there exist any solutions at all, bifurcation analysis can provide useful insight into the existence properties of the system. Also, there are situations, such as computing the critical points of thermodynamic phase diagrams, in which the locations of bifurcation points are themselves of direct interest.

search_bifurcation.m searches for a bifurcation point along a user-defined linear path in parameter space

$$\Theta(\lambda) = (1 - \lambda)\Theta_0 + \lambda\Theta_1 \qquad (2.159)$$

The program is invoked for the simple quadratic example above by

```
a = 1; % initial lambda value to try as initial guess
b = 0; % final lambda value to try as initial guess
num_lambda = 10; % # of lambda values to try as initial guess
theta_0 = 0; % parameter value at lambda = 0
theta_1 = 4; % parameter value at lambda = 1
x0 = -1.1; % initial guess of solution
fun_name = 'calc_f_quad';
[x_b,theta_b,lambda_b,ifound_b] = search_bifurcation(fun_name, ...
    theta_0,theta_1,a,b,x0,num_lambda),
```

where the routine for the function vector of the nonlinear system is

```
function f = calc_f_quad(x,theta);
f = x.^2 + theta.*x + 1;
return;
```

MATLAB summary

The main routine for solving nonlinear algebraic systems $f(x) = 0$ is **fsolve**,

```
x = fsolve(fun_name, x0, Options, P1, P2, ... );
```

fun_name is the name of a routine,

```
function f = fun_name(x, P1, P2, ... );
```

that returns the function vector for input x for the system under consideration. x0 is the initial guess, and P1, P2, . . . are optional model parameters. Options is a data structure managed by **optimset** that controls the operation of **fsolve**, e.g.

```
Options = optimset('LargeScale', 'off', 'Display', 'off');
```

If the Jacobian matrix can be computed analytically, we use

```
Options = optimset(Options, 'Jacobian', 'on');
```

and return the Jacobian as a second argument,

```
function [f, Jac] = fun_name(x, P1, P2, ... );
```

Other options that will be found to be useful in Chapter 6 are

```
Options = optimset(Options,' JacobPattern', S);
```

and

```
Options = optimset(Options, 'JacobMult', Jfun_name);
```

In the former option, the user supplies a sparse matrix S whose sparsity pattern (location of nonzero elements) matches that of the Jacobian. That is, even though the Jacobian may be difficult to compute analytically, the user can at least specify that only a small subset of Jacobian elements are known to be nonzero. **fsolve** can use this information to reduce the computational burden and memory requirement when generating an approximate Jacobian. With "JacobMult", the user supplies the name of a routine that returns the product of the Jacobian matrix with an input vector. The usefulness of this option will become clearer after our discussion of iterative methods for solving linear algebraic systems in Chapter 6.

fsolve is part of the Optimization Toolkit, and so is not available in all installations of MATLAB. If **fsolve** is unavailable, the provided routine reduced_Newton.m can be used in its place. For a single nonlinear algebraic equation, one may also use **fzero**, that is called similarly to **fsolve**,

x = fzero(fun_name, x0, Options, P1, P2, . . .);

Problems

2.A.1. Consider the following system of two nonlinear algebraic equations:

$$\begin{aligned} f_1(x_1, x_2) &= x_1^3 - 3x_1x_2^2 - x_2 + 18 = 0 \\ f_2(x_1, x_2) &= x_1^2 - 4x_1^2x_2 + x_2^3 - 2x_2^2 + 28 = 0 \end{aligned} \tag{2.160}$$

By hand, compute the Jacobian matrix of this system, and generate the refined estimate $x^{[1]}$ starting from an initial guess of $x^{[0]} = [2 \ 1]^{\mathrm{T}}$. In the reduced-step Newton method, would this estimate $x^{[1]}$ be accepted?

2.A.2. Write a MATLAB program to solve the system of Problem 2.A.1.

2.A.3. The terminal velocity v_t of a falling sphere in a Newtonian fluid is

$$v_t = \sqrt{\frac{4g(\rho_p - \rho_f)D_p}{3C_D\rho_f}} \tag{2.161}$$

ρ_f is the density of the fluid, ρ_p is the density of the particle, D_p is the particle diameter, g is the acceleration due to gravity, and C_D is a dimensionless drag coefficient that is a function of the Reynolds' number

$$Re = \frac{\rho_f v_t D_p}{\mu_f} \tag{2.162}$$

μ_f is the viscosity of the fluid. Table 5-22 in Perry & Green (1984) presents values of C_D for various Re, and also provides the correlations

$$Re < 0.1 \quad C_D = \frac{24}{Re}$$

$$0.1 < Re < 1000 \quad C_D = \left(\frac{24}{Re}\right)[1 + 0.14(Re)^{0.70}]$$

$$1000 < Re < 350\,000 \quad C_D \approx 0.445 \tag{2.163}$$

$$Re > 1\,000\,000 \quad C_D = 0.19 - \frac{8 \times 10^4}{Re}$$

Table 2.2 Drag coefficient for flow around a smooth sphere at Reynolds' numbers between 350 000 and 1 000 000. From Table 5–22 of Perry & Green (1984)

Re	C_D
350 000	0.396
400 000	0.0891
500 000	0.0799
700 000	0.0945
1 000 000	0.110

Between Reynolds' numbers of 350 000 and 1 000 000, there is a transient drop in the drag constant as the boundary layer first becomes turbulent, delaying separation on the downstream side. In this regime, we may use interpolation (MATLAB function **interp1**) of the data in Table 2.2.

Estimate the terminal velocity of a smooth spherical iron particle (density 7850 kg/m^3) in water (density 1000 kg/m^3, viscosity 0.001 Pa s) as a function of the particle diameter (in meters) over the range 0.1 mm to 10 cm.

2.B.1 Consider a CSTR, with the single gas-phase elementary reaction

$$A + B \rightarrow C \tag{2.164}$$

that has a rate constant k of 0.05 l/(mol/s). The feed stream to the reactor, at a volumetric flow rate v_0 in liters per second, is at a total pressure $P_0 = 5$ atm. It comprises reactants A and B, and a nonreacting diluent gas. The partial pressures of the reactants are $P_{A0} = 0.5$ atm and $P_{B0} = 0.5$ atm. Assume that the ideal gas law holds, and that the reactor is operated isothermally at 350 K. Assume that the inlet stream is at the same temperature and pressure as the reactor contents. For a reactor volume of 1000 l, plot the conversion of A as a function of v_0. Make sure to account for the fact that the total number of moles in the gas phase decreases with increasing conversion, so that the outlet volumetric flow rate will be somewhat smaller than v_0. For further discussion, consult Fogler (1999).

2.B.2. Consider a 1000-l CSTR in which the following reactions are taking place:

$$\begin{aligned} A + B &\rightarrow C \quad r_{R1} = k_1 c_A c_B \\ C + B &\rightarrow D \quad r_{R2} = k_2 c_B c_C \\ A &\rightarrow E \quad r_{R3} = k_3 c_A \end{aligned} \tag{2.165}$$

We have the following kinetic data

$$\begin{aligned} k_1(298\,\text{K}) &= 2.1 \times 10^{-2}\ \frac{1}{\text{mol s}} \quad & k_1(315\,\text{K}) &= 3.6 \times 10^{-2}\ \frac{1}{\text{mol s}} \\ k_2(298\,\text{K}) &= 1.5 \times 10^{-2}\ \frac{1}{\text{mol s}} \quad & k_2(315\,\text{K}) &= 4.5 \times 10^{-2}\ \frac{1}{\text{mol s}} \\ k_3(298\,\text{K}) &= 0.00012\ \text{s}^{-1} \quad & k_3(315\,\text{K}) &= 0.00026\ \text{s}^{-1} \end{aligned} \tag{2.166}$$

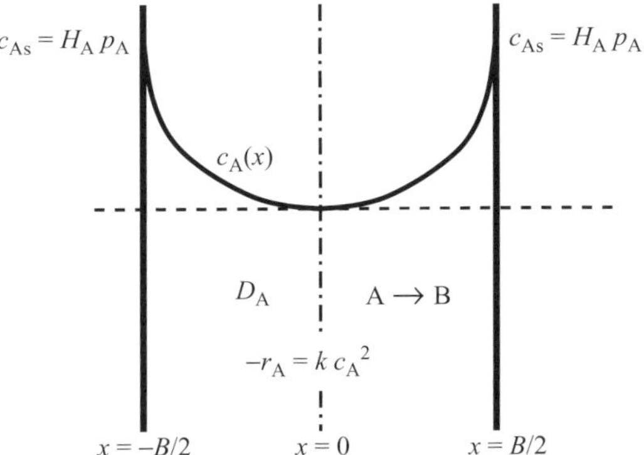

Figure 2.20 Reaction and diffusion in a catalyst slab.

We feed a stream at a volumetric flow rate of 0.1 l/s containing A, B, and a diluent solvent, with $c_{A0} = 0.5$ M and varying $\gamma = c_{B0}/c_{A0}$. Assuming isothermal operation, and neglecting any volume change due to reaction, plot the conversion of A as a function of temperature and γ.

2.B.3. Consider the problem of reaction and diffusion in a slab of catalyst of thickness B (Figure 2.20). The reaction $A \rightarrow B$ proceeds within the slab with apparent second-order kinetics, $-r_A = kc_A^2$. At the surface of the slab at $x = \pm B/2$, the concentration of A is in equilibrium with the external gas phase, $c_A(\pm B/2) = c_{As} = H_A p_A$. Within the slab, the steady-state concentration field is governed by the reaction–diffusion equation

$$0 = D_A \frac{d^2 c_A}{dx^2} - kc_A^2 \qquad (2.167)$$

Convert this problem to dimensionless form to reduce the number of independent parameters. Then, use the finite difference method to convert this boundary value problem into a set of nonlinear algebraic equations and solve with **MATLAB**. Plot the dimensionless concentration as a function of the remaining adjustable dimensionless parameter(s). To speed up your calculations, have your function routine return the Jacobian matrix.

2.C.1. In this problem, we consider the thermodynamics of a mixture of n_2 moles of linear polymer chains, each containing x segments, and n_1 moles of solvent, each of which is comparable in size to a single polymer segment. Let $n_0 = n_1 + xn_2$, so that the volume fractions of solvent and polymer are

$$\phi_1 = \frac{n_1}{n_1 + xn_2} \qquad \phi_2 = \frac{xn_2}{n_1 + xn_2} \qquad (2.168)$$

According to Flory–Huggins lattice theory (Flory, 1953), the free energy of mixing for such a mixture is

$$\Delta G_{mix} = \Delta G_{mix}^{contact} - T\Delta S_{mix}^{ideal} \qquad (2.169)$$

The ideal entropy of mixing is

$$\Delta S_{mix}^{ideal} = -R[n_1 \ln \phi_1 + n_2 \ln \phi_2] \qquad (2.170)$$

The contact free energy of mixing is modeled as

$$\Delta G_{\text{mix}}^{\text{contact}} = RT n_1 \phi_2 \chi \tag{2.171}$$

χ is a temperature-dependent dimensionless parameter that is specific to the solvent/polymer pair, where $\chi k_b T$ is a measure of the free energy penalty paid whenever a solvent molecule is in contact with a polymer segment, k_b being the Boltzmann constant. Usually $\chi = a + b/T$, where the contributions are from nonideal entropic and enthalpic effects respectively. As χ increases, the solvent and polymer segments dislike each other more, resulting eventually in phase separation.

From this lattice theory, the free energy per mole of solvents and segments, $\Delta g_{\text{mix}} = \Delta G_{\text{mix}}/n_0$, is

$$\Delta g_{\text{mix}} = RT\left[\phi_1\phi_2\chi - \phi_1 \ln\phi_1 - \frac{\phi_2}{\gamma} \ln\phi_2\right] \tag{2.172}$$

We see that as the chain length x increases, the polymer contribution to the entropy of mixing decreases, due to the constraints that neighboring segments always must be next to each other.

The chemical potentials, with respect to the pure reference states, at constant temperature and pressure, are

$$\mu_i - \mu_i^0 = \frac{\partial}{\partial n_i}\Delta G_{\text{mix}}\bigg|_{T,P,n_{j\neq i}} \tag{2.173}$$

Using the free energy expression above, we have

$$\frac{\mu_1 - \mu_1^0}{RT} = \ln\phi_1 + \left(1 - \frac{1}{x}\right)\phi_2 + \chi\phi_2^2 \qquad \frac{\mu_2 - \mu_2^0}{RT} = \ln\phi_2 - (x-1)\phi_1 + x\chi\phi_1^2 \tag{2.174}$$

At $\chi = 0$ we have only the positive ideal entropy of mixing and $\Delta G_{\text{mix}} < 0$. As χ increases, we eventually reach a point where phase separation into two coexisting phases, (I) and (II), occurs. We wish to compute the properties of the coexisting phases. Let the volume fractions of solvent in each phase be $\phi_1^{(I)}$ and $\phi_1^{(II)}$. From these values, we compute the volume fraction of the system occupied by phase (I) from the lever rule

$$\Phi^{(I)} = \frac{\phi_1 - \phi_1^{(II)}}{\phi_1^{(I)} - \phi_1^{(II)}} \tag{2.175}$$

To compute $\phi_1^{(I)}$ and $\phi_1^{(II)}$, we equate the chemical potentials in each phase,

$$(\mu_i^{(I)} - \mu_i^0)/(RT) = (\mu_i^{(II)} - \mu_i^0)/(RT) \quad i = 1,2 \tag{2.176}$$

This yields the following system of two nonlinear algebraic equations

$$\ln\phi_1^{(I)} + \left(1 - \frac{1}{x}\right)(1 - \phi_1^{(I)}) + \chi(1 - \phi_1^{(I)})^2 = \ln\phi_1^{(II)} + \left(1 - \frac{1}{x}\right)(1 - \phi_1^{(II)}) + \chi(1 - \phi_1^{(II)})^2$$

$$\ln(1 - \phi_1^{(I)}) - (x-1)\phi_1^{(I)} + x\chi[\phi_1^{(I)}]^2 = \ln(1 - \phi_1^{(II)}) - (x-1)\phi_1^{(II)} + x\chi[\phi_1^{(II)}]^2 \tag{2.177}$$

Clearly $\phi_1^{(\mathrm{I})} = \phi_1^{(\mathrm{II})}$ is always a solution; however, for fixed x, as χ increases, there may arise additional solutions for which $\phi_1^{(\mathrm{I})} \neq \phi_1^{(\mathrm{II})}$. If the free energy of this heterogeneous system,

$$\Delta g_{\mathrm{mix}}^{\mathrm{heterogeneous}} = \Phi^{(I)} \Delta g_{\mathrm{mix}}^{(\mathrm{I})} + \left(1 - \Phi^{(I)}\right) \Delta g_{\mathrm{mix}}^{(\mathrm{II})} \tag{2.178}$$

is less than the free energy Δg_{mix} for a homogeneous system, phase separation occurs. Generate a phase diagram, plotting $\phi_1^{(\mathrm{I})}$ and $\phi_1^{(\mathrm{II})}$ on the x-axis and χ on the y-axis for chain lengths of $x = 1, 10, 100, 1000$.

2.C.2. For the system of Problem 2.C.1, use the concept of a bifurcation point to directly compute χ_c, the critical value of χ, above which phase separation occurs, given an input value of x. Plot χ_c vs. x.

2.C.3. In the polycondensation reactor example above, we reduced the number of equations by deriving moment equations. This required a closure approximation to estimate the value of λ_3 given λ_0, λ_1, λ_2. Test this approximation by solving the complete set of population balance equations

$$0 = F^{(\mathrm{in})}[\mathrm{P}_m]^{(in)} - F[\mathrm{P}_m] + M r_{\mathrm{P}_m} \tag{2.179}$$

where the net rate of generation of m-mer is the sum of (2.120) and (2.124). Solve this set of nonlinear algebraic equations for $m = 1, 2, \ldots, M_{\mathrm{max}}$, and increase M_{max} until you no longer see an effect upon the polydispersity. Note that the required values of M_{max} might be very large, as even a small number of moles of such very high-molecular-weight polymer chains contribute greatly to higher-order moments. Generate the plots of Figure 2.18 for this full set of population balances and note when the closure approximation (2.127) fails to give adequate results.

3 Matrix eigenvalue analysis

We now resume our discussion of linear algebra, which previously focused upon the solution of linear systems $Ax = b$ by Gaussian elimination. The interpretation of A as a linear transformation was found useful in understanding the existence and uniqueness of solutions. Here, we consider a powerful tool in analyzing the transformational properties of a matrix, *eigenvalue analysis*, based upon identifying for a matrix A the *eigenvectors* w and corresponding scalar *eigenvalues* λ such that

$$Aw = \lambda w \tag{3.1}$$

We shall encounter numerous situations in which eigenvalue analysis provides insight into the behavior and performance of an algorithm, or is itself of direct use, as when estimating the vibrational frequencies of a structure or when calculating the states of a system in quantum mechanics. The related method of singular value decomposition (SVD), an extension of eigenvalue analysis to nonsquare matrices, is also discussed.

Orthogonal matrices

We begin our discussion of eigenvalue analysis by demonstrating how it may be used to diagnose the transformational properties of a matrix. Here, we consider a 3×3 real matrix Q that rotates vectors in $\Re^3$. We specify the particular rotation that it performs by designating an orthonormal basis set $\{u^{[1]}, u^{[2]}, u^{[3]}\}$ that is obtained from the orthonormal basis

$$e^{[1]} = \begin{bmatrix} 1 \\ 0 \\ 0 \end{bmatrix} \quad e^{[2]} = \begin{bmatrix} 0 \\ 1 \\ 0 \end{bmatrix} \quad e^{[3]} = \begin{bmatrix} 0 \\ 0 \\ 1 \end{bmatrix} \quad e_j^{[m]} = \delta_{mj} \tag{3.2}$$

by transformation under Q (Figure 3.1),

$$u^{[k]} = Qe^{[k]} \qquad u^{[j]} \cdot u^{[k]} = \delta_{jk} \tag{3.3}$$

Note that we rotate the vectors, *not* the coordinate system. We now ask:

If we only know the new basis vectors, can we determine the elements of Q?

Given only Q, is there any way that we can recognize that it performs a rotation, and if so, can we extract from Q any information about the particular rotation that it represents?

To answer the first question, we write $u^{[k]} = Qe^{[k]}$ explicitly,

$$u^{[k]} = \begin{bmatrix} Q_{11} & Q_{12} & Q_{13} \\ Q_{21} & Q_{22} & Q_{23} \\ Q_{31} & Q_{32} & Q_{33} \end{bmatrix} \begin{bmatrix} e_1^{[k]} \\ e_2^{[k]} \\ e_3^{[k]} \end{bmatrix} = \begin{bmatrix} Q_{11}e_1^{[k]} + Q_{12}e_2^{[k]} + Q_{13}e_3^{[k]} \\ Q_{21}e_1^{[k]} + Q_{22}e_2^{[k]} + Q_{23}e_3^{[k]} \\ Q_{31}e_1^{[k]} + Q_{32}e_2^{[k]} + Q_{33}e_3^{[k]} \end{bmatrix} \tag{3.4}$$

Using $e_j^{[m]} = \delta_{mj}$, we have

$$u^{[1]} = \begin{bmatrix} Q_{11} \\ Q_{21} \\ Q_{31} \end{bmatrix} \qquad u^{[2]} = \begin{bmatrix} Q_{12} \\ Q_{22} \\ Q_{32} \end{bmatrix} \qquad u^{[3]} = \begin{bmatrix} Q_{13} \\ Q_{23} \\ Q_{33} \end{bmatrix} \tag{3.5}$$

Taking the dot products of each $e^{[j]}$ with each $u^{[k]}$ yields

$$Q_{jk} = e^{[j]} \cdot u^{[k]} \tag{3.6}$$

If we want instead to rotate the coordinate system without changing the vectors, we apply the inverse rotation Q^{-1}. To derive the elements of Q^{-1}, we have merely to swap $e^{[j]} \Leftrightarrow u^{[j]}$ to obtain

$$Q_{jk}^{-1} = u^{[j]} \cdot e^{[k]} = e^{[k]} \cdot u^{[j]} = Q_{kj} = Q_{jk}^{\mathrm{T}} \tag{3.7}$$

We thus see that for this "rotation matrix" Q,

$$Q^{-1} = Q^{\mathrm{T}} \tag{3.8}$$

As a rotation transforms one orthogonal basis into another, a matrix satisfying (3.8) is said to be *orthogonal*. Note that matrices that perform improper rotations, i.e. that combine a rotation with a mirror reflection, also map orthogonal bases into other ones, and also satisfy (3.8).

A specific example of an orthogonal matrix

We now consider a specific Q that performs a counter-clockwise rotation (in the 1–2 plane) by an angle α around $e^{[3]}$ (Figure 3.1),

$$Q = \begin{bmatrix} \cos\alpha & -\sin\alpha & 0 \\ \sin\alpha & \cos\alpha & 0 \\ 0 & 0 & 1 \end{bmatrix} \tag{3.9}$$

The effect of Q on each $v \in \Re^3$ is

$$Qv = \begin{bmatrix} \cos\alpha & -\sin\alpha & 0 \\ \sin\alpha & \cos\alpha & 0 \\ 0 & 0 & 1 \end{bmatrix} \begin{bmatrix} v_1 \\ v_2 \\ v_3 \end{bmatrix} = \begin{bmatrix} v_1\cos\alpha - v_2\sin\alpha \\ v_1\sin\alpha + v_2\cos\alpha \\ v_3 \end{bmatrix} \tag{3.10}$$

In particular, the basis set obtained by rotating the basis (3.2) is

$$u^{[1]} = Qe^{[1]} = \begin{bmatrix} \cos\alpha \\ \sin\alpha \\ 0 \end{bmatrix} \qquad u^{[2]} = Qe^{[2]} = \begin{bmatrix} -\sin\alpha \\ \cos\alpha \\ 0 \end{bmatrix} \qquad u^{[3]} = Qe^{[3]} = \begin{bmatrix} 0 \\ 0 \\ 1 \end{bmatrix} \tag{3.11}$$

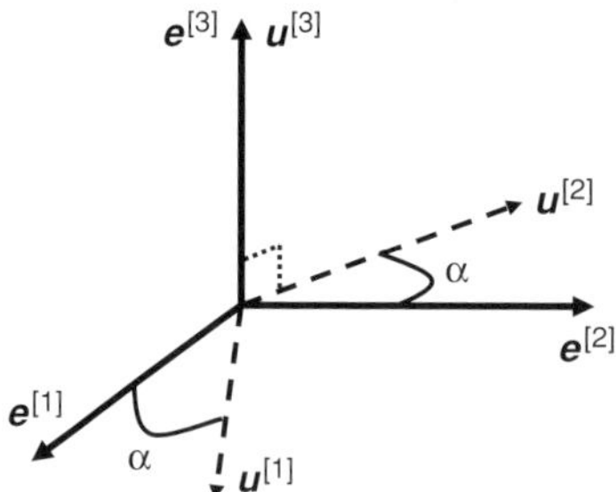

Figure 3.1 A counter-clockwise rotation in the 1–2 plane performed by Q.

To obtain the inverse rotation Q^{-1}, we obviously set $\alpha \to -\alpha$,

$$Q^{-1} = \begin{bmatrix} \cos(-\alpha) & -\sin(-\alpha) & 0 \\ \sin(-\alpha) & \cos(-\alpha) & 0 \\ 0 & 0 & 1 \end{bmatrix} = \begin{bmatrix} \cos\alpha & \sin\alpha & 0 \\ -\sin\alpha & \cos\alpha & 0 \\ 0 & 0 & 1 \end{bmatrix} = Q^{\mathrm{T}} \tag{3.12}$$

For this Q, $e^{[3]}$ has the special property that it is unchanged by the rotation,

$$Q e^{[3]} = e^{[3]} \tag{3.13}$$

Obviously, this is because Q defines a rotation about $e^{[3]}$. However, if we did not know this but rather only knew the matrix elements of Q, recognizing that $Q e^{[3]} = e^{[3]}$ immediately tells us something useful about the nature of the transformation done by Q. *This is the role of eigenvalue analysis.*

Eigenvalues and eigenvectors defined

If A is an $N \times N$ matrix, and if w is acted upon by A simply as if it were multiplied by a scalar λ,

$$Aw = \lambda w \tag{3.14}$$

then w is said to be a *characteristic vector* of A with the *characteristic value* λ. This field was developed, to a large extent, by German mathematicians, and the German word for characteristic value is *"eigenwert,"* *eigen* meaning characteristic, individual, or unique, and *wert* meaning value or worth. It is now common, if confusing, practice to use the *halb Deutsch*/half English terms *eigenvalue* and *eigenvector*.

As $Iv = v$, the "eigenpair" (w, λ) satisfies the linear system

$$(A - \lambda I)w = 0 \tag{3.15}$$

For this to hold true for $w \neq 0$, the matrix $(A - \lambda I)$ must be singular, providing a *characteristic polynomial* of degree N,

$$p(\lambda) = \det(A - \lambda I) = 0 \tag{3.16}$$

whose N roots are the eigenvalues of A.

Eigenvalues/eigenvectors of a 2 × 2 real matrix

Before discussing general eigenvector properties, let us consider the case of a 2×2 real matrix, for which analytical calculation of eigenvalues is easy,

$$\det(A - \lambda I) = \begin{vmatrix} a_{11} - \lambda & a_{12} \\ a_{21} & a_{22} - \lambda \end{vmatrix} = (a_{11} - \lambda)(a_{22} - \lambda) - a_{21}a_{12}$$
$$= \lambda^2 - (a_{11} + a_{22})\lambda + (a_{11}a_{22} - a_{21}a_{12}) \tag{3.17}$$

As the *trace* (the sum of the diagonal values) and the *determinant* of A are

$$\text{tr}(A) = a_{11} + a_{22} \qquad \det(A) = a_{11}a_{22} - a_{21}a_{12} \tag{3.18}$$

then, using the shorthand $T = \text{tr}(A)$, $D = \det(A)$,

$$\det(A - \lambda I) = \lambda^2 - T\lambda + D \qquad \lambda_{1,2} = \frac{T \pm \sqrt{T^2 - 4D}}{2} \tag{3.19}$$

In terms of the elements of A, the eigenvalues are

$$\lambda_{1,2} = \tfrac{1}{2}(a_{11} + a_{22}) \pm \tfrac{1}{2}\sqrt{(a_{11} + a_{22})^2 - 4(a_{11}a_{22} - a_{21}a_{12})} \tag{3.20}$$

For any 2×2 real matrix A, the trace equals the sum of the eigenvalues,

$$\lambda_1 + \lambda_2 = a_{11} + a_{22} = tr(A) \tag{3.21}$$

Also, we have the following properties for a real 2×2 matrix:

1 if $T^2 - 4D > 0$, then both eigenvalues are real;
2 if $T^2 - 4D = 0$, $\lambda_1 = \lambda_2 = \tfrac{1}{2}(a_{11} + a_{22})$;
3 if $T^2 - 4D < 0$, both eigenvalues are complex and $\lambda_2 = \overline{\lambda}_1$.

Also, note that the determinant equals the product of the eigenvalues,

$$\lambda_1\lambda_2 = \frac{1}{4}(T + \sqrt{T^2 - 4D})(T - \sqrt{T^2 - 4D}) = \frac{1}{4}[T^2 - (T^2 - 4D)] = D \tag{3.22}$$

Thus, A is singular if it has one or more zero eigenvalues.

We now consider some special cases of 2×2 real matrices.

A is triangular

$$A = \begin{bmatrix} a_{11} & a_{12} \\ 0 & a_{22} \end{bmatrix} \qquad \det(A) = a_{11}a_{12} \tag{3.23}$$

The eigenvalues are

$$\lambda_{1,2} = \frac{(a_{11} + a_{22}) \pm \sqrt{(a_{11} + a_{22})^2 - 4a_{11}a_{22}}}{2} = \{a_{11}, a_{22}\} \tag{3.24}$$

For any upper triangular matrix, the eigenvalues are found on the principal diagonal. This also is true for lower triangular and diagonal matrices,

$$A = \begin{bmatrix} a_{11} & 0 \\ a_{21} & a_{22} \end{bmatrix} \qquad A = \begin{bmatrix} a_{11} & 0 \\ 0 & a_{22} \end{bmatrix} \tag{3.25}$$

A is real, symmetric ($A^{\mathsf{T}} = A$)

As $a_{12} = a_{21}$, $D = a_{11}a_{22} - a_{12}^2$, and

$$\lambda_{1,2} = \tfrac{1}{2}(a_{11} + a_{22}) \pm \sqrt{(a_{11} - a_{22})^2 + 4a_{12}^2} \tag{3.26}$$

Since $(a_{11} - a_{22})^2 + 4a_{12}^2 \geq 0$, both eigenvalues are always real in this case.

Analytical computation of eigenvectors

If $w^{[j]}$ is an eigenvector of A for λ_j, then so is $cw^{[j]}$, for any $c \in C$, since

$$A w^{[j]} = \lambda_j w^{[j]} \quad \Rightarrow \quad A c w^{[j]} = \lambda_j c w^{[j]} \tag{3.27}$$

Thus, $w^{[j]}$ can have any length and still be an eigenvector for λ_j. This lack of uniqueness results from the singularity of $(A - \lambda_j I)$. The eigenvector $w^{[j]}$ must satisfy

$$(A - \lambda_j I)w = \begin{bmatrix} (a_{11} - \lambda_j) & a_{12} \\ a_{21} & (a_{22} - \lambda_j) \end{bmatrix} \begin{bmatrix} w_1^{[j]} \\ w_2^{[j]} \end{bmatrix} = \begin{bmatrix} 0 \\ 0 \end{bmatrix} = \mathbf{0} \tag{3.28}$$

which yields the two equations

$$\begin{aligned} (a_{11} - \lambda_j)w_1^{[j]} + a_{12}w_2^{[j]} &= 0 \\ a_{21}w_1^{[j]} + (a_{22} - \lambda_j)w_2^{[j]} &= 0 \end{aligned} \tag{3.29}$$

Because $\det(A - \lambda_j I) = 0$, these two equations are dependent, so we pick only one to satisfy, say the first one. The second then must be satisfied automatically. As we have flexibility in setting the length of the eigenvector, we are free to choose $|w^{[j]}| = 1$, which would yield the two equations

$$\begin{aligned} (a_{11} - \lambda_j)w_1^{[j]} + a_{12}w_2^{[j]} &= 0 \\ \left(w_1^{[j]}\right)^2 + \left(w_2^{[j]}\right)^2 &= 1 \end{aligned} \tag{3.30}$$

As the second equation is nonlinear, rather than finding a unit length eigenvector, we instead try to find one with $w_1^{[j]} = 1$ that satisfies

$$\begin{aligned} (a_{11} - \lambda_j)w_1^{[j]} + a_{12}w_2^{[j]} &= 0 \\ w_1^{[j]} &= 1 \end{aligned} \tag{3.31}$$

This linear system has a unique solution if

$$\det \begin{bmatrix} (a_{11} - \lambda_j) & a_{12} \\ 1 & 0 \end{bmatrix} = (a_{11} - \lambda_j)(0) - a_{12} = -a_{12} \neq 0 \tag{3.32}$$

in which case we obtain $w_2^{[j]}$ from the first equation,

$$(a_{11} - \lambda_j)(1) + a_{12}w_2^{[j]} = 0 \quad \Rightarrow \quad w_2^{[j]} = -\frac{(a_{11} - \lambda_j)}{a_{12}} \tag{3.33}$$

If $a_{12} = 0$, we instead set $w_2^{[j]} = 1$. Once we have computed $w^{[j]}$, we can renormalize to unit length if desired. Consider the example

$$A = \begin{bmatrix} 2 & 1 \\ 1 & 3 \end{bmatrix} \qquad \begin{aligned} D &= \det(A) = (2)(3) - (1)(1) = 6 - 1 = 5 \\ T &= \operatorname{tr}(A) = 2 + 3 = 5 \end{aligned} \tag{3.34}$$

with the eigenvalues

$$\lambda_{1,2} = \frac{1}{2}T \pm \frac{1}{2}\sqrt{T^2 - 4D} = \frac{1}{2}(5) \pm \frac{1}{2}\sqrt{25 - (4)(5)} = \frac{5}{2} \pm \frac{\sqrt{5}}{2} \tag{3.35}$$

An eigenvector $w^{[1]}$ for

$$\lambda_1 = \frac{5}{2} + \frac{\sqrt{5}}{2}$$

must satisfy

$$(a_{11} - \lambda_1)w_1^{[1]} + a_{12}w_2^{[1]} = 0 \Rightarrow -\left(\frac{1}{2} + \frac{\sqrt{5}}{2}\right)w_1^{[1]} + w_2^{[1]} = 0 \tag{3.36}$$

If we set $w_1^{[1]} = 1$, and compute $w_2^{[1]}$ from (3.36), we obtain

$$w^{[1]} = \begin{bmatrix} 1 & \left(\frac{1}{2} + \frac{\sqrt{5}}{2}\right) \end{bmatrix}^{\mathrm{T}} \tag{3.37}$$

Above, we have selected a unique eigenvector by arbitrarily setting the value of one component. More generally, we may select some vector v, say at random, and choose as a normalization, $v \cdot w^{[j]} = 1$. We replace row m of the matrix $(A - \lambda_j I)$ with the elements of v to obtain a modified matrix $(A - \lambda_j I)^{++}$. We then obtain the eigenvector by solving the linear system

$$(A - \lambda_j I)^{++}w^{[j]} = e^{[m]} \qquad e_j^{[m]} = \delta_{mj} \tag{3.38}$$

Shortly, we consider numerical methods to compute λ_k and $w^{[k]}$ without the need for tedious analytical calculations.

Multiplicity and formulas for the trace and determinant

For an $N \times N$ matrix A, the characteristic polynomial

$$p(\lambda) = \det(A - \lambda I) = \sum_{i_1=1}^{N}\sum_{i_2=1}^{N} \cdots \sum_{i_N=1}^{N} \varepsilon_{i_1,i_2,\ldots,i_N}(a_{i_1,1} - \lambda\delta_{i_1,1})\ldots(a_{i_N,N} - \lambda\delta_{i_N,N})$$

$$\tag{3.39}$$

is of degree N and thus has N roots, so an $N \times N$ matrix has N eigenvalues. However, these may not all be distinct; i.e., we may have $\lambda_j = \lambda_k$ for some $j \neq k$. As an example, consider the matrix

$$A = \begin{bmatrix} 2 & 3 & 4 \\ 0 & 2 & 5 \\ 0 & 0 & 1 \end{bmatrix} \qquad \begin{aligned} p(\lambda) &= (2 - \lambda)^2(1 - \lambda) \\ \lambda_1 &= 2 \quad \lambda_2 = 2 \quad \lambda_3 = 1 \end{aligned} \tag{3.40}$$

In general if $P \leq N$ is the number of distinct roots of (3.39), we write

$$p(\lambda) = \det(A - \lambda I) = (\lambda_1 - \lambda)^{m_1}(\lambda_2 - \lambda)^{m_2} \cdots (\lambda_P - \lambda)^{m_P} \tag{3.41}$$

m_k is the *(algebraic) multiplicity* of λ_k, i.e., the number of times that it is repeated as a root of (3.39). The multiplicities must sum to N,

$$m_1 + m_2 + \cdots + m_P = N \tag{3.42}$$

From $p(\lambda = 0)$, we find the determinant to be the product of the eigenvalues,

$$\det(A) = \lambda_1^{m_1} \lambda_2^{m_2} \ldots \lambda_P^{m_P} \tag{3.43}$$

It may also be shown that the trace equals the sum of the eigenvalues,

$$\mathrm{tr(A)} = a_{11} + a_{22} + \cdots + a_{NN} = \lambda_1 + \lambda_2 + \cdots + \lambda_N \tag{3.44}$$

A proof of this result is found in the supplemental material in the accompanying website.

Eigenvalues and the existence/uniqueness properties of linear systems

We now consider the existence and uniqueness properties of $Ax = b$ from the viewpoint of eigenvalue analysis. Let A be an $N \times N$ matrix, with the P distinct eigenvalues $\lambda_1, \lambda_2, \ldots,$ λ_P for eigenvectors $w^{[1]}, w^{[2]}, \ldots, w^{[P]}$,

$$Aw^{[k]} = \lambda_k w^{[k]} \tag{3.45}$$

Let these P distinct eigenvalues be ordered by increasing modulus,

$$|\lambda_1| \leq |\lambda_2| \leq \cdots \leq |\lambda_P| \tag{3.46}$$

We use the $\leq$ sign, even though the eigenvalues are distinct, because with complex eigenvalues we may have the distinct, but equal modulus, values

$$\lambda_k = a + ib \quad \lambda_{k+1} = a - ib \quad a, b \in \Re \tag{3.47}$$

We now examine the effect of A on an eigenvector associated with λ_1,

$$Aw^{[1]} = \lambda_1 w^{[1]} \tag{3.48}$$

If any eigenvalue is zero, it will be λ_1, as we have ordered the eigenvalues by increasing modulus. Also, if $\lambda_1 = 0$, A is singular, as

$$\det(A) = \lambda_1^{m_1} \lambda_2^{m_2} \cdots \lambda_P^{m_P} = 0 \tag{3.49}$$

Thus, if $\lambda_1 = 0$, the null space of A is not empty and there exists some $w \in K_A, w \neq 0$ such that $Aw = 0$. But, when $\lambda_1 = 0$, as

$$Aw^{[1]} = \lambda_1 w^{[1]} = 0 \tag{3.50}$$

any eigenvector $w^{[1]}$ for $\lambda_1 = 0$ is in the null space of A and $\dim(K_A) = m_1$.

Estimating eigenvalues; Gershgorin's theorem

With the significant effort required to find the roots of

$$\det(A - \lambda I) = (\lambda_1 - \lambda)^{m_1}(\lambda_2 - \lambda)^{m_2} \cdots (\lambda_P - \lambda)^{m_P} \tag{3.51}$$

it would be convenient if we could just look at a matrix and be able to "tell" what are its eigenvalues. Unfortunately, we cannot do this in general, although for a few cases it is possible. For triangular and diagonal matrices

$$U = \begin{bmatrix} U_{11} & U_{12} & & U_{1N} \\ & U_{22} & \cdots & U_{2N} \\ & & \cdot & \vdots \\ & & & U_{NN} \end{bmatrix} \qquad L = \begin{bmatrix} L_{11} & & & \\ L_{21} & L_{22} & & \\ \vdots & & \ddots & \\ L_{N1} & L_{N2} & \cdots & L_{NN} \end{bmatrix}$$

$$D = \begin{bmatrix} D_{11} & & & \\ & D_{22} & & \\ & & \ddots & \\ & & & D_{NN} \end{bmatrix} \tag{3.52}$$

the determinant equals the product of the elements along the diagonal,

$$\det(U) = U_{11}U_{22}U_{33}\ldots U_{NN} \qquad \det(L) = L_{11}L_{22}L_{33}\cdots L_{NN} \tag{3.53}$$

Thus, the characteristic equation is already factored,

$$\det(U - \lambda I) = (U_{11} - \lambda)(U_{22} - \lambda)(U_{33} - \lambda)\cdots(U_{NN} - \lambda) \tag{3.54}$$

The eigenvalues of a triangular (diagonal) matrix lie along the diagonal

$$\lambda_1 = U_{11} \qquad \lambda_2 = U_{22} \qquad \cdots \qquad \lambda_N = U_{NN} \tag{3.55}$$

For matrices that are not triangular, we cannot determine the eigenvalues by inspection, but we can obtain upper and lower bounds using Gershgorin's theorem. Let A be an $N \times N$ matrix,

$$A = \begin{bmatrix} a_{11} & a_{12} & a_{13} & \cdots & a_{1N} \\ a_{21} & a_{22} & a_{23} & \cdots & a_{2N} \\ a_{31} & a_{32} & a_{33} & \cdots & a_{3N} \\ \vdots & \vdots & \vdots & & \vdots \\ a_{N1} & a_{N2} & a_{N3} & \cdots & a_{NN} \end{bmatrix} \tag{3.56}$$

In row k, the diagonal element is a_{kk}, and the sum of the magnitudes of the off-diagonal elements is

$$\Gamma_k = |a_{k1}| + |a_{k2}| + \cdots + |a_{k,k-1}| + |a_{k,k+1}| + \cdots + |a_{kN}| \tag{3.57}$$

As the eigenvalues of A are, in general, complex, we make a graph of the complex plane and place the eigenvalue $\lambda = a + ib$ at (a, b) (Figure 3.2). On this graph, we add a circle for each row k of the matrix. The center of the circle is placed at the location of the diagonal element a_{kk} in the complex plane, and the radius of the circle is the sum of the magnitudes of the

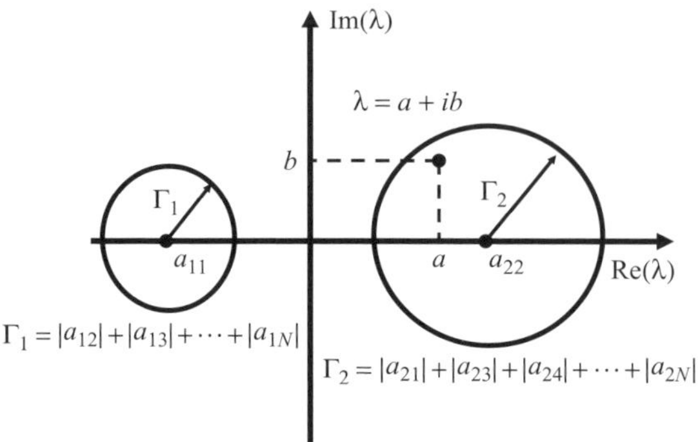

Figure 3.2 The complex plane, showing that an eigenvalue must lie within one of the Gershgorin circles.

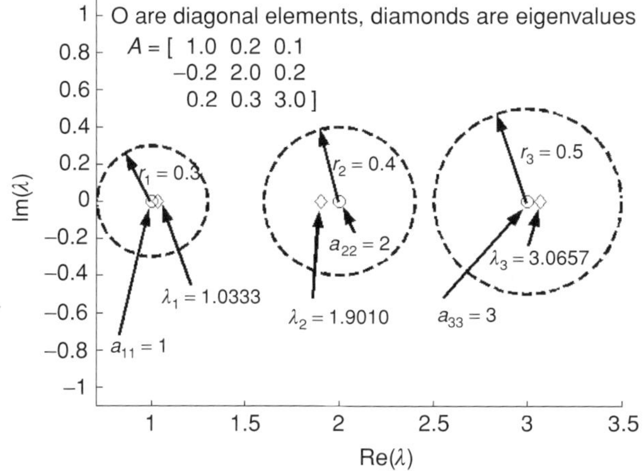

Figure 3.3 Test of Gershgorin's theorem for an example 3×3 matrix.

off-diagonal elements in the row, Γ_k. *Gershgorin's theorem* states that every eigenvalue λ must be located within one of these circles; i.e.,

$$|\lambda - a_{kk}| \leq \Gamma_k = \sum_{\substack{j=1 \\ j \neq k}}^{N} |a_{kj}| \quad \text{for some } k \in [1, N] \tag{3.58}$$

A proof, relying upon the concepts of matrix norm and spectral radius introduced in the next section, is provided in the supplemental material in the accompanying website. Figure 3.3 demonstrates the application of Gershgorin's theorem to a 3×3 matrix. **test-Gershgorin.m** generates this plot for any input matrix.

Gershgorin's theorem does not tell us what the eigenvalues are exactly, but at least it provides information about where they can be located. The eigenvalues tend to be clustered around the diagonal values when the diagonal elements are much larger in magnitude than the off-diagonal elements.

Matrix norm, spectral radius, and condition number

We have defined the norm of a vector as a rule that assigns to every $v \in C^N$ a scalar $\|v\| \in \mathfrak{R}$ that represents the "size" of v and that satisfies $\|v\| \geq 0$, where $\|v\| = 0$ if and only if $v = 0$. For each particular vector norm $\|v\|$, we can generate a corresponding *matrix norm* $\|A\|$,

$$\|A\| = \max_{v \neq 0} \frac{\|Av\|}{\|v\|} \tag{3.59}$$

How is $\|A\|$ related to the eigenvalues $\lambda_1, \lambda_2, \ldots, \lambda_N$ of A? Let $\{w^{[1]}, \ldots, w^{[N]}\} \in C^N$ be unit-length eigenvectors with $Aw^{[k]} = \lambda_k w^{[k]}$ and let $S_W = \operatorname{span}\{w^{[1]}, \ldots, w^{[N]}\}$. We can decompose any $v \in C^N$ into a component $u \in S_W$ and a component $y \notin S_W$,

$$v = u + y = c_1 w^{[1]} + c_2 w^{[2]} + \cdots + c_N w^{[N]} + y$$

$$y \notin S_W = \operatorname{span}\{w^{[1]}, \ldots, w^{[N]}\} \tag{3.60}$$

The matrix norm can then be expressed as

$$\|A\| = \max_{u \in S_W} \max_{\substack{y \notin S_W \\ u + y \neq 0}} \frac{\|Au + Ay\|}{\|u + y\|} \tag{3.61}$$

We thus must have

$$\|A\| \geq \max_{\substack{u \in S_W \\ u \neq 0}} \frac{\|Au\|}{\|u\|} \tag{3.62}$$

where the equality holds if the set of eigenvectors of A completely spans C^N; i.e., if the set $\{w^{[1]}, \ldots, w^{[N]}\}$ is linearly independent, $S_W = C^N$. We define the quantity on the right-hand side of (3.62) as the *spectral radius*, $\rho(A)$,

$$\rho(A) = \max_{\substack{u \in S_W \\ u \neq 0}} \frac{\|Au\|}{\|u\|} = \max_{\substack{\{c_1, \ldots, c_N\} \\ u \neq 0}} \frac{\|A(c_1 w^{[1]} + c_2 w^{[2]} + \cdots + c_N w^{[N]})\|}{\|c_1 w^{[1]} + c_2 w^{[2]} + \cdots + c_N w^{[N]}\|}$$

$$\rho(A) = \max_{\substack{\{c_1, \ldots, c_N\} \\ u \neq 0}} \frac{\|c_1 \lambda_1 w^{[1]} + c_2 \lambda_2 w^{[2]} + \cdots + c_N \lambda_N w^{[N]}\|}{\|c_1 w^{[1]} + c_2 w^{[2]} + \cdots + c_N w^{[N]}\|} \tag{3.63}$$

As the maximum is attained when u points in the direction of an eigenvector for the eigenvalue of largest modulus,

$$\rho(A) = \max\{|\lambda_1|, |\lambda_2|, \ldots, |\lambda_N|\} \tag{3.64}$$

The spectral radius provides a lower bound on the matrix norm,

$$\|A\| \geq \rho(A) \tag{3.65}$$

The *condition number*, κ, the ratio of the largest and smallest eigenvalue magnitudes, is

$$\kappa = \frac{\lambda_{\max}}{\lambda_{\min}} \quad \lambda_{\max} = \rho(A) \quad \lambda_{\min} = \min\{|\lambda_1|, |\lambda_2|, \ldots, |\lambda_N|\} \tag{3.66}$$

A matrix with a large condition number is said to be *ill-conditioned*.

Condition numbers are computed in MATLAB using **cond** and **condest**. Vector and matrix norms are computed by **norm** and **normest**.

Applying Gershgorin's theorem to study the convergence of iterative linear solvers

As a demonstration of the usefulness of Gershgorin's theorem, we generate a convergence criterion for the Jacobi iterative method of solving $Ax = b$. This example is typical of the use of eigenvalues in numerical analysis, and also shows why the questions of eigenvector basis set existence raised in the next section are of such importance.

We develop here a simple alternative to Gaussian elimination for solving a linear system $Ax = b$. It is based on forming an initial guess of the solution, $x^{[0]}$, and iteratively refining it to form a sequence $x^{[1]}, x^{[2]}, \ldots$ that hopefully converges to a solution; i.e.,

$$\lim_{k \to \infty} \left\| Ax^{[k]} - b \right\| = 0 \tag{3.67}$$

We add to each side of $Ax = b$ the term Bx, for some non singular B,

$$Ax + Bx = b + Bx \tag{3.68}$$

Upon rearrangement, this yields the identity

$$Bx = b + (B - A)x \tag{3.69}$$

and suggests a rule for forming a new guess $x^{[k+1]}$ from $x^{[k]}$,

$$Bx^{[k+1]} = b + (B - A)x^{[k]} \tag{3.70}$$

If $B = A$, this rule yields the exact solution after only one iteration, but it is equivalent to solving $Ax = b$ directly in the first place. We therefore want to choose some B that approximates A such that: (1) $\|B - A\|$ is small and (2) the update linear system is easy to solve (e.g. B is triangular). In the *Jacobi method*, we simply choose B to be the diagonal part of A,

$$B = \begin{bmatrix} a_{11} & & & \\ & a_{22} & & \\ & & \ddots & \\ & & & a_{NN} \end{bmatrix} \tag{3.71}$$

Under what conditions must the Jacobi method converge? To answer this question, let us define the error vector at iteration k as

$$\varepsilon^{[k]} \equiv x^{[k]} - x \qquad x = A^{-1}b \tag{3.72}$$

We want $\lim_{k \to \infty} \|\varepsilon^{[k]}\| = 0$. We obtain a rule for the transformation of the error vector at each iteration by subtracting (3.69) from (3.70),

$$B\varepsilon^{[k+1]} = (B - A)\varepsilon^{[k]} \qquad \Rightarrow \qquad \varepsilon^{[k+1]} = B^{-1}(B - A)\varepsilon^{[k]} \tag{3.73}$$

We now use Gershgorin's theorem to find when $\lim_{k \to \infty} \|\varepsilon^{[k]}\| = 0$. Let us *assume* that the matrix $B^{-1}(B - A)$ has a set of eigenvectors $\{w^{[1]}, w^{[2]}, \ldots, w^{[N]}\}$, satisfying $B^{-1}(B - A)w^{[j]} = \lambda_j w^{[j]}$, that is linearly independent and forms a basis for $\Re^N$. We can then express any vector as a linear combination of the eigenvectors, and in particular, the error associated with the initial guess can be written as

$$\varepsilon^{[0]} = c_1 w^{[1]} + c_2 w^{[2]} + \cdots + c_N w^{[N]} \tag{3.74}$$

After the first iteration, the error is

$$
\begin{aligned}
\varepsilon^{[1]} &= B^{-1}(B - A)\varepsilon^{[0]} = B^{-1}(B - A)\big[c_1 w^{[1]} + c_2 w^{[2]} + \cdots + c_N w^{[N]}\big] \\
&= c_1 B^{-1}(B - A)w^{[1]} + c_2 B^{-1}(B - A)w^{[2]} + \cdots + c_N B^{-1}(B - A)w^{[N]} \\
&= c_1 \lambda_1 w^{[1]} + c_2 \lambda_2 w^{[2]} + \cdots + c_N \lambda_N w^{[N]}
\end{aligned}
\tag{3.75}
$$

After the second iteration, the error is

$$
\begin{aligned}
\varepsilon^{[2]} &= B^{-1}(B - A)e^{[1]} = B^{-1}(B - A)\big[c_1 \lambda_1 w^{[1]} + c_2 \lambda_2 w^{[2]} + \cdots + c_N \lambda_N w^{[N]}\big] \\
&= c_1 \lambda_1^2 w^{[1]} + c_2 \lambda_2^2 w^{[2]} + \cdots + c_N \lambda_N^2 w^{[N]}
\end{aligned}
\tag{3.76}
$$

After k iterations, the error is

$$\varepsilon^{[k]} = c_1 \lambda_1^k w^{[1]} + c_2 \lambda_2^k w^{[2]} + \cdots + c_N \lambda_N^k w^{[N]} \tag{3.77}$$

If all eigenvalues of $B^{-1}(B - A)$ have moduli less than 1, $|\lambda_j| < 1$, then

$$1 > |\lambda_j| > |\lambda_j|^2 > |\lambda_j|^3 > \cdots \tag{3.78}$$

and $\lim_{k \to \infty} |c_j \lambda_j^k| = 0$ for finite $\{c_1, c_2, \ldots, c_N\}$, so that $\lim_{k \to \infty} \|\varepsilon^{[k]}\| = 0$. For the B of (3.71), we write $B^{-1}(B - A)$ explicitly,

$$
B^{-1}(B - A) =
\begin{bmatrix}
0 & (-a_{12}/a_{11}) & \cdots & (-a_{1N}/a_{11}) \\
(-a_{21}/a_{22}) & 0 & \cdots & (-a_{2N}/a_{22}) \\
\vdots & \vdots & & \vdots \\
(-a_{N1}/a_{NN}) & (-a_{N2}/a_{NN}) & \cdots & 0
\end{bmatrix}
\tag{3.79}
$$

By Gershgorin's theorem, each eigenvalue λ_j of $B^{-1}(B - A)$ must satisfy the following inequality for some $k = 1, 2, \ldots, N$:

$$|\lambda_j| \le \sum_{\substack{m=1 \\ m \neq k}}^{N} |-a_{km}/a_{kk}| = \frac{1}{|a_{kk}|} \sum_{\substack{m=1 \\ m \neq k}}^{N} |a_{km}| \tag{3.80}$$

Therefore, we can ensure that all $|\lambda_j| < 1$, if for *every* $k = 1, 2, \ldots, N$:

$$\frac{1}{|a_{kk}|} \sum_{\substack{m=1 \\ m \neq k}}^{N} |a_{km}| < 1 \Rightarrow \sum_{\substack{m=1 \\ m \neq k}}^{N} |a_{km}| < |a_{kk}| \tag{3.81}$$

That is, for every row of A, the magnitude of the diagonal element is greater than the sum of the magnitudes of all off-diagonal elements. A matrix for which this property holds is said to be *strictly diagonally dominant*. For such a matrix, the Jacobi method converges to a solution from any $x^{[0]}$.

$$A = \begin{bmatrix} 2 & -1 & 0 & 0 \\ -1 & 2 & -1 & 0 \\ 0 & -1 & 2 & -1 \\ 0 & 0 & -1 & 2 \end{bmatrix}$$

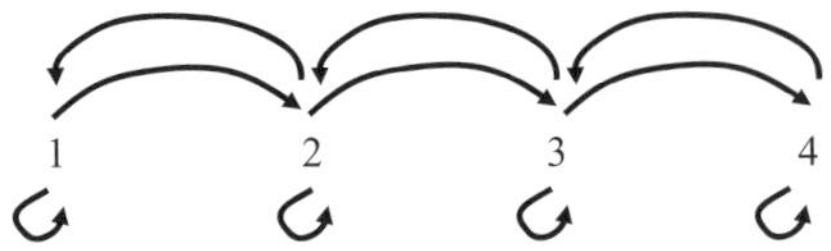

Figure 3.4 Graph for an irreducible matrix, with all nodes connected by a directed path.

Irreducible matrices

It may be shown that for matrices that are irreducible, it is sufficient only that the following, weaker inequality be satisfied,

$$\sum_{\substack{m=1 \\ m \neq k}}^{N} |a_{km}| \leq |a_{kk}| \tag{3.82}$$

A matrix A is *irreducible* if there exists no permutation matrix P, such that PAP^{T} takes the block upper triangular form

$$PAP^{\mathrm{T}} = \begin{bmatrix} A_{11} & A_{12} \\ 0 & A_{22} \end{bmatrix} \tag{3.83}$$

A_{11} and A_{22} are square submatrices. An example irreducible matrix is that obtained when discretizing the operator $-d^2/dx^2$ using finite differences,

$$A = \begin{bmatrix} 2 & -1 & & & & \\ -1 & 2 & -1 & & & \\ & -1 & 2 & \cdots & & \\ & & -1 & \cdots & -1 & \\ & & & \cdots & 2 & -1 \\ & & & & -1 & 2 \end{bmatrix} \tag{3.84}$$

As $|2| = |-1| + |-1|$, we can expect Jacobi's method to converge only if this matrix is indeed irreducible.

To show that a matrix is irreducible, we make a graph with a node for each row (column) number of the system, $k = 1, 2, \ldots, N$ (Figure 3.4 for (3.84) with $N = 4$). For each non zero element, we draw an arrow from the node of the row number to the node of the column number. If there exists a directed path that connects every pair of nodes, the matrix is irreducible.

In the derivation above, we have assumed that we can write *any* arbitrary vector $v \in \Re^N$ as a sum of the eigenvectors of $B^{-1}(B - A)$. This assumption is not always valid. Next, we derive some conditions under which a matrix is guaranteed to have a set of N linearly independent eigenvectors.

Eigenvector matrix decomposition and basis sets

For a given $N \times N$ matrix, when does the set of eigenvectors form a complete basis such that any $v \in C^N$ can be expressed as a linear combination of eigenvectors?

Can we express a matrix as a decomposition involving matrices that comprise eigenvalues and eigenvectors?

Are there special classes of matrices for which the eigenvalues can be proved to be real, or for which the set of eigenvectors is not only linearly independent, but also orthogonal?

Such questions may seem abstract, but they are in fact very important in practice. Above, our proof of convergence of Jacobi's method is based upon the assumed existence of a complete eigenvector basis for $B^{-1}(B - A)$.

Eigenvector properties of a general $N \times N$ complex matrix

Because we cannot assume that the eigenvectors and eigenvalues are real, even for a real matrix, we consider the general case where $\lambda_k \in C$, $w^{[k]} \in C^{[N]}$, and A is an $N \times N$ complex matrix,

$$
A = \begin{bmatrix}
(a_{11} + ib_{11}) & (a_{12} + ib_{12}) & \ldots & (a_{1N} + ib_{1N}) \\
(a_{21} + ib_{21}) & (a_{22} + ib_{22}) & \ldots & (a_{2N} + ib_{2N}) \\
\vdots & \vdots & & \vdots \\
(a_{N1} + ib_{N1}) & (a_{N2} + ib_{N2}) & \ldots & (a_{NN} + ib_{NN})
\end{bmatrix}
\tag{3.85}
$$

If we do not make any special assumptions about the structure of the matrix, we cannot establish generally that the eigenvectors form a complete basis for C^N. We can prove the following statements, however.

Theorem ENVG1   *Eigenvectors $w^{[j]}$ and $w^{[k]}$, satisfying $Aw^{[j]} = \lambda_j w^{[j]}$ and $Aw^{[k]} = \lambda_k w^{[k]}$ respectively, are linearly independent if $\lambda_j \neq \lambda_k$. Thus, the eigenvectors of any $N \times N$ matrix A that has N distinct eigenvalues form a complete basis for C^N.*

Proof We want to show that when $\lambda_j \neq \lambda_k$, we cannot have $cw^{[j]} = w^{[k]}$. Let us assume, contrary to the theorem, that $cw^{[j]} = w^{[k]}$. We replace $w^{[k]}$ with $cw^{[j]}$ in the first term of $Aw^{[k]} - \lambda_k w^{[k]} = 0$,

$$
Acw^{[j]} - \lambda_k w^{[k]} = \lambda_j cw^{[j]} - \lambda_k w^{[k]} = 0
\tag{3.86}
$$

and replace $w^{[k]}$ with $cw^{[j]}$ in the second term,

$$
\lambda_j cw^{[j]} - \lambda_k cw^{[j]} = c(\lambda_j - \lambda_k)w^{[j]} = 0
\tag{3.87}
$$

This poses a contradiction with $w^{[j]} \neq 0$ when $\lambda_j \neq \lambda_k$ and thus we cannot have $cw^{[j]} = w^{[k]}$. As $w^{[j]}$ and $w^{[k]}$ cannot be parallel, they are linearly independent. Thus, if A has N distinct eigenvalues, no two of $\{w^{[1]}, \ldots, w^{[N]}\}$ can be parallel and the eigenvectors form a complete basis set.   *QED*

Theorem EVG2   *For any A, it follows from the rules of matrix multiplication that we can form the following matrices from the eigenvalues and eigenvectors satisfying* $A\mathbf{w}^{[j]} = \lambda_j \mathbf{w}^{[j]}$,

$$
\Lambda = \begin{bmatrix} \lambda_1 & & & \\ & \lambda_2 & & \\ & & \ddots & \\ & & & \lambda_N \end{bmatrix}
\qquad
W = \begin{bmatrix} | & | & & | \\ \mathbf{w}^{[1]} & \mathbf{w}^{[2]} & \cdots & \mathbf{w}^{[N]} \\ | & | & & | \end{bmatrix}
\tag{3.88}
$$

and write A as

$$
AW = W\Lambda
\tag{3.89}
$$

Proof By the rules of matrix multiplication, the left-hand side of (3.89) is

$$
AW = A \begin{bmatrix} | & | & & | \\ \mathbf{w}^{[1]} & \mathbf{w}^{[2]} & \cdots & \mathbf{w}^{[N]} \\ | & | & & | \end{bmatrix}
= \begin{bmatrix} | & | & & | \\ A\mathbf{w}^{[1]} & A\mathbf{w}^{[2]} & \cdots & A\mathbf{w}^{[N]} \\ | & | & & | \end{bmatrix}
\tag{3.90}
$$

The right-hand side is

$$
\begin{aligned}
W\Lambda &= \begin{bmatrix} | & | & & | \\ \mathbf{w}^{[1]} & \mathbf{w}^{[2]} & \cdots & \mathbf{w}^{[N]} \\ | & | & & | \end{bmatrix}
\begin{bmatrix} \lambda_1 & & & \\ & \lambda_2 & & \\ & & \ddots & \\ & & & \lambda_N \end{bmatrix} \\
&= \begin{bmatrix} | & | & & | \\ \lambda_1\mathbf{w}^{[1]} & \lambda_2\mathbf{w}^{[2]} & \cdots & \lambda_N\mathbf{w}^{[N]} \\ | & | & & | \end{bmatrix}
\end{aligned}
\tag{3.91}
$$

Because $A\mathbf{w}^{[j]} = \lambda_j \mathbf{w}^{[j]}$, we see that $AW = W\Lambda$. $\qquad$ *QED*

Definition Note that while we can write any A as $AW = W\Lambda$, we *cannot* assume that W is nonsingular. Only if the eigenvectors form a complete basis for C^N will $\det(W) \neq 0$, such that W^{-1} exists. If $\det(W) \neq 0$, we say that A is *diagonalizable*, and we can write A in *Jordan form*,

$$
A = W\Lambda W^{-1}
\tag{3.92}
$$

From the Jordan form, we see that if $A\mathbf{w} = \lambda\mathbf{w}$, then $A^{-1}\mathbf{w} = \lambda^{-1}\mathbf{w}$. Using the rule $(AB)^{-1} = B^{-1}A^{-1}$, $A^{-1} = (W\Lambda W^{-1})^{-1} = W\Lambda^{-1}W^{-1}$. This is the Jordan form of A^{-1}, and thus $A^{-1}\mathbf{w} = \lambda^{-1}\mathbf{w}$.

Theorem EVG3   *Let S be some arbitrary non singular $N \times N$ complex matrix. Let A and B be $N \times N$ complex matrices related to one another by the* similarity transformation

$$
B = S^{-1}AS
\tag{3.93}
$$

Then A and B are said to be similar, *and share the same set of eigenvalues. Their eigenvectors satisfy $A\mathbf{w} = \lambda\mathbf{w}$ and $B(S^{-1}\mathbf{w}) = \lambda(S^{-1}\mathbf{w})$.*

Proof The eigenvalues of B are roots of $\det(B - \lambda I) = 0$. Substituting $B = S^{-1}AS$,

$$\det(B - \lambda I) = \det(S^{-1}AS - \lambda I) = 0 \tag{3.94}$$

We now use the identity $I = S^{-1}S = S^{-1}IS$ to obtain

$$\det(B - \lambda I) = \det(S^{-1}AS - \lambda S^{-1}IS) = \det(S^{-1}(A - \lambda I)S) = 0 \tag{3.95}$$

Now, as $\det(AB) = \det(A) \times \det(B)$,

$$\det(B - \lambda I) = \det(S^{-1}) \times \det(A - \lambda I) \times \det(S) = 0 \tag{3.96}$$

As $\det(S) \neq 0$ and $\det(S^{-1}) \neq 0$ for a nonsingular S, λ can only satisfy $\det(B - \lambda I) = 0$ if it also satisfies $\det(A - \lambda I) = 0$. Thus, A and $B = S^{-1}AS$ have the same eigenvalues.

Let $A\mathbf{w} = \lambda\mathbf{w}$. Substituting $A = SBS^{-1}$, we have $SBS^{-1}\mathbf{w} = \lambda\mathbf{w}$. Multiplying by S^{-1} yields the following relationship between eigenvectors of A and B:

$$A\mathbf{w} = \lambda\mathbf{w} \quad B(S^{-1}\mathbf{w}) = \lambda(S^{-1}\mathbf{w}) \tag{3.97}$$

$$QED$$

Definition The *Hermitian conjugate* of A, A^{H}, is obtained by taking the transpose of the matrix and its complex conjugate,

$$A_{kj} = a_{kj} + ib_{kj} \qquad \{a_{kj}, b_{kj}\} \in \Re \qquad (A^{\mathrm{H}})_{kj} = a_{jk} - ib_{jk} \tag{3.98}$$

The Hermitian conjugate of a matrix product is $(AB)^{\mathrm{H}} = B^{\mathrm{H}}A^{\mathrm{H}}$.

Definition A matrix A is said to be *Hermitian* if $A = A^{\mathrm{H}}$. A real Hermitian matrix A is *symmetric*, $A = A^{\mathrm{T}}$.

Definition A matrix U is said to be *unitary* if $U^{\mathrm{H}} = U^{-1}$. A real unitary matrix Q is *orthogonal*, $Q^{\mathrm{T}} = Q^{-1}$.

Definition A matrix A is said to be *normal* if $AA^{\mathrm{H}} = A^{\mathrm{H}}A$. Hermitian and unitary matrices are both special cases of normal matrices.

Theorem EVG4 *We can write any complex matrix A in terms of a unitary matrix U and upper triangular matrix R as the* Schur decomposition,

$$A = URU^{\mathrm{H}} \tag{3.99}$$

As $U^{\mathrm{H}} = U^{-1}$, A and R are similar, and as R is triangular, the diagonal elements of R are eigenvalues of A. In **MATLAB**, *a Schur decomposition is computed by* **schur**.

Proof We proceed by induction, showing that if (3.99) holds for $(N-1) \times (N-1)$ matrices, then it also does for $N \times N$ matrices. For $N = 1$, we have the trivial result

$$U = 1 \qquad A = R = \lambda \tag{3.100}$$

For an $N \times N$ matrix A, let $A\mathbf{w} = \lambda\mathbf{w}$, $|\mathbf{w}| = 1$, and let $\{\mathbf{w}, \mathbf{u}^{[2]}, \ldots, \mathbf{u}^{[N]}\}$ form an orthonormal basis for C^N. Then, the matrix

$$U^{[N]} = \begin{bmatrix} | & | & & | \\ \boldsymbol{w} & \boldsymbol{u}^{[2]} & \cdots & \boldsymbol{u}^{[N]} \\ | & | & & | \end{bmatrix} \qquad \left(U^{[N]}\right)^{\mathrm{H}} = \begin{bmatrix} - & \boldsymbol{w}^{\mathrm{H}} & - \\ - & \left(\boldsymbol{u}^{[2]}\right)^{\mathrm{H}} & - \\ & \vdots & \\ - & \left(\boldsymbol{u}^{[N]}\right)^{\mathrm{H}} & - \end{bmatrix} \tag{3.101}$$

is unitary, $\left(U^{[N]}\right)^{\mathrm{H}} U^{[N]} = I$, as $\{(U^{[N]})^{\mathrm{H}} U^{[N]}\}_{mn} = \boldsymbol{u}^{[m]} \cdot \boldsymbol{u}^{[n]} = \delta_{mn}$. Then,

$$A U^{[N]} = A \begin{bmatrix} | & | & & | \\ \boldsymbol{w} & \boldsymbol{u}^{[2]} & \cdots & \boldsymbol{u}^{[N]} \\ | & | & & | \end{bmatrix} = \begin{bmatrix} | & | & & | \\ \lambda\boldsymbol{w} & A\boldsymbol{u}^{[2]} & \cdots & A\boldsymbol{u}^{[N]} \\ | & | & & | \end{bmatrix} \tag{3.102}$$

and

$$\left(U^{[N]}\right)^{\mathrm{H}} A U^{[N]} = \begin{bmatrix} - & \boldsymbol{w}^{\mathrm{H}} & - \\ - & \left(\boldsymbol{u}^{[2]}\right)^{\mathrm{H}} & - \\ & \vdots & \\ - & \left(\boldsymbol{u}^{[N]}\right)^{\mathrm{H}} & - \end{bmatrix} \begin{bmatrix} | & | & & | \\ \lambda\boldsymbol{w} & A\boldsymbol{u}^{[2]} & \cdots & A\boldsymbol{u}^{[N]} \\ | & | & & | \end{bmatrix}$$

$$= \begin{bmatrix} \lambda & - & \boldsymbol{c}^{\mathrm{H}} & - \\ | & b_{22} & \cdots & b_{2N} \\ \boldsymbol{0} & \vdots & & \vdots \\ | & b_{N2} & \cdots & b_{NN} \end{bmatrix} \tag{3.103}$$

$b_{mn} = \boldsymbol{u}^{[m]} \cdot A\boldsymbol{u}^{[n]}$. That is, in terms of an $(N-1) \times (N-1)$ matrix B,

$$\left(U^{[N]}\right)^{\mathrm{H}} A U^{[N]} = \begin{bmatrix} \lambda & \boldsymbol{c}^{\mathrm{H}} \\ \boldsymbol{0} & B \end{bmatrix} \tag{3.104}$$

If the theorem holds for B, we can write $V^{\mathrm{H}} B V = T$, for $V^{\mathrm{H}} = V^{-1}$ and T upper triangular. Then, defining the $N \times N$ unitary matrix

$$U^{[N-1]} = \begin{bmatrix} 1 & \boldsymbol{0}^{\mathrm{T}} \\ \boldsymbol{0} & V \end{bmatrix} \qquad \left(U^{[N-1]}\right)^{\mathrm{H}} = \begin{bmatrix} 1 & \boldsymbol{0}^{\mathrm{T}} \\ \boldsymbol{0} & V^{\mathrm{H}} \end{bmatrix} \tag{3.105}$$

we have

$$\left(U^{[N-1]}\right)^{\mathrm{H}} \left\{\left(U^{[N]}\right)^{\mathrm{H}} A U^{[N]}\right\} U^{[N-1]} = \begin{bmatrix} 1 & \boldsymbol{0}^{\mathrm{T}} \\ \boldsymbol{0} & V^{\mathrm{H}} \end{bmatrix} \begin{bmatrix} \lambda & \boldsymbol{c}^{\mathrm{H}} \\ \boldsymbol{0} & B \end{bmatrix} \begin{bmatrix} 1 & \boldsymbol{0}^{\mathrm{T}} \\ \boldsymbol{0} & V \end{bmatrix}$$

$$= \begin{bmatrix} \lambda & \boldsymbol{c}^{\mathrm{H}} \\ \boldsymbol{0} & (V^{\mathrm{H}} B V) \end{bmatrix} \tag{3.106}$$

As $U = U^{[N]} U^{[N-1]}$ is unitary and $V^{\mathrm{H}} B V = T$, then

$$U^{\mathrm{H}} A U = R \qquad R = \begin{bmatrix} \lambda & \boldsymbol{c}^{\mathrm{H}} \\ \boldsymbol{0} & T \end{bmatrix} \tag{3.107}$$

Thus, if the theorem holds for $(N-1) \times (N-1)$ matrices, it holds for $N \times N$ matrices as well.

$$\textit{QED}$$

Special eigenvector properties of normal matrices

Normal matrices, $AA^H = A^H A$, have additional eigenvector properties.

Theorem EVN1 (spectral decomposition) *If A is a normal matrix, it is possible to find a complete orthonormal set of eigenvectors even if the matrix has eigenvalues of multiplicity greater than 1; i.e. $\det(A - \lambda I) = 0$ has repeated roots. The matrix W whose columns are these eigenvectors is unitary, and we can write A as*

$$A = W\Lambda W^H \qquad \Lambda = \operatorname{diag}(\lambda_1, \lambda_2, \ldots, \lambda_N) \tag{3.108}$$

Proof We first write A as a Schur decomposition,

$$A = URU^H \tag{3.109}$$

Taking the Hermitian conjugate,

$$A^H = (URU^H)^H = UR^H U^H \tag{3.110}$$

we then form the two matrix products

$$\begin{aligned} AA^H &= URU^H(UR^H U^H) = URR^H U^H \\ A^H A &= UR^H U^H(URU^H) = UR^H RU^H \end{aligned} \tag{3.111}$$

For A to be normal, $AA^H = A^H A$, R must be normal as well, $RR^H = R^H R$. For

$$R = \begin{bmatrix} R_{11} & R_{12} & R_{13} & \cdots & R_{1N} \\ & R_{22} & R_{23} & \cdots & R_{2N} \\ & & R_{33} & \cdots & R_{3N} \\ & & & \ddots & \vdots \\ & & & & R_{NN} \end{bmatrix} \qquad R^H = \begin{bmatrix} \bar{R}_{11} & & & & \\ \bar{R}_{12} & \bar{R}_{22} & & & \\ \bar{R}_{13} & \bar{R}_{23} & \bar{R}_{33} & & \\ \vdots & \vdots & \vdots & \ddots & \\ \bar{R}_{1N} & \bar{R}_{2N} & \bar{R}_{3N} & \cdots & \bar{R}_{NN} \end{bmatrix} \tag{3.112}$$

$RR^H = R^H R$ only if R is diagonal. As R is similar to A,

$$R = \Lambda = \begin{bmatrix} \lambda_1 & & & \\ & \lambda_2 & & \\ & & \cdots & \\ & & & \lambda_N \end{bmatrix} \tag{3.113}$$

The Schur decomposition for a normal matrix is therefore

$$A = U\Lambda U^H \tag{3.114}$$

Postmultiplication by U yields

$$AU = U\Lambda \tag{3.115}$$

The general form of the eigenvector decomposition (3.89) is $AW = W\Lambda$, where W is a matrix whose column vectors are eigenvectors of A. Therefore, for any *normal* matrix A, we can form a unitary matrix whose column vectors are eigenvectors to write A in *Jordan normal form*,

$$A = W\Lambda W^H \tag{3.116}$$

For a matrix to be unitary, its column vectors must be orthogonal, as

$$
W^{\mathrm{H}} W = \begin{bmatrix} - & \left(w^{[1]}\right)^{\mathrm{H}} & - \\ & \vdots & \\ - & \left(w^{[N]}\right)^{\mathrm{H}} & - \end{bmatrix} \begin{bmatrix} | & & | \\ w^{[1]} & \cdots & w^{[N]} \\ | & & | \end{bmatrix}
$$

$$
= \begin{bmatrix} \left(w^{[1]} \cdot w^{[1]}\right) & \cdots & \left(w^{[1]} \cdot w^{[N]}\right) \\ \vdots & & \vdots \\ \left(w^{[N]} \cdot w^{[1]}\right) & \cdots & \left(w^{[N]} \cdot w^{[N]}\right) \end{bmatrix} = I \tag{3.117}
$$

Therefore, it is always possible, for any normal matrix A, to find a complete, orthonormal basis for C^N whose members are eigenvectors of A. One can write any vector $v \in C^N$ as the *spectral decomposition*

$$
v = c_1 w^{[1]} + c_2 w^{[2]} + \cdots + c_N w^{[N]} \qquad w^{[j]} \in C^N \tag{3.118}
$$

$$
A w^{[j]} = \lambda_j w^{[j]} \quad w^{[j]} \cdot w^{[k]} = \delta_{jk} \qquad c_j = w^{[j]} \cdot v \qquad QED
$$

Corollary ENV1-1 *Let us write a normal matrix A in Jordan normal form as $A = W \Lambda W^{\mathrm{H}}$. Taking the Hermitian conjugate, $A^{\mathrm{H}} = W \Lambda^{\mathrm{H}} W^{\mathrm{H}}$. Therefore, the normal matrices A and A^{H} share the same set of eigenvectors, and the eigenvalues of A^{H} are the complex conjugates of those of A,*

$$
A w^{[j]} = \lambda_j w^{[j]} \qquad \Rightarrow \qquad A^{\mathrm{H}} w^{[j]} = \overline{\lambda}_j w^{[j]} \tag{3.119}
$$

Theorem EVN2 *If A is Hermitian, $A = A^{\mathrm{H}}$, all of its eigenvalues are real.*

Proof If $A = A^{\mathrm{H}}$, then by writing the matrices in normal form,

$$
A = W \Lambda W^{\mathrm{H}} \qquad A^{\mathrm{H}} = (W \Lambda W^{\mathrm{H}})^{\mathrm{H}} = W \Lambda^{\mathrm{H}} W^{\mathrm{H}} \tag{3.120}
$$

we see the diagonal matrix of eigenvalues must also be Hermitian,

$$
\Lambda = \Lambda^{\mathrm{H}} \tag{3.121}
$$

Thus for each $j = 1, 2, \ldots, N$, $\lambda_j = \overline{\lambda}_j$, and every eigenvalue must be real.   *QED*

Corollary ENV2-1 (spectral decomposition of $\Re^N$) *Let A be a real, symmetric matrix, $A^{\mathrm{T}} = A$, and thus Hermitian. From $A w^{[j]} = \lambda_j w^{[j]}$, as both A and λ_j are real, it is always possible to find a real set of mutually orthonormal eigenvectors for A. Therefore, we may write any vector $v \in \Re^N$ as the eigenvector expansion*

$$
v = c_1 w^{[1]} + c_2 w^{[2]} + \cdots + c_N w^{[N]} \qquad w^{[j]} \in \Re^N
$$
$$
A w^{[j]} = \lambda_j w^{[j]} \quad w^{[j]} \cdot w^{[k]} = \delta_{jk} \quad c_j = w^{[j]} \cdot v \in \Re \tag{3.122}
$$

Definition A real symmetric matrix A is said to be *positive-definite* if $v \cdot A v = v^{\mathrm{T}} A v > 0$ for all $v \in \Re^N$. Let the eigenvalues and orthonormal eigenvectors of A satisfy $A w^{[j]} = \lambda_j w^{[j]}$, $w^{[j]} \cdot w^{[k]} = \delta_{jk}$, $w^{[j]} \in \Re^N$, $\lambda_j \in \Re$. Thus, we can write any $v \in \Re^N$ as the linear combination

$$
v = \sum_{j=1}^{N} \left(w^{[j]} \cdot v\right) w^{[j]} = \sum_{j=1}^{N} c_j w^{[j]} \tag{3.123}
$$

so that

$$v \cdot Av = v \cdot A \left[\sum_{j=1}^{N} c_j w^{[j]} \right] = v \cdot \sum_{j=1}^{N} c_j A w^{[j]} = \left[\sum_{k=1}^{N} c_k w^{[k]} \right] \cdot \sum_{j=1}^{N} c_j \lambda_j w^{[j]}$$

$$= \sum_{k=1}^{N} \sum_{j=1}^{N} c_j c_k \lambda_j \left[w^{[k]} \cdot w^{[j]} \right] = \sum_{j=1}^{N} c_j^2 \lambda_j \tag{3.124}$$

Thus, a real symmetric matrix A is positive-definite if *all* of its eigenvalues are positive. If we have only that $v^{\mathrm{T}} A v \geq 0$, A is said to be *positive-semidefinite*. If $v^{\mathrm{T}} A v < 0$ or $v^{\mathrm{T}} A v \leq 0$, A is *negative-definite* or *negative-semidefinite* respectively. If no such condition holds for all $v \in \Re^N$, A is *indefinite*.

Theorem EVN3 *If U is unitary, all of its eigenvalues have moduli of 1.*

Proof We write U and U^{H} in normal form,

$$U = W \Lambda W^{\mathrm{H}} \qquad U^{\mathrm{H}} = W \Lambda^{\mathrm{H}} W^{\mathrm{H}} \tag{3.125}$$

For U to be unitary, we require

$$U U^{\mathrm{H}} = W \Lambda W^{\mathrm{H}} \left(W \Lambda^{\mathrm{H}} W^{\mathrm{H}} \right) = W \Lambda \Lambda^{\mathrm{H}} W^{\mathrm{H}} = I \tag{3.126}$$

Thus Λ must also be unitary,

$$\Lambda \Lambda^{\mathrm{H}} = W^{\mathrm{H}} I W = W^{\mathrm{H}} W = I \tag{3.127}$$

For each diagonal element of $\Lambda \Lambda^{\mathrm{H}}$, $\lambda_j \bar{\lambda}_j$, to be 1, we must have $|\lambda_j| = 1$. *QED*

Numerical calculation of eigenvalues and eigenvectors in MATLAB

We now consider the numerical calculation of eigenvalues and eigenvectors. In this section, we merely demonstrate the use of the MATLAB routines for computing eigenvalues and eigenvectors. **eig** computes all eigenvalues and eigenvectors of a matrix, and **eigs** computes only certain eigenvalues and eigenvectors of interest, e.g. those eigenvalues with the largest moduli. In the following two sections, the algorithms used by these routines are discussed in further detail.

Computing all eigenvalues and eigenvectors with eig

eig uses the iterative QR method (described below) to compute all eigenvalues and eigenvectors of a matrix. Consider the matrix

$$A = \begin{bmatrix} 1 & 2 & -1 \\ 3 & 0 & -2 \\ -1 & 1 & 4 \end{bmatrix} \tag{3.128}$$

With a single output argument, **eig** returns a vector of eigenvalues,

```
A = [1 2 -1; 3 0 -2; -1 1 4];
e = eig(A),
```

```
e =
  -1.8284
   3.8284
   3.0000
```

With two output arguments, **eig** returns first a matrix W whose column vectors are the eigenvectors of A, and a diagonal matrix D, such that $AW = WD$,

```
[W,D] = eig(A),
W =
      0.6152   -0.6198   0.7071
     -0.7527   -0.6854   0.7071
      0.2347    0.3823   0.0000
D =
     -1.8284   0        0
      0        3.8284   0
      0        0        3.0000
```

eig cannot be used with sparse-format matrices, e.g. those set by **spalloc**.

Computing extremal eigenvalues and their eigenvectors with *eigs*

Often, we need not compute all eigenvalues, but rather only certain ones, e.g. the largest or smallest in magnitude. In MATLAB this is done by **eigs** using the iterative methods discussed below. We demonstrate the routine for a positive-definite matrix A, such as is obtained by discretizing a diffusion equation in one dimension. As **eigs** is compatible with sparse-format matrices we use this option,

```
N = 25;
v = ones(N,1);
A = spdiags([-v 2*v -v], -1:1, N, N);
```

With a single output, eigs(A,k) returns the k eigenvalues of A with the largest magnitude. eigs(A) performs this calculation for $k = 6$.

```
e = eigs(A,5);
e =
    3.9854
    3.9419
    3.8700
    3.7709
    3.6460
```

We can change the types of eigenvalues computed through a third argument, SIGMA, taking the values 'LM' or 'SM' to compute the eigenvalues of largest or smallest magnitude, 'BE' to compute eigenvalues from "both ends" of the magnitude spectrum, 'LR', 'SR', 'LI', 'SI' to compute the eigenvalues of largest/smallest real or imaginary parts, and a scalar value to compute eigenvalues closest to SIGMA. For example,

```
e = eigs(A,3,'SM'),
```

```
e =
    0.1300
    0.0581
    0.0146
```

and

```
e = eigs(A,4,1.0),
```

```
e =
    1.2908
    1.0706
    0.8639
    0.6738
```

In addition to this output, **eigs** writes information about its internal calculations to the screen. To turn this off, use

```
OPTS.disp = 0;
e = eigs(A,5,'LM',OPTS);
```

Other fields in OPTS allow us to modify the behavior of the algorithm, e.g. by decreasing the tolerances. Type help eigs for more information.

With two arguments, **eigs** returns a matrix W of eigenvectors and a diagonal matrix D of eigenvalues such that $AW = WD$. The following code computes and plots the eigenvectors of largest and smallest magnitudes,

```
[W,D] = eigs(A,2,'BE',OPTS);
figure; plot(W(:,1),'–');
hold on; plot(W(:,2));
xlabel('component'); ylabel('w(k)');
title('Eigenvectors of 1-D diffusion matrix');
phrase1 = ['\ lambda = ', num2str(D(1,1))];
phrase2 = ['\ lambda = ', num2str(D(2,2))];
legend(phrase1, phrase2, 'Location', 'Best');
```

The graph generated by this code is shown in Figure 3.5.

Above, we have called **eigs** for a sparse-format matrix A. A matrix stored in full matrix format can also be used. Additionally, we can merely supply a routine that returns for each input v, the corresponding vector Av. For the matrix above, this is done by

```
function Av = diff_matrix_1D_mult(v);
```

```
N = length(v);
Av = zeros(N,1);
Av(1) = 2*v(1) - v(2);
for k = 2:(N-1)
    Av(k) = -v(k-1) +2*v(k)-v(k+1);
```

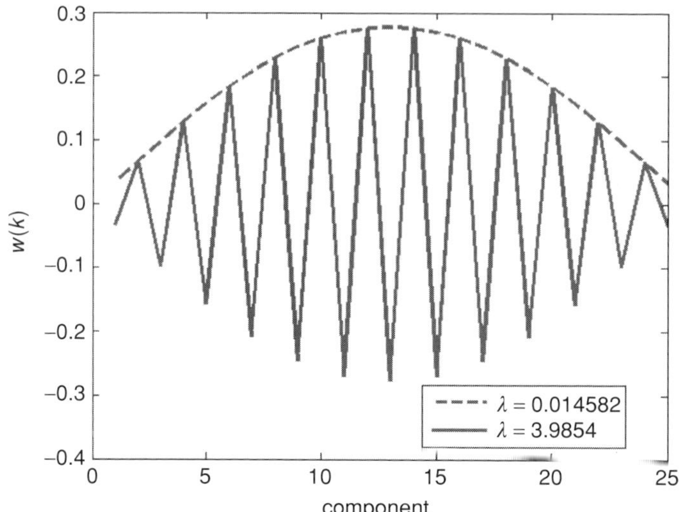

Figure 3.5 Plots of the eigenvector components for the eigenvalues of largest and smallest magnitudes for a 1-D diffusion matrix.

end
Av(N) = -v(N-1) + 2*v(N);
return;

The five eigenvalues of largest magnitude for $N = 25$ are computed by

e = eigs('diff_matrix_1D_mult',25,5,'LM',OPTS),

```
e =
    3.9854
    3.9419
    3.8700
    3.7709
    3.6460
```

Computing extremal eigenvalues

We now delve into the details of the algorithms behind **eig** and **eigs**, starting first with methods for calculating extremal eigenvalues. Let us say that we want to find the eigenvalue of largest magnitude of a matrix A, assumed diagonalizable. That is, we assume that the set of N eigenvectors of A are linearly independent, and that we can write any vector $v \in C^N$ as

$$v = c_1 w^{[1]} + c_2 w^{[2]} + \cdots + c_N w^{[N]}$$
$$A w^{[j]} = \lambda_j w^{[j]} \qquad c_j \in C \tag{3.129}$$

Let us generate at random some vector $v^{[0]}$, and write it as the linear combination above. It is highly unlikely that this random vector is an eigenvector of A, and we expect each of the coefficients $\{c_1, c_2, \ldots, c_N\}$ to be nonzero. From $v^{[0]}$, we obtain a sequence of new vectors $v^{[1]}, v^{[2]}, \ldots$ by the rule

$$v^{[k+1]} = \frac{Av^{[k]}}{\left|Av^{[k]}\right|} \tag{3.130}$$

Writing these vectors as linear combinations of eigenvectors, we have

$$Av^{[0]} = A\left[c_1 w^{[1]} + \cdots + c_N w^{[N]}\right] = c_1 \lambda_1 w^{[1]} + \cdots + c_N \lambda_N w^{[N]} \tag{3.131}$$

so that

$$v^{[1]} = \frac{Av^{[0]}}{\left|Av^{[0]}\right|} = \frac{c_1 \lambda_1 w^{[1]} + c_2 \lambda_2 w^{[2]} + \cdots + c_N \lambda_N w^{[N]}}{\left|Av^{[0]}\right|} \tag{3.132}$$

Next,

$$Av^{[1]} = \frac{A\left[c_1 \lambda_1 w^{[1]} + \cdots + c_N \lambda_N w^{[N]}\right]}{\left|Av^{[0]}\right|} = \frac{c_1 \lambda_1^2 w^{[1]} + \cdots + c_N \lambda_N^2 w^{[N]}}{\left|Av^{[0]}\right|} \tag{3.133}$$

so that

$$v^{[2]} = \frac{c_1 \lambda_1^2 w^{[1]} + c_2 \lambda_2^2 w^{[2]} + \cdots + c_N \lambda_N^2 w^{[N]}}{\left|Av^{[1]}\right| \times \left|Av^{[0]}\right|} \tag{3.134}$$

or in general,

$$v^{[k]} = \frac{c_1 \lambda_1^k w^{[1]} + c_2 \lambda_2^k w^{[2]} + \cdots + c_N \lambda_N^k w^{[N]}}{\left|c_1 \lambda_1^k w^{[1]} + c_2 \lambda_2^k w^{[2]} + \cdots + c_N \lambda_N^k w^{[N]}\right|} \tag{3.135}$$

Let us assume that the eigenvalues are ordered by decreasing modulus, and that λ_1 is both distinct and has a larger modulus than λ_2:

$$|\lambda_1| > |\lambda_2| \geq |\lambda_3| \geq \cdots \geq |\lambda_{N-1}| \geq |\lambda_N| \tag{3.136}$$

Note that this is a stricter statement than saying that λ_1 is distinct, as if $\lambda_2 = \overline{\lambda}_1$, λ_1 is distinct, but $|\lambda_2| = |\lambda_1|$. If (3.136) holds, as $k \rightarrow \infty$,

$$\left|\lambda_1^k\right| \gg \left|\lambda_2^k\right| \geq \left|\lambda_3^k\right| \geq \cdots \geq \left|\lambda_{N-1}^k\right| \geq \left|\lambda_N^k\right| \tag{3.137}$$

and for any finite $\{c_1, c_2, \ldots, c_N\}$, we eventually have

$$\left|c_1 \lambda_1^k\right| \gg \left|c_2 \lambda_2^k\right| \geq \left|c_3 \lambda_3^k\right| \geq \cdots \geq \left|c_{N-1} \lambda_{N-1}^k\right| \geq \left|c_N \lambda_N^k\right| \tag{3.138}$$

Therefore, as $k \rightarrow \infty$,

$$v^{[k]} \approx \frac{c_1 \lambda_1^k w^{[1]}}{\left|c_1 \lambda_1^k w^{[1]}\right|} \tag{3.139}$$

In this limit, as $Av^{[k]} \approx \lambda_1 v^{[k]}$,

$$\lambda_k \approx v^{[k]} \cdot Av^{[k]} \qquad w^{[1]} \approx v^{[k+1]} \tag{3.140}$$

We have assumed above that $c_1 \neq 0$; however, this is not really necessary. In the presence of round-off error, which mixes in some $w^{[1]}$ component, this algorithm will find λ_1 and $w^{[1]}$ even if initially c_1 were zero.

Convergence of power method with a degenerate leading eigenvalue

We have assumed that $|\lambda_1| > |\lambda_2|$. *What happens if* $|\lambda_1| = |\lambda_2|$? In general, the power method oscillates and the sequence $v^{[1]}$, $v^{[2]}$, ... fails to converge. However, for the special case of a Hermitian, positive-semidefinite matrix, all eigenvalues must be real and non-negative. The only way that $|\lambda_1| = |\lambda_2|$ is if $\lambda_1 = \lambda_2$. In the limit $k \to \infty$,

$$v^{[k]} \approx \frac{c_1\lambda_1^k w^{[1]} + c_2\lambda_2^k w^{[2]}}{\left|c_1\lambda_1^k w^{[1]} + c_2\lambda_2^k w^{[2]}\right|} = \frac{\lambda_1^k\left[c_1 w^{[1]} + c_2 w^{[2]}\right]}{\left|\lambda_1^k\left[c_1 w^{[1]} + c_2 w^{[2]}\right]\right|} \tag{3.141}$$

Therefore, the method converges to some linear combination of $w^{[1]}$ and $w^{[2]}$. But, if both $w^{[1]}$ and $w^{[2]}$ are eigenvectors for $\lambda_1 = \lambda_2$, $c_1 w^{[1]} + c_2 w^{[2]}$ is one as well and the power method converges to an eigenvector for $\lambda_1 = \lambda_2$, but the exact eigenvector found depends upon the initial guess.

Finding the next largest eigenvalues of a positive-semidefinite matrix

Let us say that by the power method we have found the largest magnitude eigenvalue λ_1 and its associated eigenvector $w^{[1]}$ for a positive-semidefinite matrix A. We now want to compute the next largest eigenvalue λ_2. To do so, we generate at random some new vector, that we can express as a linear combination of eigenvectors,

$$\begin{aligned}
v^{[0]} &= c_1 w^{[1]} + c_2 w^{[2]} + \cdots + c_N w^{[N]} \\
A w^{[j]} &= \lambda_j w^{[j]} \qquad c_j \in C
\end{aligned} \tag{3.142}$$

As the eigenvectors of a Hermitian matrix are orthogonal, $w^{[2]}$ is orthogonal to $w^{[1]}$. We then can project out the $w^{[1]}$ component by the operation,

$$\begin{aligned}
\tilde{v}^{[0]} &= \left[I - w^{[1]}\left(w^{[1]}\right)^{\mathrm{T}}\right]v^{[0]} = v^{[0]} - w^{[1]}\left(w^{[1]} \cdot v^{[0]}\right) \\
&= \left[c_1 w^{[1]} + c_2 w^{[2]} + \cdots + c_N w^{[N]}\right] - w^{[1]}(c_1) \\
&= c_2 w^{[2]} + \cdots + c_N w^{[N]}
\end{aligned} \tag{3.143}$$

We now take as our iteration rule,

$$\tilde{v}^{[k+1]} = \frac{\left[I - w^{[1]}\left(w^{[1]}\right)^{\mathrm{T}}\right]\left(A\tilde{v}^{[k]}\right)}{\left|\left[I - w^{[1]}\left(w^{[1]}\right)^{\mathrm{T}}\right]\left(A\tilde{v}^{[k]}\right)\right|} \tag{3.144}$$

In terms of the original expansion of $v^{[0]}$, the sequence is

$$\tilde{v}^{[k]} = \frac{c_2\lambda_2^k w^{[2]} + \cdots + c_N\lambda_N^k w^{[N]}}{\left|c_2\lambda_2^k w^{[2]} + \cdots + c_N\lambda_N^k w^{[N]}\right|} \tag{3.145}$$

If $|\lambda_2| > |\lambda_3| \geq |\lambda_4| \geq \cdots \geq |\lambda_N|$, this sequence converges to $w^{[2]}$. By continuing this process, we compute the eigenvalues and eigenvectors in order of decreasing magnitude.

Inverse inflation and shift operations to find other eigenvalues

Rather than finding the eigenvalue of largest modulus, let us find the eigenvalue λ_j of A closest to some target shift value μ. As the eigenvalues of $(A - \mu I)$ are $\lambda_j - \mu$, we only need to derive a method that finds the smallest magnitude eigenvalue of $(A - \mu I)$. We do so by applying the iterative rule

$$(A - \mu I)z^{[k]} = v^{[k]} \qquad v^{[k+1]} = \frac{z^{[k]}}{\left| z^{[k]} \right|} \tag{3.146}$$

This method can be written as

$$v^{[k+1]} = \frac{(A - \mu I)^{-1} v^{[k]}}{\left| (A - \mu I)^{-1} v^{[k]} \right|} \tag{3.147}$$

Thus, it returns the largest modulus eigenvalue of $(A - \mu I)^{-1}$. As the eigenvalues of $(A - \mu I)^{-1}$ are $(\lambda_j - \mu)^{-1}$, this method finds the eigenvalue of smallest $|\lambda_j - \mu|$. At each iteration, we must solve a linear system $(A - \mu I)z^{[k]} = v^{[k]}$, which is done efficiently by LU decomposition, so that later iterations only require solving two triangular systems by substitution.

The QR method for computing all eigenvalues

We next consider a method to compute all eigenvalues of a matrix concurrently by transforming the matrix into a similar one whose eigenvalues are easy to calculate. The transformation is done through iterative use of QR decompositions, described below.

QR decomposition of a real matrix

Just as a matrix may be factored into the product of lower and upper triangular matrices, it may also be factored into the product of an orthogonal matrix Q, $Q^{\mathrm{T}} = Q^{-1}$, and an upper triangular matrix R,

$$A = QR \tag{3.148}$$

We describe here an algorithm based on *Householder transformations (reflections)*. For any $w \in \Re^N$ with $|w| = 1$, we can generate the matrix

$$P = I - 2ww^{\mathrm{T}} = \begin{bmatrix} (1 - 2w_1w_1) & (-2w_1w_2) & \dots & (-2w_1w_N) \\ (-2w_2w_1) & (1 - 2w_2w_2) & \dots & (-2w_2w_N) \\ \vdots & \vdots & & \vdots \\ (-2w_Nw_1) & (-2w_Nw_2) & \dots & (1 - 2w_Nw_N) \end{bmatrix} \tag{3.149}$$

that negates the component of any $w \in \Re^N$ in the direction of w (Figure 3.6),

$$Px = (I - 2ww^{\mathrm{T}})x = x - 2(w \cdot x)w \tag{3.150}$$

The matrix (3.149) is symmetric and orthogonal,

$$P^{\mathrm{T}} = (I - 2ww^{\mathrm{T}})^{\mathrm{T}} = I - 2ww^{\mathrm{T}} = P \qquad P^{\mathrm{T}}P = PP = I \tag{3.151}$$

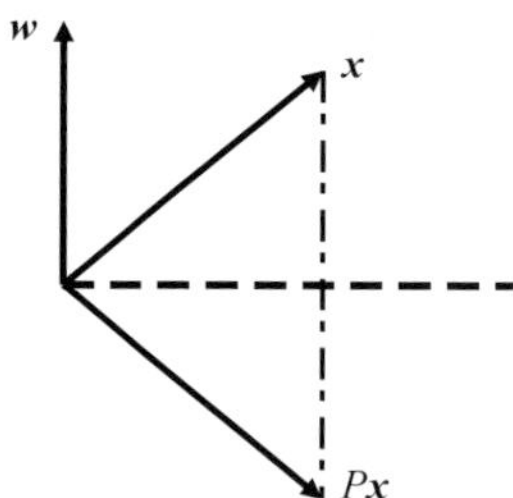

Figure 3.6 Effect of $P = I - 2\boldsymbol{w}\,\boldsymbol{w}^{\mathrm{T}}$ on an arbitrary vector $\boldsymbol{x}$.

We next demonstrate how a sequence of such Householder transformations performs the *QR factorization*,

$$A = QR = Q \begin{bmatrix} R_{11} & R_{12} & \cdots & R_{1N} \\ & R_{22} & \cdots & R_{2N} \\ & & \ddots & \vdots \\ & & & R_{NN} \end{bmatrix} \qquad QQ^{\mathrm{T}} = I \tag{3.152}$$

Let us examine the first column of R that has a nonzero element only at the first component. For any vector $\boldsymbol{x} \in \Re^{N}$, we can find a vector $\boldsymbol{v}$, generating a Householder reflection P, that zeros all but the first component of $\boldsymbol{x}$:

$$P\boldsymbol{x} = \left[I - 2\frac{\boldsymbol{v}\boldsymbol{v}^{\mathrm{T}}}{|\boldsymbol{v}|^2} \right]\boldsymbol{x} = \boldsymbol{x} - \frac{2(\boldsymbol{v}\cdot\boldsymbol{x})}{|\boldsymbol{v}|^2}\boldsymbol{v} = \begin{bmatrix} b \\ 0 \\ \vdots \\ 0 \end{bmatrix} = b\boldsymbol{e}^{[1]} \tag{3.153}$$

As $\boldsymbol{x} - 2(\boldsymbol{v}\cdot\boldsymbol{x})|\boldsymbol{v}|^{-2}\boldsymbol{v} = b\boldsymbol{e}^{[1]}$, $\boldsymbol{v}$ must be a linear combination of $\boldsymbol{x}$ and $\boldsymbol{e}^{[1]}$,

$$\boldsymbol{v} = \boldsymbol{x} + \alpha\boldsymbol{e}^{[1]} \tag{3.154}$$

As

$$\begin{aligned} |\boldsymbol{v}|^2 &= \left(\boldsymbol{x} + \alpha\boldsymbol{e}^{[1]}\right) \cdot \left(\boldsymbol{x} + \alpha\boldsymbol{e}^{[1]}\right) = |\boldsymbol{x}|^2 + 2\alpha x_1 + \alpha^2 \\ \boldsymbol{v}\cdot\boldsymbol{x} &= \left(\boldsymbol{x} + \alpha\boldsymbol{e}^{[1]}\right) \cdot \boldsymbol{x} = |\boldsymbol{x}|^2 + \alpha x_1 \end{aligned} \tag{3.155}$$

the Householder transformation acts on $\boldsymbol{x}$ as

$$\begin{aligned} P\underline{x} &= \boldsymbol{x} - \frac{2(|\boldsymbol{x}|^2 + \alpha x_1)}{|\boldsymbol{x}|^2 + 2\alpha x_1 + \alpha^2}\left(\boldsymbol{x} + \alpha\boldsymbol{e}^{[1]}\right) \\ &= \left[1 - \frac{2(|\boldsymbol{x}|^2 + \alpha x_1)}{|\boldsymbol{x}|^2 + 2\alpha x_1 + \alpha^2}\right]\boldsymbol{x} - \left[\frac{2\alpha(|\boldsymbol{x}|^2 + \alpha x_1)}{|\boldsymbol{x}|^2 + 2\alpha x_1 + \alpha^2}\right]\boldsymbol{e}^{[1]} \end{aligned} \tag{3.156}$$

We obtain $P\boldsymbol{x} = b\boldsymbol{e}^{[1]}$ by satisfying

$$1 = \frac{2(|\boldsymbol{x}|^2 + \alpha x_1)}{|\boldsymbol{x}|^2 + 2\alpha x_1 + \alpha^2} \Rightarrow \begin{array}{c} \alpha = \varepsilon|\boldsymbol{x}| \\ \varepsilon = \mathrm{sign}(x_1) = \begin{cases} -1, & x_1 < 0 \\ 1, & x_1 \geq 0 \end{cases} \end{array} \tag{3.157}$$

so that

$$P\boldsymbol{x} = -\varepsilon|\boldsymbol{x}|\boldsymbol{e}^{[1]} \tag{3.158}$$

Thus, we have an orthogonal matrix $P = Q^{[1]}$ such that $Q^{[1]}A$,

$$Q^{[1]}\begin{bmatrix} | & & | \\ a^{(1)} & \cdots & a^{[N]} \\ | & & | \end{bmatrix} = \begin{bmatrix} | & & | \\ (Q^{[1]}a^{[1]}) & \cdots & (Q^{[1]}a^{[N]}) \\ | & & | \end{bmatrix}$$

$$= \begin{bmatrix} R_{11} & R_{12} & \cdots & R_{1N} \\ 0 & | & & | \\ \vdots & \underline{a}^{(2,2)} & \cdots & \underline{a}^{(2,N)} \\ 0 & | & & | \end{bmatrix} \tag{3.159}$$

has all zero elements in the first column except at the $(1,1)$ position. We next generate the Householder reflection $P^{[2]}$ (in $\Re^{N-1}$) that transforms $a^{(2,2)} \in \Re^{N-1}$ into a vector with all zeros except for the first component. Then, we construct the $N \times N$ orthogonal matrix

$$Q^{[2]} = \begin{bmatrix} 1 & 0 \\ 0 & P^{[2]} \end{bmatrix} \tag{3.160}$$

such that

$$Q^{[2]}(Q^{[1]}A) = \begin{bmatrix} 1 & 0 \\ 0 & P^{[2]} \end{bmatrix} \begin{bmatrix} R_{11} & R_{12} & \cdots & R_{1N} \\ 0 & | & & | \\ \vdots & \underline{a}^{(2,2)} & \cdots & \underline{a}^{(2,N)} \\ 0 & | & & | \end{bmatrix}$$

$$= \begin{bmatrix} R_{11} & R_{12} & R_{13} & \cdots & R_{1N} \\ 0 & R_{22} & R_{23} & \cdots & R_{2N} \\ 0 & 0 & | & & | \\ \vdots & \vdots & \underline{a}^{(3,3)} & \cdots & \underline{a}^{(3,N)} \\ 0 & 0 & | & & | \end{bmatrix} \tag{3.161}$$

This process is repeated until we have

$$[Q^{[N-1]}Q^{[N-2]}\cdots Q^{[2]}Q^{[1]}]A = \begin{bmatrix} R_{11} & R_{12} & R_{13} & \cdots & R_{1N} \\ & R_{22} & R_{23} & \cdots & R_{2N} \\ & & R_{33} & \cdots & R_{3N} \\ & & & \ddots & \vdots \\ & & & & R_{NN} \end{bmatrix} \tag{3.162}$$

This yields the QR factorization after $\sim N^3$ FLOPs,

$$A = QR \qquad Q = (Q^{[1]})^{\mathrm{T}}(Q^{[2]})^{\mathrm{T}}\cdots(Q^{[N-2]})^{\mathrm{T}}(Q^{[N-1]})^{\mathrm{T}} \tag{3.163}$$

which is implemented in **MATLAB** by the function **qr**.

Iterative QR method for computing all eigenvalues

Note that the upper triangular matrix R that we obtain from QR decomposition is *not* similar to A and thus does *not* have the same eigenvalues. However, we can define a matrix $A^{[1]}$ that *is* similar to A,

$$A = QR \qquad A^{[1]} = Q^{-1}AQ = Q^{\mathrm{T}}AQ \tag{3.164}$$

Applying this similarity transform iteratively, we define the rule

$$A^{[k]} = Q^{[k]} R^{[k]} \qquad A^{[k+1]} = \left(Q^{[k]}\right)^{\mathrm{T}} A^{[k]} Q^{[k]} = R^{[k]} Q^{[k]} \tag{3.165}$$

As we demonstrate below for an example matrix, as $k \to \infty$, this sequence of matrices becomes of block upper triangular form,

$$A^{[k \to \infty]} = \begin{bmatrix} R_{11} & R_{12} & R_{13} & \cdots & R_{1P} \\ & R_{22} & R_{23} & \cdots & R_{2P} \\ & & R_{33} & \cdots & R_{3P} \\ & & & \ddots & \vdots \\ & & & & R_{PP} \end{bmatrix} \tag{3.166}$$

where the submatrices along the diagonal, $R_{11}, R_{22}, \ldots, R_{PP}$, are either 1×1 or 2×2. For the former case, this diagonal element is an eigenvalue of A. For the latter case, the 2×2 diagonal submatrix R_{jj} has two eigenvalues that are complex conjugates of each other (for real A) and that are eigenvalues of A. This *iterative QR method* concurrently yields all eigenvalues of A, and the corresponding eigenvectors can then be computed from (3.38).

Improving the efficiency of the QR method

In practice, we use a more complex algorithm than that above to reduce the computational workload. For example, it is standard to use Householder reflections first to convert A to *upper Hessenberg form*

$$A^{[0]} = \begin{bmatrix} a_{11}^{[0]} & a_{12}^{[0]} & a_{13}^{[0]} & \cdots & a_{1N}^{[0]} \\ a_{21}^{[0]} & a_{22}^{[0]} & a_{23}^{[0]} & \cdots & a_{2N}^{[0]} \\ & a_{32}^{[0]} & a_{33}^{[0]} & \cdots & a_{3N}^{[0]} \\ & & \ddots & \ddots & \vdots \\ & & & a_{N,N-1}^{[0]} & a_{NN}^{[0]} \end{bmatrix} \tag{3.167}$$

Then, the work necessary to perform each subsequent QR factorization scales only as N^2 rather than N^3. There is only one nonzero element below the diagonal that can be zeroed efficiently using Givens rotations, a transformation similar to that of Householder, but that is designed to zero only a single component. For brevity, we do not consider these modifications in detail, but refer the interested reader to Quateroni *et al.* (2000).

The convergence rate of the QR method can be improved using *single-shift QR iterations*, in which the rule is now

$$Q^{[k]} R^{[k]} = A^{[k]} - \mu I$$
$$A^{[k+1]} = R^{[k]} Q^{[k]} + \mu I \tag{3.168}$$

A common single-shift value is $\mu = a_{NN}^{[k]}$, but variable shifts are also used. Note that the QR iterations above do *not* converge to the Schur decomposition, because $A^{[k \to \infty]}$ is only block upper triangular, and still have may have nonzero elements in the diagonal immediately below the principal one. These lower nonzero elements can be removed, and convergence accelerated for the case of complex eigenvalues, by using *double-shift QR iterations*. If at iteration k, there appears to be a 2×2 diagonal submatrix R_{jj} with approximate eigenvalues $\lambda_j^{[k]}$ and $\bar{\lambda}_j^{[k]}$, we perform the two-step iteration

$$\begin{aligned}
Q^{[k]} R^{[k]} &= A^{[k]} - \lambda_j^{[k]} I \\
A^{[k+1]} &= A^{[k]} Q^{[k]} + \lambda_j^{[k]} I \\
Q^{[k+1]} R^{[k+1]} &= A^{[k+1]} - \bar{\lambda}_j^{[k]} I \\
A^{[k+2]} &= R^{[k+1]} Q^{[k+1]} + \bar{\lambda}_j^{[k]} I
\end{aligned} \tag{3.169}$$

and then resume single-shift iterations. For discussion of more complex, and efficient, shifting strategies and algorithms for special classes of matrices (e.g. Hermitian), consult Quateroni *et al.* (2000) and Stoer & Bulirsch (1993).

Example. QR method for a real 4×4 matrix

We now demonstrate the QR method for the real, anti-symmetric matrix

$$A = \begin{bmatrix}
4 & 0 & -1 & 1 \\
0 & 4 & 2 & -1 \\
1 & -2 & 4 & 1 \\
-1 & 1 & -1 & 4
\end{bmatrix} \tag{3.170}$$

that has the complex eigenvalues

$$\lambda_{1,2} = 4 \pm 2.8059i \qquad \lambda_{3,4} = 4 \pm 0.3564i \tag{3.171}$$

and is generated (along with a working copy) by

A = [4 0 -1 1; 0 4 2 -1; 1 -2 4 1; -1 1 -1 4]; N = size(A,1);
    A_w = A;

A single-shift QR iteration is performed by

mu = A_w(N,N); [Q,R] = qr(A_w - mu*eye(N));
A_w = R*Q +mu*eye(N);

We have the following first few values of $A^{[k]}$,

$$\begin{bmatrix}
4.0 & -1.0 & -2.5 & -0.3 \\
1.0 & 4.0 & 0.6 & -0.3 \\
2.5 & -0.6 & 4.0 & -0.0 \\
0.3 & 0.3 & 0.0 & 4.0
\end{bmatrix} \rightarrow
\begin{bmatrix}
4.0 & -2.7 & -0.8 & 0.0 \\
2.7 & 4.0 & 0.1 & 0.1 \\
0.9 & -0.1 & 4.0 & -0.3 \\
-0.0 & -0.1 & 0.3 & 4.0
\end{bmatrix} \rightarrow$$
$$A^{[1]} \qquad\qquad\qquad\qquad A^{[2]}$$

$$\begin{bmatrix}
4.0 & -2.8 & -0.1 & -0.0 \\
2.8 & 4.0 & 0.0 & -0.0 \\
0.1 & -0.0 & 4.0 & 0.4 \\
0.0 & 0.0 & -0.4 & 4.0
\end{bmatrix} \tag{3.172}$$
$$A^{[3]}$$

The single-shift QR iterations yield upon convergence

$$
A^{[k \to \infty]} = \begin{bmatrix} 4 & -2.8059 & 0 & 0 \\ 2.8059 & 4 & 0 & 0 \\ 0 & 0 & 4 & 0.3564 \\ 0 & 0 & -0.3564 & 4 \end{bmatrix} = \begin{bmatrix} R_{11} & R_{12} \\ 0 & R_{22} \end{bmatrix} \tag{3.173}
$$

The eigenvalues of A are then computed easily from the diagonal 2×2 submatrices R_{11} and R_{22}, where for each we have with $b \in \mathfrak{R}$,

$$
|R_{jj} - \lambda I| = \begin{vmatrix} (4 - \lambda) & -b \\ b & (4 - \lambda) \end{vmatrix} = (4 - \lambda)^2 + b^2 = 0 \tag{3.174}
$$

The two roots are $\lambda = 4 \pm bi$, yielding the eigenvalues of (3.171).

Normal mode analysis

We now provide an example in which eigenvalue analysis is of direct interest to a problem from chemical engineering practice. Let us say that we have some structure (it could be a molecule or some solid object) whose state is described by the F positional degrees of freedom $q \in \mathfrak{R}^F$ and the corresponding velocities $\dot{q}$. We have some model for the total potential energy of the system $U(q)$ and some model of the total kinetic energy $K(q, \dot{q})$. We wish to compute the vibrational frequencies of the structure. Such a *normal mode analysis* problem arises when we wish to compute the IR spectra of a molecule (Allen & Beers, 2005).

First, using the numerical optimization methods outlined in Chapter 5, we identify a state $\hat{q}$ that is a local minimum of the potential energy. That is, it has a lower potential energy than any neighboring states, and as it is an extremum, $\nabla U|_{\hat{q}} = 0$. We wish to describe the system's dynamics when it is perturbed slightly from this minimum energy state, and so define $\delta = q - \hat{q}$. Expanding $U(q)$ about $\hat{q}$ as a Taylor series, with $\partial U/\partial q_m|_{\hat{q}} = 0$, yields

$$
U(\hat{q}_1 + \delta_1, \dots, \hat{q}_F + \delta_F) \approx U(\hat{q}_1, \dots, \hat{q}_F) + \frac{1}{2} \sum_{m=1}^{F} \sum_{n=1}^{F} \delta_m \left(\frac{\partial^2 U}{\partial q_m \partial q_n} \Big|_{\hat{q}} \right) \delta_n \tag{3.175}
$$

Defining the *Hessian matrix* H, containing the second derivatives of $U(q)$,

$$
H_{mn} = \frac{\partial^2 U}{\partial q_m \partial q_n} \Big|_{\hat{q}} = \frac{\partial^2 U}{\partial q_n \partial q_m} \Big|_{\hat{q}} = H_{nm} \tag{3.176}
$$

the Taylor series for $U(q)$ in the vicinity of $\hat{q}$ becomes

$$
U(\hat{q}_1 + \delta_1, \dots, \hat{q}_F + \delta_F) \approx U(\hat{q}_1, \dots, \hat{q}_F) + \frac{1}{2} \sum_{m=1}^{F} \sum_{n=1}^{F} \delta_m H_{mn} \delta_n \tag{3.177}
$$

H, which from (3.176) is real symmetric, also must be positive-semidefinite, as for a local minimum $\hat{q}$, $U(\hat{q} + \delta) - U(\hat{q}) \approx \frac{1}{2} \delta^{\mathrm{T}} H \delta \geq 0$.

Let us assume for the moment that each degree of freedom has the same effective mass m_{eff}, so that the kinetic energy is

$$K(\boldsymbol{q}, \dot{\boldsymbol{q}}) = \frac{m_{\text{eff}}}{2} \sum_{k=1}^{F} \dot{q}_k^2 \tag{3.178}$$

The dynamics of $\boldsymbol{q}$ are governed by the equations of motion

$$m_{\text{eff}} \frac{d^2 q_j}{dt^2} = m_{\text{eff}} \frac{d^2 \delta_j}{dt^2} = -\frac{\partial}{\partial \delta_j} U(\hat{q}_1 + \delta_1, \ldots, \hat{q}_F + \delta_F) \tag{3.179}$$

which, in the vicinity of the local minimum, reduce to

$$m_{\text{eff}} \frac{d^2 \delta_j}{dt^2} = -\frac{\partial}{\partial \delta_j} \left[U(\hat{q}_1, \ldots, \hat{q}_F) + \frac{1}{2} \sum_{m=1}^{F} \sum_{n=1}^{F} \delta_m H_{mn} \delta_n \right]$$

$$m_{\text{eff}} \frac{d^2 \delta_j}{dt^2} = -\frac{1}{2} \sum_{n=1}^{F} H_{jn} \delta_n - \frac{1}{2} \sum_{m=1}^{F} \delta_m H_{mj} = -\sum_{n=1}^{F} H_{jn} \delta_n \tag{3.180}$$

This system of second-order ODEs is written more compactly as

$$m_{\text{eff}} \frac{d^2}{dt^2} \boldsymbol{q} = -H\boldsymbol{\delta} \tag{3.181}$$

The eigenvalues and eigenvectors of the Hessian matrix satisfy

$$H \boldsymbol{w}^{[j]} = \lambda_j \boldsymbol{w}^{[j]} \qquad j = 1, 2, \ldots, F \tag{3.182}$$

Since the Hessian is real symmetric, the eigenvectors are mutually orthogonal, and so form a convenient basis set for representing any vector. We therefore write a trial form of $\boldsymbol{\delta}(t)$ as the linear combination

$$\boldsymbol{\delta}(t) = c_1(t)\boldsymbol{w}^{[1]} + \cdots + c_F(t)\boldsymbol{w}^{[F]} \qquad \frac{d^2}{dt^2}\boldsymbol{q} = \ddot{c}_1 \boldsymbol{w}^{[1]} + \cdots + \ddot{c}_F \boldsymbol{w}^{[F]} \tag{3.183}$$

The dynamical equations then become

$$m_{\text{eff}} \left[\ddot{c}_1 \boldsymbol{w}^{[1]} + \cdots + \ddot{c}_F \boldsymbol{w}^{[F]} \right] = -H \left[c_1 \boldsymbol{w}^{[1]} + \cdots + c_F \boldsymbol{w}^{[F]} \right]$$

$$m_{\text{eff}} \ddot{c}_1 \boldsymbol{w}^{[1]} + \cdots + m_{\text{eff}} \ddot{c}_F \boldsymbol{w}^{[F]} = -\left[c_1 H \boldsymbol{w}^{[1]} + \cdots + c_F H \boldsymbol{w}^{[F]} \right]$$

$$= -\left[c_1 \lambda_1 \boldsymbol{w}^{[1]} + \cdots + c_F \lambda_F \boldsymbol{w}^{[F]} \right] \tag{3.184}$$

Equating the left- and right-hand sides separately in each eigenvector direction yields the following uncoupled set of equations for each $c_j(t)$,

$$m_{\text{eff}} \frac{d^2 c_j}{dt^2} = -\lambda_j c_j \tag{3.185}$$

We propose a trial form of the solution

$$c_j(t) = a_j \sin(\omega_j t) \tag{3.186}$$

and substitute it into the differential equation

$$\frac{d^2 c_j}{dt^2} = a_j \omega_j^2 \sin(\omega_j t) = -m_{\text{eff}}^{-1} \lambda_j a_j \sin(\omega_j t) = -m_{\text{eff}}^{-1} \lambda_j c_j \tag{3.187}$$

to show that the trial form of the solution is valid. The angular vibrational frequency of the normal mode is then

$$\omega_j = \sqrt{\frac{\lambda_j}{m_{\text{eff}}}} \tag{3.188}$$

lattice_2D_vib.m computes the normal modes of a 2-D lattice of point masses connected by harmonic springs. animate_2D_vib.m produces a movie of the oscillations for each mode. A derivation of this lattice model and a discussion of the results is provided in the supplemental material in the accompanying website.

Relaxing the assumption of equal masses

Above we have assumed that each degree of freedom has an equal effective mass. We now relax this assumption, using *Lagrange's equation of motion*,

$$\frac{d}{dt}\left(\frac{\partial L}{\partial \dot{q}_j}\right) = \frac{\partial L}{\partial q_j} \tag{3.189}$$

where the *Lagrangian*, the kinetic energy minus the potential energy, is

$$L(q, \dot{q}) = K(q, \dot{q}) - U(q) \tag{3.190}$$

For a system of point masses, each described by Cartesian coordinates, Lagrange's equation reduces to Newton's second law of motion. For small departures about a minimum energy state, Lagrange's equations of motion typically generate a system of the linearized form

$$M\left[\frac{d^2}{dt^2}\delta\right] = -H\delta \tag{3.191}$$

in which the *mass matrix M* is symmetric, positive-definite, but not necessarily diagonal. Since M is nonsingular, we can write

$$\frac{d^2}{dt^2}\delta = -M^{-1}H\delta \tag{3.192}$$

We thus must perform normal mode analysis on the matrix $M^{-1}H$, where

$$M^{-1}Hw^{[j]} = \lambda_j w^{[j]} \quad \Rightarrow \quad Hw^{[j]} = \lambda_j Mw^{[j]} \tag{3.193}$$

The generalized eigenvalue problem

A *generalized eigenvalue problem*, such as (3.193), is of the form

$$Aw = \lambda Bw \tag{3.194}$$

where B is a nonsingular matrix. While there exist general techniques to convert (3.194) into a standard eigenvalue problem using Schur decompositions, we here describe a simpler approach that may be used when B is positive-definite and A is real symmetric, as they are for the problem above with $B = M$ and $A = H$. We compute the Cholesky factorization of B,

$$B = LL^{\mathrm{T}} \tag{3.195}$$

and write

$$Aw = \lambda LL^{\mathrm{T}}w \tag{3.196}$$

We now define the transformation

$$z = L^{\mathrm{T}}w \qquad w = L^{\mathrm{T}(-1)}z \tag{3.197}$$

to obtain the corresponding eigenvalue problem

$$\left[L^{-1}AL^{\mathrm{T}(-1)}\right]z = \lambda z \tag{3.198}$$

If A is symmetric, so is $L^{-1}AL^{\mathrm{T}(-1)}$. As L is lower triangular, we can compute L^{-1} very quickly column-by-column by forward substitution

$$LL^{-1} = L \begin{bmatrix} | & & | \\ \tilde{l}^{[1]} & \cdots & \tilde{l}^{[N]} \\ | & & | \end{bmatrix} = I = \begin{bmatrix} | & & | \\ e^{[1]} & \cdots & e^{[N]} \\ | & & | \end{bmatrix} \qquad \begin{array}{c} L\tilde{l}^{[j]} = e^{[j]} \\ j = 1, 2, \ldots, N \end{array} \tag{3.199}$$

Once we obtain the eigenvalues λ_j and eigenvectors $z^{[j]}$ of $L^{-1}AL^{\mathrm{T}(-1)}$, we compute the corresponding generalized eigenvectors

$$w^{[j]} = L^{\mathrm{T}(-1)}z^{[j]} \tag{3.200}$$

eig and **eigs** allow the use of an optional nonsingular matrix B in the problem $Aw = \lambda Bw$ (type help eig or help eigs for further details). For

$$A = \begin{bmatrix} 2 & 1 & -1 \\ 1 & 4 & -2 \\ -1 & -2 & 6 \end{bmatrix} \qquad B = \begin{bmatrix} 2 & -1 & 0 \\ -1 & 2 & -1 \\ 0 & -1 & 2 \end{bmatrix} \tag{3.201}$$

the generalized eigenvalues satisfying $Aw = \lambda Bw$ are computed by

```
A = [2 1 -1; 1 4 -2; -1 -2 6];
B = [2 -1 0; -1 2 -1; 0 -1 2];
e = eig(A,B),
e =
    0.5664
    3.1128
    4.8207
```

Eigenvalue problems in quantum mechanics

Eigenvalue analysis lies at the heart of quantum mechanics. Here we consider only a simple example involving a single electron in one dimension, but the numerical approach is the same as that used in more realistic 3-D calculations of atoms and molecules. We wish to compute the energy states of a single electron in a 1-D external potential field that has the spatial periodicity

$$V(x + n2P) = V(x) \qquad n = 0, \pm1, \pm2, \ldots \tag{3.202}$$

Such a periodic system may be interpreted to be a 1-D "crystal."

The probability of finding an electron in $[x, x + dx]$ is $|\psi(x)|^2 dx$, where $\psi(x)$ is the *wavefunction* of the electron, satisfying the *Schrödinger equation*,

$$-\frac{\bar{h}^2}{2m_e} \frac{d^2\psi}{dx^2} + V(x)\psi(x) = E\psi(x) \tag{3.203}$$

m_e is the mass of an electron and $\bar{h} = h/2\pi$ where h is Planck's constant. E is the energy of the electron. For a more detailed discussion, consult Leach (2001) and Atkins & Friedman (1989).

Numerical solution of a differential equation eigenvalue problem

In (3.203), we have a differential equation eigenvalue problem, but we have been studying techniques to solve matrix eigenvalue problems. To convert this problem into a matrix one, we expand $\psi(x)$ as a linear combination of basis functions that satisfy the appropriate boundary conditions; here the periodicity condition $\psi(x + n2P) = \psi(x)$, since if $V(x)$ is periodic, we expect $\psi(x)$ to be also. We choose a plane wave basis set with members

$$\chi_m(x) = e^{iq_m x} = \cos(q_m x) + i \sin(q_m x) \tag{3.204}$$

The reasoning behind this choice of basis set is discussed in Chapter 9. Periodicity is satisfied if the allowable *wavenumbers* q_m satisfy

$$q_m = \frac{m\pi}{P} \qquad m = 0, \pm1, \pm2, \ldots \tag{3.205}$$

Using this basis, we write a trial form of the wavefunction as

$$\psi(x) = \sum_{m=-N}^{N} b_m \chi_m(x) = \sum_{m=-N}^{N} b_m e^{iq_m x} \tag{3.206}$$

We wish to compute the coefficients $\{b_m\}$ that satisfy (3.203) best and thus truncate the expansion to order N to make the problem dimension finite. Substituting this expansion into (3.203), we have

$$\sum_{m=-N}^{N} b_m \left[-\frac{\bar{h}^2}{2m_e} \frac{d^2}{dx^2} \chi_m(x) + V(x)\chi_m(x) \right] = E \sum_{m=-N}^{N} b_m \chi_m(x) \tag{3.207}$$

We now multiply this differential equation by the complex conjugates of each basis function, $\chi_n^*(x)$, to obtain a set of equations:

$$\sum_{m=-N}^{N} b_m \chi_n^*(x) \left[-\frac{\bar{h}^2}{2m_c} \frac{d^2}{dx^2} \chi_m(x) + V(x)\chi_m(x) \right] = E \sum_{m=-N}^{N} b_m \chi_n^*(x)\chi_m(x) \tag{3.208}$$

We next integrate each of these equations over $[0,\ 2P]$ to obtain a set of algebraic equations:

$$\sum_{m=-N}^{N} b_m h_{nm} = E \sum_{m=-N}^{N} b_m s_{nm} \tag{3.209}$$

The coefficients s_{nm} are

$$s_{nm} = \int_0^{2P} \chi_n^*(x)\chi_m(x)dx \tag{3.210}$$

and the coefficients h_{nm} are

$$h_{nm} = \int_0^{2P} \chi_n^*(x)[\hat{H}\chi_m(x)]dx \tag{3.211}$$

where the *Hamiltonian operator* is

$$\hat{H}\chi = -\frac{\bar{h}^2}{2m_{\mathrm{e}}}\frac{d^2}{dx^2}\chi(x) + V(x)\chi(x) \tag{3.212}$$

We compute these integrals shortly, but first let us consider the structure of (3.209), which we write as

$$\sum_{m=-N}^{N} h_{nm}b_m = E \sum_{m=-N}^{N} s_{nm}b_m \quad n = 0, \pm 1, \pm 2, \ldots, \pm N \tag{3.213}$$

Let us define the new indices

$$\begin{aligned} p = m + N + 1 \quad & m = 0, \pm 1, \pm 2, \ldots, \pm N \quad p = 1, 2, \ldots, 2N + 1 \\ q = n + N + 1 \quad & n = 0, \pm 1, \pm 2, \ldots, \pm N \quad q = 1, 2, \ldots, 2N + 1 \end{aligned} \tag{3.214}$$

so that, with $B = 2N + 1$, we define a $B \times B$ *overlap matrix S*, a $B \times B$ *Hamiltonian matrix H*, and a coefficient vector $c \in C^B$, such that

$$H_{pq} = h_{mn} \quad S_{pq} = s_{mn} \quad c_p = b_m \tag{3.215}$$

The system of equations (3.213) then becomes

$$\sum_{p=1}^{B} H_{qp}c_p = E \sum_{p=1}^{B} S_{qp}c_p \quad q = 1, 2, \ldots, B \tag{3.216}$$

This is the qth row of the generalized matrix eigenvalue problem

$$Hc = ESc \tag{3.217}$$

H and S are both Hermitian and S is also positive-definite. Let the energy eigenvalues be E_k with $Hc^{[k]} = E_k Sc^{[k]}$. These E_k are the allowable energy states of the system, whose corresponding normalized wavefunctions are

$$\psi^{[k]}(x) = \frac{\varphi^{[k]}(x)}{\int_0^{2P} \left|\varphi^{[k]}(x)\right|^2 dx} \quad \varphi^{[k]}(x) = \sum_{p=1}^{B} c_p^{[k]}\chi_{m=p-N-1}(x) \tag{3.218}$$

with the corresponding electron probability densities

$$\rho_{\mathrm{e}}^{[k]}(x) = \left|\psi^{[k]}(x)\right|^2 \tag{3.219}$$

We first must calculate the integrals for the matrix elements of S and H, and this typically comprises a significant fraction of the effort of quantum calculations. First, we consider the elements of the overlap matrix

$$S_{pq} = s_{mn} = \int_0^{2P} \chi_n{}^*(x)\chi_m(x)dx \tag{3.220}$$

Substituting for the plane wave basis functions,

$$S_{pq} = s_{mn} = \int_0^{2P} e^{-iq_n x} e^{iq_m x} dx = \int_0^{2P} \exp\left[i\left(\frac{\pi x}{P}\right)(m-n)\right]dx \tag{3.221}$$

Defining $\theta = \pi x/P$, this becomes

$$S_{pq} = s_{mn} = \left(\frac{P}{\pi}\right)\int_0^{2\pi} e^{i\theta(m-n)}d\theta = 2P\delta_{mn} = 2P\delta_{pq} \tag{3.222}$$

Thus, the overlap matrix is merely $S = (2P)I$, I being the identity matrix.

We next compute the elements of the Hamiltonian matrix, which we break into two contributions, $H = T + V$, where

$$T_{pq} = -\left(\frac{\bar{h}^2}{2m_e}\right)\int_0^{2P} \chi_n^*(x)\frac{d^2\chi_m}{dx^2}dx \tag{3.223}$$

and

$$V_{pq} = \int_0^{2P} \chi_n^*(x)V(x)\chi_m(x)dx \tag{3.224}$$

In general, we must compute the elements of V numerically. A discussion of numerical integration is not given until the next chapter, so here we merely note that we use the *trapezoid method* (**trapz** in MATLAB) in which we evaluate the integrand $f(x) = \chi_n{}^*(x)V(x)\chi_m(x)$ at n_G grid points,

$$x_j = (j-1)(\Delta x) \qquad \Delta x = \frac{2P}{n_G - 1} \qquad j = 1, 2, \ldots, n_G \tag{3.225}$$

We then approximate the integrals (3.224) using the quadrature rule

$$\int_0^{2P} f(x)dx = \sum_{j=1}^{n_G-1}\int_{x_j}^{x_{j+1}} f(x)dx \approx \sum_{j=1}^{n_G-1}\left[\frac{f(x_{j+1}) + f(x_j)}{2}\right](\Delta x) \tag{3.226}$$

For the selected basis functions, we can compute the elements of T analytically. Substituting into (3.223) for the basis functions,

$$T_{pq} = -\left(\frac{\bar{h}^2}{2m_e}\right)\int_0^{2P} e^{-iq_n x}\frac{d^2}{dx^2}e^{iq_m x}dx = \left(\frac{\bar{h}^2}{2m_e}\right)q_m^2\int_0^{2P} e^{-iq_n x}e^{iq_m x}dx \tag{3.227}$$

Using $q_m = m\pi/P$ and $\int_0^{2P} e^{i(q_m - q_n)x}dx = 2P\delta_{mn}$, we have

$$T_{pq} = \left(\frac{\bar{h}^2 m^2 \pi^2}{Pm_e}\right)\delta_{mn} \qquad m = p - N - 1 \quad n = q - N - 1 \tag{3.228}$$

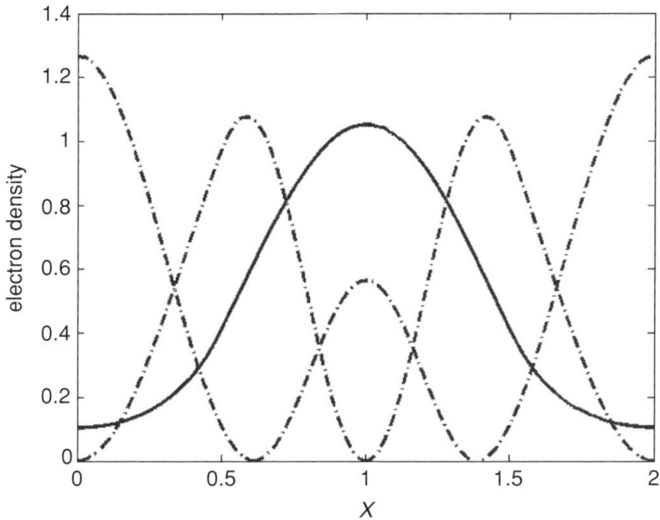

Figure 3.7 Plots of the electron densities in the ground state (*solid*) and first two excited states (*dash-dot*) for an electron in a 1-D array of square well potentials.

quantum_1D.m computes the lowest energy states for a well potential

$$
V(0 \leq x \leq 2P) = \begin{cases} -E_{\text{well}}, & \left(P - {}^w\!/_2\right) \leq x \leq \left(P + {}^w\!/_2\right) \\ 0, & \text{otherwise} \end{cases}
\tag{3.229}
$$

w is the width of the energy well, and E_{well} is its depth. Plots of the electron densities for the ground state and first two excited states with $P = 1$, $w = 1$, $E_{\text{well}} = 5$, and $N = 25$ are shown in Figure 3.7. The lowest energy eigenstates are computed using **eigs** with the 'SR' keyword.

Singular value decomposition (SVD)

With the usefulness of eigenvalue analysis, we might wonder if it can be extended to the case of nonsquare matrices of general dimension $M \times N$,

$$
A = \begin{bmatrix}
a_{11} & a_{12} & a_{13} & \ldots & a_{1N} \\
a_{21} & a_{22} & a_{23} & \ldots & a_{2N} \\
a_{31} & a_{32} & a_{33} & \ldots & a_{3N} \\
\vdots & \vdots & \vdots & & \vdots \\
a_{M1} & a_{M2} & a_{M3} & \ldots & a_{MN}
\end{bmatrix}
\tag{3.230}
$$

In fact, such an extension exists. For a $M \times N$ real matrix A, we can always generate the $N \times N$ square matrix $A^{\mathrm{T}}A$ that is symmetric and positive-semidefinite; i.e., all eigenvalues are real and are greater than or equal to zero. Let the eigenvalues of $A^{\mathrm{T}}A$ be $\mu_1, \mu_2, \ldots,$ μ_N. We then define the *singular values* of A as

$$
\sigma_1 = \sqrt{\mu_1} \qquad \sigma_2 = \sqrt{\mu_2} \quad \ldots \quad \sigma_N = \sqrt{\mu_N}
\tag{3.231}
$$

Let us assume that our matrix has more rows than columns, $M \geq N$. We then can write A in the form of a Singular Value Decomposition (SVD)

$$A = W \Sigma V^{\mathrm{T}} \tag{3.232}$$

Σ is a $M \times N$ diagonal matrix containing the singular values, W is an orthogonal $M \times M$ matrix whose columns are the M *left singular vectors* of A, and V is an orthogonal $N \times N$ matrix whose columns are the N *right singular vectors* of A.

$$\Sigma = \begin{bmatrix} \sigma_1 & & & \\ & \sigma_2 & & \\ & & \ddots & \\ & & & \sigma_N \\ & & & 0 \\ & & & \vdots \\ & & & 0 \end{bmatrix} \qquad W = \begin{bmatrix} | & | & & | \\ \boldsymbol{w}^{[1]} & \boldsymbol{w}^{[2]} & \cdots & \boldsymbol{w}^{[M]} \\ | & | & & | \end{bmatrix}$$
$$W^{\mathrm{T}} = W^{-1}$$

$$V = \begin{bmatrix} | & | & & | \\ \boldsymbol{v}^{[1]} & \boldsymbol{v}^{[2]} & \cdots & \boldsymbol{v}^{[N]} \\ | & | & & | \end{bmatrix} \tag{3.233}$$

For $M < N$, the SVD also exists, but now Σ is

$$\Sigma = \begin{bmatrix} \sigma_1 & & & \\ & \sigma_2 & & \\ & & \cdots & \\ & & & \sigma_M & 0\,0\,0 \end{bmatrix} \tag{3.234}$$

and $\sigma_{M+1} = \cdots = \sigma_N = 0$. As $A^{\mathrm{T}} A = (V \Sigma^{\mathrm{T}} W^{\mathrm{T}})(W \Sigma V^{\mathrm{T}}) = V \Sigma^{\mathrm{T}}(W^{\mathrm{T}} W) \Sigma V^{\mathrm{T}}$, we note that as W is orthogonal, $W^{\mathrm{T}} W = I_M$, I_M being the $M \times M$ identity matrix, and thus we obtain the Jordan normal form for $A^{\mathrm{T}} A$

$$A^{\mathrm{T}} A = V (\Sigma^{\mathrm{T}} \Sigma) V^{\mathrm{T}} \tag{3.235}$$

Therefore, $\Sigma^{\mathrm{T}} \Sigma = \Lambda$ is a diagonal matrix containing the eigenvalues of $A^{\mathrm{T}} A$, and the right singular vectors of A are the eigenvectors of $A^{\mathrm{T}} A$,

$$A^{\mathrm{T}} A \boldsymbol{v}^{[j]} = \sigma_j^2 \boldsymbol{v}^{[j]} \tag{3.236}$$

Previously, we have defined the rank of a square matrix as the number of linearly-independent columns (or rows); however, we can now extend this definition and provide a means for its calculation.

Definition The *rank* of an $M \times N$ matrix A is the number of its nonzero singular values.

Above we have treated the case of a real matrix A. For a complex matrix A, the SVD is $A = U \Sigma V^{\mathrm{H}}$, where now U and V are unitary. Σ contains the nonnegative singular values, as is the case when A is real.

SVD analysis and the existence/uniqueness properties of linear systems

Let us examine how SVD aids detecting the existence and uniqueness properties of linear systems. As noted in Chapter 1, the nature of the null space (kernel) of A and of the range are vitally important; however, we have not described how we may identify these subspaces for a particular matrix. Let A be a real, square $N \times N$ matrix, with the SVD $A = W \Sigma V^{\mathrm{T}}$,

$$
A = W \begin{bmatrix} \sigma_1 & & & \\ & \sigma_2 & & \\ & & \ddots & \\ & & & \sigma_N \end{bmatrix} \begin{bmatrix} - & (v^{[1]})^{\mathrm{T}} & - \\ - & (v^{[2]})^{\mathrm{T}} & - \\ & \vdots & \\ - & (v^{[N]})^{\mathrm{T}} & - \end{bmatrix}
$$

$$
= \begin{bmatrix} | & | & & | \\ w^{[1]} & w^{[2]} & \cdots & w^{[N]} \\ | & | & & | \end{bmatrix} \begin{bmatrix} - & \sigma_1(v^{[1]})^{\mathrm{T}} & - \\ - & \sigma_2(v^{[2]})^{\mathrm{T}} & - \\ & \vdots & \\ - & \sigma_N(v^{[N]})^{\mathrm{T}} & - \end{bmatrix} \tag{3.237}
$$

Therefore, we can write

$$
Ax = W \begin{bmatrix} - & \sigma_1(v^{[1]})^{\mathrm{T}} & - \\ - & \sigma_2(v^{[2]})^{\mathrm{T}} & - \\ & \vdots & \\ - & \sigma_N(v^{[N]})^{\mathrm{T}} & - \end{bmatrix} \begin{bmatrix} x_1 \\ x_2 \\ \vdots \\ x_N \end{bmatrix}
$$

$$
= \begin{bmatrix} | & | & & | \\ w^{[1]} & w^{[2]} & \cdots & w^{[N]} \\ | & | & & | \end{bmatrix} \begin{bmatrix} \sigma_1(v^{[1]} \cdot x) \\ \sigma_2(v^{[2]} \cdot x) \\ \vdots \\ \sigma_N(v^{[N]} \cdot x) \end{bmatrix} \tag{3.238}
$$

The right singular vectors $\{v^{[1]}, v^{[2]}, \ldots, v^{[N]}\}$ are orthonormal. Therefore, any vector $x \in \mathfrak{R}^N$ can be written as the linear combination

$$
x = \sum_{j=1}^{N} (v^{[j]} \cdot x) v^{[j]} \tag{3.239}
$$

Let us say the first r singular values of A are zero and that the rest are nonzero. We want to identify the null space K_A and range R_A of A. To do so, we break the linear contribution for x into two parts

$$
x = \sum_{\substack{j=1 \\ \sigma_j=0}}^{r} (v^{[j]} \cdot x) v^{[j]} + \sum_{\substack{j=r+1 \\ \sigma_j>0}}^{N} (v^{[j]} \cdot x) v^{[j]} \tag{3.240}
$$

Let us now define a second vector $y \in \mathfrak{R}^N$ that is a linear combination solely of the right singular vectors for the zero singular values,

$$
y = \sum_{\substack{j=1 \\ \sigma_j=0}}^{r} (v^{[j]} \cdot y) v^{[j]} \tag{3.241}
$$

and write $A\boldsymbol{y}$ in the form of (3.238),

$$A\boldsymbol{y} = W \begin{bmatrix} \sigma_1\big(\boldsymbol{v}^{[1]}\cdot\boldsymbol{y}\big) \\ \vdots \\ \sigma_r\big(\boldsymbol{v}^{[r]}\cdot\boldsymbol{y}\big) \\ \sigma_{r+1}\big(\boldsymbol{v}^{[r+1]}\cdot\boldsymbol{y}\big) \\ \vdots \\ \sigma_N\big(\boldsymbol{v}^{[N]}\cdot\boldsymbol{y}\big) \end{bmatrix} = W \begin{bmatrix} (0)\big(\boldsymbol{v}^{[1]}\cdot\boldsymbol{y}\big) \\ \vdots \\ (0)\big(\boldsymbol{v}^{[r]}\cdot\boldsymbol{y}\big) \\ \sigma_{r+1}(0) \\ \vdots \\ \sigma_N(0) \end{bmatrix} = W\mathbf{0} = \mathbf{0} \tag{3.242}$$

Thus, the right singular vectors for the zero singular values form an orthonormal basis for the null space (kernel) of A,

$$K_A = \text{span}\,\big\{\boldsymbol{v}^{[1]}, \ldots, \boldsymbol{v}^{[r]}\big\} \quad \sigma_1 = \ldots = \sigma_r = 0 \quad \dim(K_A) = r \tag{3.243}$$

We obtain a basis for the range of A by continuing the calculation of $A\boldsymbol{x}$,

$$A\boldsymbol{x} = \begin{bmatrix} | & & | \\ \boldsymbol{w}^{[1]} & \cdots & \boldsymbol{w}^{[N]} \\ | & & | \end{bmatrix} \begin{bmatrix} \sigma_1\big(\boldsymbol{v}^{[1]}\cdot\boldsymbol{x}\big) \\ \sigma_2\big(\boldsymbol{v}^{[2]}\cdot\boldsymbol{x}\big) \\ \vdots \\ \sigma_N\big(\boldsymbol{v}^{[N]}\cdot\boldsymbol{x}\big) \end{bmatrix}$$

$$= \begin{bmatrix} w_1^{[1]}\sigma_1\big(\boldsymbol{v}^{[1]}\cdot\boldsymbol{x}\big) + \cdots + w_1^{[N]}\sigma_N\big(\boldsymbol{v}^{[N]}\cdot\boldsymbol{x}\big) \\ w_2^{[1]}\sigma_1\big(\boldsymbol{v}^{[1]}\cdot\boldsymbol{x}\big) + \cdots + w_2^{[N]}\sigma_N\big(\boldsymbol{v}^{[N]}\cdot\boldsymbol{x}\big) \\ \vdots \\ w_N^{[1]}\sigma_1\big(\boldsymbol{v}^{[1]}\cdot\boldsymbol{x}\big) + \cdots + w_N^{[N]}\sigma_N\big(\boldsymbol{v}^{[N]}\cdot\boldsymbol{x}\big) \end{bmatrix} \tag{3.244}$$

to obtain

$$A\boldsymbol{x} = \sum_{j=1}^{N} \sigma_j\big(\boldsymbol{v}^{[j]}\cdot\boldsymbol{x}\big)\boldsymbol{w}^{[j]} \tag{3.245}$$

If $\sigma_1 = \cdots = \sigma_r = 0$ and $\sigma_j > 0$ for $j = r+1, \ldots, N$, then

$$A\boldsymbol{x} = \sum_{\substack{j=r+1 \\ \sigma_j > 0}}^{N} \sigma_j\big(\boldsymbol{v}^{[j]}\cdot\boldsymbol{x}\big)\boldsymbol{w}^{[j]} \tag{3.246}$$

Thus, any $A\boldsymbol{x} \in \Re^N$ can be written as linear combination of the left singular vectors $\{\boldsymbol{w}^{[r+1]}, \ldots, \boldsymbol{w}^{[N]}\}$, so that the range of A is

$$R_A = \text{span}\big\{\boldsymbol{w}^{[r+1]}, \ldots, \boldsymbol{w}^{[N]}\big\} \quad \sigma_{j \in [r+1,N]} > 0 \quad \dim(K_A) = N - r \tag{3.247}$$

The right and left singular vectors thus provide orthonormal basis sets for the kernel and range respectively,

$$K_A = \text{span}\big\{\boldsymbol{v}^{[j]}\big|\sigma_j = 0\big\} \qquad R_A = \text{span}\big\{\boldsymbol{w}^{[j]}\big|\sigma_j > 0\big\} \tag{3.248}$$

For $A\boldsymbol{x} = \boldsymbol{b}$, if A is singular, we can check easily for the existence of solutions using SVD. If $\boldsymbol{b} \in \Re_A$, there exist an infinite number of solutions

$$\boldsymbol{x} = \boldsymbol{z} + \sum_{\substack{j=1 \\ \sigma_j = 0}}^{N} c_j \boldsymbol{v}^{[j]} \qquad c_j \in \Re \tag{3.249}$$

z is any particular solution satisfying $Az = b$. To obtain a particular solution from the SVD, we note that if A were nonsingular, the unique solution would be

$$x = A^{-1}b = (W\Sigma V^{\mathrm{T}})^{-1}b = V^{\mathrm{T}(-1)}\Sigma^{-1}W^{-1}b = V\Sigma^{-1}W^{\mathrm{T}}b \qquad (3.250)$$

where

$$\Sigma^{-1} = \mathrm{diag}\big(\sigma_1^{-1}, \ldots, \sigma_r^{-1}, \sigma_{r+1}^{-1}, \ldots, \sigma_N^{-1}\big) \qquad (3.251)$$

Written in terms of the left and right singular vectors, the solution is

$$x = \sum_{j=1}^{N} \big(w^{[j]} \cdot b\big)\sigma_j^{-1}v^{[j]} \qquad (3.252)$$

If there exist zero singular values $\sigma_j = 0$, Σ^{-1} and $A^{-1} = V\Sigma^{-1}W^{\mathrm{T}}$ do not exist, and this formula diverges due to division by zero. We can, however, define the *pseudo (generalized) inverse* of A, in which we replace the infinite values of σ_j^{-1} with zero:

$$\tilde{A}^{-1} = V\tilde{\Sigma}^{-1}W^{\mathrm{T}} \qquad \tilde{\Sigma}^{-1} = \mathrm{diag}\big(0, \ldots, 0, \sigma_{r+1}^{-1}, \ldots, \sigma_N^{-1}\big) \qquad (3.253)$$

For, $b \in \Re_A$, b must lie in $\mathrm{span}\{w^{[j]}|\sigma_j > 0\}$, and as all $w^{[j]}$ are orthonormal, $w^{[j]} \cdot b = 0$ for all $\sigma_j = 0$. Therefore, by (3.252), we still have $Az = b$ when $b \in \Re_A$,

$$z = \sum_{\substack{j=1 \\ \sigma_j>0}}^{N} \big(w^{[j]} \cdot b\big)\sigma_j^{-1}v^{[j]} = \tilde{A}^{-1}b \qquad (3.254)$$

The general solution, for $b \in \Re_A$, is then

$$x = \sum_{\substack{j=1 \\ \sigma_j>0}}^{N} \big(w^{[j]} \cdot b\big)\sigma_j^{-1}v^{[j]} + \sum_{\substack{j=1 \\ \sigma_j=0}}^{N} c_j v^{[j]} \quad c_j \in \Re \qquad (3.255)$$

Least-squares approximate solutions

You may ask, what happens if we apply the pseudo-inverse to a vector b that is *not* in the range of A? Such is often the case for an overdetermined system with more equations than unknowns, $Ax = b, x \in \Re^N, b \in \Re^{M>N}$. Then, $z = \tilde{A}^{-1}b$ is *not* a solution, $Az \neq b$. However, it is the "closest thing to a solution," as it minimizes the residual norm,

$$|Az - b| \leq |Ay - b| \qquad \forall y \in \Re^N \qquad (3.256)$$

As this also means minimizing $|Az - b|^2 = (Az - b) \cdot (Az - b), z = \tilde{A}^{-1}b$ is said to be the *least-squares approximate solution*. The SVD solution, $z = \tilde{A}^{-1}b$, exists for systems $Ax = b$ with both square and nonsquare A matrices.

This property makes SVD very useful in statistics. Let us fit a linear model

$$y = \beta_1 x_1 + \beta_2 x_2 + \cdots + \beta_N x_N \qquad (3.257)$$

to a data set of M measurements of y, $y^{[k]}$, for known $\{x_1^{[k]}, \ldots, x_N^{[k]}\}$. From the data, let us form the $M \times N$ design matrix X, the response vector $y \in \Re^M$, and the vector of unknown parameters $\beta \in \Re^N$,

$$
X = \begin{bmatrix} x_1^{[1]} & x_2^{[1]} & \cdots x_N^{[1]} \\ x_1^{[2]} & x_2^{[2]} & \cdots x_N^{[2]} \\ \vdots & \vdots & \vdots \\ x_1^{[M]} & x_2^{[M]} & \cdots x_N^{[M]} \end{bmatrix} \quad y = \begin{bmatrix} y^{[1]} \\ y^{[2]} \\ \vdots \\ y^{[M]} \end{bmatrix} \in \Re^N \quad \beta = \begin{bmatrix} \beta_1 \\ \beta_2 \\ \vdots \\ \beta_N \end{bmatrix} \in \Re^N \tag{3.258}
$$

Using the rules of matrix multiplication, we can write the set of relationships $y^{[k]} = \beta_1 x_1^{[k]} + \beta_2 x_2^{[k]} + \cdots + \beta_N x_N^{[k]}$ as $y = X\beta$

$$
\begin{bmatrix} y^{[1]} \\ y^{[2]} \\ \vdots \\ y^{[M]} \end{bmatrix} = \begin{bmatrix} x_1^{[1]} & x_2^{[1]} & \cdots & x_N^{[1]} \\ x_1^{[2]} & x_2^{[2]} & \cdots & x_N^{[2]} \\ \vdots & \vdots & & \vdots \\ x_1^{[M]} & x_2^{[M]} & \cdots & x_N^{[M]} \end{bmatrix} \begin{bmatrix} \beta_1 \\ \beta_2 \\ \vdots \\ \beta_N \end{bmatrix}
$$

$$
= \begin{bmatrix} \beta_1 x_1^{[1]} + \beta_2 x_2^{[1]} + \cdots + \beta_N x_N^{[1]} \\ \beta_1 x_1^{[2]} + \beta_2 x_2^{[2]} + \cdots + \beta_N x_N^{[2]} \\ \vdots \\ \beta_1 x_1^{[M]} + \beta_2 x_2^{[M]} + \cdots + \beta_N x_N^{[M]} \end{bmatrix} \tag{3.259}
$$

In Chapter 1, we computed the coefficients β of a linear model by solving the system $X^T X \beta = X^T y$, obtained by premultiplying $y = X\beta$ by X^T. We also can obtain β using the pseudo-inverse from SVD,

$$
\beta = V \tilde{\Sigma}^{-1} W^T y \tag{3.260}
$$

The advantage of the SVD approach becomes evident when we do not have sufficient data to determine all coefficients β_j. Then, the right singular vectors $\{v^{[j]} | \sigma_j = 0\} \in \Re^N$ provide information about the "missing" data points x that are necessary to determine all β_j. In particular, we should add new measurements $\{x^{[N+1]}, \ldots, x^{[N+P]}\}$ such that

$$
\mathrm{span}\left\{v^{[j]} | \sigma_j = 0\right\} \subset \mathrm{span}\left\{x^{[1]}, \ldots, x^{[N]}, x^{[N+1]}, \ldots, x^{[N+P]}\right\} \tag{3.261}
$$

Then, the new design matrix with these P additional measurements will have no zero singular values, such that all coefficients β_j can be estimated. Equation (3.256) tells us that the SVD solution (3.260) is the least-squares estimate that minimizes the sum of squared errors $|X\beta - y|^2$. This subject will be discussed further in Chapter 8.

SVD in MATLAB

We wish to fit the linear model $y = \beta_0 + \beta_1 \theta_1 + \beta_2 \theta_2$ to the data in Table 3.1. Entering the corresponding design matrix and response vector,

```
X = [1 0 0; 1 1 1; 1 2 2; 1 3 3];
y = [1; 6; 11; 16];
```

we compute the SVD $X = USV^{\mathrm{H}}$

Table 3.1 Measured data for fitting linear model with SVD

measured y	θ_1	θ_2
1	0	0
6	1	1
11	2	2
16	3	3

```
[U,S,V] = svd(X),
U = 0.0547    0.8349    -0.5000    0.2236
    0.2979    0.4596     0.8333    0.0745
    0.5412    0.0843    -0.1667   -0.8199
    0.7844   -0.2909    -0.1667    0.5217
S = 5.5405    0          0
    0         1.1415     0
    0         0          0.0000
    0         0          0
V = 0.3029    0.9530     0
    0.6739   -0.2142    -0.7071
    0.6739   -0.2142     0.7071
```

We see from S that the third singular value is zero, and identify the "missing" data from the corresponding right singular vector

$$v^{[3]} = [0 - 0.7071 \quad 0.7071]^{\mathrm{T}} \tag{3.262}$$

This is clearly related to the lack of any data points in Table 3.1 that vary θ_1 and θ_2 by different amounts. We ensure (3.261) by adding a fifth data point $x^{[5]} = x^{[3]} + cv^{[3]}$, yielding for $c^{-1} = 0.7071$, $x_1 = 2 - 1 = 1$ and $x_2 = 2 + 1 = 3$. For this $x^{[5]}$, we measure $y^{[5]} = 12$, such that the new data set is

```
X = [1 0 0; 1 1 1; 1 2 2; 1 3 3; 1 1 3];
y = [1; 6; 11; 16; 12];
```

The singular values of X are now all nonzero,

```
s=svd(X),
s =
    6.3373
    1.2530
    1.1264
```

The fitted parameters are computed by

```
[U,S,V] = svd(X);
S_inv = zeros(size(S'));
S_inv(1:3,1:3) = inv(S(1:3,1:3)),
```

```
S_inv =
    0.1578      0          0         0    0
    0           0.7981     0         0    0
    0           0          0.8878    0    0
b = V*S_inv*U'*y,
b =
    1.0000
    2.0000
    3.0000
```

Thus, the fitted parameters of the linear model are $\beta_0 = 1$, $\beta_1 = 2$, $\beta_2 = 3$. This approach to the design and analysis of data sets for parameter estimation will be considered again in Chapter 8.

Computing the roots of a polynomial

The eigenvalues of a matrix A are roots of the characteristic polynomial of degree N, $\det(A - \lambda I) = 0$. Therefore, we can compute the roots of any polynomial of degree N

$$p_N(x) = a_N x^N + a_{N-1} x^{N-1} + \cdots + a_1 x + a_0 \tag{3.263}$$

by calculating the eigenvalues of a matrix A for which $\det(A - \lambda I) = p_N(\lambda)$. From expansion by minors in the first row, the *auxiliary matrix* for $p_N(x)$ is

$$A = \begin{bmatrix} (-a_{N-1}/a_N) & (-a_{N-2}/a_N) & \cdots & (-a_1/a_N) & (-a_0/a_N) \\ 1 & 0 & \cdots & 0 & 0 \\ 0 & 1 & \cdots & 0 & 0 \\ \vdots & \vdots & & \vdots & \vdots \\ 0 & 0 & \cdots & 1 & 0 \end{bmatrix} \tag{3.264}$$

For example, consider the polynomial of degree three

$$p_3(x) = x^3 - 3x^2 + x - 3 \tag{3.265}$$

The auxiliary matrix is

$$A = \begin{bmatrix} 3 & -1 & 3 \\ 1 & 0 & 0 \\ 0 & 1 & 0 \end{bmatrix} \tag{3.266}$$

The eigenvalues of this matrix are

$$\lambda_1 = 3 \qquad \lambda_2 = i \qquad \lambda_3 = -i \tag{3.267}$$

It is easy to show that these eigenvalues are in fact roots of (3.265):

$$\begin{aligned} p_3(3) &= (3)^3 - 3(3)^2 + (3) - 3 = 27 - 27 + 3 - 3 = 0 \\ p_3(i) &= (i)^3 - 3(i)^2 + (i) - 3 = -i + 3 + i - 3 = 0 \\ p_3(-i) &= (-i)^3 - 3(-i)^2 + (-i) - 3 = i + 3 - i - 3 = 0 \end{aligned} \tag{3.268}$$

In MATLAB, this technique is employed by **roots**.

MATLAB summary

The use of MATLAB to compute eigenvalues was discussed earlier in this chapter; therefore, here only a brief summary is provided. A matrix W, whose column vectors are eigenvectors of A, and a diagonal matrix D, whose principal diagonal contains the corresponding eigenvalues, are returned by

[W,D] = eig(A);

With only a single output argument, **eig** returns a vector of eigenvalues. If only a few extremal eigenvalues are desired, use **eigs**. For example, the five largest-magnitude eigenvalues of A and the corresponding eigenvectors are returned by

[W,D] = eigs(A,5, 'LM');

Other options include computing the smallest magnitude ('SM'), largest and smallest real part ('LR', 'SR'), or the eigenvalues closest to a specified target shift value. Type help eigs, or consult the earlier discussion of this chapter, for further details.

The SVD $A = USV^{\mathrm{H}}$ is computed by

[U,S,V] = svd(A);

The condition number is computed by **cond** and **condest**; the norm by **norm** and **normest**; and the rank by **rank**. Eigenvalue methods are used to compute all roots of a polynomial by **roots**.

Problems

3.A.1. From Gershgorin's theorem, derive lower and upper bounds on the possible eigenvalues of the matrix

$$A = \begin{bmatrix} 1 & 0 & 3 \\ 0 & 2 & 1 \\ 3 & 1 & -1 \end{bmatrix} \tag{3.269}$$

3.A.2. Compute by hand the eigenvalues and eigenvectors of (3.269), and check your results using MATLAB.

3.A.3. Consider the following matrices,

$$A = \begin{bmatrix} 0 & -1 & -2 & 1 \\ -1 & 2 & 0 & 4 \\ -2 & 0 & 3 & 0 \\ 1 & 4 & 0 & -1 \end{bmatrix} \qquad B = \begin{bmatrix} 6 & 2 & 1 \\ 0 & 5 & -1 \\ -1 & 3 & 2 \end{bmatrix}$$

$$C = \begin{bmatrix} 3 & 2 \\ 1 & -1 \end{bmatrix} \qquad D = \begin{bmatrix} 0 & -1 & 0 \\ 1 & 0 & 0 \\ 0 & 0 & 1 \end{bmatrix} \tag{3.270}$$

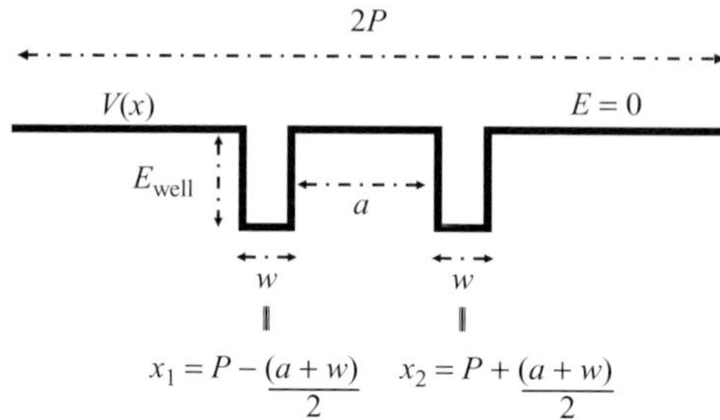

Figure 3.8 Double-well potential for a 1-D quantum mechanics problem.

(a) Without computing the actual eigenvalues, can you tell if any of the matrices above must have all real eigenvalues? Explain why you make this judgement.

(b) For each of those guaranteed to have all real eigenvalues, provide upper and lower bounds on the eigenvalues.

(c) Show that D is unitary.

(d) Compute by hand the eigenvalues and unit-length eigenvectors of C.

3.A.4. Consider a random 4×4 matrix generated by $\mathsf{A} = \mathsf{rand(4)}$. Similarly to (3.170)–(3.174) use the double-shift iterative QR method to find its eigenvalues, reporting each intermediate matrix $A^{[k]}$. Then, compute $A^T A$ and repeat the calculation, demonstrating that its eigenvalues are real. Once the eigenvalues of A and $A^T A$ have been calculated, compute their eigenvectors, and demonstrate that those of $A^T A$ are orthogonal. You may use **MATLAB** to solve linear systems and perform the QR decomposition at each iteration.

3.B.1. Modify $\mathsf{quantum\_1D.m}$ to compute the lowest-energy states for the double-well potential system shown in Figure 3.8, with the parameters

$$P = 10 \qquad a = 2 \qquad w = 1 \qquad E_{\text{well}} = 10$$

3.B.2. Consider the positive-definite matrix A, obtained by discretizing the Poisson equation $-\nabla^2 \varphi = f$ in d dimensions on a hypercube grid of N^d points, with the following nonzero elements in each row for $\Delta x_j = 1$,

$$A_{kk} = 2d \qquad A_{k,k \pm N^m} = -1 \qquad m = 0, 1, \ldots, d - 1 \tag{3.271}$$

Plot as functions of N the largest and smallest eigenvalues and the condition number for $d = 1, 2, 3$. For $d = 3$, extend the calculation to relatively large values of N by not storing the matrix (even in sparse format) but rather by merely supplying a routine that returns Av given an input value of v.

3.B.3. We wish to fit the model $y = \beta_0 + \beta_1 \theta_1 + \beta_2 \theta_2$ to the data of Table 3.2. Compute the SVD of the design matrix, and show that the data are sufficient to determine all parameters in the proposed model. Then, compute the best fit of the parameters to the data. NOTE: The "\" linear solver of **MATLAB** returns the least squares solution for overdetermined systems.

3.B.4. It is common in mechanics to describe a rotation in $\mathfrak{R}^3$ by its three *Euler angles* $(\varphi, \theta, \psi), 0 < \varphi < 2\pi, 0 < \theta < \pi, 0 < \psi < 2\pi$. The corresponding

Table 3.2 Data for fitting quadratic two-variable model

k	1	2	3	4	5	6	7	8
θ_1	0	1	1	1	2	2	2	0
θ_2	1	0	1	2	1	2	0	2
y	1.53	1.11	2.83	4.39	4.02	5.92	2.00	3.23

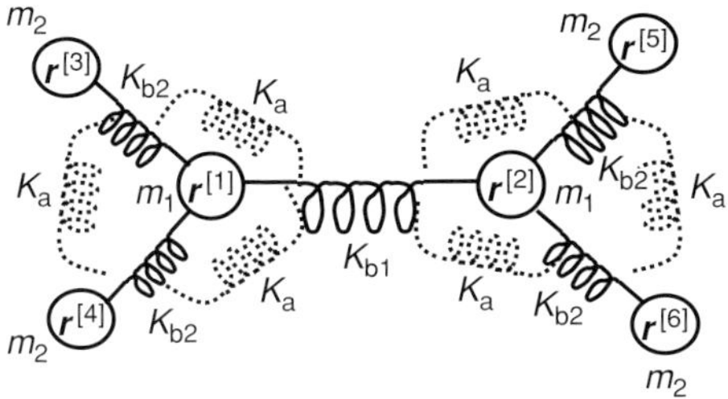

Figure 3.9 Molecule in two dimensions with bond and angle spring energy models.

transformation involves a rotation of angle φ about the z-axis, followed by a rotation of θ about the new x-axis, followed by a rotation of angle ψ about the new z-axis. Using the notation, $C_\psi = \cos_\psi$, $S_\varphi = \sin\varphi$, etc., the orthogonal matrix for this rotation is

$$Q = \begin{bmatrix} (C_\psi C_\varphi - S_\psi C_\theta S_\varphi) & (-C_\psi S_\varphi - S_\psi C_\theta S_\varphi) & (S_\psi S_\theta) \\ (S_\psi C_\varphi + C_\psi C_\theta S_\varphi) & (-S_\psi S_\varphi + C_\psi C_\theta C_\varphi) & (-C_\psi S_\theta) \\ (S_\theta S_\varphi) & (S_\theta C_\varphi) & (C_\theta) \end{bmatrix} \tag{3.272}$$

Write a MATLAB routine that takes a vector v and the set of three Euler angles, and returns the rotated vector. What are the vectors that $e^{[j]}, j = 1, 2, 3$, are transformed into by a rotation with $\varphi = \pi/4$, $\theta = \pi/4$, $\psi = \pi/4$?

3.C.1. Consider the simple ethylene-like 2-D "molecule" in Figure 3.9 whose six atoms are located at the coordinates $r^{[\alpha]} = [x_\alpha y_\alpha]^{\mathrm{T}}$, $\alpha = 1, 2, \ldots, 6$. For simplicity, we neglect here any out-of-plane distortion, and compute the vibrational frequencies of the molecule from a proposed energy model involving harmonic springs. There are two types of springs: bond springs and angle springs. For each pair of bonded atoms α and β, we have a bond-stretching contribution to the potential energy,

$$U_{\alpha\beta}^{[b]}(r^{[\alpha]}, r^{[\beta]}) = \tfrac{1}{2}K_b[r_{\alpha\beta} - l]^2 \qquad r_{\alpha\beta} = \left| r^{[\beta]} - r^{[\alpha]} \right| \tag{3.273}$$

that helps to maintain the distance at the natural bond length l. The quantity K_b is the harmonic spring constant for this bond-stretching term. In Figure 3.9, the bond-stretching springs are shown as solid lines, and there are two types, with

$$\begin{aligned} k_{b1} &= 200 & l_1 &= 1.5 \\ k_{b2} &= 100 & l_2 &= 1 \end{aligned} \tag{3.274}$$

The angle-bending springs are shown in Figure 3.9 as dotted lines, and help to preserve the

4 Initial value problems

Armed with techniques for solving linear and nonlinear algebraic systems (Chapters 1 and 2) and the tools of eigenvalue analysis (Chapter 3), we are now ready to treat more complex problems of greater relevance to chemical engineering practice. We begin with the study of *initial value problems (IVPs)* of *ordinary differential equations (ODEs)*, in which we compute the trajectory in time of a set of N variables $x_j(t)$ governed by the set of first-order ODEs

$$\frac{d}{dt}x = \dot{x} = f(x) \tag{4.1}$$

We start the simulation, usually at $t_0 = 0$, at the *initial condition*, $x(t_0) = x^{[0]}$. Such problems arise commonly in the study of chemical kinetics or process dynamics. While we have interpreted above the variable of integration to be time, it might be another variable such as a spatial coordinate.

Our task will be to develop iterative rules for updating the trajectory by taking small steps forward in time. We would like the numerical trajectory to agree with the exact solution

$$x(t) = x^{[0]} + \int_{t_0}^{t} f(x(t'))dt' \tag{4.2}$$

Therefore, this problem is closely related to that of numerically computing the values of definite integrals

$$I_F = \int_a^b f(x)dx \tag{4.3}$$

Thus, we first consider the subject of *numerical integration (quadrature)*. As we can compute I_F analytically when $f(x)$ is a polynomial,

$$f(x) = \sum_{k=0}^{N} c_k x^k \qquad I_F = \sum_{k=0}^{N} \frac{c_k}{k+1}\left[b^{k+1} - a^{k+1}\right] \tag{4.4}$$

our first topic will be *polynomial interpolation*, the representation of an arbitrary function $f(x)$ by an approximating polynomial.

Following a discussion of polynomial interpolation and numerical integration, a survey is presented of the major techniques for solving IVPs, as implemented in MATLAB. Then, the issues of numerical accuracy and stability are treated at depth for commonly-used ODE solvers. Next, we consider *differential-algebraic equation (DAE)* systems that contain both

ODEs and nonlinear algebraic equations of the general form

$$M\dot{x} = f(x) \qquad x(t_0) = x^{[0]} \tag{4.5}$$

M is a matrix, in general itself a function of x and t, and singular, as it contains a row of all zeros for each algebraic equation. Finally, we present a robust method, based upon IVP solvers, to study how the solution to a set of nonlinear algebraic equations depends upon its parameters, *parametric continuation*.

Initial value problems of ordinary differential equations (ODE-IVPs)

IVPs arise when we study the dynamics of a system governed by a set of first-order ODEs, such as the batch reactor kinetics for the network of two elementary reactions

$$\begin{aligned}
A + B &\rightarrow C \quad r_{R1} = k_1 c_A c_B \\
C + B &\rightarrow D \quad r_{R2} = k_2 c_C c_B
\end{aligned} \tag{4.6}$$

At $t_0 = 0$, we start with the initial concentrations

$$c_A(t_0) = c_{A0} \quad c_B(t_0) = c_{B0} \quad c_C(t_0) = c_D(t_0) = 0 \tag{4.7}$$

The time evolution of the system follows the set of first-order ODEs

$$\begin{aligned}
\frac{dc_A}{dt} &= -r_{R1} \qquad \frac{dc_B}{dt} = -r_{R1} - r_{R2} \\
\frac{dc_C}{dt} &= r_{R1} - r_{R2} \qquad \frac{dc_D}{dt} = r_{R2}
\end{aligned} \tag{4.8}$$

We wish to use a general notation system for IVPs, and so define a *state vector*, x, that completely describes the state of the system at any time sufficiently well to predict its future behavior; here,

$$x = \begin{bmatrix} c_A & c_B & c_C & c_D \end{bmatrix}^T \tag{4.9}$$

We then write the ODE system, substituting for the reaction rates, as

$$\begin{aligned}
\dot{x}_1 &= -k_1 x_1 x_2 = f_1(x; k_1, k_2) \quad \dot{x}_2 = -k_1 x_1 x_2 - k_2 x_3 x_2 = f_2(x; k_1, k_2) \\
\dot{x}_3 &= k_1 x_1 x_2 - k_2 x_3 x_2 = f_3(x; k_1, k_2) \quad \dot{x}_4 = k_2 x_3 x_2 = f_4(x; k_1, k_2)
\end{aligned} \tag{4.10}$$

We collect the parameters of the system into a *parameter vector*

$$\Theta = \begin{bmatrix} k_1 & k_2 \end{bmatrix}^T \tag{4.11}$$

and write (4.10) in the *standard ODE-IVP form*

$$\dot{x} = f(x; \Theta) \qquad x(t_0) = x^{[0]} \tag{4.12}$$

We next show that this problem formulation is quite general by considering the following:

How do we express the system in the form of (4.12) if the function vector is itself time-dependent?

What if we have ODEs of higher order than one?

To resolve the first question, let us say that we have a problem with time-dependent function values

$$\dot{x} = f(t, x; \Theta) \qquad x(t_0) = x^{[0]} \tag{4.13}$$

Expanding the state vector to include time,

$$y = \begin{bmatrix} x_1 \\ \vdots \\ x_N \\ t \end{bmatrix} \qquad g = \begin{bmatrix} f_1 \\ \vdots \\ f_N \\ 1 \end{bmatrix} \tag{4.14}$$

we write the system in the standard form as

$$\dot{y} = g(y; \Theta) \quad y(t_0) = y^{[0]} = \begin{bmatrix} x^{[0]} \\ t_0 \end{bmatrix} \tag{4.15}$$

To answer the second question let us say that we have a second-order ODE for $u(t)$

$$a_2(u, t)\frac{d^2 u}{dt^2} + a_1(u, t)\frac{du}{dt} + a_0(u, t) = 0 \tag{4.16}$$

We define the state vector

$$x = \begin{bmatrix} u & \dfrac{du}{dt} & t \end{bmatrix}^{\mathrm{T}} \tag{4.17}$$

and write the ODE as

$$a_2(x_1, x_3)\frac{dx_2}{dt} + a_1(x_1, x_3)x_2 + a_0(x_1, x_3) = 0 \tag{4.18}$$

As long as $a_2(x_1, x_3) \neq 0$, we can write (4.18) in the standard form

$$\dot{x} = \begin{bmatrix} \dot{x}_1 \\ \dot{x}_2 \\ \dot{x}_3 \end{bmatrix} = \begin{bmatrix} x_2 \\ -[a_1(x_1, x_3)x_2 + a_0(x_1, x_3)]/a_2(x_1, x_3) \\ 1 \end{bmatrix} \tag{4.19}$$

Thus if we can solve (4.12) by calculating (4.2), we can simulate the dynamics of a wide variety of systems.

Polynomial interpolation

Because we can evaluate the integral of a polynomial analytically, integration techniques typically employ polynomial approximations of the integrand. The general problem of *polynomial interpolation* is as follows. Let us say that we have sampled some function $f(x)$ at the $N + 1$ *support points* $\{x_0, x_1, x_2, \ldots, x_N\}$ to obtain the function values $\{f_0, f_1, f_2, \ldots, f_N\}$, $f_j = f(x_j)$. We wish to construct a polynomial of degree N

$$p(x) = a_0 + a_1 x + a_2 x^2 + \cdots + a_N x^N \tag{4.20}$$

such that $p(x)$ agrees with $f(x)$ at each support point,

$$p(x_j) = a_0 + a_1 x_j + a_2 x_j^2 + \cdots + a_N x_j^N = f(x_j) \qquad j = 0, 1, 2, \ldots, N \tag{4.21}$$

The coefficients of the polynomial therefore must satisfy the linear system

$$
X\boldsymbol{a} =
\begin{bmatrix}
1 & x_0 & x_0^2 & \cdots & x_0^N \\
1 & x_1 & x_1^2 & \cdots & x_1^N \\
1 & x_2 & x_2^2 & \cdots & x_2^N \\
\vdots & \vdots & \vdots & & \vdots \\
1 & x_N & x_N^2 & \cdots & x_N^N
\end{bmatrix}
\begin{bmatrix}
a_0 \\ a_1 \\ a_2 \\ \vdots \\ a_N
\end{bmatrix}
=
\begin{bmatrix}
f_0 \\ f_1 \\ f_2 \\ \vdots \\ f_N
\end{bmatrix}
= \boldsymbol{f}
\tag{4.22}
$$

As long as no two support points are colocated, $\det(X) \neq 0$, and there is a unique polynomial of degree N that solves the interpolation problem. However, we would like to obtain the interpolating polynomial without having to solve a system of N linear equations by elimination, requiring $\sim N^3$ FLOPs. Also, X easily may have quite a high condition number and thus round-off errors may be significant, as we multiply, add, and subtract numbers of vastly different magnitudes during elimination.

Lagrange interpolation

From the drawbacks of solving (4.22) by elimination, it would be nice to have a direct way to construct $p(x)$ from the function values. In fact, it is possible to express $p(x)$ as the linear combination

$$
p(x) = \sum_{j=0}^{N} f_j L_j(x)
\tag{4.23}
$$

The $\{L_j(x)\}$ that satisfy $p_j(x) = f_j(x)$ are the *Lagrange polynomials*

$$
\begin{aligned}
L_j(x) &= \frac{(x - x_0) \cdots (x - x_{j-1})(x - x_{j+1}) \cdots (x - x_N)}{(x_j - x_0) \cdots (x_j - x_{j-1})(x_j - x_{j+1}) \cdots (x_j - x_N)} \\
&= \prod_{\substack{k=0 \\ k \neq j}}^{N} \left[\frac{x - x_k}{x_j - x_k} \right]
\end{aligned}
\tag{4.24}
$$

$L_j(x_k) = \delta_{jk}$. In Figure 4.1, we plot the Lagrange polynomials for support points at $\{0, 0.5, 1.0\}$. We use these polynomials to approximate $f(x) = \sqrt{x}$ on $[0, 1]$ by (4.23). Note that outside of $[0, 1]$, $p(x)$ and $f(x)$ diverge rapidly. This demonstrates the drastic loss in accuracy that may occur when extrapolating outside of the range of data using polynomials.

Newton interpolation

One disadvantage of Lagrange interpolation is that if we add another support point x_{N+1}, we have to recompute all of the Lagrange polynomials again from scratch. *Newton interpolation* avoids this difficulty, and expresses the interpolating polynomial as

$$
\begin{aligned}
P_N(x) = {}& a_0 + a_1(x - x_0) + a_2(x - x_0)(x - x_1) \\
& + \cdots + a_N(x - x_0)(x - x_1)(x - x_2)\ldots(x - x_{N-1})
\end{aligned}
\tag{4.25}
$$

We can evaluate $P_N(x)$ rapidly through the factorization

$$
P_N(x) = a_0 + (x - x_0)[a_1 + (x - x_1)[a_2 + (x - x_2)[\cdots + (x - x_{N-1})a_N]]]
\tag{4.26}
$$

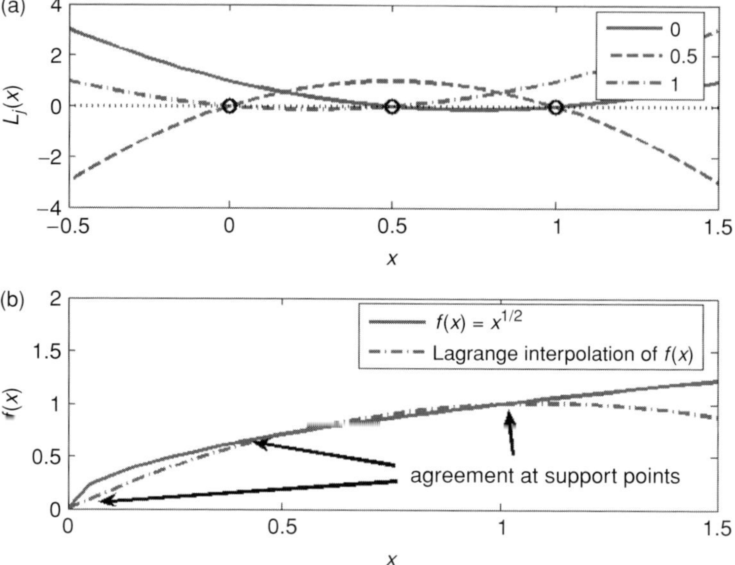

Figure 4.1 (a) Lagrange polynomials for the support points $\{0, 0.5, 1\}$ (b). Lagrange interpolation of the square root function on $[0, 1]$.

Given the function values $\{f_0, f_1, \ldots, f_N\}$ at $\{x_0, x_1, \ldots, x_N\}$, how do we obtain the coefficients $\{a_0, a_1, \ldots, a_N\}$ of the Newton interpolating polynomial?

We sequentially solve the equations

$$\begin{aligned}
f(x_0) &= a_0 \\
f(x_1) &= a_0 + (x_1 - x_0)a_1 \\
f(x_2) &= a_0 + (x_2 - x_0)a_1 + (x_2 - x_0)(x_2 - x_1)a_2
\end{aligned} \tag{4.27}$$

First, $a_0 = f(x_0)$. We next obtain a_1 from

$$f(x_1) = f(x_0) + (x_1 - x_0)a_1 \quad \Rightarrow \quad a_1 = \frac{f(x_1) - f(x_0)}{(x_1 - x_0)} \tag{4.28}$$

To get a_2, we subtract $f(x_1)$ from $f(x_2)$,

$$\begin{aligned}
f(x_2) - f(x_1) &= [a_0 + (x_2 - x_0)a_1 + (x_2 - x_0)(x_2 - x_1)a_2] - [a_0 + (x_1 - x_0)a_1] \\
f(x_2) - f(x_1) &= (x_2 - x_1)a_1 + (x_2 - x_0)(x_2 - x_1)a_2 \\
\frac{f(x_2) - f(x_1)}{(x_2 - x_1)} &= a_1 + (x_2 - x_0)a_2
\end{aligned} \tag{4.29}$$

Defining the *first-order divided differences*

$$f[x_j, x_k] = \frac{f(x_j) - f(x_k)}{(x_j - x_k)} = \frac{f(x_k) - f(x_j)}{(x_k - x_j)} \tag{4.30}$$

Table 4.1 Values of the cube root function in [0.4, 0.6]

x	$f(x) = x^{1/3}$
0.4	0.7368
0.5	0.7937
0.6	0.8434

this yields

$$\frac{f[x_1, x_2] - f[x_0, x_1]}{(x_2 - x_0)} = a_2 \equiv f[x_0, x_1, x_2] \tag{4.31}$$

We define *divided differences* of higher orders recursively

$$f[x_0, x_1, \ldots, x_k] = \frac{f[x_1, \ldots, x_k] - f[x_0, \ldots, x_{k-1}]}{(x_k - x_0)} \tag{4.32}$$

and find that the coefficients a_j are simply divided differences of the $\{f_k\}$:

$$a_j = f[x_0, x_1, x_2, \ldots, x_j] \tag{4.33}$$

Comparing the polynomial $P_N(x)$ for support points $\{x_0, x_1, \ldots, x_N\}$,

$$P_N(x) = a_0 + a_1(x - x_0) + \cdots + a_N(x - x_0)(x - x_1)\ldots(x - x_{N-1}) \tag{4.34}$$

to that, $P_{N+1}(x)$, obtained by adding an additional support point x_{N+1},

$$\begin{aligned} P_{N+1}(x) = {} & a_0 + a_1(x - x_0) + \cdots + a_N(x - x_0)(x - x_1)\ldots(x - x_{N-1}) \\ & + a_{N+1}(x - x_0)(x - x_1)\ldots(x - x_N) \end{aligned} \tag{4.35}$$

we see that

$$P_{N+1}(x) = P_N(x) + a_{N+1}(x - x_0)(x - x_1)\ldots(x - x_N) \tag{4.36}$$

Thus, all previously-computed coefficients $\{a_0, \ldots, a_N\}$ remain unchanged.

As a demonstration, we use the Newton method to interpolate $f(x) = x^{1/3}$ in [0.4, 0.6] using the support points $\{0.4, 0.5, 0.6\}$ (Table 4.1). We wish to predict $f(0.55) = 0.8193$. First, we compute $f[x_j] = f(x_j)$,

$$a_0 = f[x_0] = 0.7368 \qquad f[x_1] = 0.7937 \qquad f[x_2] = 0.8434 \tag{4.37}$$

Next, we obtain the first-order divided differences $f[x_0, x_1]$ and $f[x_1, x_2]$,

$$a_1 = f[x_0, x_1] = \frac{f[x_1] - f[x_0]}{x_1 - x_0} = \frac{0.7937 - 0.7368}{0.5 - 0.4} = 0.5689$$

$$f[x_1, x_2] = \frac{f[x_2] - f[x_1]}{x_2 - x_1} = \frac{0.8434 - 0.7937}{0.6 - 0.5} = 0.4973 \tag{4.38}$$

From these, we get the second-order divided difference $f[x_0, x_1, x_2]$,

$$a_2 = f[x_0, x_1, x_2] = \frac{f[x_1, x_2] - f[x_0, x_1]}{x_2 - x_0} = \frac{0.4973 - 0.5689}{0.6 - 0.4}$$
$$= -0.3581 \qquad (4.39)$$

The Newton interpolating polynomial, and its value at $x = 0.55$, are

$$P_2(x) = a_0 + (x - x_0)[a_1 + (x - x_1)a_2]$$
$$= (0.7648) + (0.55 - 0.4)[(0.5689) + (0.55 - 0.5)(-0.3581)] = 0.8195 \qquad (4.40)$$

Thus, the approximation of $f(0.55) = 0.8193$ is quite accurate.

Hermite Interpolation

We have interpolated a function based on its values at the support points; however, we may wish to include as well information about the leading order derivatives at some or all of the support points. In *Hermite interpolation*, we find the polynomial $p(x)$ of degree N that satisfies the following $N + 1$ conditions at the $M + 1$ points $x_0 < x_1 < \cdots < x_M$,

$$p^{(k)}(x_j) = \left.\frac{d^k p}{dx^k}\right|_{x_j} = \left.\frac{d^k f}{dx^k}\right|_{x_j} = f_j^{(k)} \qquad \begin{array}{l} j = 0, 1, \ldots, M \\ k = 0, 1, \ldots, n_j - 1 \\ \Sigma_j n_j = N + 1 \end{array} \qquad (4.41)$$

That is, at each support point x_j, $p(x)$ matches the values of $f(x)$ and its derivatives up to order $n_j - 1$. The interpolating polynomial is

$$p(x) = \sum_{j=0}^{M} \sum_{k=0}^{n_j - 1} f_j^{(k)} L_{jk}(x) \qquad (4.42)$$

where the $L_{jk}(x)$ are computed from auxiliary polynomials $l_{jk}(x)$ by

$$l_{jk}(x) = \frac{(x - x_j)^k}{k!} \prod_{\substack{h=0 \\ h \neq j}}^{M} \left(\frac{x - x_h}{x_j - x_h}\right)^{n_h} \qquad (4.43)$$

$$L_{j, n_j - 1}(x) = l_{j, n_j - 1}(x) \qquad j = 0, 1, \ldots, M \qquad (4.44)$$

$$L_{jk}(x) = l_{jk}(x) - \sum_{v=k+1}^{n_j - 1} l_{jk}^{(v)}(x_j) L_{jv}(x) \qquad k = n_j - 2, n_j - 3, \ldots, 0 \qquad (4.45)$$

We demonstrate this method by approximating a function $f(x)$ in the domain $[a, b]$ from the values of $f(x)$ and $f^{(1)}(x)$ at either end. The auxiliary polynomials for $k = 0, 1$ with $x_0 = a, x_1 = b$ are

$$l_{0k}(x) = \frac{(x - a)^k}{k!}\left(\frac{x - b}{a - b}\right)^2 \qquad l_{1k}(x) = \frac{(x - b)^k}{k!}\left(\frac{x - a}{b - a}\right)^2 \qquad (4.46)$$

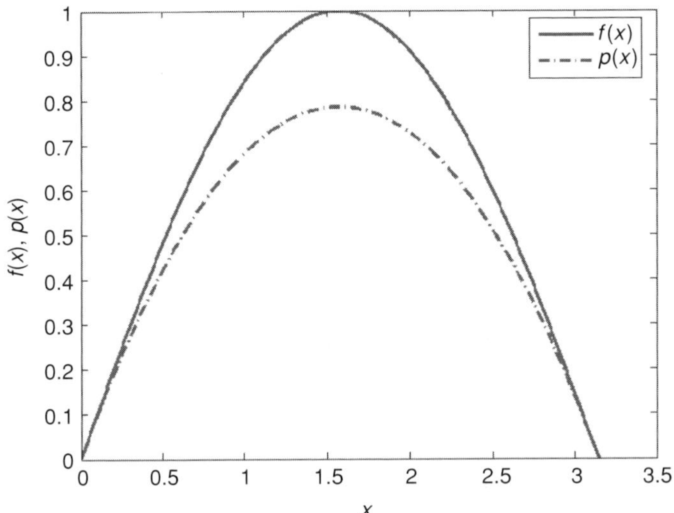

Figure 4.2 Plot of the sine function $f(x)$ at its Hermite interpolating polynomial $p(x)$, fitting the function and first derivative values at each end point.

Thus, we have from (4.44)

$$L_{01}(x) = l_{01}(x) = (x - a)\left(\frac{x - b}{a - b}\right)^2 \qquad L_{11}(x) = l_{01}(x) = (x - b)\left(\frac{x - a}{b - a}\right)^2 \quad (4.47)$$

and from (4.45), $L_{j0}(x) = l_{j0}(x) - l_{j0}^{(1)}(x_j)L_{j0}(x)$, yielding

$$L_{00}(x) = \left(\frac{x - b}{a - b}\right)^2 - \left[\frac{2}{(a - b)}\right]\left[(x - a)\left(\frac{x - b}{a - b}\right)^2\right]$$

$$(4.48)$$

$$L_{01}(x) = \left(\frac{x - a}{b - a}\right)^2 - \left[\frac{2}{(b - a)}\right]\left[(x - b)\left(\frac{x - a}{b - a}\right)^2\right]$$

Using hermite_ex.m, we approximate the sine function on $[0, \pi]$ by

hermite_ex('sin',0,pi);

to generate Figure 4.2.

Other types of interpolation

Here, we have considered interpolation using only polynomials to match function values at a set of support points; however, many other types of interpolation exist, e.g. with trigonometric functions instead of polynomials. For brevity, we do not consider these methods here, but refer the interested reader to Press *et al.* (1992) and Quateroni *et al.* (2000). The interpolation methods introduced above are sufficient to meet our immediate needs of computing the values of definite integrals. In MATLAB, various options for polynomial interpolation are available in **interp1**.

Newton–Cotes integration

We can now address the problem of *numerical integration (quadrature)*.

We want to integrate $f(x)$ over $[a, b]$. Interpolating $f(x)$ at the support points $a = x_0 < x_1 < \cdots < x_N = b$, we obtain

$$\int_a^b f(x)dx \approx \int_a^b \left[\sum_{j=0}^N f_j L_j(x)\right] dx = \sum_{j=0}^N f_j \int_a^b L_j(x)dx = \sum_{j=0}^N w_j f_j \tag{4.49}$$

In *Newton–Cotes integration*, we place the support points uniformly,

$$x_j = a + jh \qquad j = 0, 1, \ldots, N \qquad h = (b - a)/N \tag{4.50}$$

to yield the weights

$$w_j = h\alpha_j \qquad \alpha_j = \int_0^N L_j(t)dt \qquad L_j(t) = \prod_{k=0,k\neq j}^N \left(\frac{t - k}{j - k}\right) \tag{4.51}$$

Common leading-order methods are

$$\int_a^b f(x)dx = \frac{b - a}{2}\{f_0 + f_1\} + O(|b - a|^3) \quad \text{trapezoid rule} \tag{4.52}$$

$$\int_a^b f(x)dx = \frac{b - a}{6}\{f_0 + 4f_1 + f_2\} + O(|b - a|^4) \qquad \text{Simpson's rule} \tag{4.53}$$

$$\int_a^b f(x)dx = \frac{b - a}{8}\{f_0 + 3f_1 + 3f_2 + f_3\} + O(|b - a|^5) \qquad \text{3/8 rule} \tag{4.54}$$

To improve the accuracy of the integral estimate, we can either move to a higher interpolation order, or more conveniently, subdivide the integration domain into a number of subdomains, $a = x_0 < x_1 < \cdots < x_N = b$,

$$\int_a^b f(x)dx = \int_{x_0}^{x_1} f(x)dx + \int_{x_1}^{x_2} f(x)dx + \cdots + \int_{x_{N-1}}^{x_N} f(x)dx \tag{4.55}$$

and apply to each subdomain one of the low-order Newton–Cotes rules. By this approach, we obtain the popular *composite trapezoid* integration rule, implemented in MATLAB as **trapz**,

$$\int_a^b f(x)dx \approx \sum_{j=1}^N \frac{(x_j - x_{j-1})}{2}[f(x_j) + f(x_{j-1})] \qquad \text{error} \propto \frac{1}{N^2} \tag{4.56}$$

An efficient means to improve accuracy is to estimate the error in each interval separately, and adaptively add new support points only to those intervals where rapid changes in $f(x)$ yield the highest errors. We can further increase the accuracy by computing several approximate values of the definite integral with different values of $h = (b - a)/N$, and extrapolating the results to the limit $h \to 0$. In MATLAB, adaptive refinement is applied with Simpson's rule in **quad**.

Use of *trapz* and *quad*

We demonstrate the use of these quadrature functions by computing the integral of $f(x) = e^{-x}$ over [0, 1], for which

$$I_F = \int_0^1 e^{-x}dx = -e^{-x}\Big|_0^1 = e^{-x}\Big|_1^0 = 1 - e^{-1} = 0.6321 \tag{4.57}$$

Using **trapz**, the integral and the corresponding error are computed by

```
N = 50;
x = linspace(0,1,N);
f = exp(-x);
int_trapz = trapz(x,f),
   int_trapz = 0.6321
int_exact = 1 – exp(-1),
   int_exact = 0.6321
error_trapz = int_exact – int_trapz,
   error_trapz = -2.1939e-005
```

The accuracy of the calculation is improved by increasing the number of support points N. This integral can also be computed using **quad**. First, we write a routine that returns the integrand function,

```
function f = calc_integrand_nexp(x);
f = exp(-x);
return;
```

and then compute the integral using **quad**,

```
int_quad = quad('calc_integrand_nexp',0,1),
   int_quad = 0.6321
error_quad = int_exact − int_quad,
   error_quad = -1.3768e-009
```

quad expects the integrand routine to return a vector of function values for an input vector of argument values. The accuracy can be increased by adjusting the tolerance through an additional argument,

```
int_quad_2 = quad('calc_integrand_nexp',0,1,1e-14),
   int_quad_2 = 0.6321
error_quad_2 = int_exact – int_quad_2,
   error_quad_2 = 0
```

Gaussian quadrature

In Newton–Cotes integration, the support points are equally spaced; however, we might ask whether there is a more accurate choice of support points. We examine this issue for the *weighted integral*

$$I_w = \int_a^b w(x) f(x) dx \tag{4.58}$$

$[a, b]$ may be finite or infinite (e.g. $[0, \infty]$, $[-\infty, \infty]$), and $w(x)$ is a nonnegative *weight function*, $w(x) \geq 0$, all of whose moments are finite,

$$\mu_k = \int_a^b x^k w(x) dx \quad k = 0, 1, 2, \ldots \tag{4.59}$$

As examples of weighted integrals, consider

$$\text{Cartesian} \quad \int_a^b f(x) dx \quad w(x) = 1 \tag{4.60}$$

$$\text{radial in two dimensions} \quad \int_0^R f(r) 2\pi r \, dr \quad w(r) = 2\pi r \tag{4.61}$$

$$\text{radial in three dimensions} \quad \int_0^R f(r) 4\pi r^2 dr \quad w(r) = 4\pi r^2 \tag{4.62}$$

We now examine the optimal placement of support points for the rule

$$I_w = \int_a^b w(x) f(x) dx \approx \sum_{j=0}^N w_j f(x_j) \tag{4.63}$$

Here for brevity, we discuss only the major results. The supplemental material in the accompanying website presents a more extensive discussion, with proofs of all theorems.

Preliminary definitions

Let us define the following sets of polynomials:

$$\begin{aligned} \Pi_j &\equiv \{p(x) | \text{degree}(p) \leq j\} = \text{all polynomials of degree j or less} \\ {}_1\Pi_j &\equiv \{p(x) | p(x) = x^j + a_1 x^{j-1} + \cdots + a_j\} \end{aligned} \tag{4.64}$$

That is, ${}_1\Pi_j$ is the set of polynomials that have degree j exactly, and that also have a leading coefficient of 1. Thus, ${}_1\Pi_j \subset \Pi_j$.

Definition A function $f(x)$ is said to be *square-integrable* over $[a, b]$ for $w(x)$ if the following definite integral exists and is finite:

$$\|f\|_2^2 = \int_a^b w(x) [f(x)]^2 dx \geq 0 \tag{4.65}$$

Definition The *scalar product* of two real square-integrable functions is

$$\langle f | g \rangle \equiv \int_a^b w(x) f(x) g(x) dx \tag{4.66}$$

Definition Two square-integrable functions are said to be **orthogonal** if their scalar product is zero,

$$\langle f | g \rangle = \int_a^b w(x) f(x) g(x) dx = 0 \Rightarrow f \perp g \tag{4.67}$$

Orthogonal polynomials

We now identify some useful properties of orthogonal polynomials.

Theorem GQ1 *For the interval [a, b] and a nonnegative weight function $w(x)$, there exists a set of orthogonal polynomials of leading coefficient one, $p_j(x) \in {}_1\Pi_j, j = 0, 1, 2, \ldots,$ such that $\langle p_j | p_k \rangle = 0$ for $j \neq k$. These polynomials are defined uniquely by the recursion formula*

$$p_{j+1}(x) = (x - \delta_{j+1})p_j(x) - \gamma_{j+1}^2 p_{j-1}(x) \tag{4.68}$$

where

$$p_{-1}(x) = 0 \quad p_0(x) = 1$$
$$\delta_{j+1} = \langle xp_j | p_j \rangle / \langle p_j | p_j \rangle \quad j = 1, 2, \ldots$$
$$\gamma_{j+1}^2 = \begin{cases} 0, j = 0 \\ \langle p_j | p_j \rangle / \langle p_{j-1} | p_{j-1} \rangle, \quad j = 1, 2, \ldots \end{cases} \tag{4.69}$$

A proof of this theorem is provided in the supplemental material.

Theorem GQ2 *The N roots $\{x_1, x_2 \ldots, x_N\}$ of the orthogonal polynomial $p_N(x)$ are real, distinct and lie in (a, b), i.e. $a < x_1 < x_2 \cdots < x_N < b$. The roots of $p_N(x)$ may be computed by the eigenvalue technique of Chapter 3.*

A proof of this theorem is provided in the supplemental material.

Theorem GQ3 *The $N \times N$ matrix*

$$A = \begin{bmatrix} p_0(x_1) & p_0(x_2) & \cdots & p_0(x_N) \\ p_1(x_1) & p_1(x_2) & \cdots & p_1(x_N) \\ \vdots & \vdots & & \vdots \\ p_{N-1}(x_1) & p_{N-1}(x_2) & \cdots & p_{N-1}(x_N) \end{bmatrix} \tag{4.70}$$

is nonsingular for any mutually distinct arguments $\{x_1, x_2, \ldots, x_N\}$.

A proof of this theorem is provided in the supplemental material.

Gaussian quadrature

We next find that the roots $x_1, x_2, \ldots, x_N$ of the orthogonal polynomial $p_N(x)$ provide a set of support points that yield highly accurate estimates of weighted integrals on $[a, b]$.

Theorem GQ4 *Let $\{x_1, x_2 \ldots, x_N\}$ be the N distinct roots of $p_N(x)$ in the open interval (a, b). Let $\{w_1, w_2, \ldots, w_N\}$ be the solution of the linear system*

$$\sum_{k=1}^{N} p_j(x_k)w_k = \begin{cases} \langle p_0 | p_0 \rangle, & \text{if } j = 0 \\ 0, & \text{if } j = 1, 2, \ldots, N - 1 \end{cases} \tag{4.71}$$

whose matrix (4.70) is nonsingular by Theorem GQ3. Then $w_j > 0$ for all $j = 1, 2, \ldots,$

N, and for all $p(x) \in \Pi_{2N-1}$,

$$\int_a^b w(x)p(x)dx = \sum_{j=1}^{N} w_j p(x_j) \qquad (4.72)$$

That is, if we place the support points at the zeros of the orthogonal polynomial $p_N(x)$, we obtain the exact *value of the definite integral for* any *polynomial integrand of degree $2N - 1$ or less. This is* much better *than we can usually expect, since with only N support points, we generally can represent exactly only polynomials of degree $N - 1$ or less.*

A proof of this theorem is provided in the supplemental material.

Gaussian quadrature with $w(x) = 1$ and the use of **quadl**

To compute the integral $\int_a^b f(x)dx$, with $w(x) = 1$, we define ξ such that

$$x(\xi) = \left[\frac{a+b}{2}\right] + \left[\frac{b-a}{2}\right]\xi \qquad x(-1) = a \qquad x(1) = b \qquad (4.73)$$

and thus

$$\int_a^b f(x)dx = \left[\frac{b-a}{2}\right]\int_{-1}^{1} f[x(\xi)]dx \qquad (4.74)$$

We use the set of orthogonal polynomials for $[a, b] = [1, 1]$, $w(x) = 1$, the *Legendre polynomials*. We select an integration order through our choice of N, and compute the roots of $p_N(x)$, e.g. by the eigenvalue method. Then, we solve for the weights, and evaluate the integral. The node locations and quadrature weights can be stored for repeated use.

quadl applies Gaussian quadrature adaptively. Actually, **quadl** uses as a default ***Lobatto quadrature***, which adds two more support points at a and b to the roots of $p_N(x)$ within (a, b). However, sometimes we want all support points to lie in the interior of the domain. Let us say that we wish to integrate a function with a singularity at $a < x_s < b$; i.e., $f(x_s)$ blows up to $\pm\infty$.

Let us say that this singularity is integrable so that $\int_a^b f(x)dx$ exists and is finite. Then, we compute the integral by splitting (a, b) as

$$\int_a^b f(x)dx = \int_a^{x_s} f(x)dx + \int_{x_s}^b f(x)dx \qquad (4.75)$$

and applying Gaussian quadrature to each integral to avoid evaluating $f(x_s)$.

We demonstrate **quadl** to integrate $f(x) = e^{-x}$ over $[0, 1]$. Continuing the calculations that demonstrated **trapz** and **quad**,

```
int_quadl = quadl('calc_integrand_nexp',0,1),
   int_quadl = 0.6321
```

Like **quad**, **quadl** expects the integrand routine to return a vector of function values for an input vector of argument values.

Multidimensional integrals

The techniques introduced above are extended easily to multidimensional integrals. For example, consider the double integral

$$I_D = \int_a^b \int_{l(x)}^{u(x)} f(x, y)\,dy\,dx \tag{4.76}$$

Let us define the function of x alone,

$$F(x) = \int_{l(x)}^{u(x)} f(x, y)\,dy \tag{4.77}$$

such that $I_D = \int_a^b F(x)\,dx$. Applying the methods of 1-D integration,

$$\int_a^b F(x)\,dx \approx \sum_{k=0}^{N} w_k^{\langle x \rangle} F(x_k) \tag{4.78}$$

We approximate each $F(x_k)$ using support points $y_{kj} \in [l(x_k), u(x_k)]$,

$$F(x_k) = \int_{l(x_k)}^{u(x_k)} f(x_k, y)\,dy \approx \sum_{j=0}^{M_k} w_{kj}^{\langle y \rangle} f(x_k, y_{kj}) \tag{4.79}$$

such that

$$I_D = \int_a^b \int_{l(x)}^{u(x)} f(x, y)\,dy\,dx \approx \sum_{k=0}^{N} \sum_{j=0}^{M_k} \left[w_k^{\langle x \rangle} w_{kj}^{\langle y \rangle} \right] f(x_k, y_{kj}) \tag{4.80}$$

dblquad applies this approach to integration domains that are rectangles in two dimensions. For nonrectangular domains, we must compute the integral over a rectangular region encompassing the domain of integration, and then set the integrand to zero outside of the integration domain,

$$\int_a^b \int_{l(x)}^{u(x)} f(x, y)\,dy\,dx = \int_a^b \int_{l_{\min}}^{u_{\max}} f(x, y)\Omega(x, y)\,dy\,dx$$
$$u_{\max} = \max_{x \in [a, b]} u(x) \qquad\qquad l_{\min} = \min_{x \in [a, b]} l(x)$$
$$\Omega(x, y) = \begin{cases} 1, & \text{if } l(x) \le y \le u(x) \\ 0, & \text{otherwise} \end{cases} \tag{4.81}$$

In three dimensions, triple integrals are evaluated using **triplequad**. To integrate $f(x, y) = x^2 + 2y^2 - 2xy$ over the unit circle, $x^2 + y^2 \le 1$,

$$I_D = \int_{x^2+y^2 \le 1} [x^2 + 2y^2 - 2xy]\,dx\,dy \tag{4.82}$$

we write a routine that returns $f(x, y)$ for (x, y) within the circle and a value of zero for (x, y) outside of it. **dblquad** expects the function to accept a vector of x arguments and a scalar y argument. Here, the function is

```
function f = calc_integrand_2D(x,y);
f = zeros(size(x));
```

```
for k = 1:length(f)
   var1 = x(k)^2 + y^2;
   if(var1 > 1)
      f(k) = 0;
   else
      f(k) = x(k)^2 + 2*y^2 − 2*x(k)*y;
   end
end
return;
```

The value of the integral is computed by

```
int_2D = dblquad('calc_integrand_2D',-1,1,-1,1),
   int_2D = 2.3562
```

Monte Carlo integration

For integrations over very complex domains or in spaces of very high dimension, we use *Monte Carlo integration*, a stochastic technique in which points are generated at random, and contribute to the integral if they lie within the domain of integration. A more efficient implementation of Monte Carlo integration, based on importance sampling, is discussed in Chapter 7, and applications of this method to Bayesian statistical analysis are considered in Chapter 8. Here, we present a simple Monte Carlo algorithm to compute (4.82). We define the *indicator function* for the unit circle

$$\Omega_{r \leq 1}(x, y) = \begin{cases} 1, & x^2 + y^2 \leq 1 \\ 0, & \text{otherwise} \end{cases} \tag{4.83}$$

Then, for $f(x, y) = x^2 + 2y^2 - 2xy$, the integral is

$$I_D = \int_{-1}^{1} \int_{-1}^{1} f(x, y)\Omega_{r \leq 1}(x, y)dxdy \tag{4.84}$$

Using **rand**, a function that returns a uniformly-distributed random number in $[0, 1]$, we generate a long sequence of values $(x^{[k]}, y^{[k]})$ and compute the average value over this sequence of the integrand of (4.84),

$$\langle f\Omega \rangle_s = \frac{1}{N_s} \sum_{k=1}^{N_s} f\left(x^{[k]}, y^{[k]}\right)\Omega_{r \leq 1}\left(x^{[k]}, y^{[k]}\right) \tag{4.85}$$

For large N_s, this sampled average should agree with the exact value

$$\langle f\Omega \rangle_{\text{exact}} = \frac{1}{A_{sd}} \int_{-1}^{1} \int_{-1}^{1} f(x, y)\Omega_{r \leq 1}(x, y)dxdy \tag{4.86}$$

where the area A_{sd} of the sampled domain is

$$A_{sd} = \int_{-1}^{1} \int_{-1}^{1} (1)dxdy = 4 \tag{4.87}$$

Thus, as $\langle f\Omega \rangle_s \approx \langle f\Omega \rangle_{\text{exact}}$, we have the approximate value for the integral

$$\int_{-1}^{1} \int_{-1}^{1} f(x, y)\Omega_{r \leq 1}(x, y)\,dx\,dy \approx A_{\text{sd}}\langle f\Omega \rangle_s$$

$$= \frac{A_{\text{sd}}}{N_s} \sum_{k=1}^{N_s} f\left(x^{[k]}, y^{[k]}\right)\Omega_{r \leq 1}\left(x^{[k]}, y^{[k]}\right) \qquad (4.88)$$

The following code uses this method to obtain an approximate value for the integral of (4.82) in good agreement with that obtained from **dblquad**.

```
NumIter = 5000000; % increase for greater accuracy
f_sum = 0;
for iter = 1:NumIter
   x = -1 + 2*rand; y = -1 + 2*rand;
   r = x^2 + y^2;
   if(r <= 1)
      f_sum = f_sum + x^2 + 2*y^2 - 2*x*y;
   end
end
f_avg = f_sum/NumIter;
area = 4;
int_2D_MC = f_avg*area,
   int_2D_MC = 2.3563
```

Linear ODE systems and dynamic stability

We now resume our discussion of IVPs, starting with a system of linear first-order ODEs $\dot{x} = Ax$,

$$
\begin{array}{l}
\dot{x}_1 = a_{11}x_1 + a_{12}x_2 + \cdots + a_{1N}x_N \\
\dot{x}_2 = a_{21}x_1 + a_{22}x_2 + \cdots + a_{2N}x_N \\
\vdots \\
\dot{x}_N = a_{N1}x_1 + a_{N2}x_2 + \cdots + a_{NN}x_N
\end{array}
\begin{bmatrix} \dot{x}_1 \\ \dot{x}_2 \\ \vdots \\ \dot{x}_N \end{bmatrix}
=
\begin{bmatrix}
a_{11} & a_{12} & \cdots & a_{1N} \\
a_{21} & a_{22} & \cdots & a_{2N} \\
\vdots & \vdots & & \vdots \\
a_{N1} & a_{N2} & \cdots & a_{NN}
\end{bmatrix}
\begin{bmatrix} x_1 \\ x_2 \\ \vdots \\ x_N \end{bmatrix}
$$

$$(4.89)$$

that can be solved analytically. Hence, we use it to test the performance of numerical integration algorithms. For a single equation, the solution is

$$x(t) = e^{at}x^{[0]} \qquad \dot{x} = ae^{at}x^{[0]} = ax \qquad (4.90)$$

For multiple ODEs the solution is of a similar form

$$x(t) = e^{At}x^{[0]} \qquad (4.91)$$

where we define the *matrix exponential function* as

$$e^{A} = I + A + \frac{1}{2!}AA + \frac{1}{3!}AAA + \frac{1}{4!}AAAA + \cdots \qquad (4.92)$$

We can show that (4.91) is indeed a solution, as

$$\dot{x} = \frac{d}{dt}\left[I + At + \frac{t^2}{2!}AA + \frac{t^3}{3!}AAA + \frac{t^4}{4!}AAAA + \cdots\right]x^{[0]}$$

$$= \left[A + tAA + \frac{t^2}{2!}AAA + \frac{t^3}{3!}AAAA + \cdots\right]x^{[0]}$$

$$= A\left[I + At + \frac{t^2}{2!}AA + \frac{t^3}{3!}AAA + \cdots\right]x^{[0]} = Ae^{At}x^{[0]} = Ax(t) \qquad (4.93)$$

Stability of the steady state of a linear system

Obviously, $\dot{x} = Ax$ has a steady state at $x_s = 0$ where $\dot{x} = 0$, but is it a stable steady state?

Definition   A *steady state* of a dynamic system is one in which the time derivatives of each state variable are zero. A steady state x_s is *stable*, if following every infinitesimal perturbation away from x_s, the system returns to x_s. A steady state x_s is *unstable*, if any infinitesimal perturbation causes the system to move away from x_s. A steady state for which a perturbation neither grows nor decays with time is said to be *neutrally stable*. Stability is a property of the particular steady state and not of the differential equation.

We now determine the stability of $x_s = 0$ for $\dot{x} = Ax$. Let us assume that A is diagonalizable, so that any $v \in \Re^N$ can be written as the linear combination

$$v = c_1 w^{[1]} + c_2 w^{[2]} + \cdots + c_N w^{[N]} \quad c_j \in C$$
$$A w^{[j]} = \lambda_j w^{[j]} \quad \lambda_j \in C \qquad (4.94)$$

To determine the stability of $x_s = 0$, we compute the response, starting at $x^{[0]} = \varepsilon$, and examine whether $x(t)$ returns to $x_s = 0$ or it diverges. That is, if the system is stable, we must have

$$\lim_{t\to\infty}\|x(t)\| = \lim_{t\to\infty}\|e^{At}\varepsilon\| = 0 \qquad (4.95)$$

We expand e^{At} and $\varepsilon = c_1 w^{[1]} + c_2 w^{[2]} + \cdots + c_N w^{[N]}$ to write $x(t) = e^{At}\varepsilon$ as

$$x(t) = \left[I + At + \frac{t^2}{2!}AA + \cdots\right]\left[\sum_{j=1}^{N} c_j w^{[j]}\right]$$

$$= \sum_{j=1}^{N} c_j \left[I + At + \frac{t^2}{2!}AA + \cdots\right]w^{[j]}$$

$$= \sum_{j=1}^{N} c_j \left[1 + t\lambda_j + \frac{t^2}{2!}\lambda_j^2 + \cdots\right]w^{[j]} = \sum_{j=1}^{N} c_j e^{\lambda_j t} w^{[j]} \qquad (4.96)$$

In general, the eigenvalues of A are complex

$$\lambda_j = a_j + ib_j \quad a_j = \text{Re}(\lambda_j) \quad b_j = \text{Im}(\lambda_j) \qquad (4.97)$$

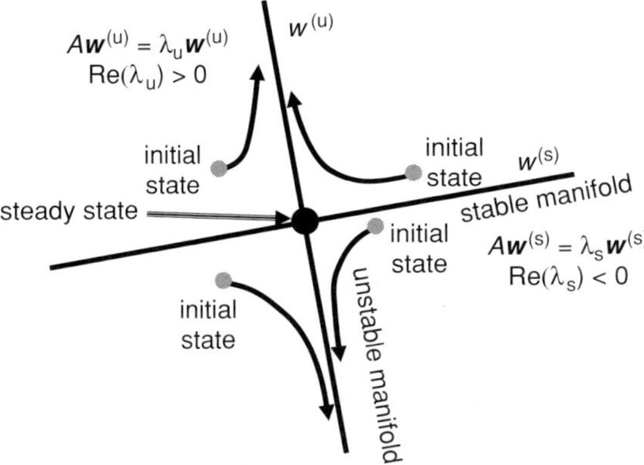

Figure 4.3　Phase plot of trajectories of 2-D system from various initial states with respect to the stable and unstable manifolds.

so that the response of the system is

$$x(t) = \sum_{j=1}^{N} c_j e^{\lambda_j t} w^{[j]} = \sum_{j=1}^{N} c_j e^{a_j t}[\cos(b_j t) + i\,\sin(b_j t)]w^{[j]} \tag{4.98}$$

If $b_j \neq 0$, the system oscillates during its response; however, as long as $a_j = Re(\lambda_j) < 0$ for *all* eigenvalues, $\lim_{t \to \infty} e^{a_j t} = 0$. Therefore,

If *all* eigenvalues λ_j of A have Re $(\lambda_j) < 0$, then for all $x^{[0]}$, $\lim_{t \to \infty} \|x(t)\| = 0$, and the steady state is *stable*.

If *any* of the eigenvalues λ_j of A have Re $(\lambda_j) > 0$, the steady state is *unstable*.

When *every* eigenvalue λ_j of A satisfies $Re(\lambda_j) \leq 0$ but there is at least *one* with Re $(\lambda_j) = 0$, the system does not return always to the steady state, but it does not diverge. The steady state is *neutrally stable*.

Definition　The span of all eigenvectors corresponding to the eigenvalues with real parts less than zero is the *stable manifold* of the system,

$$W^{(s)} = \text{span}\{w^{[j]}\,|\,Re(\lambda_j) < 0\} \tag{4.99}$$

The span of all eigenvectors corresponding to eigenvalues with real parts greater than zero is the *unstable manifold* of the system,

$$W^{(u)} = \text{span}\{w^{[j]}\,|\,Re(\lambda_j) > 0\} \tag{4.100}$$

The span of all eigenvectors corresponding to eigenvalues with real parts equal to zero is the *center manifold* of the system,

$$W^{(c)} = \text{span}\{w^{[j]}\,|\,Re(\lambda_j) = 0\} \tag{4.101}$$

The trajectory of the state vector approaches a steady state along its stable manifold and diverges along its unstable manifold (Figure 4.3).

Stability of a steady state of a nonlinear system

We now generalize the stability results to nonlinear systems, $\dot{x} = f(x; \Theta)$, with a steady state x_s, where $f(x_s; \Theta) = 0$. x_s is stable if $\lim_{t \to \infty} \|x(t) - x_s\| = 0$ following any infinitesimal perturbation ε away from x_s. If we only perturb the system slightly, we can represent each function near x_s as

$$f_j(x_s + \varepsilon) \approx f_j(x_s) + \sum_{k=1}^{N} \left.\frac{\partial f_j}{\partial x_k}\right|_{x_s} \varepsilon_k = \sum_{k=1}^{N} \left.\frac{\partial f_j}{\partial x_k}\right|_{x_s} \varepsilon_k \tag{4.102}$$

Defining the *Jacobian matrix*, whose elements are functions of x and Θ,

$$J(x; \Theta) = \begin{bmatrix} (\partial f_1/\partial x_1) & (\partial f_1/\partial x_2) & \cdots & (\partial f_1/\partial x_N) \\ (\partial f_2/\partial x_1) & (\partial f_2/\partial x_2) & \cdots & (\partial f_2/\partial x_N) \\ \vdots & \vdots & & \vdots \\ (\partial f_N/\partial x_1) & (\partial f_N/\partial x_2) & \cdots & (\partial f_N/\partial x_N) \end{bmatrix} \qquad J_{mn}(x; \Theta) = \left.\frac{\partial f_m}{\partial x_n}\right|_{(x;\Theta)} \tag{4.103}$$

the function vector in the vicinity of the steady state is approximately

$$f_j(x_s + \varepsilon) \approx \sum_{k=1}^{N} J_{jk}(x_s; \Theta)\varepsilon_k \quad \Rightarrow \quad f(x_s + \varepsilon; \Theta) \approx J(x_s; \Theta)\varepsilon \tag{4.104}$$

As x_s is fixed, $d(x_s + \varepsilon)/dt = \dot{\varepsilon}$, and the dynamics near x_s are described by

$$\dot{\varepsilon} = J(x_s; \Theta)\varepsilon \tag{4.105}$$

Thus, we can apply the stability analysis presented above, using

$$J(x_s; \Theta)w^{[k]} = \lambda_k w^{[k]} \tag{4.106}$$

The steady state x_s of $\dot{x} = f(x; \Theta)$ is *stable with respect to infinitesimal perturbations* if *every* eigenvalue of the Jacobian matrix $J(x_s; \Theta)$ has a real part less than zero; $\mathrm{Re}(\lambda_k) < 0$, for all k.

If *any* eigenvalue of $J(x_s; \Theta)$ has a positive real part, x_s is *unstable with respect to infinitesimal perturbations;* $\mathrm{Re}(\lambda_k) > 0$ for any k.

If *all* eigenvalues of $J(x_s; \Theta)$ have nonnegative real parts, but there is at least *one* with a zero real part, x_s is *neutrally stable with respect to infinitesimal perturbations;* $\mathrm{Re}(\lambda_k) \leq 0$ for all k, $\mathrm{Re}(\lambda_m) = 0$ for some m.

Note that in each of the statements above, we have included the restriction *"with respect to infinitesimal perturbations."* The system may very well respond unstably to large, finite perturbations even if all eigenvalues of the Jacobian have negative real parts.

Example. Stability of steady states for nonlinear ODE systems

Consider the system of two nonlinear ODEs

$$\begin{aligned} \dot{x}_1 &= -2(x_1 - 1)^2 - 2(x_1 - 1) + (x_2 - 1) = -2x_1^2 + 2x_1 + x_2 - 1 = f_1 \\ \dot{x}_2 &= -(x_1 - 1) - 3(x_2 - 1)^2 - 4(x_2 - 1) = -x_1 - 3x_2^2 + 2x_2 + 2 = f_2 \end{aligned} \tag{4.107}$$

which clearly has a steady state at (1, 1). The Jacobian matrix

$$
J(x) = \begin{bmatrix} \dfrac{\partial f_1}{\partial x_1} & \dfrac{\partial f_1}{\partial x_2} \\[2ex] \dfrac{\partial f_2}{\partial x_1} & \dfrac{\partial f_2}{\partial x_2} \end{bmatrix} = \begin{bmatrix} (-4x_1 + 2) & (1) \\ (-1) & (-6x_2 + 2) \end{bmatrix} \tag{4.108}
$$

evaluated at the steady state is

$$
J(x_s) = \begin{bmatrix} (-4(1) + 2) & (1) \\ (-1) & (-6(1) + 2) \end{bmatrix} = \begin{bmatrix} -2 & 1 \\ -1 & -4 \end{bmatrix} \tag{4.109}
$$

The eigenvalues of the Jacobian are

```
Jac = [-2 1; -1 -4];
eig(Jac)',
   ans = -3 -3
```

Both eigenvalues have negative real parts and thus the steady state (1, 1) is stable. As the eigenvalues are real, we do not expect the response to be oscillatory. Using the routine,

```
function f = stable_calc_f(t,x);
f = zeros(2,1);
f(1) = -2*x(1)^2 + 2*x(1) + x(2) -1;
f(2) = -x(1) - 3*x(2)^2 + 2*x(2) + 2;
return;
```

the following code (**ode45** is explained in further detail below) plots the system response for a small perturbation away from the steady state,

```
% set initial state with small perturbation
x_0 = [1;1] + 0.1*randn(2,1);
% compute response
[t_traj,x_traj] = ode45('stable_calc_f',[0 2],x_0);
% make plot
figure; subplot(2,1,1); plot(t_traj,x_traj(:,1));
xlabel('t'); ylabel('x_1(t)'); title('Response to small perturbation');
subplot(2,1,2); plot(t_traj,x_traj(:,2));
xlabel('t'); ylabel('x_2(t)');
```

A sample response is shown in Figure 4.4.

Let us next consider another system with a steady state at $x_s = (1, 1)$, but that has a different function f_1,

$$
\begin{aligned}
\dot{x}_1 &= (x_1 - 1)^2 + 3(x_1 - 1) - (x_2 - 1) = x_1^2 + x_1 - x_2 - 1 = f_1 \\
\dot{x}_2 &= -(x_1 - 1) - 3(x_2 - 1)^2 - 4(x_2 - 1) = -x_1 - 3x_2^2 + 2x_2 + 2 = f_2
\end{aligned} \tag{4.110}
$$

The Jacobian matrix at the steady state x_s is now

$$
J(x_s) = \begin{bmatrix} (2x_{s1} + 1) & (-1) \\ (-1) & (-6x_{s2} + 2) \end{bmatrix} = \begin{bmatrix} 3 & -1 \\ -1 & -4 \end{bmatrix} \tag{4.111}
$$

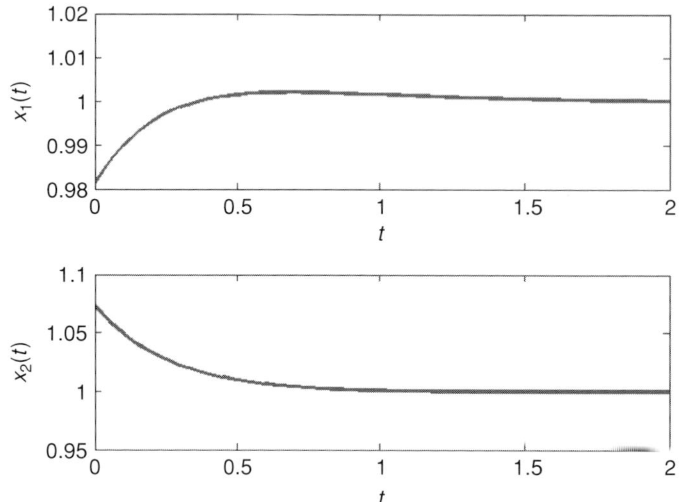

Figure 4.4 Response of a nonlinear system to small perturbation for a stable steady state.

and has eigenvalues,

```
Jac = [3 -1; -1 -4];
eig(Jac)',
ans = -4.1401 3.1401
```

Now, the steady state is unstable. With the routine,

```
function f = unstable_calc_f(t,x);
f = zeros(2,1);
f(1) = x(1)^2 + x(1) - x(2) - 1;
f(2) = -x(1) - 3*x(2)^2 + 2*x(2) + 2;
return;
```

we compute the response to a small perturbation as above, but now substitute 'unstable_calc_f' in the call to **ode45**. An example unstable response is shown in Figure 4.5. Here, for this particular random perturbation, the trajectory approaches a second steady state at $x'_s = [-2.1761 \ 1.5595]^T$. x'_s is a stable steady state as its Jacobian matrix

$$J(x'_s) = \begin{bmatrix} (2x'_{s1} + 1) & (-1) \\ (-1) & (-6x'_{s2} + 2) \end{bmatrix} = \begin{bmatrix} -3.3520 & -1 \\ -1 & -7.3570 \end{bmatrix} \tag{4.112}$$

has all eigenvalues with negative real parts, as is evident from Gershgorin's theorem. Non-linear ODE systems can have multiple steady states, each with different stability properties. For other guesses of the initial perturbation, the response blows up to infinity.

generate_phase_plots_ex1.m plots $x_2(t)$ vs. $x_1(t)$ for random initial states, with the trajectories shown as lines emanating from circles at the initial guesses (Figure 4.6). The manifolds for each steady state are shown as solid lines for stable eigenvalues and

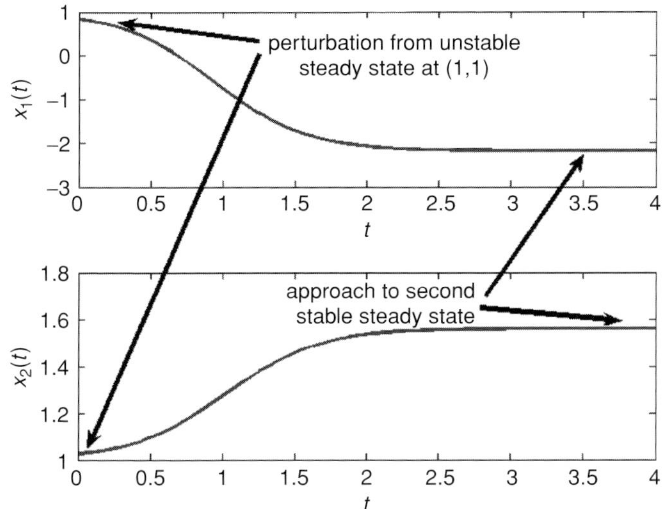

Figure 4.5 Response of nonlinear system to small perturbation for an unstable steady state showing the approach to another, stable steady state.

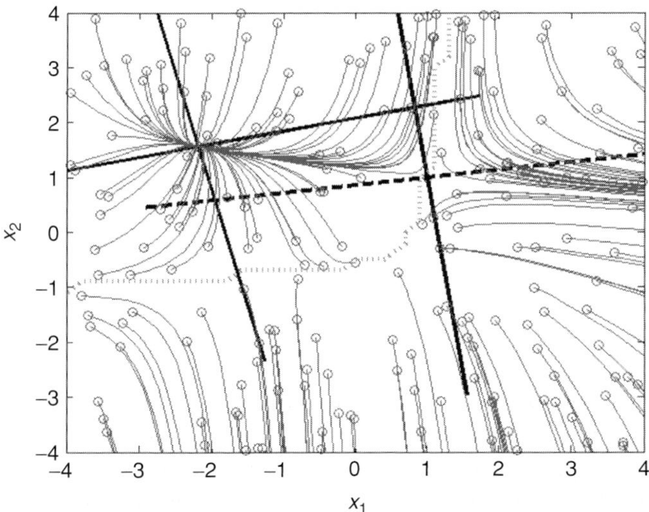

Figure 4.6 Trajectories of an unstable system in phase space, relative to the stable and unstable manifolds of each steady state. The boundary of the domain of attraction for stable steady state is shown as a dotted line.

dashed lines for unstable eigenvalues. Figure 4.6 also shows as a dotted line the boundary between the points that converge to the stable steady state x'_s (its *domain of attraction*) and those that do not. This boundary passes through the unstable steady state at $(1, 1)$, explaining why some random perturbations converge to the stable steady state while others do not.

Overview of ODE-IVP solvers in MATLAB

We next provide an overview of ODE-IVP solvers. The contents of this section provide a sufficient background to solve problems of the form

$$\dot{x} = f(x; \Theta) \qquad x(t_0) = x^{[0]} \tag{4.113}$$

First, we describe the basic time-marching approach of ODE-IVP solvers and contrast explicit and implicit, single-step and multistep solvers. Then, we demonstrate the use of the explicit single-step solver **ode45** and the implicit multistep solver **ode15s**.

Time-marching ODE-IVP solvers

ODE solvers update $x(t)$ in discrete time steps of size Δt to compute $x(t_k)$ at times $t_0 < t_1 < t_2 < \cdots$. For a constant time step, $t_k = t_0 + k(\Delta t)$; but often Δt varies throughout the course of the simulation. Δt is smaller for greater accuracy when $x(t)$ changes rapidly and is larger for increased simulation speed when $x(t)$ changes slowly. Over each time step, the exact update of the state vector is

$$x(t_k + \Delta t) - x(t_k) = \int_{t_k}^{t_k + \Delta t} \dot{x}(t)dt = \int_{t_k}^{t_k + \Delta t} f(x(t); \Theta)dt \tag{4.114}$$

If $x^{[k]} = x(t_k)$, the new state at $t_{k+1} = t_k + \Delta t$ can be estimated by a rule that approximates (4.114), of the form

$$x^{[k+1]} - x^{[k]} = (\Delta t)F\left[x^{[k]}, x^{[k+1]}, f(x; \Theta)\right] \qquad x^{[k+1]} \approx x(t_{k+1}) \tag{4.115}$$

$F[x^{[k]}, x^{[k+1]}, f(x; \Theta)]$ is a rule involving the old and new states that defines the method. At each step, there is introduced a new *local error*, proportional to some power of Δt, that is the difference between the approximate update (4.115) and the exact update (4.114).

As (4.115) uses information only about the state values at the beginning, $x^{[k]}$, and end, $x^{[k+1]}$, of the current time step, it is said to define a *single-step integration method*. For example, in the *Crank–Nicholson method*

$$F_{\text{CN}}\left[x^{[k]}, x^{[k+1]}, f(x; \Theta)\right] \equiv \tfrac{1}{2}\left[f(x^{[k]}; \Theta) + f(x^{[k+1]}; \Theta)\right] \tag{4.116}$$

and in the *implicit (backward) Euler method*

$$F_{\text{BE}}\left[x^{[k]}, x^{[k+1]}, f(x; \Theta)\right] \equiv f(x^{[k+1]}; \Theta) \tag{4.117}$$

Because $f(x; \Theta)$ generally is nonlinear, (4.115) often cannot be rearranged to provide a direct expression for $x^{[k+1]}$. Then, (4.115) is said to generate an *implicit integration method* that requires a nonlinear algebraic system to be solved at each time step.

Explicit single-step methods

By contrast, an *explicit single-step integration method*

$$x^{[k+1]} - x^{[k]} = (\Delta t)F\left[x^{[k]}, f(x; \Theta)\right] \tag{4.118}$$

yields the new state $x^{[k+1]}$ directly without solving an algebraic system. For example, in the *explicit (forward) Euler method*

$$F_{FE}\big[x^{[k]}, \, f(x; \Theta)\big] \equiv f\big(x^{[k]}; \Theta\big) \qquad x^{[k+1]} = x^{[k]} + (\Delta t) f\big(x^{[k]}; \Theta\big) \tag{4.119}$$

The explicit Euler method is not very accurate and is not often used. A more popular single-step explicit rule is the *fourth-order Runge–Kutta (RK4) method*

$$k^{(1)} = (\Delta t) f\big(x^{[k]}\big) \qquad k^{(2)} = (\Delta t) f\big(x^{[k]} + k^{(1)}/2\big)$$
$$k^{(3)} = (\Delta t) f\big(x^{[k]} + k^{(2)}/2\big) \qquad k^{(4)} = (\Delta t) f\big(x^{[k]} + k^{(3)}\big)$$
$$x^{[k+1]} - x^{[k]} = \tfrac{1}{6}\big[k^{(1)} + 2k^{(2)} + 2k^{(3)} + k^{(4)}\big] \tag{4.120}$$
$$\text{local error} \sim O(\Delta t^5) \qquad \text{global error} \sim O(\Delta t^4)$$

The RK4 method is said to be fourth-order as the *global error*, the net difference between the numerical and true solutions over the course of a simulation, is proportional to the fourth power of the time step Δt. For a method of order p, after some period of simulation time $t_N = t_0 + N(\Delta t)$,

$$x(t_N) - x(t_0) = \int_{t_0}^{t_N} \dot{x}(t)\,dt = x^{[N]} - x^{[0]} + O[(N\Delta t)^p] \tag{4.121}$$

We do not derive (4.120) here, but rather the simpler RK2 method, yet the path to higher-order RK methods becomes clear. We want to update $\dot{x} = f(x)$ from t_k to $t_{k+1} = t_k + \Delta t$ by approximating (4.114). With the explicit Euler method, we neglect the time variation of $f(x(t))$ to obtain the rule

$$x^{[k+1]} - x^{[k]} = (\Delta t) f\big(x^{[k]}\big) \equiv k^{(1)} \tag{4.122}$$

But, once we have computed $k^{(1)}$, we can use it to approximate the integrand of (4.114) more accurately. We take the mid-point of the full explicit Euler step, $x^{[k]} + k^{(1)}/2$, and evaluate the function at this point

$$k^{(2)} = (\Delta t) f\big(x^{[k]} + k^{(1)}/2\big) \approx (\Delta t) f(x(t_k + \Delta t/2)) \tag{4.123}$$

We now form a linear approximation to $f(x(t))$ over the time step,

$$f(x(t)) = f(x(t_k)) \left[\frac{(\Delta t/2) - (t - t_k)}{(\Delta t)/2}\right] + f(x(t_k + \Delta t/2)) \left[\frac{(t - t_k)}{(\Delta t)/2}\right] + O(\Delta t^2) \tag{4.124}$$

which we write in terms of $k^{(1)}$ and $k^{(2)}$,

$$f(x(t)) = k^{(1)} \left[\frac{(\Delta t/2) - (t - t_k)}{(\Delta t)^2/2}\right] + k^{(2)} \left[\frac{(t - t_k)}{(\Delta t)^2/2}\right] + O(\Delta t^2) \tag{4.125}$$

We substitute this approximation into (4.114),

$$x^{[k+1]} - x^{[k]} = \int_{t_k}^{t_k + \Delta t} \left\{ k^{(1)} \left[\frac{(\Delta t/2) - (t - t_k)}{(\Delta t)^2/2}\right] + k^{(2)} \left[\frac{(t - t_k)}{(\Delta t)^2/2}\right] + O(\Delta t^2) \right\} dt \tag{4.126}$$

Integrating, we obtain the *second-order Runge–Kutta (RK2) method*

$$x^{[k+1]} - x^{[k]} = k^{(2)} + O(\Delta t^3)$$
$$k^{(1)} = (\Delta t)f(x^{[k]}) \qquad k^{(2)} = (\Delta t)f(x^{[k]} + k^{(1)}/2) \qquad (4.127)$$
$$\text{local error} \sim O(\Delta t^3) \qquad \text{global error} \sim O(\Delta t^2)$$

Higher-order Runge–Kutta methods are generated by using $k^{(1)}$ and $k^{(2)}$ to construct even more accurate approximations for $f(x(t))$.

In MATLAB, **ode45** implements the *Runge–Kutta–Fehlberg 4-5 (RKF45) method*, based upon a fifth-order Runge–Kutta rule, from which a fourth-order update formula is extracted using the same function evaluations. If the two methods agree closely, the time step can be increased safely, but if they disagree, the time step is too large and should be reduced. Thus, the user can specify a desired level of accuracy, and let **ode45** adjust the step size as needed. Use of **ode45** is demonstrated below.

Implicit multistep methods

In a *multistep integration method*, the update rule also depends upon the values of x at times in the immediate past, up to a horizon length m_h. Implicit multi-step integration methods are of the general form

$$\alpha_{-1}x^{[k+1]} + \alpha_0 x^{[k]} = (\Delta t)\beta_{-1}f(x^{[k+1]}; \Theta) + U^{[k]} \qquad (4.128)$$

α_{-1}, α_0, and β_{-1} are dimensionless scalars and we have some rule for $U^{[k]}$ involving the current and old values of the state vector,

$$U^{[k]} = F\left[\Delta t, f(x; \Theta); t_k, x^{[k]}, t_{k-1}, x^{[k-1]}, t_{k-2}, x^{[k-2]}, \ldots, t_{k-m_h}, x^{[k-m_h]}\right]$$
$$(4.129)$$

For example, we can use polynomial interpolation of the past state and function data to approximate the exact update (4.114),

$$x(t_k + \Delta t) - x^{[k]} = \int_{t_k}^{t_k + \Delta t} f(x(t))dt = \int_0^1 \left(\frac{dx}{d\tau}\right)d\tau \qquad \tau \equiv \frac{t - t_k}{\Delta t} \qquad (4.130)$$

We have available at times $\{t_k, t_{k-1}, \ldots, t_{k-m_h}\}$ in the recent past the values of the state vector $\{x^{[k]}, x^{[k-1]}, \ldots, x^{[k-m_h]}\}$ and the time derivative function $\{f^{[k]}, f^{[k-1]}, \ldots, f^{[k-m_h]}\}$. We define the scaled time quantities

$$\tau_j = \frac{t_{k-j} - t_k}{\Delta t} \qquad \dot{x}(\tau_j) = \frac{dx}{d\tau}\bigg|_{x^{[k-j]}} = \left(\frac{dt}{d\tau}\right)\frac{dx}{dt}\bigg|_{x^{[k-j]}} = (\Delta t)f(x^{[k-j]}) \qquad (4.131)$$

We shall assume that the time step is constant in the recent past. If not, we can interpolate the values for nonuniform Δt to estimate the values of x and f at times in the past that *are* uniformly spaced. Thus, we have available for interpolation of $x(\tau)$ the data

$$
\begin{array}{ccccc}
\text{past} & \text{past} & \text{past} & \text{current} & \text{future} \\
\tau_{m_h} = -m_h & \tau_2 = -2 & \tau_1 = -1 & \tau_0 = 0 & \tau_{-1} = 1 \\
x(\tau_{m_h}) & \cdots \quad x(\tau_2) & x(\tau_1) & x(\tau_0) & x(\tau_{-1}) \\
\left.\dfrac{dx}{d\tau}\right|_{\tau_{m_h}} & \left.\dfrac{dx}{d\tau}\right|_{\tau_2} & \left.\dfrac{dx}{d\tau}\right|_{\tau_1} & \left.\dfrac{dx}{d\tau}\right|_{\tau_0} & \left.\dfrac{dx}{d\tau}\right|_{\tau_{-1}}
\end{array}
\tag{4.132}
$$

For uniform Δt, $x(\tau_j) = x^{[k-j]}$ and with (4.131), we use the data of (4.132) to approximate $x(\tau)$ by Hermite interpolation,

$$
x(\tau) \approx \pi(\tau) = \sum_{j=-1}^{m_h} \left[x^{[k-j]} L_{j0}(\tau) + (\Delta t) f^{[k-j]} L_{j1}(\tau) \right]
\tag{4.133}
$$

Above, we use both past state and function data, but a *backward difference formula (BDF) method* uses only the states at the present and past times,

$$
x(\tau) \approx \pi(\tau) = x^{[k+1]} L_{-1,0}(\tau) + (\Delta t) f^{[k+1]} L_{-1,1}(\tau) + \sum_{j=0}^{m_h} x^{[k-j]} L_{j0}(\tau)
\tag{4.134}
$$

Upon differentiation of (4.133), we have

$$
\frac{dx}{d\tau} \approx \frac{d\pi}{d\tau} = \sum_{j=-1}^{m_h} \left[x^{[k-j]} \frac{dL_{j0}}{d\tau} + (\Delta t) f^{[k-j]} \frac{dL_{j1}}{d\tau} \right]
\tag{4.135}
$$

Substituting (4.135) into (4.130) yields the update rule

$$
x^{[k+1]} - x^{[k]} = \int_0^1 \left(\frac{d\pi}{dt} \right) d\tau = \int_0^1 \left\{ \sum_{j=-1}^{m_h} \left[x^{[k-j]} \frac{dL_{j0}}{d\tau} + (\Delta t) f^{[k-j]} \frac{dL_{j1}}{d\tau} \right] \right\} d\tau
\tag{4.136}
$$

which can be written as

$$
\alpha_{-1} x^{[k+1]} + \sum_{j=0}^{m_h} \alpha_j x^{[k-j]} = (\Delta t)\beta_{-1} f^{[k+1]} + (\Delta t) \sum_{j=0}^{m_h} \beta_j f^{[k-j]}
\tag{4.137}
$$

where the coefficients are

$$
\alpha_{-1} = 1 - \int_0^1 \frac{dL_{-1,0}}{d\tau} d\tau \quad \alpha_0 = -1 - \int_0^1 \frac{dL_{00}}{d\tau} d\tau \quad \alpha_{j \in [1,m_h]} = -\int_0^1 \frac{dL_{j0}}{d\tau} d\tau
$$

$$
\beta_j = \int_0^1 \frac{dL_{j1}}{d\tau} d\tau \qquad j = -1,\, 0,\, 1,\, \ldots,\, m_h
\tag{4.138}
$$

Equation (4.137) is of the form of (4.128) with

$$
U^{[k]} = (\Delta t) \sum_{j=0}^{m_h} \beta_j f^{[k-j]} - \sum_{j=1}^{m_h} \alpha_j x^{[k-j]}
\tag{4.139}
$$

We now discuss the numerical solution of (4.128). We first generate an initial guess of the new state vector, $x^{[k+1,0]}$ from an explicit rule such as (4.118). Then, we use Newton's method to generate a sequence of refined estimates $x^{[k+1,1]}, x^{[k+1,2]}, \ldots$ that should converge to the solution of (4.128) if Δt is small enough. Such an approach, using an explicit rule to make an initial guess of $x^{[k+1]}$ followed by iterative solution of an implicit rule, is known as a *predictor/corrector method*.

Let $x^{[k+1,q]}$ be our current estimate of $x^{[k+1]}$ at the qth Newton iteration, with the function and Jacobian values $f^{[k+1,q]}$ and $J^{[k+1,q]}$. We approximate $f(x; \Theta)$ in the vicinity of $x^{[k+1,q]}$ as

$$f(x; \Theta) \approx f^{[k+1,q]} + J^{[k+1,q]}\left(x - x^{[k+1,q]}\right) \tag{4.140}$$

Substituting (4.140) into (4.128) for $f(x^{[k+1]}; \Theta)$, we have

$$\alpha_{-1}x^{[k+1]} + \alpha_0 x^{[k]} \approx (\Delta t)\beta_{-1}\left\{ f^{[k+1,q]} + J^{[k+1,q]}\left(x^{[k+1]} - x^{[k+1,q]}\right)\right\} + U^{[k]} \tag{4.141}$$

This yields the following linear algebraic system for $x^{[k+1, q+1]} \approx x^{[k+1]}$,

$$A^{[k+1,q]}x^{[k+1,q+1]} = b^{[k+1,q]}$$
$$A^{[k+1,q]} = \alpha_{-1}I - (\Delta t)\beta_{-1}J^{[k+1,q]} \tag{4.142}$$
$$b^{[k+1,q]} = -\alpha_0 x^{[k]} + (\Delta t)\beta_{-1}\left\{ f^{[k+1,q]} - J^{[k+1,q]}x^{[k+1,q]}\right\} + U^{[k]}$$

These Newton iterations are repeated until (4.128) is satisfied, yielding $x^{[k+1]}$. The entire process then is repeated for the next time step.

Clearly, an implicit method requires significantly more work per time step than an explicit method; however, some tricks are available to reduce the computational burden. When Δt is small, the changes in x from one iteration to the next are slight, as the difference between $x^{[k]}$ and $x^{[k+1]}$ is proportional to Δt. Thus, $A = \alpha_{-1}I - (\Delta t)\beta_{-1}J$ varies little and may be held constant, at least temporarily. LU factorization allows us to solve a large fraction of the systems of (4.142) without the need for elimination each time.

Stiffness and the choice of integration method

We might wonder why we bother to discuss implicit methods at all, if explicit methods allow much easier and faster updates at each time step. The reason is that for many problems, explicit methods may require, for reasons of numerical stability, the use of very small time steps. Then, for a given overall simulation time, the number of updates performed with an explicit method may be so much greater than with an implicit method that the implicit method becomes favored.

Implicit methods are favored for IVPs that are *stiff*, in which the condition number of the Jacobian matrix (the ratio of the largest and smallest eigenvalue moduli) is very large. Stiff systems are by no means rare and so we must be prepared to use both explicit and implicit methods. As a general rule, if we have no reason to expect that a problem is stiff, we first try an explicit method. If it fails, i.e., it keeps running with no end in sight, we try an implicit method. A more detailed treatment of stiffness and a comparison of the numerical stability properties of explicit and implicit methods are provided following the demonstration of the MATLAB ODE solvers.

ODE solvers in MATLAB

A list of ODE solvers (and of other routines that act on functions) is returned by help funfun, and a documentation window is opened by doc funfun. The two main routines of interest are **ode45**, an explicit single-step integrator, and **ode15s**, an implicit multistep integrator that works well for stiff systems. We demonstrate the use of **ode45** and **ode15s** for a simple batch reactor with the two elementary reactions $A + B \rightarrow C$ and $C + B \rightarrow D$

$$\frac{dc_A}{dt} = -r_{R1} \quad \frac{dc_B}{dt} = -r_{R1} - r_{R2} \quad \frac{dc_C}{dt} = r_{R1} - r_{R2} \quad \frac{dc_D}{dt} = r_{R2}$$

$$r_{R1} = k_1 c_A c_B \quad r_{R2} = k_2 c_C c_B \tag{4.143}$$

$$c_A(0) = c_{A0} \quad c_B(0) = c_{B0} \quad c_C(0) = 0 \quad c_D(0) = 0$$

To use **ode45** and **ode15s**, we must provide at least a routine that returns for input values of t and x, the value of $\dot{x} = f(t, x)$,

function f = calc_f(t, x, P1, P2, . . .);

P1, P2, . . . are optional fixed parameters. For (4.143), the routine is

```
function f = batch_reactor_ex_calc_f(t, x, k1, k2);
f = zeros(size(x));
% extract concentrations from state vector
cA = x(1); cB = x(2); cC = x(3); cD = x(4);
% compute rates of each reaction
r1 = k1*cA*cB; r2 = k2*cC*cB;
f(1) = -r1; % mole balance on A
f(2) = -r1 -r2; % mole balance on B
f(3) = r1 - r2; % mole balance on C
f(4) = r2; % mole balance on D
return;
```

For $k_1 = 1, k_2 = 0.5, c_{A0} = 1, c_{B0} = 2$, (4.143) is simulated until an end time $t_{end} = 10$ by calling **ode45** with the code

```
% set initial state
cA_0 = 1; cB_0 = 2; x_0 = [cA_0; cB_0; 0; 0];
k_1 = 1; k_2 = 0.5; % set fixed parameters
t_end = 10;  % set end time
% call ode45 to perform simulation
[t_traj,x_traj] = ode45(@batch_reactor_ex_calc_f, ...
   [0 t_end], x_0, [ ], k_1, k_2);
% plot results
figure; plot(t_traj,x_traj(:,1));
hold on; plot(t_traj,x_traj(:,2),'--');
plot(t_traj,x_traj(:,3),'-.'); plot(t_traj,x_traj(:,4),':');
xlabel('time t'); ylabel('concentration c_j(t)');
```

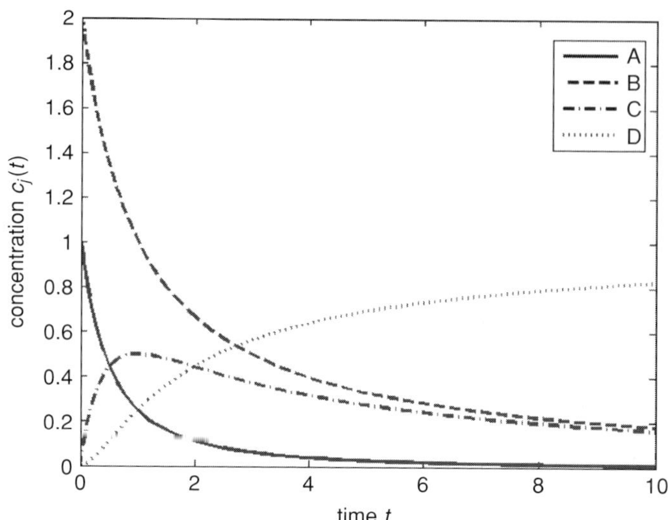

Figure 4.7 Dynamic concentrations of species in a batch chemical reactor ($A + B \rightarrow C$, $C + B \rightarrow D$).

title('Batch reactor, A + B ==> C, C + B ==> D');
legend('A','B','C','D','Location','Best');

The plot is shown in Figure 4.7. The general syntax for **ode45** is

[t_traj, x_traj] = ode45(fun_name, t_span, x_0, OPTIONS, P1, P2, . . .);

fun_name is the name of the routine that returns the time derivative vector. t_span is the set of times at which the state values are desired. If only a start and an end time are given (as above), **ode45** returns state values at several times within the time span. x_0 is the initial state. OPTIONS is a structure, set by **odeset**, that allows the user to override the default behavior of **ode45**. To accept the default arguments, use the empty set [ ] at this position. P1, P2, . . . are fixed parameters that are passed as arguments to fun_name. t_traj is a vector of times at which the state vector is returned, and the corresponding values of $x_k(t)$ are located in the kth column of x_traj.

The use of **ode15s** is analogous to that of **ode45**; however, if only the function values are returned by the user-supplied routine, **ode15s** has to evaluate the Jacobian matrix itself, which is costly for large systems. Therefore, for large ODE systems, it is helpful to supply a second routine that returns the Jacobian matrix for input t and x. While for (4.143) the effort is not necessary, we use it to demonstrate the procedure. The Jacobian is

$$J = \begin{bmatrix} (-k_1 c_B) & (-k_1 c_A) & 0 & 0 \\ (-k_1 c_B) & (-k_1 c_A - k_2 c_C) & (-k_2 c_B) & 0 \\ (k_1 c_B) & (k_1 c_A - k_2 c_C) & (-k_2 c_B) & 0 \\ 0 & (k_2 c_C) & (k_2 c_B) & 0 \end{bmatrix} \tag{4.144}$$

We supply a routine that computes the Jacobian,

```
function Jac = batch_reactor_ex_calc_Jac(t, x, k1, k2);
% allocate memory for the Jacobian matrix
N = length(x); Jac = zeros(N,N);
% extract state variables
cA = x(1); cB = x(2); cC = x(3); cD = x(4);
% set non zero values of the Jacobian
% row 1 – mole balance on A
Jac(1,1) = -k1*cB; Jac(1,2) = -k1*cA;
% row 2 – mole balance on B
Jac(2,1) = -k1*cB; Jac(2,2) = -k1*cA-k2*cC; Jac(2,3) = -k2*cB;
% row 3 – mole balance on C
Jac(3,1) = k1*cB; Jac(3,2) = k1*cA-k2*cC; Jac(3,3) = -k2*cB;
% row 4 - mole balance on D
Jac(4,2) = k2*cC; Jac(4,3) = k2*cB;
return;
```

and inform **ode15s** that this routine has been provided through **odeset**,

```
OPTIONS = odeset('Jacobian', 'batch_reactor_ex_calc_Jac');
[t_traj, x_traj] = ode15s(@batch_reactor_ex_calc_f, ...
   [0 t_end], x_0, OPTIONS, k_1, k_2);
```

Other flags in OPTIONS allow us to modify the tolerances or to provide an Events function that identifies when specified functions approach zero, and optionally stop the simulation whenever this happens. Type help odeset for more information.

Example. Stiffness and the QSSA in chemical kinetics

We contrast the behavior of the explicit **ode45** and implicit **ode15s** ODE solvers for a simple chemical system that exhibits stiffness for certain parameter values. Consider a reaction mechanism for $A \rightarrow B$ in which A first collides with an energetic third body M to form an activated species A^*. This activated species then decomposes to the product B. The mechanism

$$\begin{aligned} A + M &\rightarrow A^* + M & r_{R1} &= k_1 c_M c_A \\ A^* &\rightarrow B & r_{R2} &= k_2 c_{A^*} \end{aligned} \tag{4.145}$$

yields the batch kinetics

$$\frac{dc_A}{dt} = f_1 = -k_1 c_M c_A \quad c_{A0} = 1$$

$$\frac{dc_{A^*}}{dt} = f_2 = k_1 c_M c_A - k_2 c_{A^*} \quad c_{A^*0} = 0 \tag{4.146}$$

$$\frac{dc_B}{dt} = f_3 = k_2 c_{A^*} \quad c_{B0} = 0$$

*Table 4.2 Comparison of **ode45** and **ode15s** for kinetics system with an activated mechanism that becomes stiff with increasing k_2. Here, the condition number equals k_2*

k_2	CPU time with **ode45** (s)	CPU time with **ode15s** (s)
1	0.0200	0.0401
10	0.0200	0.0901
100	0.2103	0.0801
1000	1.8226	0.0701
10 000	24.8648	0.0701

As $d[c_A + c_{A^*} + c_B]/dt = 0$, $c_B = c_{A0} - c_A - c_{A^*}$ and thus we only need to simulate the first two equations. The Jacobian of the system for $\{c_A(t), c_{A^*}(t)\}$ is

$$J = \begin{bmatrix} \dfrac{\partial f_1}{\partial c_A} & \dfrac{\partial f_1}{\partial c_{A^*}} \\[2mm] \dfrac{\partial f_2}{\partial c_A} & \dfrac{\partial f_2}{\partial c_{A^*}} \end{bmatrix} = \begin{bmatrix} -k_1 c_M & 0 \\ k_1 c_M & -k_2 \end{bmatrix} \tag{4.147}$$

The eigenvalues of J are $-k_1 c_M$ and $-k_2$. When the activated species is very reactive, $k_2 \gg k_1 c_M$ and the system is stiff. Let us examine what happens to the performances of **ode45** and **ode15s**. QSSA_ex.m uses **cputime** to compare the CPU times required to solve the ODE-IVP with the two solvers when $k_1 c_M = 1$ and k_2 is increased from a value of 1 (Table 4.2). As the system becomes stiff, **ode45** requires more CPU time to simulate the response, due to a need to use very small time steps to preserve numerical stability. **ode15s** performs much better when the system is stiff, showing little change in performance. The concentrations of A, A*, and B are plotted for the nonstiff case $k_2 = 1$ in Figure 4.8. As expected, c_{A^*} initially grows as it is produced by the first reaction, and then decreases later as it is consumed by the second reaction.

For the stiff case $k_2 = 100$, the dynamic concentration profiles are very different (Figure 4.9). Except for a very short initial period where c_{A^*} increases rapidly from $c_{A^*0} = 0$, it appears that c_{A^*} remains proportional to c_A. Such an observation leads to the *quasi-steady state approximation* (*QSSA*), which states that the concentration of a very active species such as A* is in dynamic equilibrium between generation and destruction; i.e.,

$$\frac{dc_{A^*}}{dt} = k_1 c_M c_A - k_2 c_{A^*} \approx 0 \quad \Rightarrow \quad c_{A^*} \approx \left(\frac{k_1 c_M}{k_2}\right) c_A \tag{4.148}$$

With the QSSA, the system reduces to a single ODE,

$$\frac{dc_A}{dt} = -k_1 c_M c_A \quad c_{A0} = 1 \quad c_{A^*} = \left(\frac{k_1 c_M}{k_2}\right) c_A \quad c_B = c_{A0} - c_A - c_{A^*} \tag{4.149}$$

that is solved by **ode45** with little difficulty. For $k_2 = 10^4$, **ode45** requires 25 CPU seconds to solve the full ODE system (4.146) but only 0.03 CPU seconds to solve the single QSSA ODE (4.149). Inspection of Figure 4.9 and Figure 4.10 shows the QSSA to be quite accurate for $k_2/(k_1 c_M) = 100$.

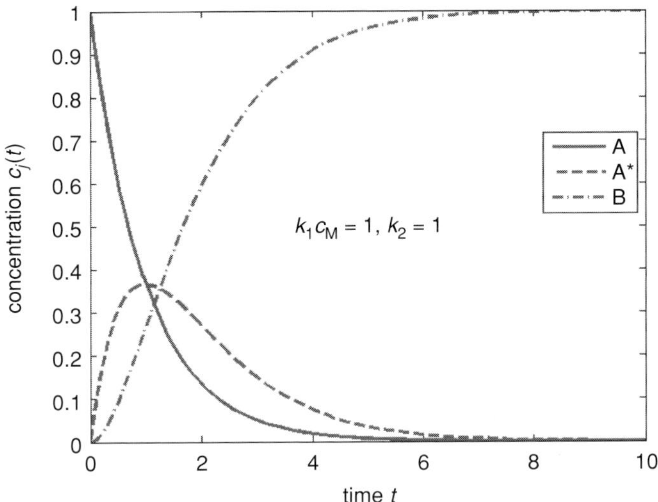

Figure 4.8 Dynamic concentration profiles for an activated reaction mechanism in the nonstiff case, $k_2 = 1$ ($A + M \rightarrow A^* + M$, $A^* \rightarrow B$).

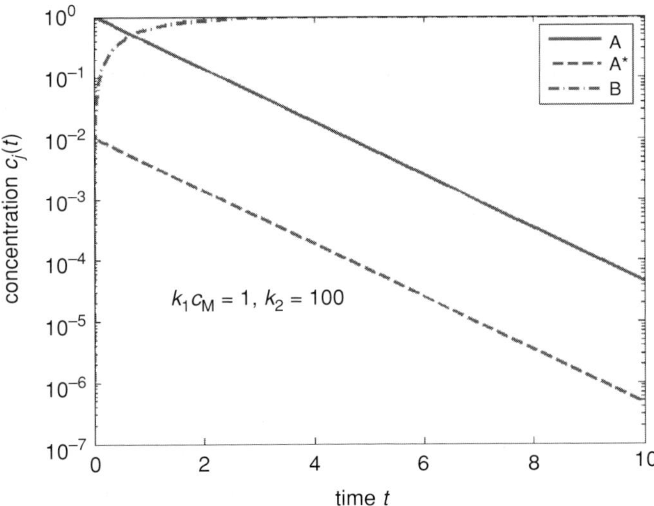

Figure 4.9 Dynamic concentration profiles for the activated reaction mechanism in the stiff case, $k_2 = 100$ ($A + M \rightarrow A^* + M$, $A^* \rightarrow B$).

Accuracy and stability of single-step methods

Above we have demonstrated the use of MATLAB solvers, with little discussion of their performance. Here, we address these issues for the restricted class of single-step integrators

$$x^{[k+1]} = x^{[k]} + (\Delta t)\left[(1 - \theta)f(x^{[k]}) + \theta f(x^{[k+1]})\right] \tag{4.150}$$

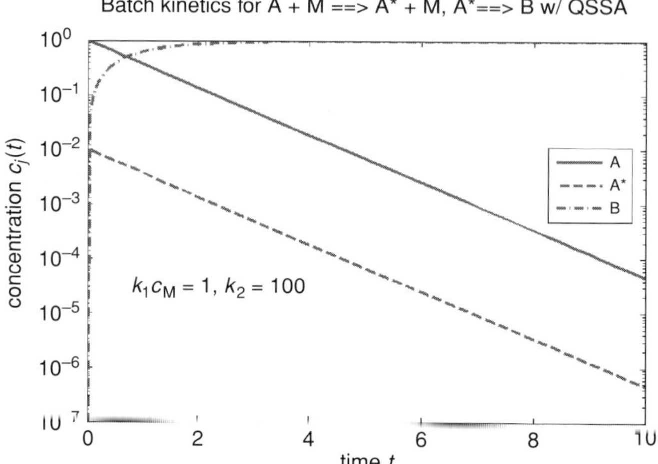

Figure 4.10 Dynamic concentration profiles for the activated reaction mechanism with $k_2 = 100$, obtained using the QSSA ($A + M \rightarrow A^* + M$, $A^* \rightarrow B$).

This includes as special cases the *explicit (forward) Euler method* for $\theta = 0$, the *implicit (backward) Euler method* for $\theta = 1$, and the *Crank–Nicholson method* for $\theta = 1/2$.

Numerical accuracy and the order of an integration method

We begin first with a discussion of accuracy. Let us form Taylor series expansions around $x(t_k)$ and $x(t_{k+1})$ in the forward and reverse time directions respectively, with $t_{k+1} = t_k + \Delta t$,

$$x(t_{k+1}) = x(t_k) + (\Delta t)f(x^{[k]}) + \frac{(\Delta t)^2}{2}\ddot{x}(t_k) + \frac{(\Delta t)^3}{6}\dddot{x}(t_k) + \cdots \tag{4.151}$$

$$x(t_k) = x(t_{k+1}) - (\Delta t)f(x^{[k+1]}) + \frac{(\Delta t)^2}{2}\ddot{x}(t_{k+1}) - \frac{(\Delta t)^3}{6}\dddot{x}(t_{k+1}) + \cdots \tag{4.152}$$

Multiplying (4.151) by $(1 - \theta)$, (4.152) by $(-\theta)$, and adding the resulting two equations, we obtain

$$x(t_{k+1}) - x(t_k) = (\Delta t)\big[(1 - \theta)f(x^{[k]}) + \theta f(x^{[k+1]})\big] + LE \tag{4.153}$$

where the *local error* introduced by truncating the two expansions is

$$LE = \frac{(\Delta t)^2}{2}[(1 - \theta)\ddot{x}(t_k) - \theta\ddot{x}(t_{k+1})] + \frac{(\Delta t)^3}{6}[(1 - \theta)\dddot{x}(t_k) + \theta\dddot{x}(t_{k+1})] + \cdots \tag{4.154}$$

As the leading term in the local error is proportional to $(\Delta t)^2$, we write

$$x(t_{k+1}) = x(t_k) + (\Delta t)\big[\theta f(x^{[k]}) + (1 - \theta)f(x^{[k+1]})\big] + O[(\Delta t)^2] \tag{4.155}$$

By this analysis, we see that at every time step, there is a new local error, proportional to $(\Delta t)^2$, introduced into our numerical state trajectory $\{x^{[1]}, x^{[2]}, \ldots\}$. Over the course of a simulation of duration t_{sim}, the number of time steps taken is $t_{sim}/(\Delta t)$. If at each of these steps, we introduce an additional local error proportional to $(\Delta t)^2$, then the *global error*, the net difference between the numerical trajectory and the true trajectory $x(t)$, is proportional to $[t_{sim}/(\Delta t)](\Delta t)^2 = t_{sim}(\Delta t)$. As the global error is proportional to the first power of Δt, (4.150) is said to be a *first-order accurate integration method*. We usually wish to use integration methods of high orders of accuracy, as then a reduction in Δt yields a dramatic decrease in the global error. Note that the most accurate choice of θ is $\theta = 1/2$, the Crank–Nicholson method, for which the $(\Delta t)^2$ terms nearly cancel out.

Absolute stability of an integration method

Accuracy is not the only consideration when choosing an integration method. Usually, a property of more practical importance is *numerical stability*. We examine here questions of stability for the linear test equation

$$\dot{x} = Ax \tag{4.156}$$

where we assume that the matrix A is diagonalizable. We wish to perturb the system by a small amount ε away from its steady state $x_s = 0$ and then compare the resulting numerical response with a time step Δt to the true response $x(t) = e^{At}\varepsilon$. As A is assumed to be diagonalizable, we can write

$$\varepsilon = x^{[0]} = c_1^{[0]}w^{[1]} + c_2^{[0]}w^{[2]} + \cdots + c_N^{[0]}w^{[N]} \quad c_j^{[0]} \in C$$
$$Aw^{[j]} = \lambda_j w^{[j]} \quad \lambda_j = a_j + ib_j \tag{4.157}$$

The true response of the system to this perturbation is

$$x(t) = e^{At}\varepsilon = \sum_{j=1}^{N} c_j^{[0]} e^{\lambda_j t} w^{[j]} \tag{4.158}$$

If $x_s = 0$ is stable, $\mathrm{Re}(\lambda_j) < 0$ for all j, and the true response $x(t)$ returns to $x_s = 0$ for all perturbations. Thus, if we find that the numerical response does not behave similarly, we obviously have a problem.

Definition *Absolute stability*. Let us generate a sequence of values $x^{[k]}$ that are meant to approximate the response of the system $\dot{x} = Ax$ at times $t_k = k(\Delta t)$ for $k = 0, 1, 2, \ldots$. We expand each member of the sequence as a linear combination of the eigenvectors of A, $x^{[k]} = \Sigma_{j=1}^{N} c_j^{[k]} w^{[j]}$. We say that the rule generating this sequence is *absolutely stable* if, for every eigenvalue of A with $\mathrm{Re}(\lambda_j) < 0$, $|c_j^{[k+1]}| \leq |c_j^{[k]}|$. That is, the numerical response of the system does not grow along the stable manifold of A.

From (4.150), the numerical response for $\dot{x} = Ax$ is generated by the rule

$$x^{[k+1]} = x^{[k]} + (\Delta t)\big[(1 - \theta)Ax^{[k]} + \theta Ax^{[k+1]}\big] \tag{4.159}$$

We expand the old and new states in eigenvectors of A,

$$x^{[k]} = \sum_{j=1}^{N} c_j^{[k]} w^{[j]} \quad x^{[k+1]} = \sum_{j=1}^{N} c_j^{[k+1]} w^{[j]} \tag{4.160}$$

and substitute these expansions into (4.159),

$$\sum_{j=1}^{N} c_j^{[k+1]} \boldsymbol{w}^{[j]} = \sum_{j=1}^{N} c_j^{[k]} \boldsymbol{w}^{[j]} + (\Delta t) \left[(1-\theta) A \sum_{j=1}^{N} c_j^{[k]} \boldsymbol{w}^{[j]} + \theta A \sum_{j=1}^{N} c_j^{[k+1]} \boldsymbol{w}^{[j]} \right] \tag{4.161}$$

Using $A \boldsymbol{w}^{[j]} = \lambda_j \boldsymbol{w}^{[j]}$, collecting terms yields

$$0 = \sum_{j=1}^{N} \left\{ -c_j^{[k+1]} + c_j^{[k]} + c_j^{[k]}(\Delta t)(1-\theta)\lambda_j + c_j^{[k+1]}(\Delta t)\theta\lambda_j \right\} \boldsymbol{w}^{[j]} \tag{4.162}$$

Setting to zero each component of (4.162) yields

$$\frac{c_j^{[k+1]}}{c_j^{[k]}} = \frac{1 + (\Delta t)\lambda_j (1-\theta)}{1 - (\Delta t)\lambda_j \theta} \tag{4.163}$$

For simplicity, let us assume that each λ_j of interest for the determination of absolute stability is real. Then, $\lambda_j < 0$ and absolute stability requires

$$\left| \frac{c_j^{[k+1]}}{c_j^{[k]}} \right| = \left| \frac{1 - \omega_j(1-\theta)}{1 + \theta\omega_j} \right| \le 1 \qquad \omega_j = -(\Delta t)\lambda_j > 0 \tag{4.164}$$

Let us consider the three cases of interest:

$$\text{explicit Euler} \qquad \left| \frac{c_j^{[k+1]}}{c_j^{[k]}} \right|_{\theta=0} = \left| \frac{1 - \omega_j}{1} \right| \le 1 \quad \text{only when } \omega_j \le 2 \tag{4.165}$$

$$\text{implicit Euler} \qquad \left| \frac{c_j^{[k+1]}}{c_j^{[k]}} \right|_{\theta=1} = \left| \frac{1}{1 + \omega_j} \right| \le 1 \quad \forall \omega_j > 0 \tag{4.166}$$

$$\text{Crank–Nicholson} \left| \frac{c_j^{[k+1]}}{c_j^{[k]}} \right|_{\theta=1/2} = \left| \frac{1 - \frac{1}{2}\omega_j}{1 + \frac{1}{2}\omega_j} \right| \le 1 \quad \forall \omega_j > 0 \tag{4.167}$$

For both the implicit Euler and the Crank–Nicholson method, we find that for any $\omega_j > 0$, (4.164) is satisfied and the methods are absolutely stable.

Definition A method is *A-stable* if it is absolutely stable for all positive time steps.

Thus, the implicit Euler and Crank–Nicholson methods, or more generally (4.150) whenever $\theta \ge 1/2$, are A-stable. Here, we have only shown this to be true when all stable eigenvalues are real. For four values of θ, Figure 4.11 plots the modulus of the growth coefficient $\mu_j = c_j^{[k+1]}/c_j^{[k]}$ as a function of ω_j in the complex plane. The region of absolute stability, $|\mu_j| \le 1$, lies within the contour $|\mu_j| = 1$. When the rules are biased more towards the future state than the old state, $\theta \ge 1/2$, the method is A-stable.

By contrast, the explicit Euler method loses absolute stability whenever $\omega_j > 2$ for *any* stable eigenvalue. To see what happens in this case, Figure 4.12 plots the numerical trajectories for $\dot{x} = -x$, $x(0) = 1$, for which the true solution is $x(t) = e^{-t}$. The explicit

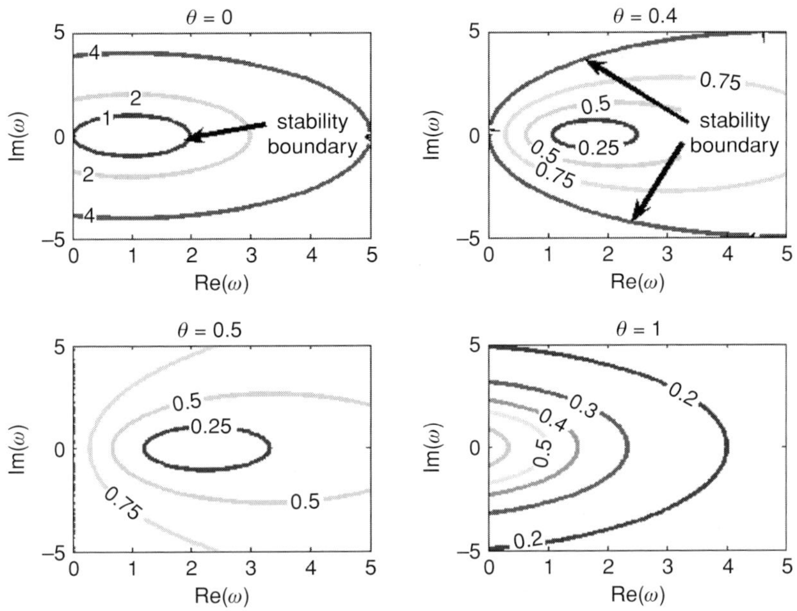

Figure 4.11 Magnitude of the growth coefficient vs. real and imaginary parts of the dimensionless time step. The region of absolute stability lies within the contour of 1.

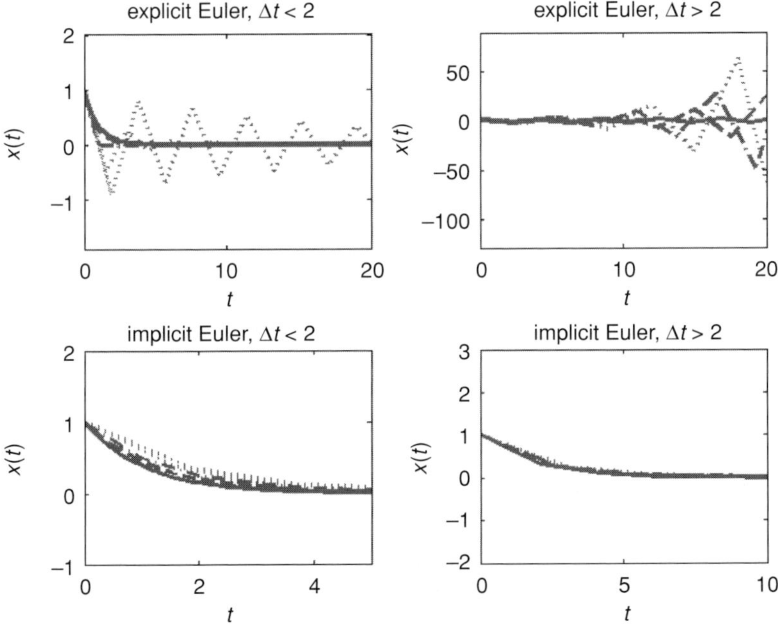

Figure 4.12 Numerical trajectories for $dx/dt = -x$, $x(0) = 1$ for explicit Euler and implicit Euler at various time steps less than 2 and greater than 2. *Upper right* shows loss of absolute stability for the explicit method. Time steps below 2 are $\{0.1, 0.5, 1.0, 1.5, 1.9\}$. Time steps above 2 are $\{2.1, 2.5, 2.75, 3\}$.

Euler method becomes unstable for $\Delta t > 2$ resulting in spurious oscillations and divergence. By contrast, the A-stable implicit Euler method remains well-behaved, though not accurate, even for large time steps.

Time step restrictions for stiff systems

We can now see why explicit methods have difficulty with stiff systems. For $\dot{x} = Ax$, $Aw^{[j]} = \lambda_j w^{[j]}$, (4.159) yields the numerical response

$$x^{[k]} = \sum_{j=1}^{N} c_j^{[0]} \mu_j^k w^{[j]} \qquad \mu_j = \frac{1 - \omega_j(1 - \theta)}{1 + \theta \omega_j} \qquad \omega_j = -(\Delta t)\lambda_j \tag{4.168}$$

For absolute stability, we must have $|\mu_j| \leq 1$ for *every* stable eigenvalue with $\mathrm{Re}(\omega_j) > 0$; however, when $\theta < 1/2$, we have an upper limit on the allowable time step (Figure 4.11). If $\lambda_{\max}$ is the largest eigenvalue magnitude, for absolute stability we must have

$$\omega_{\max} = (\Delta t)\lambda_{\max} \leq \omega_{\mathrm{c}} \qquad \omega_{\mathrm{c}} = 2 \text{ for explicit Euler} \tag{4.169}$$

Thus, the time step must be smaller than $(\Delta t)_{\mathrm{c}} = \omega_{\mathrm{c}}/\lambda_{\max}$. But, inspection of (4.158) shows that the time required for $x(t)$ to relax back to the steady state $x_{\mathrm{s}} = 0$ is governed by the *smallest* eigenvalue $\lambda_{\min}$, as $c_j(t) = c_j^{[0]} e^{\lambda_j t}$. Thus, we shall have to continue the simulation until $t_{\mathrm{end}} \approx \lambda_{\min}^{-1}$. To maintain absolute stability, the number of time steps has to be greater than

$$N_{\min} \approx \frac{t_{\mathrm{end}}}{(\Delta t)_{\mathrm{c}}} = \frac{\lambda_{\min}^{-1}}{\omega_{\mathrm{c}}/\lambda_{\max}} = \frac{(\lambda_{\max}/\lambda_{\min})}{\omega_{\mathrm{c}}} = \frac{\kappa(A)}{\omega_{\mathrm{c}}} \tag{4.170}$$

$\kappa(A)$ is the condition number of A. Therefore, for stiff systems with high condition numbers, explicit methods are forced to take a very large number of time steps, requiring many function evaluations and much CPU time. The restriction (4.169) is not present for A-stable integrators.

Error rejection

There is another benefit that accrues from absolute stability. Let us start a simulation at an initial state slightly perturbed from the previous one at $x^{[0]} = \Sigma_{j=1}^{N}(c_j^{[0]} + \delta c_j^{[0]})w^{[j]}$, to generate the numerical trajectory

$$x^{[k]} = \sum_{j=1}^{N} \left(c_j^{[0]} + \delta c_j^{[0]}\right) \mu_j^k w^{[j]} \tag{4.171}$$

As in (4.168), μ_j is the growth coefficient. The difference between the original trajectory (4.168) and the perturbed trajectory (4.171) is

$$\delta x^{[k]} = \sum_{j=1}^{N} \left(\delta c_j^{[0]}\right) \mu_j^k w^{[j]} \tag{4.172}$$

When all $|\mu_j| < 1$, the effect of this perturbation decays to zero. This property, *error rejection*, is important and desirable in numerical simulation, as then the round-off errors that are continually introduced do not accumulate but remain manageable, as their effects become less and less important at later times. By contrast, if any $|\mu_j| > 1$, round-off errors grow exponentially and may drown out the true response with random noise.

Stiff systems from discretized PDEs

These issues of stability and error rejection become very important for stiff systems, as the time step must be chosen to accommodate the largest eigenvalue (fastest mode). We have seen above an example of a stiff system from chemical kinetics, but another important source of stiff systems is the simulation of time-dependent PDEs such as the diffusion equation

$$\frac{\partial \varphi}{\partial t} = \frac{\partial^2 \varphi}{\partial x^2} \tag{4.173}$$

The finite difference method yields the linear ODE system, $\varphi_j \equiv \varphi(x_j)$,

$$\frac{d\varphi_j}{dt} = \frac{\varphi_{j-1} - 2\varphi_j + \varphi_{j+1}}{(\Delta x)^2} \qquad \dot{\varphi} = -\frac{1}{(\Delta x)^2} A\varphi$$

$$A = \begin{bmatrix} 2 & -1 & & & \\ -1 & 2 & -1 & & \\ & -1 & 2 & \dots & \\ & & \dots & \dots & -1 \\ & & & -1 & 2 \end{bmatrix} \tag{4.174}$$

The eigenvalues and eigenvectors $A w^{[k]} = \lambda_k w^{[k]}$ of this diffusion matrix are

$$\lambda_k = 4 \sin^2\left[\frac{k\pi}{2(N+1)}\right] \qquad w_j^{[k]} = \sin\left[\frac{k\pi j}{N+1}\right] \qquad k = 1, 2, \dots, N \tag{4.175}$$

The largest (fastest mode) eigenvalue occurs for $k = N$,

$$\lambda_N = 4 \sin^2\left[\frac{N\pi}{2(N+1)}\right] \approx 4 \qquad w_j^{[N]} = \sin\left[\frac{N\pi j}{N+1}\right] \approx \sin(\pi j) \tag{4.176}$$

The wavelength of this eigenvector is the spacing of the grid; i.e., it describes variations on the smallest length scale that can be resolved by the grid (Figure 3.5). The smallest (slowest) eigenvalue occurs for $k = 1$,

$$\lambda_1 = 4 \sin^2\left[\frac{\pi}{2(N+1)}\right] \approx 4\left[\frac{\pi}{2(N+1)}\right]^2 = \frac{\pi}{N^2} \qquad w_j^{[1]} = \sin\left[\frac{\pi j}{N+1}\right] \tag{4.177}$$

The slowest mode eigenvector describes variations across the entire domain on the largest length scale that can be resolved by the grid.

The condition number in the limit of large N is

$$\kappa = \frac{\lambda_N}{\lambda_1} \approx \frac{4N^2}{\pi} \tag{4.178}$$

Unless the number of grid points is very small, the set of first order ODEs obtained from discretizing a partial differential equation is very stiff.

Stiff stability of BDF methods

Above, we have studied in detail the stability properties of simple single-step methods of the class (4.150). More generally, explicit methods always have time restrictions of the form (4.169), and the more a method makes use of past data to predict the future, the more strict the time step restrictions become. We consider here the stability properties of implicit multi-step BDF methods

$$\alpha_{-1} x^{[k+1]} + \sum_{n=0}^{m_h} \alpha_n x^{[k-n]} = (\Delta t) \beta_{-1} f^{[k+1]} \tag{4.179}$$

The coefficients are computed from (4.138) by substituting the Hermite interpolation polynomial (4.134) into the update formula (4.130). An efficient implementation of BDF methods is presented below in the discussion of DAE systems. For the test equation $\dot{x} = Ax$, $Aw^{[j]} = \lambda_j w^{[j]}$, we again expand the states in the numerical trajectory in eigenvectors of A, $x^{[k]} = \sum_{j=1}^{N} c_j^{[k]} w^{[j]}$, and define the growth coefficient for each mode as $\mu_j = c_j^{[k+1]}/c_j^{[k]}$. Absolute stability requires $|\mu_j| \leq 1$. In terms of the $\{c^{[k]}\}$, we write the states at other times as $x^{[k-n]} = \sum_{j=1}^{N} c_j^{[k]} \mu_j^{-n} w^{[j]}$. Substitution into (4.179) with $f^{[k+1]} = Ax^{[k+1]}$ yields

$$\sum_{n=-1}^{m_h} \alpha_n \left\{ \sum_{j=1}^{N} c_j^{[k]} \mu_j^{-n} w^{[j]} \right\} = (\Delta t)\beta_{-1} A \left\{ \sum_{j=1}^{N} c_j^{[k]} \mu_j w^{[j]} \right\} \tag{4.180}$$

Using $Aw^{[j]} = \lambda_j w^{[j]}$ and collecting terms, yields

$$0 = \sum_{j=1}^{N} c_j^{[k]} \left\{ \sum_{n=-1}^{m_h} \alpha_n \mu_j^{-n} - \beta_{-1}(\Delta t)\lambda_j \mu_j \right\} w^{[j]} \tag{4.181}$$

Equating the sum in the braces to zero, defining again $\omega_j = -(\Delta t)\lambda_j$, and multiplying by $\mu_j^{m_h}$, we obtain

$$0 = \sum_{n=-1}^{m_h} \alpha_n \mu_j^{m_h-n} - \beta_{-1}\omega_j \mu_j^{m_h+1} \tag{4.182}$$

For absolute stability, all growth coefficients that are roots of this *stability polynomial* must have $|\mu_j| \leq 1$. Figure 4.13 shows the regions of absolute stability for BDF methods of various orders $p = m_h + 1$ in the complex plane of ω_j. As we are concerned only with $\text{Re}(\lambda_j) < 0$, a method is A-stable (absolutely stable for all $\Delta t > 0$) if the entire half-plane $\text{Re}(\omega_j) > 0$ lies within the region of absolute stability. This is the case for $p = 1$ and $p = 2$, but not for $p \geq 3$, as the methods become unstable for complex eigenvalues.

However, we see from Figure 4.13 that even though these higher-order methods are not A-stable, they do remain absolutely stable when $\text{Re}(\omega_j) = -(\Delta t)\text{Re}(\lambda_j) \gg 1$. Thus, when we use a time step much larger than the characteristic time $[\text{Re}(\lambda_j)]^{-1}$ of the mode, $|c_j^{[k+1]}/c_j^{[k]}| < 1$ and these modes decay to zero. Thus, these methods are said to have *stiff decay*.

From our discussion of the discretized time-dependent diffusion equation (4.174), we have seen that the fastest modes have spatial wavelengths comparable to the distance between

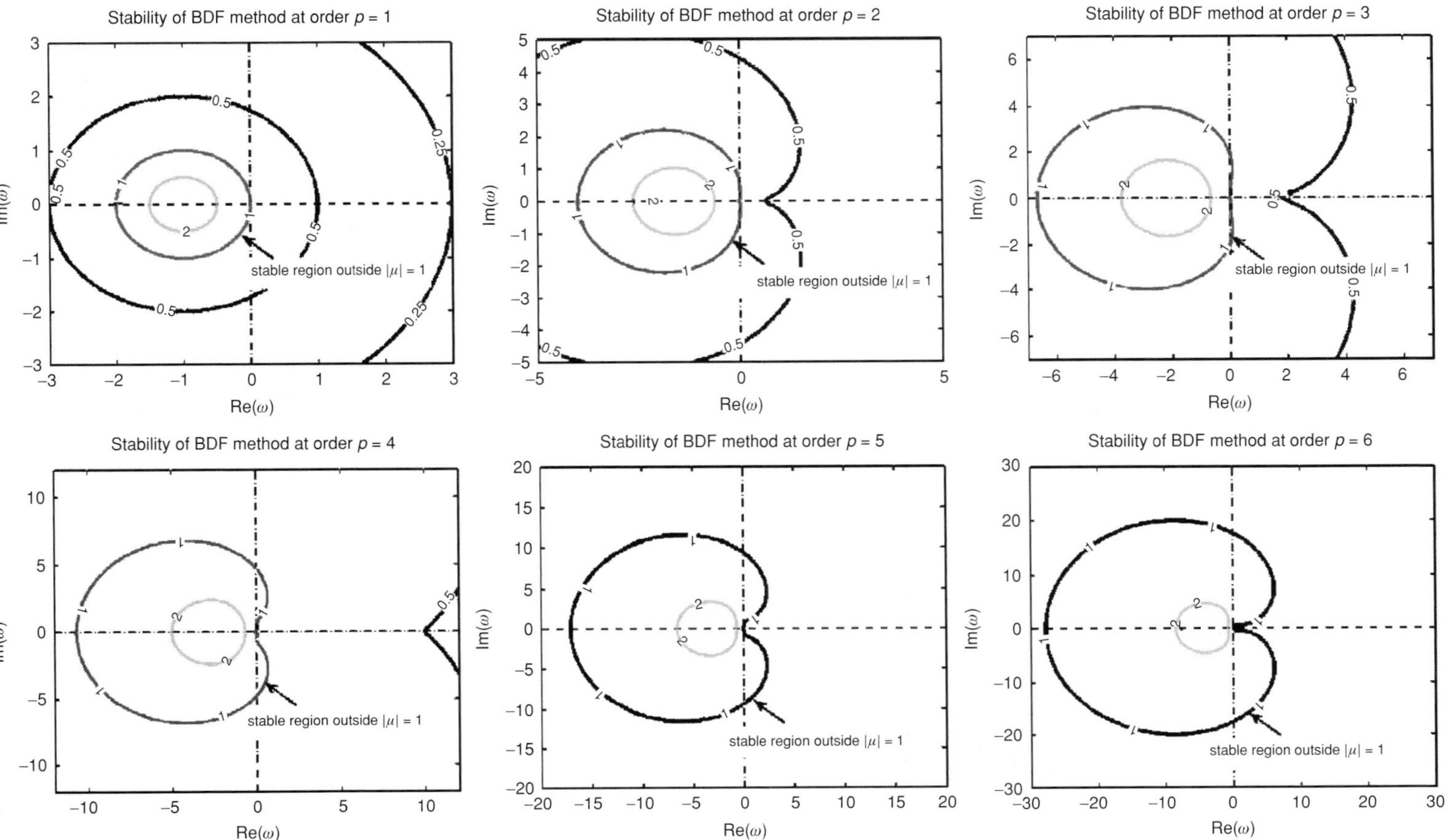

Figure 4.13 Regions of absolute stability for BDF methods of various orders.

grid points, and that the slowest modes have longer wavelengths comparable to the dimensions of the physical domain. Often, the real physical behavior occurs on these long wavelengths and the true solution has little variation from one grid point to the next. Thus, for the true solution, the fast modes with $[\mathrm{Re}(\lambda_j)]^{-1} \ll (\Delta t)$ *should* have $c_j \approx 0$, and we only observe significant activity in these modes, $c_j \neq 0$, due to the exponential growth of round-off errors when $|\mu_j| \gg 1$. If we use a BDF method with stiff decay, even though we are *not* simulating the fast mode dynamics accurately, they do correctly decay to zero in the true solution. The fast modes therefore do not corrupt the dynamics of interest taking place on the slow time scales. Therefore, integrators with stiff decay can simulate the dynamics occurring on the slow time scales of a stiff system using time steps that are very large on the scale of the fast modes. Essentially, a QSSA is made on the fast mode dynamics.

Symplectic methods for classical mechanics

The integration methods discussed above have been compared in terms of stability and accuracy; however, no integration method with a finite time step is perfectly accurate. Over long simulation periods, the discrepancy between the predicted and actual system behavior may be significant.

This is of importance for applications such as celestial mechanics and molecular dynamics, in which we simulate the motion of a number of interacting particles of masses m_α, positions $\boldsymbol{r}_\alpha$, velocities $\boldsymbol{v}_\alpha$, with a total potential energy function $V(\boldsymbol{r}_1, \boldsymbol{r}_2, \ldots, \boldsymbol{r}_N)$. The motion of each point mass is governed by Newton's second law of motion

$$m_\alpha \frac{d\boldsymbol{v}_\alpha}{dt} = \boldsymbol{F}_\alpha \qquad \boldsymbol{F}_\alpha = -\frac{\partial}{\partial \boldsymbol{r}_\alpha} V(\boldsymbol{r}_1, \boldsymbol{r}_2, \ldots, \boldsymbol{r}_N) \tag{4.183}$$

In the absence of an external potential or dissipation (frictional forces), the total system energy is constant,

$$E = \sum_{\alpha=1}^{N} \tfrac{1}{2} m_\alpha (\boldsymbol{v}_\alpha \cdot \boldsymbol{v}_\alpha) + V(\boldsymbol{r}_1, \boldsymbol{r}_2, \ldots, \boldsymbol{r}_N) = \text{constant} \tag{4.184}$$

Due to integration errors, this property is generally not satisfied by the numerical trajectory of the system; however, for a special class of integrators, total energy is conserved (approximately) even over long simulations.

From Noether's theorem (Arnold, 1989), it is known that the conservation of energy is related to the invariance of Newton's equation of motion to time reversal. That is, if we follow a conservative system for some period of time and then reverse the direction of time, the system will exactly retrace, in reverse, its previous trajectory. It may be shown that integration rules that are *symplectic* (i.e., symmetric with respect to the direction of time) have favorable energy conservation properties that make them most suitable for simulating the classical mechanics of conservative systems. A more complete discussion of symplectic integrators is found in Frenkel & Smit (2001). Here, we merely provide a popular symplectic integrator, the *velocity Verlet rule,*

$$r_\alpha \leftarrow r_\alpha + v_\alpha(\Delta t) + \frac{(\Delta t)^2}{2m_\alpha} F_\alpha \quad \alpha = 1, 2, \ldots, N$$

$$v_\alpha \leftarrow v_\alpha + \frac{F_\alpha}{2m_\alpha}(\Delta t) \quad \alpha = 1, 2, \ldots, N$$

$$\text{compute new forces} \quad F_\alpha \quad \alpha = 1, 2, \ldots, N$$

$$v_\alpha \leftarrow v_\alpha + \frac{F_\alpha}{2m_\alpha}(\Delta t) \quad \alpha = 1, 2, \ldots, N$$

(4.185)

Differential-algebraic equation (DAE) systems

In the previous sections, we have treated systems described by a set of ODEs. We now consider the addition of algebraic equations to obtain a *DAE system*. We show how the BDF method can be modified to accommodate a system of mixed differential and algebraic equations. Consider the DAE system

$$M(x)\dot{x} = f(x)$$

(4.186)

$M(x)$ is a state-dependent *mass matrix*. If $M(x)$ is nonsingular, we can decouple the set of equations into standard ODE form

$$\dot{x} = M^{-1}f(x)$$

(4.187)

We consider here the case where $M(x)$ is singular, as when the system is modeled by a combination of differential and algebraic equations. As a particular example, consider the system

$$\dot{y} = F(y, z)$$
$$0 = G(y, z)$$

(4.188)

System (4.188) is expressed in the DAE form (4.186) as

$$M\dot{x} = \begin{bmatrix} I & 0 \\ 0 & 0 \end{bmatrix} \begin{bmatrix} \dot{y} \\ \dot{z} \end{bmatrix} = \begin{bmatrix} F(y, z) \\ G(y, z) \end{bmatrix} = f(x)$$

(4.189)

BDF method for DAE systems of index one

We now show how the BDF method can be modified to simulate a DAE system when $M(x)$ is singular (Ascher & Petzold, 1998). The first step is to generate an explicit *predictor polynomial* that extrapolates the state behavior at past times (not necessarily uniformly spaced) into the future,

$$\pi^{(p)}(\tau_j) = x(\tau_j) = x^{[k-j]} \quad \tau_j = \frac{t_{k-j} - t_k}{\Delta t} \quad j = 0, 1, 2, \ldots, m_h$$

(4.190)

This polynomial is constructed easily with Newton interpolation,

$$\pi^{(p)}(\tau) = a_0 + a_1(\tau - \tau_{m_h}) + a_2(\tau - \tau_{m_h})(\tau - \tau_{m_h-1})$$
$$+ \cdots + a_{m_h+1}(\tau - \tau_{m_h})(\tau - \tau_{m_h-1})\cdots(\tau - \tau_1)(\tau)$$

(4.191)

where $a_k = x[\tau_{m_h}, \tau_{m_h-1}, \ldots, \tau_{m_h-k}]$. We then generate a *corrector polynomial* that agrees with the predictor at uniformly spaced times in the recent past,

$$\pi^{(c)}(-j) = \pi^{(p)}(-j) \qquad j = 0, 1, 2, \ldots, m_h \tag{4.192}$$

and that also matches the time derivative at $t_{k+1} = t_k + \Delta t$,

$$\left.\frac{d\pi^{(c)}}{d\tau}\right|_{\tau=1} = (\Delta t)\left.\frac{dx}{dt}\right|_{t_{k+1}} \tag{4.193}$$

With $x^{[k+1]} = \pi^{(c)}(1)$ and $M^{[k+1]} = M(x^{[k+1]})$, this yields the equation

$$M^{[k+1]}\dot{\pi}^{(c)}(1) = (\Delta t)f\left(x^{[k+1]}\right) \tag{4.194}$$

that we solve for the state $x^{[k+1]}$ at the new time $t_{k+1} = t_k + \Delta t$.

Rather than construct $\pi^{(c)}(\tau)$, we find it easier to form the polynomial

$$\delta\pi(\tau) \equiv \pi^{(c)}(\tau) - \pi^{(p)}(\tau) \qquad \delta\pi(-j) = 0 \qquad j = 0, 1, 2, \ldots, m_h \tag{4.195}$$

Using Newton interpolation,

$$\begin{aligned}
\delta\pi(\tau) = \; & \delta\pi[1] + \delta\pi[1, 0](\tau - 1) + \delta\pi[1, 0, -1](\tau - 1)(\tau) \\
& + \delta\pi[1, 0, -1, -2](\tau - 1)(\tau)(\tau + 1) + \cdots \\
& + \delta\pi[1, 0, -1, -2, \ldots, -m_h](\tau - 1)(\tau)(\tau + 1)\ldots(\tau + m_h - 1)
\end{aligned} \tag{4.196}$$

Starting with $\delta\pi[1] = \delta\pi(1) = \pi^{(c)}(1) - \pi^{(p)}(1)$, from $\delta\pi(0) = 0$ we have

$$\delta\pi[1, 0] = \frac{\delta\pi[0] - \delta\pi[1]}{(0 - 1)} = \frac{0 - \delta\pi(1)}{(-1)} = \delta\pi(1) \tag{4.197}$$

Next, we use $\delta\pi(0) = \delta\pi(-1) = 0$ to compute

$$\delta\pi[1, 0, -1] = \frac{\delta\pi[0, -1] - \delta\pi[1, 0]}{(-1 - 1)} = \frac{0 - \delta\pi(1)}{(-2)} = \frac{1}{2}\delta\pi(1) \tag{4.198}$$

Similarly, as $\delta\pi(0) = \delta\pi(-1) = \delta\pi(-2) = 0$,

$$\delta\pi[1, 0, -1, -2] = \frac{\delta\pi[0, -1, -2] - \delta\pi[1, 0, -1]}{(-2 - 1)} = \frac{0 - \frac{1}{2}\delta\pi(1)}{(-3)} = \frac{\delta\pi(1)}{3!} \tag{4.199}$$

Continuing this approach, we find the general result

$$\delta\pi[1, 0, -1, -2, \ldots, -j] = \frac{\delta\pi(1)}{j!} \tag{4.200}$$

The difference polynomial is then

$$\delta\pi(\tau) = \delta\pi(1)\left\{1 + \sum_{j=1}^{m_h+1}\left(\frac{1}{j!}\right)\left[\prod_{k=-1}^{j-2}(\tau + k)\right]\right\} \tag{4.201}$$

Taking the derivative of this polynomial at $\tau = 1$,

$$\delta\dot{\pi}(1) = \delta\pi(1)\left\{\sum_{j=1}^{m_h+1}\left(\frac{1}{j!}\right)\sum_{k=-1}^{j-2}\left[\prod_{\substack{p=-1 \\ p\neq k}}^{j-2}(1 + p)\right]\right\} \tag{4.202}$$

In the sum over k, all terms with $k \neq -1$ evaluate to zero, because in the product over p, a factor for $p = -1$ contributes a value of $1 + p = 0$. As

$$\sum_{\substack{k=-1}}^{j-2} \left[\prod_{\substack{p=-1 \\ p \neq k}}^{j-2} (1 + p) \right] = \prod_{p=0}^{j-2} (1 + p) = \prod_{l=1}^{j-1} l = (j - 1)! \tag{4.203}$$

we find that

$$\delta\dot{\pi}(1) = \delta\pi(1) \left\{ \sum_{j=1}^{m_h+1} \left(\frac{1}{j!} \right) (j - 1)! \right\} = \delta\pi(1)\alpha_{-1} \qquad \alpha_{-1} = \sum_{j=1}^{m_h+1} \left(\frac{1}{j} \right) \tag{4.204}$$

Substituting $\dot{\pi}^{(c)}(1) = \delta\dot{\pi}(1) + \dot{\pi}^{(p)}(1)$ into (4.194) and using (4.204),

$$M^{[k+1]} \left\{ \alpha_{-1}\delta\pi(1) + \dot{\pi}^{(p)}(1) \right\} = (\Delta t) f\left(x^{[k+1]} \right) \tag{4.205}$$

As $x^{[k+1]} = \pi^{(c)}(1) = \delta\pi(1) + \pi^{(p)}(1)$, we substitute for $\delta\pi(1)$ in (4.205) to obtain a nonlinear algebraic equation for the new state $x^{[k+1]}$,

$$0 = g\left(x^{[k+1]} \right) = \alpha_{-1} M^{[k+1]} x^{[k+1]} - (\Delta t) f\left(x^{[k+1]} \right) - M^{[k+1]} U^{[k]}$$
$$U^{[k]} = \alpha_{-1}\pi^{(p)}(1) - \dot{\pi}^{(p)}(1) \tag{4.206}$$

Starting with an initial guess of $x^{[k+1,0]} = \pi^{(p)}(1)$, we solve (4.206) using Newton's method, for which the Jacobian matrix is

$$B^{[k+1]} = \alpha_{-1} M\left(x^{[k+1]} \right) - (\Delta t) J\left(x^{[k+1]} \right) \qquad B = \frac{\partial g}{\partial x^{\mathrm{T}}} \qquad J = \frac{\partial f}{\partial x^{\mathrm{T}}} \tag{4.207}$$

As x changes little from one iteration to the next and α_{-1} is fixed by m_h (*fixed leading-coefficient BDF method*), B varies slowly and we can save much CPU time through LU factorization. This algorithm marches forward in time similarly to an ODE system; however, for the Newton iterations to be successful, the matrix $B^{[k+1]}$ must be nonsingular. This is unfortunately not always the case. We can identify the condition that must be met for $B^{[k+1]}$ to be invertible for the special case of (4.188),

$$M(x)\dot{x} = f(x) \qquad M = \begin{bmatrix} I & 0 \\ 0 & 0 \end{bmatrix} \qquad x = \begin{bmatrix} y \\ z \end{bmatrix} \qquad f(x) = \begin{bmatrix} F(y, z) \\ G(y, z) \end{bmatrix} \tag{4.208}$$

The Jacobian matrix of $f(x)$ takes the partitioned form

$$J(x) = \begin{bmatrix} \left(\dfrac{\partial F}{\partial y^{\mathrm{T}}} \right) & \left(\dfrac{\partial F}{\partial z^{\mathrm{T}}} \right) \\[2mm] \left(\dfrac{\partial G}{\partial y^{\mathrm{T}}} \right) & \left(\dfrac{\partial G}{\partial z^{\mathrm{T}}} \right) \end{bmatrix} \tag{4.209}$$

The Newton update matrix (4.207) for this system is then

$$B(x) = \begin{bmatrix} \left[\alpha_{-1} I - (\Delta t)\dfrac{\partial F}{\partial y^{\mathrm{T}}} \right] & \left[-(\Delta t)\left(\dfrac{\partial F}{\partial z^{\mathrm{T}}} \right) \right] \\[3mm] \left[-(\Delta t)\left(\dfrac{\partial G}{\partial y^{\mathrm{T}}} \right) \right] & \left[-(\Delta t)\left(\dfrac{\partial G}{\partial z^{\mathrm{T}}} \right) \right] \end{bmatrix} = \begin{bmatrix} B_{11} & B_{12} \\ B_{21} & B_{22} \end{bmatrix} \tag{4.210}$$

If $(\partial \boldsymbol{G}/\partial \boldsymbol{z}^{\mathrm{T}})$ is singular, then as $\Delta t \to 0$, $B(\boldsymbol{x})$ also becomes singular. To see this, consider a system with two differential and two algebraic equations. Then, as $\Delta t \to 0$, writing $b_{ij} = (\Delta t)a_{ij}$, we have

$$
B \approx
\begin{bmatrix}
\alpha_{-1} & [(\Delta t)a_{12}] & [(\Delta t)a_{13}] & [(\Delta t)a_{14}] \\
[(\Delta t)a_{21}] & \alpha_{-1} & [(\Delta t)a_{23}] & [(\Delta t)a_{24}] \\
[(\Delta t)a_{31}] & [(\Delta t)a_{32}] & [(\Delta t)a_{33}] & [(\Delta t)a_{34}] \\
[(\Delta t)a_{41}] & [(\Delta t)a_{42}] & [(\Delta t)a_{43}] & [(\Delta t)a_{44}]
\end{bmatrix}
\tag{4.211}
$$

The determinant of this matrix is

$$
\det(B) = \sum_{i_1=1}^{4}\sum_{i_2=1}^{4}\sum_{i_3=1}^{4}\sum_{i_4=1}^{4} \varepsilon_{i_1,i_2,i_3,i_4} b_{1,i_1} b_{2,i_2} b_{3,i_3} b_{4,i_4}
\tag{4.212}
$$

As $\Delta t \to 0$, the terms with $i_1 = 1$, $i_2 = 2$ dominate the others, and

$$
\det(B) \approx \sum_{i_3=1}^{4}\sum_{i_4=1}^{4} \varepsilon_{1,2,i_3,i_4}(\alpha_{-1})(\alpha_{-1}) b_{3,i_3} b_{4,i_4} = \alpha_{-1}^{2} \sum_{i_3=1}^{4}\sum_{i_4=1}^{4} \varepsilon_{i_3,i_4} b_{3,i_3} b_{4,i_4}
\tag{4.213}
$$

Thus, $\det(B) \approx \alpha_{-1}^{2} \det(B_{22})$, and if $B_{22} = (\partial \boldsymbol{G}/\partial \boldsymbol{z}^{\mathrm{T}})$ is singular, so is $B(\boldsymbol{x})$ and Newton's method (4.207) fails, even in the limit as $\Delta t \to 0$.

If, however, $(\partial \boldsymbol{G}/\partial \boldsymbol{z}^{\mathrm{T}})$ is nonsingular, then $B(\boldsymbol{x})$ will *not* be singular as $\Delta t \to 0$, and the linear system at each Newton update has a unique solution. If $(\partial \boldsymbol{G}/\partial \boldsymbol{z}^{\mathrm{T}})$ is nonsingular, it is also easy to determine a set of *consistent initial conditions*, i.e., one that satisfies all algebraic equations. We set $\boldsymbol{y}$, and then compute $\boldsymbol{z}$ using Newton's method to solve

$$
\boldsymbol{0} = \boldsymbol{G}(\boldsymbol{y}, \boldsymbol{z})
\tag{4.214}
$$

If $(\partial \boldsymbol{G}/\partial \boldsymbol{z}^{\mathrm{T}})$ is nonsingular, we can take a single time derivative,

$$
\frac{d}{dt}\begin{bmatrix}\dot{\boldsymbol{y}} = \boldsymbol{F}(\boldsymbol{y}, \boldsymbol{z}) \\ \boldsymbol{0} = \boldsymbol{G}(\boldsymbol{y}, \boldsymbol{z})\end{bmatrix} \Rightarrow \quad \ddot{\boldsymbol{y}} = \left(\frac{\partial \boldsymbol{F}}{\partial \boldsymbol{y}^{\mathrm{T}}}\right)\dot{\boldsymbol{y}} + \left(\frac{\partial \boldsymbol{F}}{\partial \boldsymbol{z}^{\mathrm{T}}}\right)\dot{\boldsymbol{z}} \quad \boldsymbol{0} = \left(\frac{\partial \boldsymbol{G}}{\partial \boldsymbol{y}^{\mathrm{T}}}\right)\dot{\boldsymbol{y}} + \left(\frac{\partial \boldsymbol{G}}{\partial \boldsymbol{z}^{\mathrm{T}}}\right)\dot{\boldsymbol{z}}
\tag{4.215}
$$

to obtain an equivalent system with a nonsingular mass matrix,

$$
\begin{bmatrix} I & 0 \\ 0 & \left(\dfrac{\partial \boldsymbol{G}}{\partial \boldsymbol{z}^{\mathrm{T}}}\right) \end{bmatrix}\begin{bmatrix}\dot{\boldsymbol{y}} \\ \dot{\boldsymbol{z}}\end{bmatrix} = \begin{bmatrix} \boldsymbol{F}(\boldsymbol{y}, \boldsymbol{z}) \\ -\left(\dfrac{\partial \boldsymbol{G}}{\partial \boldsymbol{y}^{T}}\right)\boldsymbol{F}(\boldsymbol{y}, \boldsymbol{z}) \end{bmatrix}
\tag{4.216}
$$

Definition The *index* of a DAE system is the number of differentiations required to convert it into an ODE system with a nonsingular mass matrix.

For the example above with nonsingular $(\partial \boldsymbol{G}/\partial \boldsymbol{z}^{\mathrm{T}})$, the index is 1. High-index DAE systems are not uncommon, and one must perform the necessary reduction in index to 1 through differentiations in order to use the BDF method presented above. Also, for high-index problems, it can be challenging to identify a consistent set of initial questions.

In MATLAB, **ode15s** can simulate index-1 DAE systems. In other languages, the package DASSL by Petzold and coworkers is popular and performs well. A further discussion of DAEs may be found in Ascher & Petzold (1998).

Example. Dynamics on the 2-D unit circle

We demonstrate using **ode15s** to solve a DAE-IVP for a simple system in which an algebraic equation constrains a 2-D trajectory to the unit circle,

$$\frac{dx_1}{dt} = f_1 = -[x_1 - \cos\theta_{\text{end}}] + [x_2 - \sin\theta_{\text{end}}]$$
$$0 = f_2 = x_1^2 + x_2^2 - 1 \tag{4.217}$$

The trajectory is a simple relaxation to $(\cos\theta_{\text{end}},\ \sin\theta_{\text{end}})$ along the unit circle. The following routine returns the function vector:

```
function f = calc_f(t,x);
f = zeros(2,1);
theta_end = 10/180*pi;
f(1) = -(x(1) - cos(theta_end)) + (x(2) - sin(theta_end));
f(2) = x(1)^2 + x(2)^2 - 1;
return;
```

We solve this DAE-IVP from an initial guess with $x_1^{[0]} = 0$ by

```
x_0 = [0; 0.8]; % set initial guess
M = [1 0; 0 0];  % set mass matrix
Options = odeset('Mass',M,'MassSingular','yes', ...
    'MState Dependence','none'); % inform ode15s of mass matrix
[t_traj,x_traj] = ode15s(@calc_f, [0 10], x_0, Options);
```

Here, we have a constant mass matrix, but we also could have provided a function that returns $M(t, x)$ were the mass matrix state-dependent (see help odeset for further details). Note that for $x_1^{[0]} = 0$, the two possible values of x_2 that satisfy the algebraic equation $f_2 = 0$ are $x_2^{[0]} = \pm 1$, but here the initial guess is $x_2^{[0]} = 0.8$. The resulting trajectory (Figure 4.14) shows that the trajectory starts on the unit circle, satisfying $f_2 = 0$ and demonstrates that **ode15s** first computes a consistent initial state.

Example. Heterogeneous catalysis in a packed bed reactor

We demonstrate the DAE-IVP use of **ode15s** for a more realistic chemical engineering problem of modeling an isothermal packed bed reactor with a solid-catalyzed gas-phase reaction $A \Leftrightarrow B + C$. First, A adsorbs reversibly to a vacant site S on the catalyst. The adsorbed species $A \cdot S$ then reacts to form an adsorbed $B \cdot S$ and a free C molecule in the gas phase. The adsorbed product $B \cdot S$ must then desorb, freeing again the active site.

$$A + S \Leftrightarrow A \cdot S \qquad \hat{r}_{\text{aA}} = k_{\text{aA}}p_A c_v - k_{\text{dA}}c_{A\cdot S}$$
$$A \cdot S \Leftrightarrow B \cdot S + C \qquad \hat{r}_s = k_s\left[c_{A\cdot S} - \frac{c_{B\cdot S}\,p_C}{K_s}\right] \qquad \cdot \tag{4.218}$$
$$B \cdot S \Leftrightarrow B + S \qquad \hat{r}_{\text{dB}} = k_{\text{dB}}c_{B\cdot S} - k_{\text{aB}}p_B c_v$$

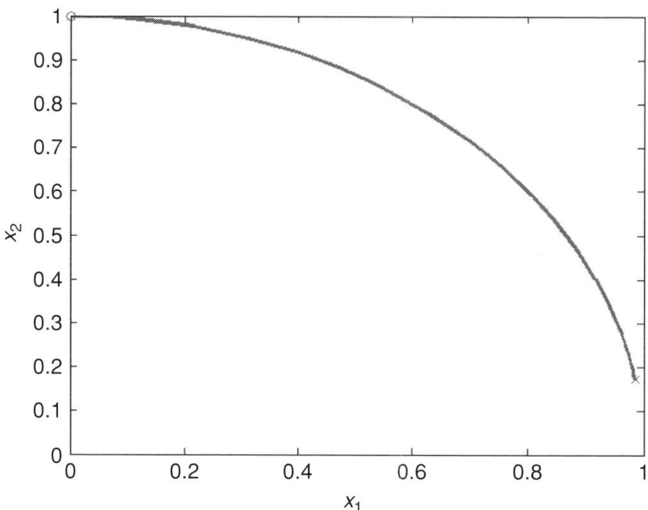

Figure 4.14 Trajectory for DAE-IVP along the unit circle.

$c_{A \cdot S}$, $c_{B \cdot S}$, and c_v are the concentrations (moles per unit mass of catalyst) of adsorbed A, adsorbed B, and vacant sites S. A site balance yields

$$c_{tot} = c_v + c_{A \cdot S} + c_{B \cdot S} \tag{4.219}$$

where c_{tot} is the total concentration of active sites. p_A, p_B, and p_C are the partial pressures of A, B, and C in the gas phase. If we assume that the surface reaction step is rate limiting, the reaction rate (moles per unit time per unit mass of catalyst) is

$$\hat{r}_R \cong \hat{r}_s = k_s \left[c_{A \cdot S} - \frac{c_{B \cdot S} p_C}{K_s} \right] \tag{4.220}$$

and the absorption/desorption steps are in equilibrium,

$$\begin{aligned}
\hat{r}_{aA} &\cong 0 \quad \Rightarrow \quad 0 = K_{aA} p_A c_v - c_{A \cdot S} \quad K_{aA} = k_{aA}/k_{dA} \\
\hat{r}_{aB} &\cong 0 \quad \Rightarrow \quad 0 = K_{aB} p_B c_v - c_{B \cdot S} \quad K_{aB} = k_{aB}/k_{dB}
\end{aligned} \tag{4.221}$$

We obtain K_{aA}, K_{aB}, and c_{tot} from the adsorption isotherms of A and B.

We conduct the reaction in a packed bed reactor, assuming that the heat transfer is sufficiently fast for the reactor to be isothermal. The mass W of catalyst in a region of volume V in the reactor is $W = \rho_s(1 - \phi)V$, where ρ_s is the density of the solid catalyst and ϕ is the void fraction of the bed. Let $F_A(W)$ be the flow rate (moles per unit time) of A passing through the particular surface in the reactor for which the mass of catalyst in the region between this surface and the inlet is W. The mole balance on A for the region between W and $W + \delta W$ is

$$0 = F_A(W) - F_A(W + \delta W) - (\delta W)\hat{r}_R \tag{4.222}$$

As $\delta W \to 0$ we obtain an ODE for $F_A(W)$ (and likewise for B and C),

$$\frac{dF_A}{dW} = -\hat{r}_R \quad \frac{dF_B}{dW} = \hat{r}_R \quad \frac{dF_C}{dW} = \hat{r}_R \tag{4.223}$$

Let the feed stream be a gas mixture of A and a nonreactive diluent gas G. Then, for a specified volumetric flow rate υ_0, total pressure P_0, and inlet temperature T_0, ideal-gas behavior yields

$$F_{A0} = \frac{p_{A0}\upsilon_0}{R T_0} \qquad F_{G0} = \frac{(P_0 - p_{A0})\upsilon_0}{R T_0} \tag{4.224}$$

Similarly, from the molar flow rates, local pressure P, and local temperature T, we can compute the local partial pressures and volumetric flow rate,

$$\upsilon = \frac{F_{tot} R T}{P} \qquad F_{tot} = F_A + F_B + F_C + F_{G0} \tag{4.225}$$

$$p_j = \frac{F_j R T}{\upsilon} = \left(\frac{F_j}{F_{tot}}\right) P \qquad j = A, B, C \tag{4.226}$$

We compute the local pressure using the Ergun equation to model the pressure drop across a packed bed (Fogler, 1999). For a bed of cross-sectional area A_c, catalyst solid density ρ_s, and void fraction ϕ,

$$\frac{dP}{dW} = -\left[\frac{\beta_0}{A_c(1-\phi)\rho_s}\right]\left(\frac{T}{T_0}\right)\left(\frac{P_0}{P}\right)\left(\frac{F_{tot}}{F_{tot,0}}\right) \tag{4.227}$$

where

$$\beta_0 = \frac{\gamma(1-\phi)}{\rho_0 g_c D_p \phi^3}\left[\frac{150(1-\phi)\mu}{D_p} + 1.75\gamma\right] \qquad \gamma = \frac{\rho_0 \upsilon_0}{A_c} \tag{4.228}$$

D_p is the particle diameter, ρ_0 is the inlet gas density, μ is the gas viscosity, and in SI units the conversion factor g_c is 1.

Above we have a set of governing equations, some differential and some algebraic. Here, we could manipulate the equations analytically to obtain a set of purely differential equations, but this may not always be possible. Thus, we simulate the system as a DAE-IVP with the state vector

$$x = \begin{bmatrix} F_A & F_B & F_C & P & c_{A\cdot S} & c_{B\cdot S} & c_v \end{bmatrix}^T \tag{4.229}$$

For the DAE format $M\dot{x} = f(x)$, the mass matrix is

$$M = \begin{bmatrix} I_{4\times 4} & O_{4\times 3} \\ O_{3\times 4} & O_{3\times 3} \end{bmatrix} \tag{4.230}$$

where I and O are respectively the identity and zero matrices of the specified sizes. The functions in the DAE model are

$$f_1 = -\hat{r}_R \qquad f_2 = \hat{r}_R \qquad f_3 = \hat{r}_R$$
$$f_4 = -\left[\frac{\beta_0}{A_c(1-\phi)\rho_s}\right]\left(\frac{P_0}{P}\right)\left(\frac{F_{tot}}{F_{tot,0}}\right) \tag{4.231}$$
$$f_5 = K_{aA} p_A c_v - c_{A\cdot S} \qquad\qquad f_6 = K_{aB} p_B c_v - c_{B\cdot S}$$
$$f_7 = c_{tot} - c_v - c_{A\cdot S} - c_{B\cdot S}$$

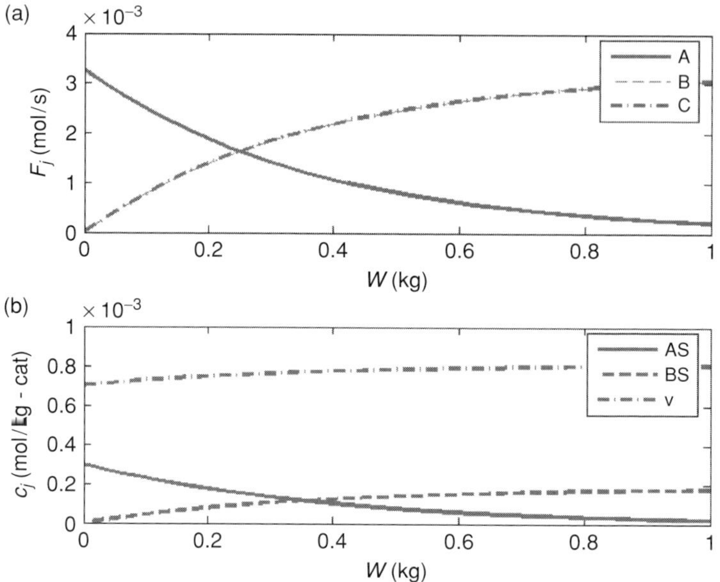

Figure 4.15 (a) Molar flow rate and (b) catalyst site concentrations as functions of catalyst mass in a packed bed reactor.

To evaluate these functions, we first have to compute some intermediate quantities from a set of auxiliary equations

$$F_{tot} = F_A + F_B + F_C + F_{G0}$$

$$p_A = \left(\frac{F_A}{F_{tot}}\right) P \qquad p_B = \left(\frac{F_B}{F_{tot}}\right) P \qquad p_C = \left(\frac{F_C}{F_{tot}}\right) P \tag{4.232}$$

$$\hat{r}_R = k_s \left[c_{A \cdot S} - \frac{c_{B \cdot S} p_C}{K_s} \right]$$

For our system, we use the parameters

$$P_0 = 1\,\text{atm} \qquad p_{A0} = 0.1\,\text{atm} \qquad T_0 = T = 373\,\text{K}$$

$$c_{tot} = 10^{-3}\ \text{moles of sites per kilogram of catalyst}$$

$$\rho_0 = 2.9\,\text{kg/m}^3 \qquad \mu = 2 \times 10^{-5}\,\text{Pa s} \qquad \upsilon_0 = 0.001\,\text{m}^3/\text{s}$$

$$D_p = 5\,\text{mm} \quad \phi = 0.64 \quad A_c = 0.0079\,\text{m}^2 \quad \rho_s = 900\,\text{kg/m}^3 \tag{4.233}$$

$$K_{aA} = 4.2 \times 10^{-5}\,\text{Pa}^{-1} \qquad K_{aB} = 2.5 \times 10^{-5}\,\text{Pa}^{-1}$$

$$k_s = 30\ \text{s}^{-1} \qquad K_s = 9.12 \times 10^5\,\text{Pa}$$

PBR_DAE_sim.m simulates the packed bed reactor for a user-specified total mass of catalyst in the reactor (example results are shown in Figure 4.15).

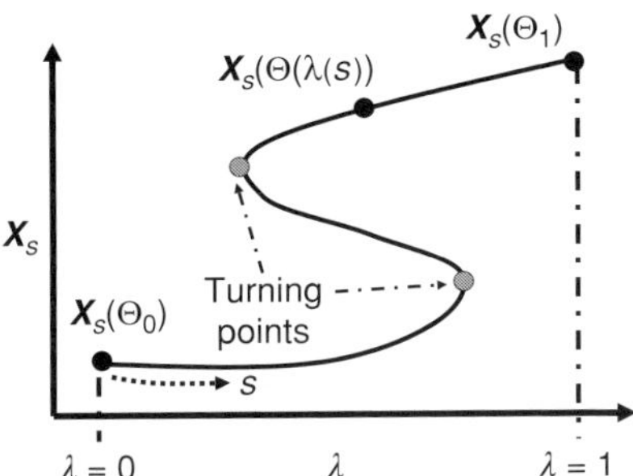

Figure 4.16 Arc length continuation of a nonlinear algebraic system, showing solution path passing through a turning point.

Parametric continuation

Above, we have focused upon simulating dynamic systems $M\dot{x} = f(x; \Theta)$; however, we can use these techniques to study how the solution $x_s(\Theta)$ of an algebraic system $f(x; \Theta) = 0$ depends upon Θ. Let the parameters vary along some path $\Theta(\lambda)$, $0 \leq \lambda \leq 1$, such as

$$\Theta(\lambda) = (1 - \lambda)\Theta_0 + \lambda\Theta_1 \tag{4.234}$$

We wish to study how the solution $x_s(\Theta(\lambda))$ varies along this path. An obvious approach would be to derive a system of ODEs for $d\Theta/d\lambda$ and $dx_s/d\lambda$; however, as shown in the supplemental material, the resulting ODE system encounters difficulties at turning points where the slope of $dx_s/d\lambda$ diverges (Figure 4.16). By contrast, if we define s to be the arc length along the path in (x_s, λ) space, and solve a system of ODEs for dx_s/ds and $d\lambda/ds$, turning points pose no difficulty as both derivatives remain finite. An algorithm for this *arc length continuation* method is presented in the supplemental material. arclength_continuation.m constructs the curve (x_s, λ) for a specified linear path (4.234). The syntax is

```
[x_c, Param_c, lambda_c, fnorm_c, stab_c] = ...
    arclength_continuation(fun_name, ...
    Param_0, Param_1, x0, AcrLenParam);
```

fun_name is the function that returns the function vector,

```
function f = fun_name(x, theta);
```

Param_0 and Param_1 are Θ_0 and Θ_1. The initial guess of the solution for $0 = f(x; \Theta_0)$ is x0. ArcLenParam is an optional structure that varies the performance of the algorithm; see the help utility for its description. x_c and Param_c are arrays in which each column vector contains the state $x_s(\lambda)$ or parameter vector $\Theta(\lambda)$ for a point along the solution curve $0 = f(x_s(\lambda); \Theta(\lambda))$. lambda_c contains the value of λ for each point. fnorm_c contains the function norms for each point (should be near zero). stab_c stores a value of 1 if the point is stable or critically stable and a value of 0 if the point is unstable. This routine traces only a single curve. For a more complex routine that can handle branching, see the AUTO package (indy.cs.concordia.ca/auto/).

Example. Multiple steady states in a nonisothermal CSTR

We demonstrate the use of arclength_continuation.m for the study of multiple steady states in a nonisothermal CSTR. Consider a perfectly-mixed stirred-tank reactor with A reacting to form B

$$A \rightarrow B \quad r = k_1(T)c_A \tag{4.235}$$

The temperature dependence of the rate constant is

$$k_1(T) = k_1(T_{\text{ref}}) \exp\left[-\frac{E_a}{R}\left(\frac{1}{T} - \frac{1}{T_{\text{ref}}}\right)\right] \tag{4.236}$$

For constant volume CSTR with no volume change due to reaction, the mole balances for a volumetric feed rate v are

$$\frac{d}{dt}\{Vc_A\} = v(c_{A,\text{in}} - c_A) - Vk_1(T)c_A$$
$$\frac{d}{dt}\{Vc_B\} = v(c_{B,\text{in}} - c_B) + Vk_1(T)c_A \tag{4.237}$$

Assuming constant density ρ and specific heat capacity $\hat{C}_p$ of the reaction medium, the enthalpy balance on the reactor is

$$\frac{d}{dt}\{V_\rho \hat{C}_p T\} = v\rho \hat{C}_p(T_{\text{in}} - T) - V(\Delta H)k_1(T)c_A - UA(T - T_c) \tag{4.238}$$

ΔH is the heat of reaction (negative for exothermic reactions), UA is the product of the heat transfer coefficient and area for a coolant jacket, through which flows at high rate a coolant of temperature T_c.

Dividing by V and setting the time derivatives to zero, we have the following three algebraic equations for the steady-state behavior

$$\frac{dc_A}{dt} = \frac{v}{V}(c_{A,\text{in}} - c_A) - k_1(T_{\text{ref}}) \exp\left[-\frac{E_a}{R}\left(\frac{1}{T} - \frac{1}{T_{\text{ref}}}\right)\right] c_A = 0$$

$$\frac{dc_B}{dt} = \frac{v}{V}(c_{B,\text{in}} - c_B) + k_1(T_{\text{ref}}) \exp\left[-\frac{E_a}{R}\left(\frac{1}{T} - \frac{1}{T_{\text{ref}}}\right)\right] c_A = 0 \tag{4.239}$$

$$\frac{dT}{dt} = \frac{v}{V}\rho \hat{C}_p(T_{\text{in}} - T) - (\Delta H)k_1(T)c_A - \frac{UA}{V}(T - T_c) = 0$$

Defining the dimensionless time, concentrations, and temperature

$$\tau = \frac{tv}{V} \qquad \varphi_A = \frac{c_A}{c_{A,\text{in}}} \qquad \varphi_B = \frac{c_B}{c_{A,\text{in}}} \qquad \theta = \frac{T}{T_{\text{in}}} \tag{4.240}$$

the dimensionless *Damköhler number*

$$Da = \frac{k_1(T_{\text{in}})V}{v} \tag{4.241}$$

and the scaled heat of reaction, cooling efficiency, and activation energy

$$\beta = \frac{(\Delta H)c_{A,\text{in}}}{\rho \hat{C}_p T_{\text{in}}} \qquad \chi = \frac{UA}{v\rho \hat{C}_p} \qquad \gamma = \frac{E_a}{R T_{\text{in}}} \tag{4.242}$$

the governing equations (4.239) can be written in dimensionless form

$$\frac{d\varphi_A}{d\tau} = 1 - \varphi_A - (Da)\exp\left[\frac{\gamma(\theta - 1)}{\theta}\right]\varphi_A = 0$$

$$\frac{d\varphi_B}{d\tau} = \varphi_B^{(in)} - \varphi_B + (Da)\exp\left[\frac{\gamma(\theta - 1)}{\theta}\right]\varphi_A = 0 \qquad (4.243)$$

$$\frac{d\theta}{d\tau} = 1 - \theta - \beta(Da)\exp\left[\frac{\gamma(\theta - 1)}{\theta}\right]\varphi_A - \chi(\theta - \theta_c) = 0$$

This is a set of three nonlinear algebraic equations with the six dimensionless parameters $\varphi_B^{(in)}$, Da, β, χ, γ, θ_c. We set $\varphi_B^{(in)} = 0$, and note that φ_B can be obtained directly from the values of φ_A and θ,

$$\varphi_B = (Da)\exp\left[\frac{\gamma(\theta - 1)}{\theta}\right]\varphi_A \qquad (4.244)$$

Therefore, we can remove it from the list of unknowns and compute it whenever needed; i.e., we make it an *auxiliary variable*. Moreover, we note that φ_B does not appear in either the equation for φ_A or that for θ, and thus we do not need to compute its value until the solution is found. We therefore reduce the system to two equations,

$$f_1(\varphi_A, \theta) = 1 - \varphi_A - (Da)\exp\left[\frac{\gamma(\theta - 1)}{\theta}\right]\varphi_A = 0$$

$$\qquad (4.245)$$

$$f_2(\varphi_A, \theta) = 1 - \theta - \beta(Da)\exp\left[\frac{\gamma(\theta - 1)}{\theta}\right]\varphi_A - \chi(\theta - \theta_c) = 0$$

with the five dimensionless parameters Da, β, χ, γ, θ_c.

We now use arc-length continuation to draw curves of the dependence of φ_A and θ upon Da for fixed, β, χ, γ and $\theta_c = 1$. We start our calculations at $Da = 10^{-2}$, for which good initial guesses are $\varphi_A = \theta = 1$. Defining the parameter vector as

$$\Theta = [\log(Da) \quad \beta \quad \chi \quad \gamma \quad \theta_c]^{\mathrm{T}} \qquad (4.246)$$

using the fixed values β, χ, γ, $\theta_c = 1$, we vary Da from $Da_0 = 10^{-2}$ at $\lambda = 0$ to $Da_1 = 10^2$ at $\lambda = 1$. nonisothermal_CSTR_Da_scan.m performs this calculation using nonisothermal_CSTR_calc_f.m.

We present results with $\beta = -1$ (exothermic reaction), $\chi = 0$ (no heat transfer to coolant jacket), and $\theta_c = 1$, for various values of γ, the dimensionless activation energy. When $\gamma = 1$, the temperature increases smoothly from a low Da limit of 1 to an upper Da limit of 2 (Figure 4.17). As we increase the activation energy to $\gamma = 5$, the upturn becomes more pronounced (Figure 4.18). At $\gamma = 8$ the curve becomes vertical, and the change in temperature is very sudden (Figure 4.19).

At higher activation energies, e.g. $\gamma = 12$, the system exhibits multiple steady states (Figure 4.20). As we initially increase Da from a small value, we reach a turning point at which the steady state becomes unstable. Further increase in Da at this point results in a sudden jump to a different steady state with a higher temperature (ignition). If we were then

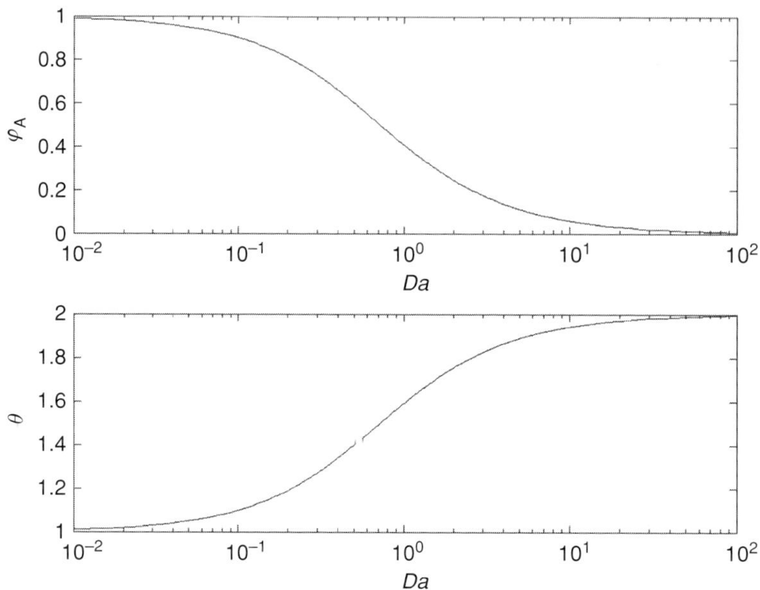

Figure 4.17 Nonisothermal CSTR behavior for $\gamma = 1$ ($\beta = -1$, $\chi = 0$, $\theta_c = 1$).

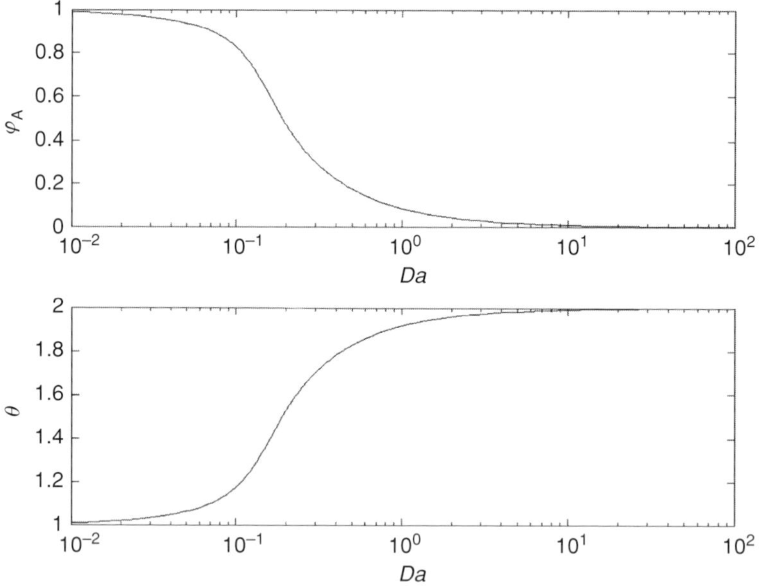

Figure 4.18 Nonisothermal CSTR behavior for $\gamma = 5$ ($\beta = -1$, $\chi = 0$, $\theta_c = 1$).

to decrease Da, we would reach another turning point at which the system jumps back to the low-temperature steady state (extinction). For Da values between the turning points, the system has three steady states, but only the upper and lower ones are stable. Clearly, it is best to design the system to operate safely away from this region of rapid, difficult-to-control, transitions.

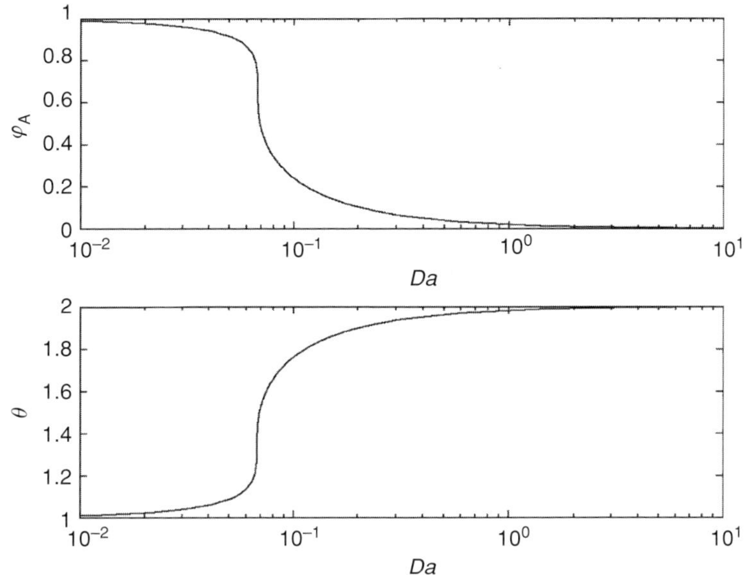

Figure 4.19 Nonisothermal CSTR behavior for $\gamma = 8$ ($\beta = -1$, $\chi = 0$, $\theta_c = 1$).

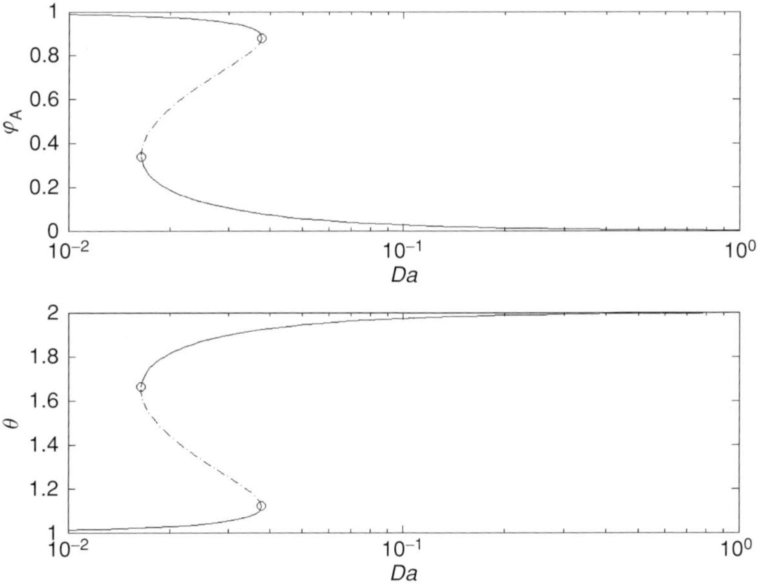

Figure 4.20 Nonisothermal CSTR behavior for $\gamma = 12$ ($\beta = -1$, $\chi = 0$, $\theta_c = 1$).

MATLAB summary

The two main routines for solving ODE-IVPs are **ode45** and **ode15s**. For nonstiff problems, the explicit **ode45** method is recommended. For stiff systems, the implicit **ode15s** is preferred. A list of available ODE solvers is returned by help funfun. For the ODE system $\dot{x} = f(t, x; \Theta)$, we must at a minimum supply a routine that returns $f(t, x; \Theta)$ with the syntax

function f = calc_f(t, x, P1, P2, . . .);

P1, P2, . . . are optional fixed parameters. The IVP is solved by

[t_traj, x_traj] = ode45(@calc_f, t_span, x_0, Options, P1, P2, . . .);

t_span contains the start and end times of the simulation, or optionally a list of discrete times at which the states are needed. x_0 is the initial state and **Options** is a structure, managed by **odeset**, that allows the user to modify the solver behavior (use [ ] to accept the default options). P1, P2, . . . are optional fixed parameters passed to the function routine, here calc_f.

Implicit methods such as **ode15s** require the Jacobian matrix $J = \partial f / \partial x^{\mathrm{T}}$. If only a routine such as calc_f is supplied, **ode15s** estimates the Jacobian by finite differences. For large, sparse systems, the work associated with Jacobian estimation can be reduced significantly by supplying a matrix with the same sparsity pattern as the Jacobian, through the "JPattern" field of **Options**. Even better, we can supply a routine that computes the Jacobian,

function Jac = calc_Jac(t, x, P1, P2, . . .);

and set the "Jacobian" field of **Options** to its name, here 'calc_Jac'.

For a DAE system $M\dot{x} = f(t, x; \Theta)$, **ode15s** may be used if the system is of index one, supplying either a constant mass matrix or a function,

function M = calc_Mass(t, x);

and informing **ode15s** of its identity through the "Mass" field of **Options**, along with the appropriate specifications of "MStateDependence" and "MassSingular." More general DAE systems of the fully implicit form $F(t, x, \dot{x}) = 0$ can be solved by **ode15i**. The low-order implicit methods **ode23s** and **ode23tb** can be useful for very stiff systems, such as discretized PDEs (discussed later in Chapter 6).

odeplot plots the state trajectory, and **odeprint** prints it to the screen. Related functions for 2-D and 3-D plots are **odephas2** and **odephas3**.

Problems

4.A.1. Consider the following ODE system with a steady state at $(1, 2)$,

$$\dot{x}_1 = f_1(x_1, x_2) = -2(x_1 - 1)(x_2 - 2)^2 + (x_1 - 1)(x_2 - 2) - (x_2 - 2) - 3(x_1 - 1)$$
$$\dot{x}_2 = f_2(x_1, x_2) = -(x_1 - 1) + (x_1 - 1)^2(x_2 - 2) \tag{4.247}$$

Derive the linearized ODE system that describes the response of the system to small perturbations around the steady state. Is the steady state stable? Will the steady state exhibit an oscillatory response to small perturbations? Confirm your results by writing a program to simulate the response to a random, small perturbation.

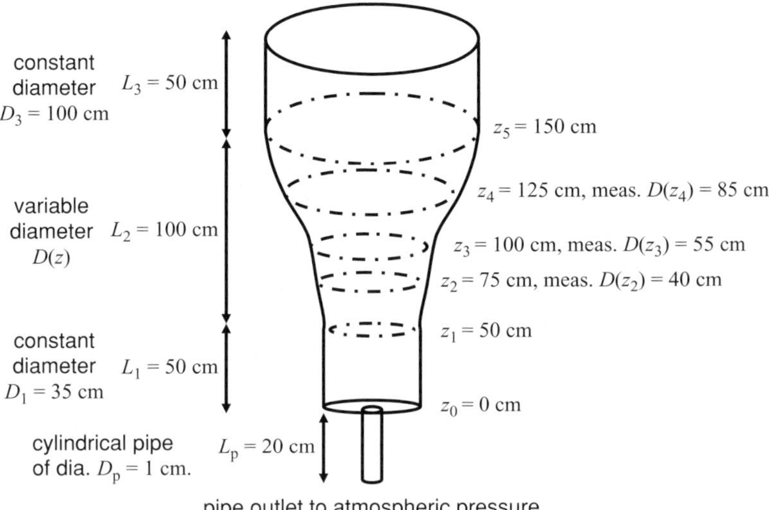

Figure 4.21 Tank of variable cross-sectional area.

4.A.2. Compute the value of the following definite integral using both **dblquad** and Monte Carlo integration,

$$I_D = \int_1^2 \int_0^{\sqrt{x}} [(x-1)^2 + y^2]\, dy\, dx \tag{4.248}$$

4.A.3. Consider the 1-D motion of a point mass, connected to the origin by a harmonic spring, that experiences a frictional drag force and a time-dependent external force. The equation of motion is

$$m\frac{d^2x}{dt^2} = -Kx - \zeta\frac{dx}{dt} + F_{\text{ext}}(t) \tag{4.249}$$

Let the mass be 1 kg, and let the harmonic frequency $\omega_c = \sqrt{K/m}$ be 2π rad/s. Let the external force be $F_{\text{ext}}(t) = (10^{-2}\text{ N})\sin(\omega t)$, where ω is set at $0.01\omega_c$, $0.1\omega_c$, $0.5\omega_c$, $0.9\omega_c$ $0.99\omega_c$, and $0.999\omega_c$. Plot $x(t)$ for each of these cases when $\zeta = 0$ and describe what happens as $\omega \to \omega_c$. Repeat for non zero ζ, and describe how the behavior changes as ζ increases.

4.B.1. Consider the axisymmetric tank shown in Figure 4.21. We want to estimate the time that it takes for the tank to drain if filled with water (density 1000 kg/m^3, viscosity 0.001 Pa s). Let $h(t)$ be the height of water in the tank, $h(0) = 2m$. $v_p(t)$. The mean velocity in the drain pipe and the volumetric flow rate out of the tank is

$$v_{\text{out}}(t) = \frac{\pi}{4}D_p^2 v_p = \frac{\pi}{4}[D(h)]^2 \left|\frac{dh}{dt}\right| \tag{4.250}$$

To obtain v_p, we use the engineering Bernoulli equation to relate the velocity and pressure at $z = h(t)$ to the values at the pipe outlet at $z = -L_p$,

$$\left[\frac{1}{2}\rho\left(\frac{dh}{dt}\right)^2 + p_{atm} + \rho g h\right] - \left[\frac{1}{2}\rho v_p^2 + p_{atm} + \rho g(-L_p)\right] = \text{net viscous loss} > 0 \tag{4.251}$$

The net viscous loss is dominated by the entrance flow at the pipe outlet and the viscous dissipation within the outlet pipe itself,

$$\text{net viscous loss} \approx K_L\left(\frac{1}{2}\rho v_p^2\right) + f_D\left(\frac{L_p}{D_p}\right)\left(\frac{1}{2}\rho v_p^2\right) \tag{4.252}$$

For a well-rounded entrance, $K_L \approx 0.2$. The friction factor f_D is a function of the Reynolds' number in the pipe,

$$Re = \frac{\rho v_p D_p}{\mu} \tag{4.253}$$

For laminar flow when $Re < 2100$, $f_D = 64/Re$. For $Re > 4000$, the friction factor can be estimated from the empirical Colebrook equation

$$\frac{1}{\sqrt{f_D}} = -2\log_{10}\left[\frac{(e/D_p)}{3.7} + \frac{2.51}{Re\sqrt{f_D}}\right] \tag{4.254}$$

e is the characteristic surface roughness length of the pipe wall, and for commercial steel is 0.045 mm. In the intermediate range $2100 < Re < 4000$, it is difficult to predict the friction factor.

Plot the height of water and the volume of water in the tank as functions of time. How long does it take for the tank to empty?

4.B.2. For the test equation, $\dot{x} = \lambda x$, $Re(\lambda) < 0$, plot the region of absolute stability for the RK4 method in the complex plane of $\omega = -(\Delta t)\lambda$.

4.B.3. Consider the calculation of radial integrals in three dimensions,

$$I_F = \int_0^R f(r)4\pi r^2 dr = R^3 \int_0^1 f(\xi R)4\pi \xi^2 d\xi \tag{4.255}$$

Write a routine that computes the Gaussian quadrature weights and nodes $\{\xi_0, \xi_1, \ldots, \xi_N\}$ for $N = 3, 5, 7$. For $f(r) = 1 + r + r^2 + \cdots + r^m$, $R = 1$, plot the error in the computed integral as a function of degree m. Compare the errors to those obtained from Newton Cotes integration with $N + 1$ support points.

4.C.1. Consider again the tank in Problem 4.B.1. We want to maintain the height of water at $h_{set} = 1.5\,\text{m}$ by adding an input water stream of volumetric flowrate v_{in}. Compute the inlet flow rate $v_{in,set}$ that maintains h_{set} at steady state.

Next, let us say that there is a second inlet whose flow rate $v_{in,2}(t)$ we cannot control. We then must use feedback control to vary $v_{in}(t)$ in order to maintain a constant height. Let the error of the height from its set point be $e(t) = h(t) - h_{set}$ and let $v_{in}(t) = v_{in,set} + u(t)$, where for a proportional integral derivative (PID) controller

$$u(t) = K_C\left[e(t) + \frac{1}{\tau_I}\int_0^t e(s)ds + \tau_D\frac{de}{dt}\right] \tag{4.256}$$

For various choices of the controller tuning parameters $\{K_C, \tau_I, \tau_D\}$, simulate the closed-loop response for several perturbations. Start at steady state with $u(0) = e(0) = v_{in,2}(0) = 0$ and immediately change $v_{in,2}$ to a constant value chosen at random within $0 \leq v_{in,2} \leq (0.1)v_{in,set}$. For each response, a measure of how well the controller rejects the disturbance is

$$F(K_C, \tau_I, \tau_D) = \int_0^{t_H} |e(t)|^2 dt \tag{4.257}$$

t_H is a horizon time suitably long for the response to be measured; for example, the time required for the height to approach within 99% of its new steady-state value if we were to take no control action for $v_{in,2} = (0.1)v_{in,set}$. Of all the $\{K_C, \tau_I, \tau_D\}$ sets that you try, report the one with the best $F(K_C, \tau_I, \tau_D)$, averaged over many disturbances. If at any time during a simulation, $h(t)$ exceeds 2 or drops below 0, stop and reject the controller design.

5 Numerical optimization

Many problems in chemical engineering are expressed mathematically as optimization problems, and involve finding the particular x that minimizes some *cost function* $F(x)$. Each component of x may vary either continuously or discretely. In this chapter, we assume that each x_j varies continuously. In Chapter 7, we consider stochastic techniques that can be used with discretely-varying parameters.

An optimization problem may be *unconstrained*, in which case each x_j can take any real value, or it can be *constrained*, such that an allowable x must satisfy some collection of equality and inequality constraints

$$g(x) = 0 \qquad \text{or} \qquad h(x) \geq 0 \tag{5.1}$$

We consider first unconstrained problems, and then treat constraints. Here, the focus is upon methods that identify local minima; i.e., points that are lower in cost function than their neighbors. The stochastic methods of Chapter 7 return (eventually) global minima; therefore, the reader is referred to that discussion if identifying the global minimum is necessary.

Numerical optimization problems arise in many contexts. To predict the geometry of a molecule, we find the conformation of its atoms with the lowest potential energy. In process design x contains parameters such as equipment sizes, flow rates, temperatures, etc., and the cost function is a measure of the economic cost of operating the process. We fit a mathematical model for a system by minimizing the sum of squared differences between the model predictions and experimental data. In optimal control, we choose the best set of control inputs to maintain a process at the desired set point.

In addition to a discussion of the basic techniques for identifying local minima in continuous parameter space, the use of optimization routines in MATLAB is demonstrated. As these routines are part of an optional optimization toolkit, an alternative routine is provided that can be used without the toolkit.

Local methods for unconstrained optimization problems

We begin by considering iterative techniques that start at some initial guess $x^{[0]}$, and generate a sequence of estimates $x^{[1]}, x^{[2]}, \ldots$ that (hopefully) converges to a *local minimum* $x_{\min}$ of $F(x)$. That is, for x within $|x - x_{\min}| \leq \varepsilon, \varepsilon > 0, F(x) \geq F(x_{\min})$. If we envision $F(x)$ to be a physical elevation, a local minimum lies at the bottom of a "valley," and we move to it from $x^{[0]}$ by descending "downhill" until we cannot decrease $F(x)$ any further.

To use these methods, we must supply at least a routine that returns the cost function value $F(x)$ for an input x. Gradient methods in addition require knowledge of the *gradient vector* $\gamma(x) = \nabla F|_x$, and Newton methods require knowledge (or an approximation) of the *Hessian matrix*

$$H = \nabla^2 F = H^{\mathrm{T}} \qquad H_{jk}(x) = \left.\frac{\partial^2 F}{\partial x_j \partial x_k}\right|_x = \left.\frac{\partial \gamma_k}{\partial x_j}\right|_x = \left.\frac{\partial^2 F}{\partial x_k \partial x_j}\right|_x = H_{kj}(x) \qquad (5.2)$$

While these derivatives can be estimated by finite differences, for large systems it is helpful to supply them directly for an input x. With knowledge of higher-order derivatives, an optimization routine has a better understanding of the local shape of the cost function surface, and can search more efficiently for a local minimum.

The simplex method

The simplex method, implemented as **fminsearch** in the optional MATLAB optimization toolkit, requires only a routine that returns $F(x)$. While simplex methods are used commonly for *linear programming* problems with linear cost functions and constraints (Nocedal & Wright, 1999), for unconstrained optimization with nonlinear cost functions, the gradient and Newton methods discussed below are preferred. Thus, we provide here only a cursory description, and refer the interested reader to the supplemental material in the accompanying website for further details.

The basic idea is easiest to see for a 2-D problem (Figure 5.1). In $\Re^2$, we form a triangle with vertices at $x^{[0]}$ and the two points

$$x^{[1]} = x^{[0]} + l_1 e^{[1]} \qquad x^{[2]} = x^{[0]} + l_2 e^{[2]} \qquad (5.3)$$

We want to move $\{x^{[0]}, x^{[1]}, x^{[2]}, \ldots\}$ so that the triangle comes to enclose a local minimum $x_{\min}$ while shrinking in size. In Figure 5.1, a typical sequence of simplex moves is shown, if the farther away a point is from $x_{\min}$, the higher is its cost function. First (upper left), the vertex of highest cost function value is moved in the direction of the triangle's center towards a region where we expect the cost function values to be lower (upper right). After a sequence of such moves (lower right), the triangle eventually contains in its interior a local minimum (we hope). At this point, the size of the triangle is reduced until it tightly bounds the local minimum (lower left). In practice, both vertex moves and triangle shrinking occur throughout the calculation, as the vertex moves are most likely to succeed when the triangle is small enough that $F(x)$ varies linearly over the triangle region. In $\Re^N$ the triangle becomes a *simplex* with $N + 1$ vertices.

Gradient methods

Algorithms that use knowledge of the gradient vector $\gamma(x) = \nabla F|_x$ are more efficient, because the gradient points in the "uphill" direction of steepest increase in cost function

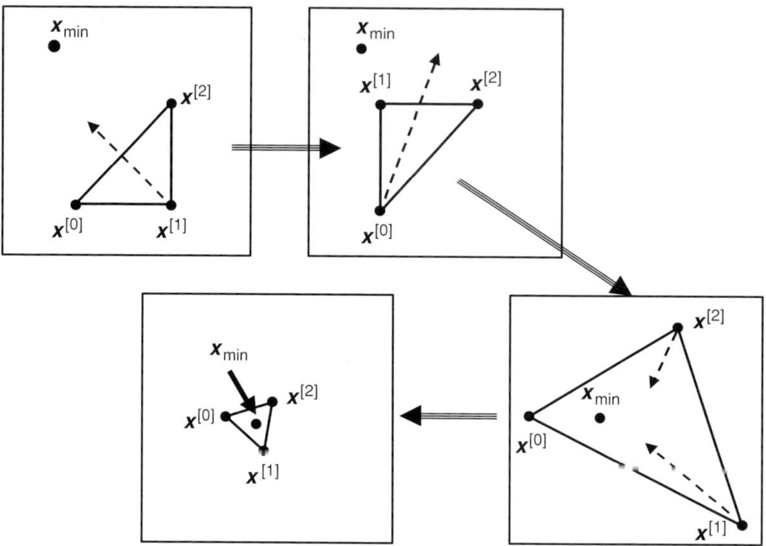

Figure 5.1 The simplex method moves the vertices of the simplex (in two dimensions, a triangle) until it contains the local minimum, and shrinks the simplex to find the local minimum.

(Figure 5.2). Thus, the opposing *direction of steepest descent* $d(x) = -\gamma(x)$ points directly "downhill" and any vector p with $p \cdot \gamma(x) < 0$ also points downhill. If $x^{[k]}$ is the current estimate of the minimum, we generate a *search direction* $p^{[k]}$ such that $p^{[k]} \cdot \gamma(x^{[k]}) < 0$. Then, we move a step length $a^{[k]} > 0$ in this direction to a new estimate $x^{[k+1]} = x^{[k]} + a^{[k]}p^{[k]}$ with a lower cost function value

$$F\left(x^{[k+1]} = x^{[k]} + a^{[k]}p^{[k]}\right) < F\left(x^{[k]}\right) \tag{5.4}$$

As $p^{[k]} \cdot \gamma(x^{[k]}) < 0$, this is possible to achieve, at least for very small $a^{[k]}$, when $|\gamma(x^{[k]})| \neq 0$ as

$$F\left(x^{[k]} + \varepsilon p^{[k]}\right) = F\left(x^{[k]}\right) + \varepsilon\left[p^{[k]} \cdot \gamma(x^{[k]})\right] + O[\varepsilon^2] \tag{5.5}$$

When $|\gamma(x^{[k]})| = 0$, $x^{[k]}$ is an extremum ("flat point"), and as we move downhill in cost function at each step, $x^{[k]}$ is likely a local minimum. In practice (5.4) is insufficient to ensure convergence, as it may happen that $|F(x^{[k+1]}) - F(x^{[k]})| \to 0$ too quickly. Thus, we require $a^{[k]}$ to also satisfy a criterion for a minimum acceptable rate of descent (Figure 5.3),

$$\frac{\left|F\left(x^{[k]}\right) - F\left(x^{[k]} + a^{[k]}p^{[k]}\right)\right|}{a^{[k]}} \geq \chi_1\left[-\gamma^{[k]} \cdot p^{[k]}\right] \qquad \chi_1 \in (0, 1) \tag{5.6}$$

and a sufficient decrease in steepness (Figure 5.4),

$$\left|p^{[k]} \cdot \nabla F\right|_{x^{[k]}+a^{[k]}p^{[k]}} \leq \chi_2\left|p^{[k]} \cdot \gamma^{[k]}\right| \qquad \chi_2 \in (\chi_1, 1) \tag{5.7}$$

The algorithm for a gradient minimizer is easy to program,

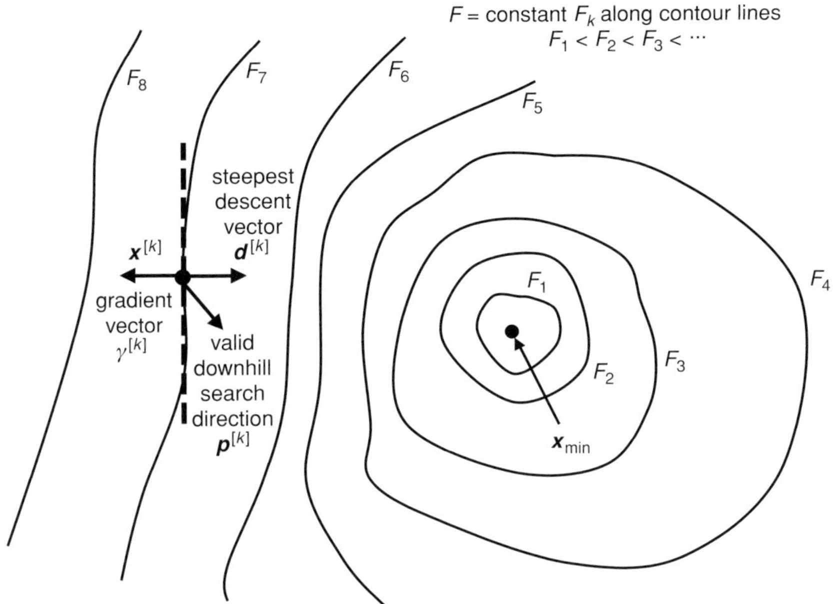

Figure 5.2 Contour plot of $F(\boldsymbol{x})$ showing the gradient and steepest descent directions at $\boldsymbol{x}^{[k]}$. Acceptable search directions point downhill.

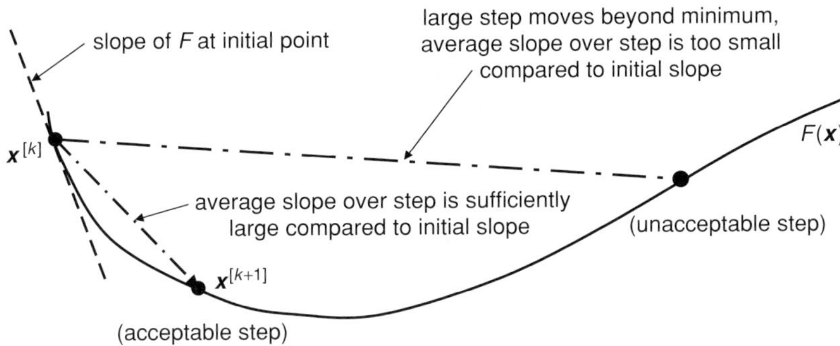

Figure 5.3 Taking too large a step in the search direction violates the criterion of sufficient rate of descent.

Make initial guess $\boldsymbol{x}^{[0]}$

Compute initial cost function, gradient $F(\boldsymbol{x}^{[0]})$, $\gamma^{[0]} = \gamma(\boldsymbol{x}^{[0]})$

for $k = 0, 1, 2, \ldots, k_{\max}$

    if $|\gamma^{[k]}| \leq \delta_{\mathrm{abs}}$, STOP and accept $\boldsymbol{x}^{[k]}$ as local minimum

    Generate search direction $\boldsymbol{p}^{[k]}$ such that $\boldsymbol{p}^{[k]} \cdot \gamma^{[k]} < 0$

    Find $a^{[k]} > 0$ such that descent criteria are met,

$$F\left(\boldsymbol{x}^{[k]} + a^{[k]}\boldsymbol{p}^{[k]}\right) < F\left(\boldsymbol{x}^{[k]}\right)$$

$$\frac{\left|F\left(\boldsymbol{x}^{[k]}\right) - F\left(\boldsymbol{x}^{[k]} + a^{[k]}\boldsymbol{p}^{[k]}\right)\right|}{a^{[k]}} \geq \chi_1\left[-\gamma^{[k]} \cdot \boldsymbol{p}^{[k]}\right]$$

$$\left|\boldsymbol{p}^{[k]} \cdot \boldsymbol{\nabla} F\big|_{\boldsymbol{x}^{[k]} + a^{[k]}\boldsymbol{p}^{[k]}}\right| \leq \chi_2\left|\boldsymbol{p}^{[k]} \cdot \gamma^{[k]}\right|$$

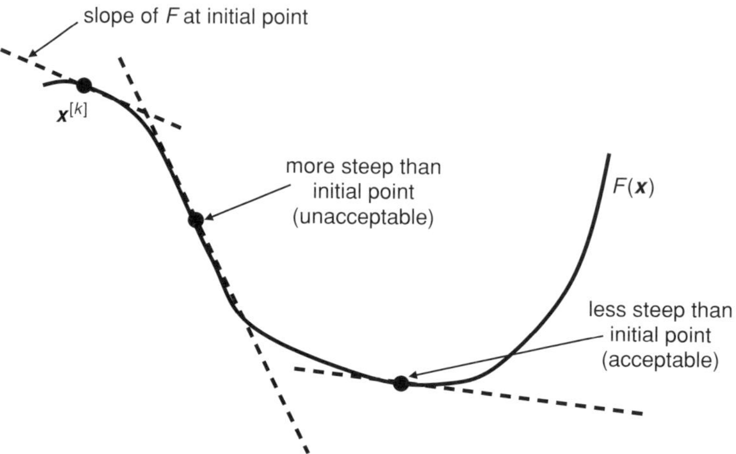

Figure 5.4 Taking too small a step in the search direction violates the condition of sufficient decrease in steepness.

$$\text{Set new estimate, } x^{[k+1]} = x^{[k]} + a^{[k]} p^{[k]}$$
$$\text{repeat for loop } k = 0, 1, 2, \ldots, k_{\max}$$

The two unspecified steps in this algorithm are: (1) the selection of a search direction $p^{[k]}$, and (2) the selection of a step size $a^{[k]}$. We discuss each in turn, starting with the choice of step size.

Strong and weak line searches

There are two general approaches to compute $a^{[k]}$. In a *strong line search*, we find the value of $a^{[k]}$ that minimizes $F(x)$ along the line emanating from $x^{[k]}$ in the direction $p^{[k]}$. However, as the search direction probably does not point directly at $x_{\min}$, much of this effort is wasted. Thus, a *weak line search*, in which we use a simple method to generate an acceptable $a^{[k]}$ without requiring it to be a line minimum, is often used instead.

Because a strong line search is a minimization in only one variable, $a^{[k]}$, we can use robust search methods that are infeasible in multiple dimensions. Essentially, we have the problem of minimizing the cost function

$$f(a) = F\big(x^{[k]} + a p^{[k]}\big) \qquad \frac{df}{da} = p^{[k]} \cdot \gamma\big(x^{[k]} + a p^{[k]}\big) \tag{5.8}$$

Because $df/da|_0 < 0$, we can increase a from zero until we find a segment in which df/da changes sign; thus, a line extremum must lie within this region. Then, either through interval-halving or by fitting a polynomial $\pi(a)$ to $f(a)$ and/or df/da and finding where $d\pi/da = 0$, we identify the $a^{[k]}$ that minimizes (5.8).

It may be of little help to identify the line minimum if the search direction itself does not point at $x_{\min}$. Thus, a weak line search is often an attractive option. A common method is the *backtrack line search*, which starts with an initial step $a_{\max}$. If $a_{\max}$ does not satisfy the

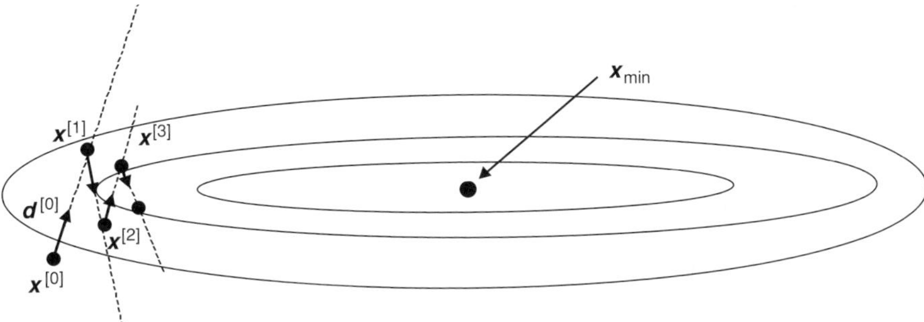

Figure 5.5 Steepest descent method results in a zig-zag trajectory in steep valleys that yields poor convergence.

descent criteria, it is halved until an acceptable value is found:

$$a^{[k]} = a_{\max}$$
$$a^{[k]} \leftarrow a^{[k]}/2, \text{ iterate until acceptable step size found} \tag{5.9}$$

$a_{\max}$ is obtained by fitting a low-order polynomial in a to $F(x^{[k]} + a p^{[k]})$,

$$F\left(x^{[k]} + a p^{[k]}\right) \approx c_0 + c_1 a + c_2 a^2 \tag{5.10}$$

The expansion coefficients are obtained at the cost of one additional function evaluation at a large step size a',

$$c_0 = F\left(x^{[k]}\right) \qquad c_1 = p^{[k]} \cdot \gamma^{[k]} < 0 \qquad c_2 = \frac{\left[F\left(x^{[k]} + a' p^{[k]}\right) - c_0 - c_1 a'\right]}{(a')^2} \tag{5.11}$$

The minimum of this quadratic polynomial in $[0, a']$ sets $a_{\max}$,

$$a_{\max} = \begin{cases} -c_1/2c_2, & c_2 > 0 \\ a', & c_2 \leq 0 \end{cases} \tag{5.12}$$

If $F(x)$ is itself quadratic, and a' is large enough, $a_{\max}$ is the line minimum, and this approach is a strong line search.

Choosing the search direction

Once an appropriate $a^{[k]}$ has been found (since we require the search direction to be one of initially decreasing cost function, such a step always exists, even if it is very small), $x^{[k+1]} = x^{[k]} + a^{[k]} p^{[k]}$ is the new estimate of the minimum. The new gradient, $\gamma^{[k+1]}$, and steepest descent, $d^{[k+1]} = -\gamma^{[k+1]}$, vectors are computed, and we then choose a new search direction $p^{[k+1]}$. It must be a descent direction, $\gamma^{[k+1]} \cdot p^{[k+1]} < 0$, but which of the infinite number of descent directions should we choose?

The most obvious choice, the *method of steepest descent*, searches always in the direction of steepest descent, $p^{[k+1]} = -\gamma^{[k+1]}$. For the first iteration, this is the logical choice. At successive iterations, however, the steepest descent method converges slowly as it often yields zig-zag trajectories when travelling down steep valleys (Figure 5.5).

In the *conjugate gradient method*, this zig-zag behavior is reduced by mixing-in a portion of the previous search direction such that the trajectory tends not to double back upon itself,

$$\boldsymbol{p}^{[k+1]} = -\boldsymbol{\gamma}^{[k+1]} + \beta^{[k]}\boldsymbol{p}^{[k]} \tag{5.13}$$

$\beta^{[k+1]}$ is commonly chosen by one of the following two formulas:

$$\beta^{[k]} = \begin{cases} \dfrac{\gamma^{[k+1]} \cdot \gamma^{[k+1]}}{\gamma^{[k]} \cdot \gamma^{[k]}}, & \text{Fletcher–Reeves (CG-FR)} \\[2em] \dfrac{\gamma^{[k+1]} \cdot \left(\gamma^{[k+1]} - \gamma^{[k]}\right)}{\gamma^{[k]} \cdot \gamma^{[k]}}, & \text{Polak–Ribiere (CG-PR)} \end{cases} \tag{5.14}$$

The CG-FR formula yields search directions with very favorable convergence properties for quadratic cost functions, as is explained below. The CG-PR formula gives the same search direction as CG-FR for quadratic cost functions, but its extra term biases the search direction towards the direction of steepest descent when the cost function is far from quadratic. As steepest descent is more robust in this case, CG-PR is preferred.

A gradient minimizer routine

gradient_minimizer.m implements either the CG-PR method as the default or the steepest descent method. It uses a weak line search starting from an initial step based on quadratic approximation. Thus, when the cost function is (locally) quadratic, a strong line search is conducted. The syntax is

[x, F, grad, iflag, x_traj] = gradient_minimizer(. . .
   func_name, x0, OptParam, ModelParam);

func_name is the name of a routine that returns $F(\boldsymbol{x})$ and $\boldsymbol{\gamma}(\boldsymbol{x})$,

function [F, grad, iOK] = func_name(x, ModelParam);

Here, iOK is 1 if the cost function was evaluated correctly.

ModelParam is a structure that contains the fixed parameters of the cost function. x0 is the initial guess, and OptParam is an optional structure that controls the behavior of the minimizer. As output, x is the local minimum, F and grad are its cost function and gradient, iflag is 1 if the minimizer converged. x_traj is an optional array recording the progress of the minimizer.

The steepest descent and gradient methods are demonstrated in Figure 5.6 and Figure 5.7 for the cost function

$$F(\boldsymbol{x}) = (x_1 - 1)^2 + 10(x_2 - 2)^2 + c(x_1 - 1)^4 + c(x_2 - 2)^4 \tag{5.15}$$

The following routine returns the cost function and its gradient:

function [F, grad, iOK] = simple_cost_func(x, ModelParam);
iOK = 0; c = ModelParam.c;
dx1 = x(1) - 1; dx2 = x(2) - 2;

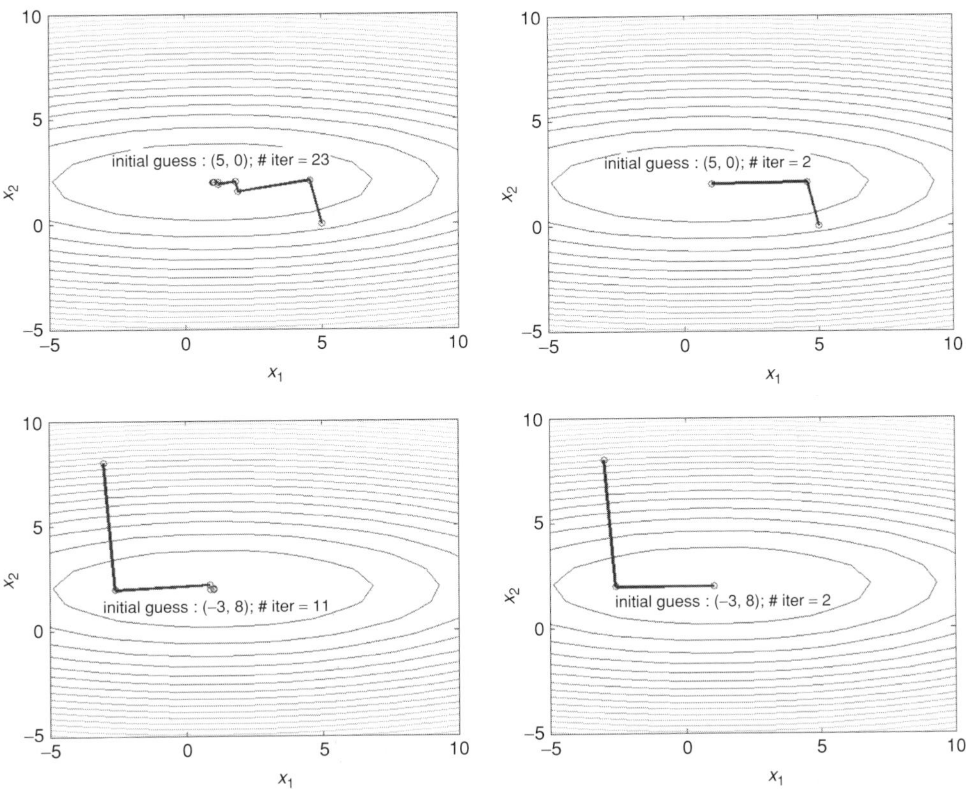

Figure 5.6 Gradient minimizers applied to a quadratic cost function. Results at left from steepest descent method and those at right from conjugate gradient method.

```
grad = zeros(size(x));
F = dx1^2 + 10*dx2^2 + c*(dx1^4) + c*(dx2^4);
grad(1) = 2*dx1 + 4*c*dx1^3; grad(2) = 2*10*dx2 + 4*c*dx2^3;
iOK = 1;
return;
```

For the quadratic case ($c = 0$) with the default (CG-PR) method, the minimization is performed by

```
ModelParam.c = 0;
x0 = [5; 0];
[x, F, grad, iflag, x_traj] = gradient_minimizer('simple_cost_func', ...
    x0, [], ModelParam);
```

To use the steepest descent method, we type

```
OptParam.method = 0;
[x, F, grad, iflag, x_traj] = gradient_minimizer('simple_cost_func', ...
    x0, OptParam, ModelParam);
```

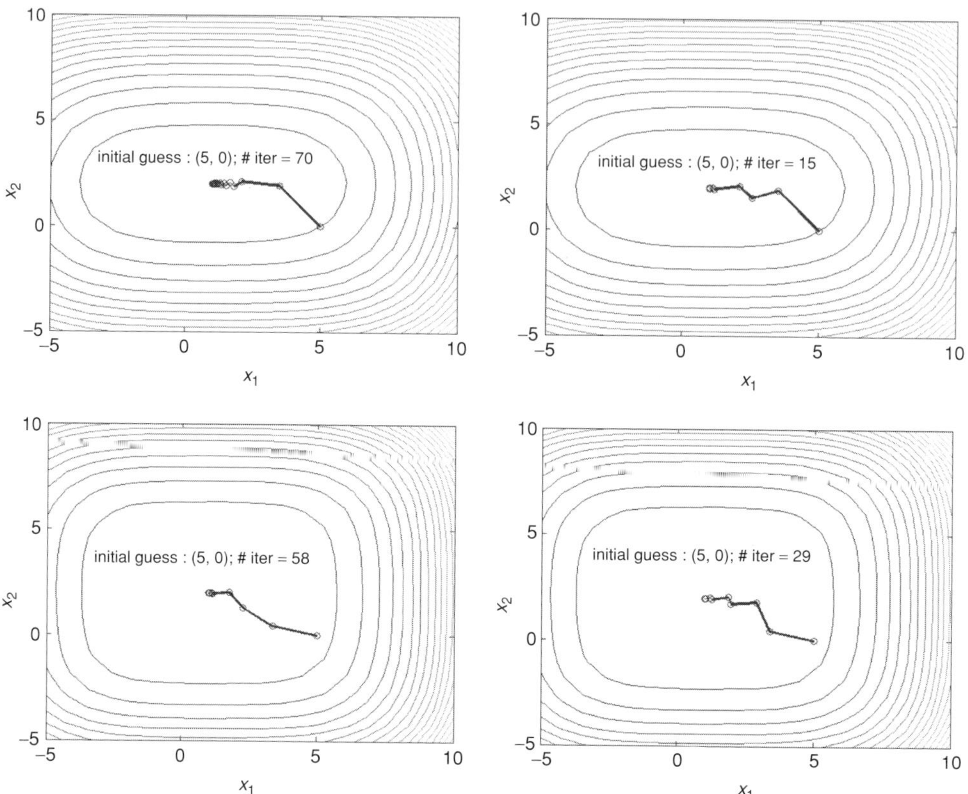

Figure 5.7 Gradient minimizers applied to a nonquadratic cost function. Steepest descent (left) and conjugate gradient (right) with $c = 0.1$ (upper) and $c = 1$ (lower).

Figure 5.6 compares the conjugate gradient and steepest descent methods for the quadratic case, $c = 0$. While the steepest descent method generates a zig-zag trajectory, the conjugate gradient method avoids this and reaches the minimum after only two iterations (we note that 2 is also the dimension of x). Figure 5.7 compares the two methods with nonzero quartic terms. The conjugate gradient method is no longer able to find the minimum after only two iterations, yet it is still more efficient than the steepest descent method.

Conjugate gradient method applied to quadratic cost functions

We now consider the origin of the excellent performance of the conjugate gradient method for a quadratic cost function, defined in terms of a symmetric, positive-definite matrix A and a vector b,

$$F(x) = \frac{1}{2}x^\mathrm{T} Ax - b^\mathrm{T}x \qquad \gamma(x) = Ax - b \tag{5.16}$$

As the cost function is minimized when $Ax = b$, this method is often used to solve linear systems when elimination methods are too costly (more on this subject in Chapter 6). Let

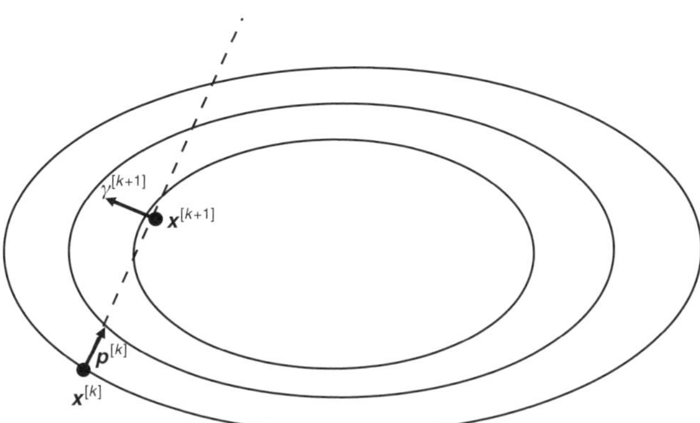

Figure 5.8 At the minimum along the line, the local gradient is perpendicular to the search direction.

the current estimate be $x^{[k]}$. We perform a line search in the direction $p^{[k]}$, and as $F(x)$ is quadratic, we can compute analytically the line minimum $a^{[k]}$ where the gradient is perpendicular to the search direction (Figure 5.8),

$$p^{[k]} \cdot \gamma\left(x^{[k]} + a^{[k]} p^{[k]}\right) = p^{[k]} \cdot \left\{A\left(x^{[k]} + a^{[k]} p^{[k]}\right) - b\right\} = 0 \tag{5.17}$$

Using $\gamma^{[k]} = Ax^{[k]} - b$, this yields the update

$$x^{[k+1]} = x^{[k]} + a^{[k]} p^{[k]} \qquad a^{[k]} = -\frac{p^{[k]} \cdot \gamma^{[k]}}{p^{[k]} \cdot A p^{[k]}} \tag{5.18}$$

We next select a new search direction $p^{[k+1]}$ to identify the next estimate

$$x^{[k+2]} = x^{[k+1]} + a^{[k+1]} p^{[k+1]} \tag{5.19}$$

How should we choose the new search direction $p^{[k+1]}$? We see from Figure 5.8 that we have already minimized the cost function in the direction $p^{[k]}$. It would be nice if, when we do the subsequent line searches in the directions $p^{[k+1]}, p^{[k+2]}, \ldots$, we do nothing to "mess up" the fact that we have found an optimal coordinate in the direction $p^{[k]}$. If so, and if the set of search directions is linearly independent, then after at most N iterations, we are guaranteed to have found the exact position of the minimum, in the absence of round-off error.

We thus choose $p^{[k+1]}$ such that $x^{[k+2]}$ remains optimal in the direction $p^{[k]}$,

$$\frac{d}{d\alpha} F\left(x^{[k+2]} + \alpha p^{[k]}\right)\Big|_{\alpha=0} = p^{[k]} \cdot \gamma\left(x^{[k+2]}\right) = 0 \tag{5.20}$$

Writing $\gamma^{[k+2]} = Ax^{[k+2]} - b$ and using (5.19), we have

$$p^{[k]} \cdot \left\{Ax^{[k+2]} - b\right\} = p^{[k]} \cdot \left\{\left[Ax^{[k+1]} - b\right] + a^{[k+1]} A p^{[k+1]}\right\} = 0$$
$$p^{[k]} \cdot \gamma^{[k+1]} + a^{[k+1]}\left(p^{[k]} \cdot A p^{[k+1]}\right) = 0 \tag{5.21}$$

As we have already established that $p^{[k]} \cdot \gamma^{[k+1]} = 0$, if we choose the new search direction to be *A-conjugate* to the old one,

$$p^{[k]} \cdot A p^{[k+1]} = 0 \tag{5.22}$$

then $p^{[k]} \cdot \gamma(x^{[k+2]}) = 0$, and performing a line minimization along $p^{[k+1]}$ will do nothing to alter the fact that we have found the optimal coordinate in the direction $p^{[k]}$. The N vectors $p^{[1]}, p^{[2]}, \ldots, p^{[N]}$ generated by this method are linearly independent. As after N iterations we will have found the correct coordinates of the minimum in each of these directions, we are guaranteed to have found the exact position of the minimum. Thus, in the absence of round-off error, the conjugate gradient algorithm terminates after at most N iterations. Note that this method is somewhat misnamed, as by (5.22) it is the search directions $\{p^{[k]}\}$ that are conjugate, not the gradients $\{\gamma^{[k]}\}$, which in fact may be shown to be orthogonal.

To implement the A-conjugacy condition, $p^{[k]} \cdot A p^{[k+1]} = 0$, we write the new search direction as the linear combination

$$p^{[k+1]} = -\gamma^{[k+1]} + \beta^{[k]} p^{[k]} \tag{5.23}$$

Multiplying by A and enforcing A-conjugacy yields

$$p^{[k]} \cdot A p^{[k+1]} = 0 = -p^{[k]} \cdot A \gamma^{[k+1]} + \beta^{[k]} p^{[k]} \cdot A p^{[k]}$$

$$\beta^{[k]} = \frac{\gamma^{[k+1]} \cdot A p^{[k]}}{p^{[k]} \cdot A p^{[k]}} = \frac{\gamma^{[k+1]} \cdot \gamma^{[k+1]}}{\gamma^{[k]} \cdot \gamma^{[k]}} \tag{5.24}$$

The latter form for $\beta^{[k]}$ is obtained from the first through use of conditions that follow from A-conjugacy. If this formula for $\beta^{[k]}$ is applied to a nonquadratic cost function, the CG-FR method is obtained. Because the relations above assume the cost function to be quadratic, the CG-FR method is not guaranteed to terminate in N iterations; however, once the algorithm is close enough to the minimum for quadratic approximation to be accurate, the self-terminating aspects of the conjugate gradient method become useful. The CG-FR and CG-PR formulas agree for a quadratic cost function (as $\gamma^{[k]} \cdot \gamma^{[k+1]} = 0$). Whenever the cost function is far from quadratic, the orthogonality of the gradients will be lost, and the additional term in the CG-PR formula pushes the search direction towards the steepest descent direction. This usually improves the performance, and the CG-PR method is the recommended choice of gradient method.

The conjugate gradient algorithm to minimize $F(x) = \frac{1}{2} x^{\mathrm{T}} A x - b^{\mathrm{T}} x$, solving $Ax = b$, is

make initial guess $x^{[0]}$; set initial gradient, $\gamma^{[0]} = A x^{[0]} - b$
set initial search direction, $p^{[0]} = -g^{[0]}$
for $k = 0, 1, \ldots, (1 + \varepsilon)N$, where $\varepsilon \geq 0$ allows for round-off error
    if $\|\gamma^{[k]}\| \leq \delta_{\mathrm{tol}}$, STOP and ACCEPT $x^{[k]}$ as solution
    perform line search.

$$y^{[k]} = A p^{[k]}, \, a^{[k]} = -\left(p^{[k]} \cdot \gamma^{[k]}\right) / \left(p^{[k]} \cdot y^{[k]}\right)$$
$$x^{[k+1]} = x^{[k]} + a^{[k]} p^{[k]}$$

    compute new gradient, $\gamma^{[k+1]} = \gamma^{[k]} + a^{[k]} y^{[k]}$

get new search direction, $\beta^{[k]} = \left(\gamma^{[k+1]} \cdot \gamma^{[k+1]}\right) / \left(\gamma^{[k]} \cdot \gamma^{[k]}\right)$,

$$p^{[k+1]} = -\gamma^{[k+1]} + \beta^{[k]} p^{[k]}$$

end for $k = 0, 1, \ldots (1 + \varepsilon)N$

An alternative expression for the step length, equivalent to (5.18), is

$$a^{[k]} = \left(\gamma^{[k]} \cdot \gamma^{[k]}\right) / \left(p^{[k]} \cdot y^{[k]}\right) \tag{5.25}$$

In MATLAB this algorithm is used by **pcg** (more on this routine in Chapter 6). Note that at each CG iteration, we only need to multiply the current solution estimate by A, a quick procedure when A is sparse. Note also, that as no fill-in occurs, this method is well suited to large, sparse systems.

Newton line search methods

The gradient vector provides information about the local slope of the cost function surface. Further improvement in efficiency is gained by using knowledge of the local curvature of the surface, as encoded in the real, symmetric *Hessian matrix*, with the elements

$$H_{jk} = \frac{\partial \gamma_k}{\partial x_j} = \frac{\partial^2 F}{\partial x_j \partial x_k} = \frac{\partial^2 F}{\partial x_k \partial x_j} = H_{kj} \tag{5.26}$$

Again, we use an iterative method with a line search direction $p^{[k]}$,

$$x^{[k+1]} = x^{[k]} + a^{[k]} p^{[k]} \tag{5.27}$$

We use the Hessian to make better selections of the search directions and step lengths. For small p, Taylor expansion of the gradient yields

$$\gamma\left(x^{[k]} + p\right) - \gamma\left(x^{[k]}\right) \approx H\left(x^{[k]}\right)p + O[|p|^2] \tag{5.28}$$

As $\gamma(x_{\min}) = 0$, we set $\gamma(x^{[k]} + p) = 0$ to yield a linear system for $p^{[k]}$:

$$H\left(x^{[k]}\right)p^{[k]} = -\gamma\left(x^{[k]}\right) \tag{5.29}$$

This appears to be equivalent to solving $\gamma(x) = 0$ by Newton's method, but there is an important difference. Newton's method just looks for a point where the gradient is zero, but we specifically want to find a *minimum* of $F(x)$. Yet, the gradient is zero at maxima and at saddle points as well. Thus, we must ensure that $p^{[k]} \cdot \gamma^{[k]} < 0$, so that we can use a backtrack line search, starting from $a^{[k]} = 1$, to enforce

$$F\left(x^{[k+1]}\right) = F\left(x^{[k]} + a^{[k]} p^{[k]}\right) < F\left(x^{[k]}\right) \tag{5.30}$$

How can we be assured that the search direction generated by (5.29) is indeed a descent direction?

To make things general, let us use not the exact Hessian, but some real-symmetric approximation to it:

$$B^{[k]} \approx H\big(x^{[k]}\big) \qquad \big(B^{[k]}\big)^{\mathrm{T}} = B^{[k]} \tag{5.31}$$

The search direction is obtained by solving

$$B^{[k]} p^{[k]} = -\gamma^{[k]} \tag{5.32}$$

For small $a^{[k]}$, we approximate the change in $F(x)$ as

$$F\big(x^{[k]} + a^{[k]} p^{[k]}\big) - F\big(x^{[k]}\big) \approx a^{[k]}\big(p^{[k]} \cdot \gamma^{[k]}\big) + O\big[\big(a^{[k]}\big)^2\big] \tag{5.33}$$

But, from our update rule $\gamma^{[k]} = -B^{[k]} p^{[k]}$; therefore,

$$F\big(x^{[k]} + a^{[k]} p^{[k]}\big) - F\big(x^{[k]}\big) \approx -a^{[k]}\big(p^{[k]} \cdot B^{[k]} p^{[k]}\big) + O\big[\big(a^{[k]}\big)^2\big] \tag{5.34}$$

We first try $a^{[k]} = 1$, and iteratively halve its value until we find that the descent criteria are satisfied. Thus, $a^{[k]}$ eventually will be small enough for (5.34) to be valid, so that $F(x)$ will reduced if $p^{[k]} \cdot B^{[k]} p^{[k]} > 0$, i.e., if $B^{[k]}$ is positive-definite (all of its eigenvalues are positive).

Let $x_{\min}$ be a local minimum such that no neighboring points have a lower cost function. Then, as for very small p,

$$F(x_{\min} + p) - F(x_{\min}) \approx \frac{1}{2} p^{\mathrm{T}} H(x_{\min})p + O[|p|^2] \Rightarrow p^{\mathrm{T}} H(x_{\min})p \geq 0 \tag{5.35}$$

It is possible for $H(x_{\min})$ to have one or more zero eigenvalues and for $x_{\min}$ still to be a local minimum, but there can be no negative eigenvalues. Thus, at (and very nearby) a minimum, the Hessian is at least positive-semidefinite. Away from the near vicinity of a local minimum, however, it is quite possible for the Hessian to have negative eigenvalues, so that the search direction generated by $H(x^{[k]})p^{[k]} = -\gamma(x^{[k]})$ is *not* a descent direction.

It is not necessary, however, that we use the exact Hessian when computing $p^{[k]}$. As $\gamma(x_{\min}) = 0$, $B^{[k]} p^{[k]} = -\gamma(x_{\min})$ has for any nonsingular $B^{[k]}$ the unique solution $p^{[k]} = 0$, so that $x^{[k+1]} = x_{\min}$. We can get the same eventual solution even if $B^{[k]} \neq H^{[k]}$. Thus, if $B^{[k]}$ is (nearly) singular, we could add a value $\tau > 0$ to each diagonal element, such that by Gershgorin's theorem $(B^{[k]} + \tau I)$ *is* positive-definite. Then, we solve $(B^{[k]} + \tau I)p^{[k]} = -\gamma^{[k]}$ to generate a search direction that is guaranteed to point downhill.

Often we construct $B^{[k]}$ from past gradient evaluations. As the Hessian is the Jacobian of the gradient, we generate $B^{[k+1]}$ from $B^{[k]}$ so that it satisfies the *secant condition*

$$B^{[k+1]} \Delta x = \Delta \gamma \qquad \Delta x = x^{[k+1]} - x^{[k]} \qquad \Delta \gamma = \gamma^{[k+1]} - \gamma^{[k]} \tag{5.36}$$

or, a corresponding condition for the Hessian inverse

$$\big(B^{[k+1]}\big)^{-1} \Delta \gamma = \Delta x \tag{5.37}$$

The smallest update consistent with (5.37) is achieved by the *Broyden–Fletcher–Goldfarb–Shanno (BFGS) formula*, which ensures that if $B^{[0]}$ is positive-definite (e.g., $B^{[0]} = I$), so is $B^{[1]}, B^{[2]}, \ldots,$

$$B^{[k+1]} = B^{[k]} + \frac{\Delta\gamma(\Delta\gamma)^{\mathrm{T}}}{(\Delta\gamma)^{\mathrm{T}}\Delta x} - \frac{B^{[k]}\Delta x(\Delta x)^{\mathrm{T}}B^{[k]}}{(\Delta x)^{\mathrm{T}}B^{[k]}\Delta x} \tag{5.38}$$

Equivalent update formulas exist for the approximate Hessian inverse or Cholesky factor, which are more commonly used in practice than (5.38). For further details of such quasi-Newton methods consult Nocedal & Wright (1999).

Trust-region Newton method

Newton line search algorithms perform badly when $B^{[k]}$ is nearly singular, as the search directions become erratic. The *trust-region Newton method*, in which step length and search direction are chosen concurrently, is more robust in this case. For small p, the cost function may be approximated in the vicinity of $x^{[k]}$ by a quadratic model function

$$F(x^{[k]} + p) \approx F(x^{[k]}) + \gamma^{[k]} \cdot p + \frac{1}{2}p \cdot B^{[k]}p \equiv m^{[k]}(p) \tag{5.39}$$

We assume that $m^{[k]}(p)$ agrees well with $F(x)$ within a *trust region* $|p| < \Delta^{[k]}$. The *trust radius* $\Delta^{[k]}$ is modified from one iteration to the next based on the observed level of disagreement between the true and model cost functions. We obtain $\Delta x^{[k]}$ by finding (at least approximately) a minimum, within the trust region, of the model function $m^{[k]}(p)$.

The advantage of the trust-region method is that even if $B^{[k]}$ is nearly singular, $m^{[k]}(p)$ still has a useful minimum within the trust region. As long as $F(x)$ is reduced, $x^{[k+1]} = x^{[k]} + \Delta x^{[k]}$ is accepted, else the old value is recycled and the trust radius is reduced. If the agreement between $F(x^{[k+1]}) - F(x^{[k]})$ and $m^{[k]}(\Delta x^{[k]}) - m^{[k]}(0)$ is good (poor), the trust radius is increased (decreased) for the next iteration.

The dogleg method

The *dogleg method* solves the trust-region subproblem approximately, assuming that $B^{[k]}$ is positive-definite, although perhaps nearly singular. We say that this method solves the problem *approximately*, because we do not bother to find the true minimum, but merely an easily-found point within the trust region that lowers $m^{[k]}(p)$ more than the steepest descent method does.

The dogleg method is based upon an analysis of how the trust-region minimum changes as a function of the trust radius $\Delta^{[k]}$. When $\Delta^{[k]}$ is very large, we have effectively an unconstrained problem and the trust-region minimum is the full Newton step $p^{(n)}$, the global minimum of $m^{[k]}(p)$,

$$B^{[k]}p^{(n)} = -\gamma^{[k]} \tag{5.40}$$

On the other hand, when $\Delta^{[k]}$ is very small, the curvature of $m^{[k]}(p)$ may be neglected, and the minimum is found by moving in the steepest descent direction as far as is allowed,

$$p^{(d)} = -\Delta^{[k]}\gamma^{[k]}/|\gamma^{[k]}| \tag{5.41}$$

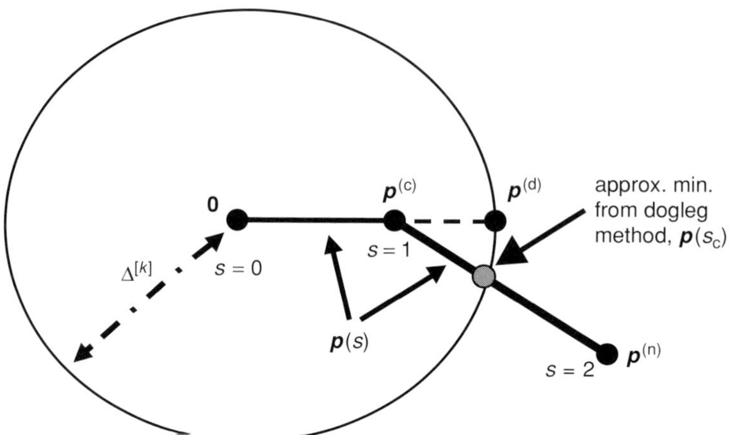

Figure 5.9 The trust-region Newton dogleg method finds the position on the two connected line segments that minimizes the model function while lying within the trust region.

When $B^{[k]}$ is positive-definite, the local model function value is greater than that predicted by neglecting the curvature. Thus, the actual minimum of $m^{[k]}(p)$ along the line connecting the origin and $p^{(d)}$ is found at some intermediate point $p^{(c)}$, the *Cauchy point*,

$$p^{(c)} = \alpha p^{(d)} \qquad 0 \le \alpha \le 1 \tag{5.42}$$

where

$$0 = \frac{d}{d\alpha} m^{[k]}(\alpha p^{(d)}) = \frac{d}{d\alpha} \left\{ F(x^{[k]}) + \alpha \gamma^{[k]} \cdot p^{(d)} + \frac{\alpha^2}{2} p^{(d)} \cdot B^{[k]} p^{(d)} \right\} \tag{5.43}$$

such that

$$\alpha = -\left[\gamma^{[k]} \cdot p^{(d)} \right] / \left[p^{(d)} \cdot B^{[k]} p^{(d)} \right] \tag{5.44}$$

For any intermediate $\triangle^{[k]}$, we first compute the full Newton step. As $p^{(n)}$ is a global minimum for $m^{[k]}(p)$, if it lies within the trust region, $|p^{(n)}| \le \triangle^{[k]}$, we accept it as the update, $\triangle x^{[k]} = p^{(n)}$. Otherwise, we must approximate the minimum, accounting for the constraint, as follows.

We form the "dogleg" curve $p(s)$ from the connected line segments running from 0 to $p^{(c)}$ and from $p^{(c)}$ to $p^{(n)}$ (Figure 5.9),

$$p(s) = \begin{cases} s p^{(c)}, & \text{if } 0 \le s \le 1 \\ p^{(c)} + (s-1)(p^{(n)} - p^{(c)}), & \text{if } 1 \le s \le 2 \end{cases} \tag{5.45}$$

The constrained minimum lies along this curve when $\triangle^{[k]}$ is very large or very small, and thus for intermediate $\triangle^{[k]}$ it poses a convenient 1-D restricted problem that is quickly solved,

$$\text{minimize } m^{[k]}(p(s)) \text{ subject to} |p(s)| \le \triangle^{[k]} \tag{5.46}$$

It may be shown that along this curve, $|p(s)|$ increases monotonically and $m^{[k]}(p(s))$

decreases monotonically,

$$\frac{d}{ds}|p(s)|^2 > 0 \qquad \frac{d}{ds}m^{[k]}(p(s)) < 0 \tag{5.47}$$

As we are considering the case where $p^{(n)}$ lies outside of the trust region and $p^{(d)}$ lies within it, the minimum along the curve occurs at $1 \le s_c \le 2$ where $p(s)$ crosses the trust region boundary,

$$|p(s_c)|^2 = \left|p^{(c)} + (s_c - 1)(p^{(n)} - p^{(c)})\right|^2 = \left(\Delta^{[k]}\right)^2 \tag{5.48}$$

In this case, the accepted new estimate is

$$x^{[k+1]} = x^{[k]} + \Delta x^{[k]} \qquad \Delta x^{[k]} = p(s_c) = p^{(c)} + (s_c - 1)(p^{(n)} - p^{(c)}) \tag{5.49}$$

The dogleg method allows us to identify quickly a point within the trust region that lowers the model cost function at least as much as the Cauchy point. The advantage over the Newton line search procedure is that the full Newton step is not automatically accepted as the search direction, avoiding the problems inherent in its erratic size and direction when $B^{[k]}$ is nearly singular. Further methods to find approximate trust-region minimums are discussed in Nocedal & Wright (1999).

Newton methods for large problems

In the line search and trust-region Newton methods, we must obtain the full Newton step $p^{(n)}$ by solving a linear system at each Newton iteration,

$$B^{[k]} p^{(n)} = -\gamma^{[k]} \qquad B^{[k]} \approx H(x^{[k]}) \qquad B^{[k]} > 0 \tag{5.50}$$

As $B^{[k]}$ is real-symmetric/positive-definite, the conjugate gradient method can be used. As discussed in Chapter 1, fill-in during elimination for large, sparse systems yields very long computation times and massive memory requirements; however, in the conjugate gradient method, no elimination is necessary (see algorithm above). We only need to compute at each conjugate gradient iteration the product of $B^{[k]}$ with the current estimate of $p^{(n)}$. When $B^{[k]}$ is sparse, this product is fast to compute and we only need to store the nonzero elements of $B^{[k]}$.

BFGS can be applied to large problems when the Hessian is sparse if the update formula (5.38) is provided with the sparsity pattern so that only the nonzero positions are stored and updated. Alternatively, only the most recent gradient vectors may be retained and used in (5.38). Both approaches allow the construction of approximate Hessians with limited memory usage. For a more detailed discussion of memory-efficient BFGS methods, consult Nocedal & Wright (1999).

The need in Newton's method to store $B^{[k]}$ and to solve a linear system (5.50) at each iteration poses a significant challenge for large optimization problems. For large problems with Hessians that are dense or whose sparsity patterns are unknown, the tricks above cannot be used. Instead, the nonlinear conjugate gradient method, which does not require any curvature knowledge, is recommended.

Unconstrained minimizer fminunc in MATLAB

The routine **fminunc**, part of the optional MATLAB optimization toolkit, uses either a gradient or a Newton algorithm to solve an unconstrained, continuous optimization problem, with the choice of method depending upon the size of the problem and upon the level of information about the gradient and Hessian provided by the user. It is called with the syntax

[x, F, exitflag, output] = fminunc (fun, x0, OPTIONS, P1, P2, . . .);

fun is the name of a routine that returns the cost function value,

function F = fun(x, P1, P2, . . .);

x0 is the initial guess, and P1, P2, . . . are optional parameters to be passed through **fminunc** to fun. OPTIONS is set by **optimset**. If fun supplies the value of the gradient as well, use

OPTIONS = optimset('GradObj', 'on');

and give fun the syntax,

function [F, grad] = fun(x, P1, P2, . . .);

If in addition, you supply the Hessian matrix, add to OPTIONS,

OPTIONS = optimset(OPTIONS, 'Hessian', 'on');

and write fun as

function [F, grad, Hessian] = fun(x, P1, P2, . . .);

The output arguments include the local minimum x and its cost function value F.

For large problems, **fminunc** can be informed of the sparsity pattern of the Hessian through the 'HessPattern' field. Instead of supplying the Hessian itself, one can merely pass the name of a routine that computes the product of an input vector with the Hessian through the 'HessMult' field. **fminunc** can be told explicitly to use algorithms effective for large problems by setting the 'LargeScale' field to 'on'. The 'Display' field controls the level of detail printed to the screen about the progress of the minimization and 'Diagnostics' can be set to 'on' to gain further information about the optimizer performance.

Example. A simple cost function

We demonstrate **fminunc** for the simple cost function (5.15) used to compare the performance of the steepest descent and conjugate gradient methods. The routine simple_cost_func.m that returns the value of this cost function was presented previously following (5.15). The code below uses **fminunc** to find the minimum at $(1, 2)$,

```
x0 = [5;0];
ModelParam.c = 1;
[x,F] = fminunc (@simple_cost_func, x0, [], ModelParam),
Warning: Gradient must be provided for trust-region method;
    using line-search method instead.
> In fminunc at 241
```

*Optimization terminated: relative infinity-norm of gradient less than
    options.TolFun.*
x = 1; 2
F = 0

To let **fminunc** know that the routine returns the gradient vector, use the 'GradObj' field of
optimset,

Options = optimset('GradObj', 'on');
[x,F] = fminunc (@simple_cost_func, x0, Options, ModelParam),
*Optimization terminated: first-order optimality less than
    OPTIONS. TolFun, and no negative/zero curvature detected in
    trust region model.*
x = 1; 2
F = 0

To suppress the printing of messages on the screen, use the 'Display' field,

Options = optimset(Options, 'Display','off');
[x,F] = fminunc (@simple_cost_func, x0, Options, ModelParam),
*   x = 1; 2*
*   F = 0*

For this problem, the steepest descent method required 58 iterations and the conjugate
gradient method 29. From the additional output arguments,

[x,F,exitflag,output] = fminunc (@simple_cost_func, . . .
**   x0, Options, ModelParam),**
x̃ = 1; 2
F̃ = 0
exitflag = 1
output =
*   iterations: 4*
*   funcCount: 5*
*   cgiterations: 4*
*   firstorderopt: 0*
*   algorithm: 'large-scale: trust-region Newton'*
*   message: [1x137 char]*

we see that **fminunc** was quite efficient at finding the minimum. To avoid using a large-scale
algorithm on this small problem, type

Options = optimset(Options,'LargeScale','off');
[x,F,exitflag,output] = fminunc (@simple_cost_func, x0, Options, ModelParam),
x = 1; 2
F = 0
exitflag = 1
output =

iterations: 1
funcCount: 2
stepsize: 0.0417
firstorderopt: 0
algorithm: 'medium-scale: Quasi-Newton line search'
message: [1x85 char]

Example. Fitting a kinetic rate law to time-dependent data

There need not always be an analytical expression for the cost function. Often, the cost function itself is computed by a numerical calculation. For example, let us say that we are studying the enzymatic conversion of a substrate S to a product P in a batch bioreactor. We expect the rate of conversion, in units of micromoles converted per minute per milligram of enzyme, to be described by Michaelis–Menten kinetics, with possibly substrate inhibition,

$$-\hat{r}_s = \frac{V_m[S]}{K_m + [S] + K_{si}^{-1}[S]^2} \tag{5.51}$$

For a bioreactor of volume V_R, the number of micromoles of substrate, N_S, is related to the substrate molar concentration [S] by

$$N_s = \alpha_c V_R[S] \tag{5.52}$$

α_c is 10^6 µmol/mol. Thus, the mole balance on the substrate in a bioreactor containing m_E mg of enzyme is

$$\frac{d[S]}{dt} = \left(\frac{m_E}{\alpha_c V_R}\right)\hat{r}_s = -\left(\frac{m_E}{\alpha_c V_R}\right)\left[\frac{V_m[S]}{K_m + [S] + K_{si}^{-1}[S]^2}\right] \tag{5.53}$$

For a reactor of volume 100 ml containing 10 mg of enzyme, Table 5.1 records the substrate concentration as a function of time, starting from an initial concentration of 2 M. We wish to fit $\theta = [V_m\ K_m\ K_{si}]^T$ by minimizing the cost function

$$F_c(\theta) = \tfrac{1}{2}\sum_{k=1}^{N_d}[S_{pred}(t_k;\theta) - S_{obs}(t_k)]^2 \tag{5.54}$$

At each time t_k, S_{obs} is the observed [S], and S_{pred} is the predicted value from (5.53). Here, there is no analytical expression for the cost function, as we must solve the initial value problem for [S] as a function of time numerically. fit_enzyme_batch_sim1.m uses **ode45** to simulate the batch kinetics for input values of the rate law parameters in order to evaluate the cost function. Either **fminsearch** or **fminunc** is used to perform the optimization. Here, we rely upon the optimizer to estimate the gradient through finite difference approximations. The agreement between the fitted equation and the data is shown in Figure 5.10.

Table 5.1 Measured substrate concentration (in M) in batch enzymatic reactor

time (min)	[S] (M)
30	1.87
60	1.73
90	1.58
120	1.43
180	1.07
240	0.63
300	0.12
315	0.04
330	0.01

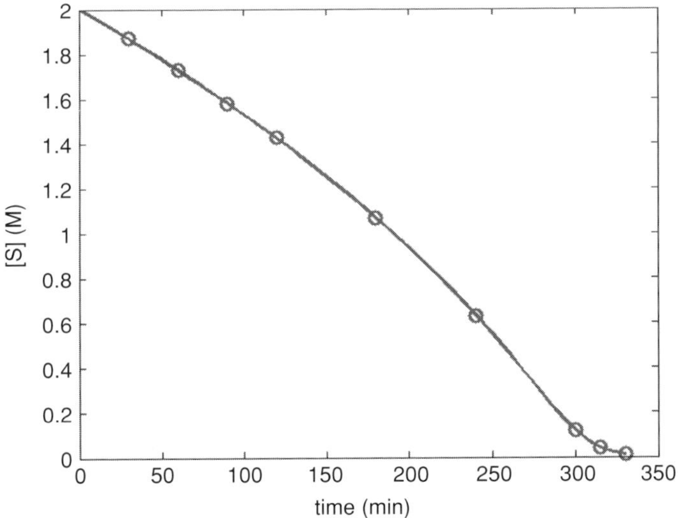

Figure 5.10 Measured substrate concentrations vs. time and predictions from fitted rate law with $V_m = 200\ \mu\text{mol}/(\text{min}/\text{mg}_E)$, $K_m = 0.201$ M, and $K_{si} = 0.5616$ M.

Lagrangian methods for constrained optimization

In the preceding sections, we considered only unconstrained optimization problems in which x may take any value. Here, we extend these methods to constrained minimization problems, where to be acceptable (or *feasible*), x must satisfy a number n_e of equality constraints $g_i(x) = 0$ and a number n_i of inequality constraints $h_j(x) \geq 0$, where each $g_i(x)$ and $h_j(x)$ are assumed to be differentiable nonlinear functions. This constrained optimization problem is expressed mathematically as

$$\text{minimize } F(x)$$
$$\text{subject to } \begin{array}{ll} g_i(x) = 0 & i = 1, 2, \ldots, n_e \\ h_j(x) \geq 0 & j = 1, 2, \ldots, n_i \end{array} \tag{5.55}$$

One approach to solve (5.55), the *penalty method*, is to add to $F(x)$ penalty terms that push x away from regions in which constraints are violated,

$$F_\mu(x) = F(x) + \frac{1}{2\mu} \left\{ \sum_{i=1}^{n_e} [g_i(x)]^2 + \sum_{j=1}^{n_i} H(-h_j(x))[h_j(x)]^2 \right\} \tag{5.56}$$

The *Heaviside step function* $H(x)$ is

$$H(x) = \begin{cases} 1, x \geq 0 \\ 0, x < 0 \end{cases} \tag{5.57}$$

The problem with this approach is that to enforce the constraints exactly, we must take the limit $\mu \to 0$. But as this limit is approached, the penalty terms come to dominate the actual cost function of interest, and so the numerical minimization of $F_\mu(x)$ in this limit is difficult.

Here we consider the *augmented Lagrangian method*, which converts the constrained problem into a sequence of unconstrained minimizations. We first treat equality constraints, and then extend the method to include inequality constraints.

Optimization with equality constraints

For simplicity, we first restrict our discussion to the case where we have only equality constraints,

$$\text{minimize } F(x)$$
$$\text{subject to } \quad g_i(x) = 0 \quad i = 1, 2, \ldots, n_e \tag{5.58}$$

To make things even simpler, we consider at first only a single constraint,

$$\text{minimize } F(x)$$
$$\text{subject to } \quad g(x) = 0 \tag{5.59}$$

For an unconstrained problem, a necessary condition for $x_{\min}$ to be a minimum is that the local gradient be zero,

$$\nabla F|_{x_{\min}} = 0 \tag{5.60}$$

What is the similar necessary condition for a point to be a constrained minimum in the presence of an equality constraint?

Let $x_{\min}$ be a constrained minimum. Then, the curve $g(x) = 0$ and the contours of $F(x)$ will look something like Figure 5.11. If $x_{\min}$ is a constrained minimum, we cannot move in either direction along the curve $g(x) = 0$ and decrease the cost function.

If t is a tangent vector to the curve at $x_{\min}$, a first-order Taylor approximation in the tangent direction gives

$$F(x_{\min} + \varepsilon t) - F(x_{\min}) = \varepsilon \nabla F|_{x_{\min}} \cdot t \tag{5.61}$$

Thus, at a constrained minimum, ∇F must be perpendicular to any tangent direction of the

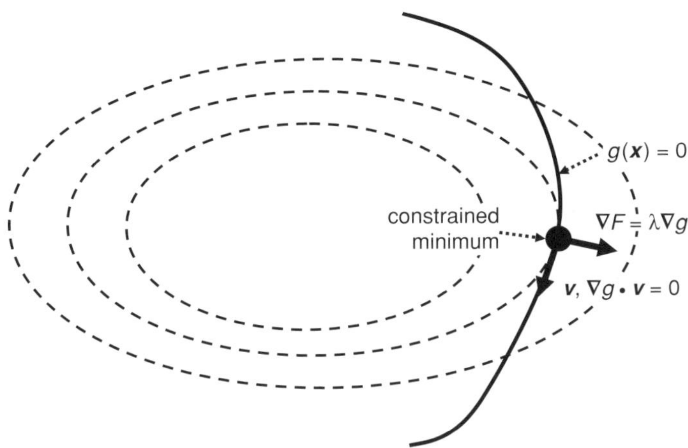

Figure 5.11 First-order optimality conditions for a constrained minimum of $F(x)$ along $g(x) = 0$.

constraint curve $g(x) = 0$,

$$\nabla F|_{x_{\min}} \cdot t = 0 \tag{5.62}$$

If we use a similar Taylor approximation for $g(x)$, we obtain

$$g(x_{\min} + \varepsilon t) - g(x_{\min}) = \varepsilon \nabla g|_{x_{\min}} \cdot t \tag{5.63}$$

As in the limit $\varepsilon \to 0$, both $g(x_{\min} + \varepsilon t)$ and $g(x_{\min})$ lie along $g(x) = 0$,

$$\nabla g|_{x_{\min}} \cdot t = 0 \tag{5.64}$$

At the constrained minimum, both ∇F and ∇g are perpendicular to any tangent vector of the curve. Does that mean that ∇F and ∇g are parallel?

We can always write (through orthogonal projection) any vector p as $p = \lambda \nabla g + v$, the sum of a vector parallel to ∇g and a vector v that is perpendicular to ∇g. Applying this to ∇F,

$$\nabla F|_{x_{\min}} = \lambda \nabla g|_{x_{\min}} + v \qquad \nabla g \cdot v = 0 \tag{5.65}$$

After a move εv, the changes in $F(x)$ and $g(x)$ are

$$F(x_{\min} + \varepsilon v) - F(x_{\min}) = \varepsilon \nabla F|_{x_{\min}} \cdot v$$
$$g(x_{\min} + \varepsilon v) - g(x_{\min}) = \varepsilon \nabla g|_{x_{\min}} \cdot v = 0 \tag{5.66}$$

If there is indeed some $v \neq 0$ contributing to ∇F, we can decrease $F(x)$ by moving in the direction of $\pm v$ and still satisfy $g(x) = 0$, as then

$$F(x_{\min} + \varepsilon v) - F(x_{\min}) = \varepsilon [\lambda \nabla g|_{x_{\min}} + v] \cdot v = \varepsilon (v \cdot v) \tag{5.67}$$

Thus, at a constrained minimum, ∇F must be parallel to ∇g,

$$\nabla F|_{x_{\min}} = \lambda \nabla g|_{x_{\min}} \tag{5.68}$$

This yields our generalization of the condition $\nabla F = 0$ to accommodate an equality constraint. Rather than finding a flat point in the cost function, we search for a flat point of the

Lagrangian

$$L(x; \lambda) \equiv F(x) - \lambda g(x) \tag{5.69}$$

where at the constrained minimum,

$$\nabla L|_{x_{\min}} = \nabla F|_{x_{\min}} - \lambda \nabla g|_{x_{\min}} = 0 \tag{5.70}$$

λ is known as a *Lagrange multiplier*, and its value must be chosen so that at the minimum of the Lagrangian, $g(x) = 0$.

The Lagrangian gives us a powerful technique to convert a constrained optimization problem into a familiar unconstrained one, but there remains the problem of computing the Lagrange multiplier used to define $L(x; \lambda)$. In the *augmented Lagrangian method*, we add a quadratic penalty to the Lagrangian that serves to enforce $g(x) = 0$ whenever we use the incorrect value of λ. Starting with an initial guess $\lambda^{[0]}$ of the multiplier, we choose some $\mu^{[0]} > 0$ and define the augmented Lagrangian

$$L_A^{[0]}\big(x; \lambda^{[0]}, \mu^{[0]}\big) \equiv F(x) - \lambda^{[0]} g(x) + \frac{1}{2\mu^{[0]}} [g(x)]^2 \tag{5.71}$$

We then find a point $x^{[0]}$ minimizing $L_A^{[0]}$, where

$$\nabla L_A^{[0]}\Big|_{x^{[0]}} = 0 = \nabla F|_{x^{[0]}} - \left[\lambda^{[0]} - \frac{g\big(x^{[0]}\big)}{\mu^{[0]}}\right] \nabla g|_{x^{[0]}} \tag{5.72}$$

This yields the following update of the Lagrange multiplier estimate

$$\lambda^{[1]} = \lambda^{[0]} - \frac{g\big(x^{[0]}\big)}{\mu^{[0]}} \tag{5.73}$$

With this updated multiplier, we define a new augmented Lagrangian and optionally make the penalty stronger, $0 < \mu^{[1]} \leq \mu^{[0]}$. We take $x^{[0]}$ as the initial guess in a minimization of the new Lagrangian

$$L_A^{[1]}\big(x; \lambda^{[1]}, \mu^{[1]}\big) \equiv F(x) - \lambda^{[1]} g(x) + \frac{1}{2\mu^{[1]}} [g(x)]^2 \tag{5.74}$$

We iterate this process until convergence to the proper multiplier value is reached; i.e., until at the unconstrained minimum of the augmented Lagrangian, the equality constraint is satisfied.

How do we treat multiple equality constraints? If $x_{\min}$ satisfies all constraints, $g_i(x_{\min}) = 0, i = 1, \ldots, n_e$, then we can only move away from $x_{\min}$ in a direction v that is tangent to *all* constraints,

$$\nabla g_i|_{x_{\min}} \cdot v = 0 \quad \forall i = 1, 2, \ldots, n_e \tag{5.75}$$

If $x_{\min}$ is a constrained minimum, we cannot decrease the cost function by moving in any such tangent direction, so that for any v satisfying (5.75),

$$\nabla F|_{x_{\min}} \cdot v = 0 \tag{5.76}$$

We can always write $\nabla F|_{x_{\min}}$ as a linear combination of the $\nabla g_i|_{x_{\min}}$ plus some tangent vector v satisfying (5.75),

$$\nabla F|_{x_{\min}} = \sum_{i=1}^{n_e} \lambda_i \nabla g_i|_{x_{\min}} + v \tag{5.77}$$

Taking the dot product with $\boldsymbol{v}$, at a constrained minimum (5.76) requires that $\boldsymbol{v} = \boldsymbol{0}$; hence,

$$\nabla F|_{\boldsymbol{x}_{\min}} = \sum_{i=1}^{n_e} \lambda_i \, \nabla g_i|_{\boldsymbol{x}_{\min}} \tag{5.78}$$

For multiple equality constraints we thus define the Lagrangian as

$$L(\boldsymbol{x}; \boldsymbol{\lambda}) = F(\boldsymbol{x}) - \sum_{i=1}^{n_e} \lambda_i g_i(\boldsymbol{x}) \tag{5.79}$$

and use the same technique as before. Note that the existence of a constrained minimum is not guaranteed, for there may exist no points that satisfy all constraints simultaneously.

Example. Finding the closest points on two ellipses

We consider here a simple example in which we wish to find the points on two ellipses that are nearest to each other. Let (x_1, y_1) and (x_2, y_2) be the coordinates of the points on the first and second ellipses, centered at $(x_{c1}, y_{c1}) = (0, 0)$ and $(x_{c2}, y_{c2}) = (5, 5)$ respectively. For (x_1, y_1) and (x_2, y_2) to lie on the ellipses, they must satisfy the two equality constraints

$$\begin{aligned} a_1(x_1 - x_{c1})^2 + b_1(y_1 - y_{c1})^2 &= 1 \qquad a_1 = 0.5 \qquad b_1 = 0.3 \\ a_2(x_2 - x_{c2})^2 + b_2(y_2 - y_{c2})^2 &= 1 \qquad a_2 = 0.2 \qquad b_2 = 0.4 \end{aligned} \tag{5.80}$$

ellipse_min.m uses the augmented Lagrangian method to minimize

$$F(x_1, y_1, x_2, y_2) = (x_1 - x_2)^2 + (y_1 - y_2)^2 \tag{5.81}$$

subject to (5.80). The initial guesses are the ellipse centers and the Lagrange multipliers are set initially to zero.

Figure 5.12 shows the trajectories at each multiplier iteration of (x_1, y_1) (circles) and (x_2, y_2) (diamonds), with a line connecting the final optimal points. **fminunc** is used to minimize the augmented Lagrangian at each multiplier iteration. In practice, the MATLAB constrained minimizer **fmincon** would be preferred; ellipse_min.m is meant only to demonstrate the augmented Lagrangian technique. Both **fminunc** and **fmincon** are part of the optional MATLAB optimization toolkit. If your installation of MATLAB does not include this toolkit, you must apply the augmented Lagrangian procedure directly, as in ellipse_min.m, and use the provided routine gradient_minimizer.m to perform each uncon-strained minimization. The code for solving this problem with **fmincon** is presented below, following our discussion of inequality constraints.

Treatment of inequality constraints

We now consider methods to solve the problem

$$\text{minimize } F(\boldsymbol{x})$$

$$\text{subject to} \quad \begin{aligned} g_i(\boldsymbol{x}) &= 0 \qquad i = 1, 2, \ldots, n_e \\ h_j(\boldsymbol{x}) &\geq 0 \qquad j = 1, 2, \ldots, n_i \end{aligned} \tag{5.82}$$

Example. Optimal steady-state design of a CSTR

We wish to design a 1-l CSTR to produce C from A and B by the reactions

$$
\begin{aligned}
A + B &\to C & r_{R1} &= k_1(T)c_A c_B \\
A &\to S_1 & r_{R2} &= k_2(T)c_A \\
C &\to S_2 & r_{R3} &= k_3(T)c_C
\end{aligned}
\tag{5.129}
$$

The rate constants depend upon the temperature as

$$
k_j(T) = k_j(T_{\text{ref}}) \exp\left\{ \frac{E_{aj}}{R T_{\text{ref}}} \left[1 - \frac{T_{\text{ref}}}{T} \right] \right\}
\tag{5.130}
$$

where

$$
\begin{aligned}
\text{reaction 1} \quad & k_1(T_{\text{ref}} = 298\,\text{K}) = 2.3\,\frac{1}{\text{mol h}} & \frac{E_{a1}}{R T_{\text{ref}}} &= 15.4 \\
\text{reaction 2} \quad & k_2(T_{\text{ref}} = 298\,\text{K}) = 0.2\,\text{h}^{-1} & \frac{E_{a2}}{R T_{\text{ref}}} &= 17.9 \\
\text{reaction 3} \quad & k_3(T_{\text{ref}} = 298\,\text{K}) = 0.1\,\text{h}^{-1} & \frac{E_{a3}}{R T_{\text{ref}}} &= 22.3
\end{aligned}
\tag{5.131}
$$

We wish to vary the volumetric flow rate v, the concentrations c_{A0} and c_{B0} of A and B in the inlet stream, and the temperature T to achieve various design goals. The vector of adjustable design parameters is

$$
\boldsymbol{\theta} = [v \; c_{A0} \; c_{B0} \; T]^{\mathsf{T}}
\tag{5.132}
$$

To maximize the production of C, the cost function is

$$
F_1(\boldsymbol{\theta}) = -v c_C
\tag{5.133}
$$

Equation (5.133) does not take into account the loss of A and C to side product formation. Let us say that we can easily separate the products and recycle the unreacted A and B. For simplicity, let us also neglect any heating/cooling or power costs. The net economic cost of operating the reactor is then

$$
F_2(\boldsymbol{\theta}) = (-v)[-\$_A(c_{A0} - c_A) - \$_B(c_{B0} - c_B) + \$_C c_C - \$_{D_{S_1}} c_{S_1} - \$_{D_{S_2}} c_{S_2}]
\tag{5.134}
$$

$\$_A$, $\$_B$, $\$_C$ are the prices per mole of A, B, and C, and $\$_{D_{S_1}}$, $\$_{D_{S_2}}$ are the disposal costs of the side products. By minimizing this cost function, we get the maximal economic benefit. Let us assume the prices

$$
\begin{aligned}
\$_A &= 4.5 & \$_B &= 1.1 & \$_C &= 8.2 \\
\$_{D_{S_1}} &= 1.0 & \$_{D_{S_2}} &= 1.0 &&
\end{aligned}
\tag{5.135}
$$

A and B are fed into the reactor in a carrier solvent, with initial concentrations subject to the constraints

$$
c_{A0} \geq 10^{-4}\,\text{M} \quad c_{B0} \geq 10^{-4}\,\text{M} \quad c_{A0} + c_{B0} \leq 2\,\text{M}
\tag{5.136}
$$

The volumetric flow rate is constrained to be in the interval

$$
10^{-4}\,l/h \leq v \leq 360\,l/h
\tag{5.137}
$$

and the temperature is constrained to lie within the interval

$$298 \text{ K} \leq T \leq 360 \text{ K} \tag{5.138}$$

To compute the steady-state concentrations, a dynamic model of the CSTR in overflow mode is integrated until steady state is reached,

$$\frac{dc_A}{dt} = \frac{\upsilon(c_{A0} - c_A)}{V_R} - r_{R1} - r_{R2} \qquad \frac{dc_B}{dt} = \frac{\upsilon(c_{B0} - c_B)}{V_R} - r_{R1}$$

$$\frac{dc_C}{dt} = -\frac{\upsilon c_C}{V_R} + r_{R1} - r_{R3} \qquad \frac{dc_{S_1}}{dt} = -\frac{\upsilon c_{S_1}}{V_R} + r_{R2} \qquad \frac{dc_{S_2}}{dt} = -\frac{\upsilon c_{S_2}}{V_R} + r_{R3} \tag{5.139}$$

Initially, the concentrations equal the inlet values. Time integration ensures that we design only for stable steady states. The constrained optimization is performed by optimal_design_CSTR.m. Starting at the initial guess

$$\upsilon^{(\text{guess})} = 1 \quad c_{A0}^{(\text{guess})} = 0.5 \quad c_{B0}^{(\text{guess})} = 0.5 \quad T^{(\text{guess})} = 310 \tag{5.140}$$

maximizing the production of C by minimizing (5.133) yields

$$\begin{aligned}
\text{initial } F_1 &= -0.182 \quad \text{final } F_1 = -27.148 \\
\upsilon &= 360 \quad c_{A0} = 1 \quad c_{B0} = 1 \quad T = 360 \\
c_A &= 0.913 \quad c_B = 0.924 \quad c_C = 0.075 \\
c_{S_1} &= 0.011 \quad c_{S_2} = 9.8 \times 10^{-4}
\end{aligned} \tag{5.141}$$

Thus, the optimal design is found at the limits of the constraints. A high flow rate yields a low residence time and thus a small concentration of C in the outlet stream (note that we maximize the product of this concentration with the volumetric flow rate). The A and B inlet concentrations are maximized to yield the fastest rate of the first reaction, and the temperature is at the upper limit. The loss of C by the third reaction is slight, as the short residence time and low C concentration make the rate of this reaction small compared to the first one.

Next, we maximize the economic benefit, by minimizing (5.134), starting at the design (5.141), to yield the optimal design

$$\begin{aligned}
\text{initial } F_2 &= -46.37 \quad \text{final } F_2 = -48.36 \\
\upsilon &= 360 \quad c_{A0} = 0.832 \quad c_{B0} = 1.168 \quad T = 360 \\
c_A &= 0.749 \quad c_B = 1.094 \quad c_C = 0.073 \\
c_{S_1} &= 0.009 \quad c_{S_2} = 9.5 \times 10^{-4}
\end{aligned} \tag{5.142}$$

Here, at the cost of producing slightly less C, we have increased the economic benefit by reducing the side product formation and the consumption of A by enriching the feed stream with B.

Optimal control

Let us consider a dynamic system, described by the *state vector* $x(t) \in \mathfrak{R}^N$, and governed by the ODE system

$$\dot{x} = f(t, x(t), u(t); \theta) \tag{5.143}$$

$u(t) \in \mathfrak{R}^M$ is a set of tunable *control inputs* and θ is a set of fixed system parameters. For simplicity, we drop explicit reference to the latter and write the ODE system as $\dot{x} = f(t, x, u)$. At time t_0, we start at the initial state $x(t_0) = x^{[0]}$, and wish to determine the trajectory of control inputs $u(t)$ within $t \in [t_0, t_H]$ that minimize a *cost functional*

$$F\left[u(t); x^{[0]}\right] = \int_{t_0}^{t_H} \sigma(s,\, x(s),\, u(s))ds + \pi(x(t_H)) \tag{5.144}$$

t_H is the *horizon time* for this *optimal control* problem. For example, we may wish to maintain the system at some set point x_{set}, in which case a suitable choice of cost functional is

$$F\left[u(t); x^{[0]}\right] = \int_{t_0}^{t_H} \{|x(s) - x_{\mathrm{set}}|^2 + C_U|u(s) - u_{\mathrm{set}}|^2\}ds + C_H|x(t_H) - x_{\mathrm{set}}|^2 \tag{5.145}$$

u_{set} is the input necessary to maintain the system steady at the set point and $C_U, C_H > 0$.

How do we find the "best" $u(t)$ that minimizes (5.144)? We describe first a direct approach for an open-loop problem in which we compute the entire optimal trajectory for a specific initial state. Then, we outline an alternative *dynamic programming* approach that turns the integral equation (5.144) into a corresponding time-dependent partial differential equation, and generates a closed-loop optimal feedback control law.

As a first approach, let us parameterize $u(t)$ as a piecewise-constant function by splitting $[t_0, t_H]$ into N_S subintervals separated by the time points

$$t_k = t_0 + k(\Delta t) \quad \Delta t = \frac{t_H - t_0}{N_S} \tag{5.146}$$

In the subinterval $t_{k-1} \leq t < t_k$, we hold $u(t)$ constant at $u^{[k]}$. Thus, we write

$$u(t) = \sum_{k=1}^{N_s} u^{[k]}\Omega_k(t) \quad \Omega_k(t) = \begin{cases} 1, & t_{k-1} \leq t < t_k \\ 0, & \text{otherwise} \end{cases} \tag{5.147}$$

and characterize the trajectory $u(t)$ by the vector $U \in \mathfrak{R}^{N_S M}$,

$$U^{\mathrm{T}} = \left[\left(u^{[1]}\right)^{\mathrm{T}}\left(u^{[2]}\right)^{\mathrm{T}} \ldots \left(u^{[N_s]}\right)^{\mathrm{T}}\right] \tag{5.148}$$

Let $x(t; U)$ be the solution to the ODE-IVP

$$\frac{d}{dt}x(t; U) = f(t,\, x(t; U),\, u(t; U)) \quad x(t_0) = x^{[0]} \tag{5.149}$$

where $u(t; U)$ is computed by (5.147). The cost functional (5.144) then is approximated by a cost function of U,

$$F(U; x^{[0]}) = \sum_{k=1}^{N_s} \left[\int_{t_{k-1}}^{t_k} \sigma(s,\, x(s; U),\, u^{[k]})ds\right] + \pi(x(t_H; U)) \tag{5.150}$$

We then minimize (5.150) using the techniques described above. As U can be of quite high a dimension, this optimization problem can be costly.

An open-loop optimal control routine

optimal_control.m implements the procedure described above, and is called with the syntax

function [TRAJ, iflag] = optimal_control(FUN, PARAM, SIM, TRAJ0);

FUN is a structure containing the names of the user-supplied routines that define the optimal control problem. FUN.f is the routine that returns the time derivative vector for $\dot{x} = f(t, x, u)$,

function f =FUN.f(t, x, u, PARAM);

PARAM is an optional user-specified structure of fixed system parameters. FUN.sigma returns the integrand $\sigma(t, x, u)$ of the cost functional,

function sigma =FUN.sigma(t, x, u, PARAM);

FUN.pi returns the value of $\pi(x(t_{\mathrm{H}}))$,

function pi =FUN.pi(xH, PARAM);

FUN.constraint is an optional routine that defines the set of constraints that apply to the control input in each subinterval,

$$A u^{[k]} \leq b \quad A^{(eq)} u^{[k]} = b^{(eq)} \quad L B(j) \leq u_j^{[k]} \leq U B(j) \tag{5.151}$$

and is called with the syntax

function UCON = FUN.constraint();

UCON contains the fields UCON.A, UCON.b, UCON.A_eq, UCON.b_eq, UCON.LB, and UCON.UB. FUN.u0 is a routine that returns the initial guess of the time-dependent control input $u(t)$,

function u0 =FUN.u0(t, PARAM);

SIM is a data structure that contains information about how the optimization calculation is to be performed. SIM.t0 is the initial time and SIM.tH is the horizon time. SIM.NS is the number of subintervals. SIM.x0 is the initial state. SIM.isRestart is 0 if the simulation is to start at the initial guess supplied by FUN.u0 and is nonzero if the information in the optional input parameter TRAJ0 sets the initial input trajectory. SIM.constraint is 0 if the control inputs are not constrained, and is nonzero if FUN.constraint is to be used to define a set of input control constraints. SIM.verbose is 0 if no information is to be printed to the screen and is nonzero if the status of the calculation is to be displayed.

The output is TRAJ, a data structure that contains information about the optimal trajectory. TRAJ.t is a vector of the times that separate the piecewise-constant subintervals, TRAJ.t(k) $=t_k$. The $u^{[k]}$ for that subinterval are found in row k of TRAJ.u. The state trajectory for the optimal control inputs is returned in TRAJ.t_xtraj and TRAJ.x_xtraj in the same format used by the ODE solvers. The control inputs at these times are returned in TRAJ.u_xtraj.

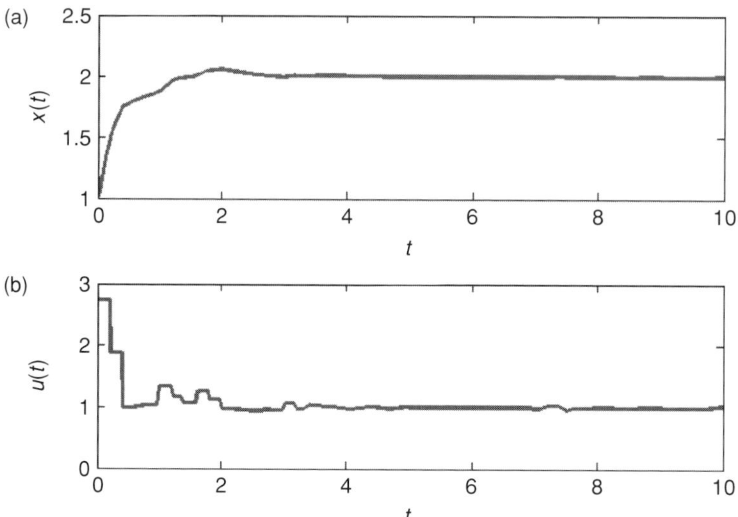

Figure 5.14 (a) Optimal state and (b) control trajectories for the set-point problem.

test_optimal_control.m demonstrates the use of this routine for the simple example of minimizing for the system

$$\dot{x} = -(x - 1) + u \qquad (5.152)$$

a cost functional that forces x to a set point $x_{set} = 2$,

$$F[u(t); x^{[0]}] = \int_{t_0}^{t_H} \{|x(s) - x_{set}|^2 + C_U|u(s) - u_{set}|^2\}ds + C_H|x(t_H) - x_{set}|^2 \qquad (5.153)$$

where $u_{set} = 1$, $t_H = 10$, $C_U = 0.1$, $C_H = 10$, and the control inputs are subject to the constraints $-10 \leq u \leq 10$. Fifty piecewise-constant subintervals are used to parameterize $u(t)$. From an initial uniform guess $u(t) = u_{set} = 1$, the optimal state and control trajectories are shown in Figure 5.14.

Dynamic programming

We revisit the optimal control problem of finding the trajectory of control inputs $\boldsymbol{u}(t)$ for $t \in [t_0, t_H]$ that minimizes the cost functional

$$F[\boldsymbol{u}(t); \boldsymbol{x}^{[0]}] = \int_{t_0}^{t_H} \sigma(s, \boldsymbol{x}(s), \boldsymbol{u}(s))ds + \pi(\boldsymbol{x}(t_H)) \qquad (5.154)$$

for a system governed by the ODE-IVP,

$$\dot{\boldsymbol{x}} = \boldsymbol{f}(t, \boldsymbol{x}, \boldsymbol{u}) \quad \boldsymbol{x}(t_0) = \boldsymbol{x}^{[0]} \qquad (5.155)$$

We introduce here a *dynamic programming* approach due to Bellman (1957), and define at each $t \in [t_0, t_H]$ the *Bellman function* $V(t, \boldsymbol{x})$ to be the optimal "cost to go" value; i.e., the additional cost accrued after time t if we were to implement the optimal inputs at all subsequent times $s \in [t, t_H]$,

$$V(t, x) = \min_{u(s>t)} \left\{ \int_t^{t_H} \sigma(s, x(s), u(s))ds + \pi(x(t_H)) \right\} \tag{5.156}$$

$V(t_0, x^{[0]})$ is the optimal value of $F[u(t); x^{[0]}]$. At the horizon time $V(t_H, x) = \pi(x)$, but how do we work backwards in time from this known result, and how from such a calculation, do we determine the optimal $u(t)$?

We obtain a differential equation for $V(t, x)$ by taking the derivative of (5.156), evaluated at the optimal control input $u(t)$,

$$\frac{d}{dt} V(t, x) = -\sigma(t, x(t), u(t)) \tag{5.157}$$

We relate this total time derivative to partial derivatives of $V(t, x)$,

$$\frac{d}{dt} V(t, x(t)) = \frac{\partial V}{\partial t} + \nabla V \cdot \frac{dx}{dt} = \frac{\partial V}{\partial t} + \nabla V \cdot f(t, x, u) \tag{5.158}$$

Thus, for the optimal control trajectory, (5.157) and (5.158) yield the partial differential equation

$$\left. \frac{\partial V}{\partial t} \right|_{\text{opt}} = -\sigma(t, x, u) - \nabla V \cdot f(t, x, u) \tag{5.159}$$

We next define the "backward" time $\tau = t_H - t$, and rewrite the Bellman function as $\varphi(\tau, x) = V(t, x)$. We have for this equation the "initial" condition $\varphi(0, x) = \pi(x)$. We then rewrite (5.159) as

$$\left. \frac{\partial \varphi}{\partial \tau} \right|_{\text{opt}} = \sigma(t_H - \tau, x, u) + \nabla \varphi \cdot f(t_H - \tau, x, u) \tag{5.160}$$

For nonoptimal trajectories, $\varphi(\tau, x)$ increases faster with increasing τ than for the optimal trajectory, so that, in general,

$$\frac{\partial \varphi}{\partial \tau} \geq \sigma(t_H - \tau, x, u) + \nabla \varphi \cdot f(t_H - \tau, x, u) \tag{5.161}$$

Thus, for the optimal trajectory, we have the following PDE for $\varphi(\tau, x)$, known as the *Hamilton–Jacobi–Bellman (HJB) equation*:

$$\frac{\partial \varphi}{\partial \tau} = \min_{u(\tau, x)} [\sigma(t_H - \tau, x, u) + \nabla \varphi \cdot f(t_H - \tau, x, u)] \tag{5.162}$$

We work backwards in time, and at each (τ, x), use the input $u(\tau, x)$ that minimizes the term in the square brackets. Note that it is possible for $\partial \varphi / \partial \tau < 0$ even if $\sigma(t_H - \tau, x, u) > 0$, as the requirement for monotonic decrease in $V(t, x)$ with increasing t is $d\varphi/d\tau = \partial \phi/\partial \tau - \nabla \varphi \cdot f > 0$.

In an *open-loop control problem*, we compute the optimal control input trajectory and then fully implement it over the entire period $t_0 \leq t \leq t_H$. For this, the direct approach of the last section is sufficient. In a *closed-loop control problem*, we only implement the initial input $\boldsymbol{u}^*(t_0) = \boldsymbol{u}(t_H - t_0, \boldsymbol{x}^{[0]})$ for some period Δt, after which we measure the resulting new state $\boldsymbol{x}_{\text{new}}$. This may not be equal to $\boldsymbol{x}^*(\Delta t)$ due to random error or inadequacy of the model. We use feedback to compensate for this, by computing a new "best current input" by minimizing the functional again, shifting $t_0 \to t_0 + \Delta t$, $t_H \to t_H + \Delta t$, and $\boldsymbol{x}^{[0]} \to \boldsymbol{x}_{\text{new}}$. Note, however, that if neither the system's time derivative function nor the cost functional integrand depend upon the absolute value of time, then we have exactly the same dynamic programming problem that we have just solved from (5.162). Therefore, we do not need in this case to redo the entire calculation, but just implement as the new control input $\boldsymbol{u}(t_H - t_0, \boldsymbol{x}_{\text{new}})$. Thus, we obtain from $\mathrm{u}(\tau = t_H - t_0, \boldsymbol{x})$ the optimal feedback control law for the system, and this is the primary advantage of the dynamic programming approach.

For further description of this subject, and its implementation in process control, consult Sontag (1990). Of course, to apply this method, we must be able to solve (5.162). The numerical solution of such partial differential equations is the subject of the next chapter.

Example. A simple 1-D optimal control problem

We consider again the problem in which we wish to control $x(t)$ at a set point $x_{\text{set}} = 2$. The system is governed by the ODE

$$\dot{x}(t) = f(x, u) = -(x - 1) + u \tag{5.163}$$

We wish to determine a control law for this system by solving the HJB equation. The cost functional that we wish to minimize is

$$F\big[u(t); x^{[0]}\big] = \int_0^{t_H} \left\{ \frac{C_U}{2}[u(s) - u_{\text{set}}]^2 + [x(s) - x_{\text{set}}]^2 \right\} ds + C_H[x(t_H) - x_{\text{set}}]^2 \tag{5.164}$$

$u_{\text{set}} = 1$ is the steady value that maintains the system at the set point. Thus, this functional penalizes both excessive departure from the set point and very large control inputs. The HJB equation for this system is

$$\frac{\partial \varphi}{\partial \tau} = \min_{u(\tau, x)} \left\{ \frac{C_U}{2}[u - u_{\text{set}}]^2 + [x - x_{\text{set}}]^2 + \frac{\partial \varphi}{\partial x}[-(x - 1) + u] \right\} \tag{5.165}$$

Note that if $C_U = 0$, the extremum condition becomes $\partial \varphi / \partial x = 0$, and so if $\partial \varphi / \partial x > 0$, u should be decreased until it reaches its lower bound, and it should be increased to its upper bound if $\partial \varphi / \partial x < 0$. Thus, $C_U > 0$ is necessary for an unconstrained minimum to exist. If so, the optimal control input is

$$u(\tau, x) = u_{\text{set}} - C_U^{-1} \frac{\partial \varphi}{\partial x} \tag{5.166}$$

We thus have the following PDE problem,

$$\frac{\partial \varphi}{\partial \tau} = \frac{C_U}{2}[u(\tau, x) - u_{set}]^2 + [x - x_{set}]^2 + \frac{\partial \varphi}{\partial x}[-(x - 1) + u(\tau, x)]$$

$$\text{initial condition } \varphi(0, x) = C_H[x(t_H) - x_{set}]^2 \tag{5.167}$$

To solve this problem numerically, we use the method of finite differences, explained in further detail in Chapter 6. We restrict the x-domain to $x_{lo} \leq x \leq x_{hi}$, where the limits are chosen to be larger than any conceivable x-value that could be encountered in practice. Then, we place a grid of N points, uniformly-spaced, in this domain,

$$x_k = x_{lo} + (k - 1)(\Delta x) \quad \Delta x = \frac{x_{hi} - x_{lo}}{N - 1} \tag{5.168}$$

At each x_k, we use a finite difference approximation to estimate $\partial \varphi / \partial x$,

$$\left.\frac{\partial \varphi}{\partial x}\right|_{x_j} = \frac{1}{\Delta x}[A_{lo}\varphi_{k-1} + A_{mid}\varphi_k + A_{hi}\varphi_{k+1}] \quad A_{lo} + A_{mid} + A_{hi} = 0 \tag{5.169}$$

where $\varphi_k(\tau) \equiv \varphi(\tau, x_k)$. For reasons that will become clear in our discussion of convection in Chapter 6, we choose here the set of one-sided differences,

$$\begin{aligned} \text{if } f(x, u) \leq 0 \quad A_{lo} = -1 \quad A_{mid} = +1 \quad A_{hi} = 0 \\ \text{else} \quad A_{lo} = 0 \quad A_{mid} = -1 \quad A_{hi} = +1 \end{aligned} \tag{5.170}$$

Note that if x_{hi} is large enough, $f(t, x, u) = -(x - 1) + u < 0$, and we have the one-sided difference pointing "into" the grid, and we have no problem applying (5.169). Similar reasoning holds at the lower boundary.

We now solve the HJB equation numerically by integrating the set of ODEs

$$\frac{d\varphi_k}{dt} = \frac{C_U}{2}[u_k(\tau) - u_{set}]^2 + [x_k - x_{set}]^2 + \left.\frac{\partial \varphi}{\partial x}\right|_{x_k} f(x_k, u_k(\tau))$$

$$f(x_k, u_k(\tau)) = -(x_k - 1) + u_k(\tau) \quad \varphi_k(0) = C_H[x_k(t_H) - x_{set}]^2 \tag{5.171}$$

$$u_k(\tau) = u_{set} - C_U^{-1}\left.\frac{\partial \varphi}{\partial x}\right|_{x_k}$$

The feedback control law $u_{con}(x)$ is then

$$u_{con}(x_k) = u_k(\tau = t_H) \tag{5.172}$$

The optimal control at any point may be computed from (5.166). control_1D_HJB.m solves this HJB equation for specified t_H, C_U, C_H. For $t_H = 10$, $C_U = 1$, and $C_H = 10$, the resulting feedback control law is shown in Figure 5.15. For this simple linear system and quadratic cost functional, the optimal control law is a simple proportional controller with a gain of $K = -0.732$. The advantage of this approach is that it can be extended (though at perhaps great numerical cost) to nonlinear systems and to systems involving input constraints.

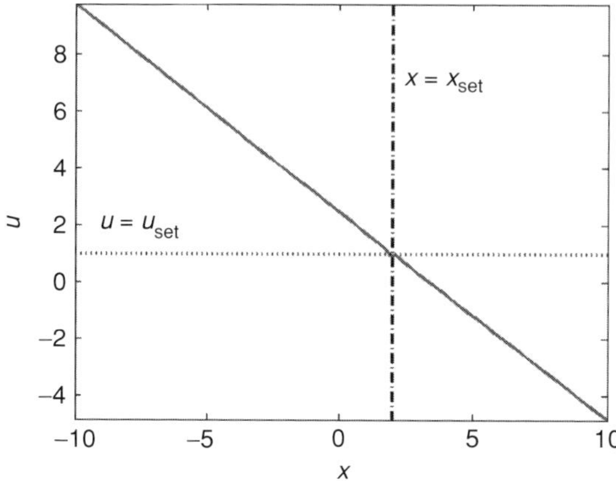

Figure 5.15 Optimal feedback control law.

MATLAB summary

The optional MATLAB optimization toolkit contains useful routines for finding local minima. **fminsearch** uses the nonlinear simplex method, but for unconstrained problems, **fminunc** is preferred. The latter uses either a gradient method or a trust-region Newton method depending upon the size of the problem and the level of information about the gradient and Hessian supplied by the user. **fmincon** finds a local minimum in the presence of linear and nonlinear equality and inequality constraints. The syntax of these routines is detailed in the corresponding sections of this chapter. If the optimization toolkit is not available, gradient_minimizer.m can be used to perform unconstrained minimizations. Constraints could then be handled explicitly using the augmented Lagrangian method described above. optimal_control.m uses a direct approach to solve open-loop optimal control problems.

Problems

5.A.1. Compute by hand a minimum of the cost function, for $x \in \Re^2$

$$F(x) = (x_1 - 3) + (x_2 - 1) + (x_1 - 1)^2 + (x_2 - 3)^2 \tag{5.173}$$

Check your results by solving the problem numerically.

5.A.2. Compute the point $x \in \Re^2$ that minimizes the cost function

$$F(x) = g \cdot x + \tfrac{1}{2} x \cdot Hx \qquad g = \begin{bmatrix} -2 \\ 1 \end{bmatrix} \qquad H = \begin{bmatrix} 3 & 1 \\ 1 & 2 \end{bmatrix} \tag{5.174}$$

Now, compute the constrained minimum subject to

$$x_1^2 + x_2^2 = 1 \tag{5.175}$$

Table 5.2 Measured data of system performance

θ_1	θ_2	$F(\theta)$
2.653	2.639	0.948
2.625	2.703	0.744
1.865	2.699	0.381
2.591	3.104	0.393
1.337	2.772	0.648
1.779	2.699	0.411
2.470	2.515	1.162
1.265	3.247	0.784

Then, compute the constrained minimum along the unit circle with the additional requirements that both x_1 and x_2 be nonnegative.

5.B.1. We wish to use the enzyme whose kinetics, described by (5.51), were studied earlier in this chapter, in an immobilized-enzyme packed bed reactor. Neglecting any internal mass transfer resistance (we assume the enzyme is immobilized in very small pellets), we compute the outlet substrate concentration by solving the ODE-IVP

$$\frac{dc_S}{dW} = -\frac{1}{\alpha_c \upsilon} \left[\frac{V_m c_S}{K_m + c_S + K_{si}^{-1} c_S^2} \right] \qquad c_S(W = 0) = c_{S0} \tag{5.176}$$

c_{S0} is the substrate concentration in moles, and is constrained to lie in $[10^{-4}, 2]$. W is the mass of enzyme in the reactor in milligrams, and we integrate (5.176) to the total mass $W_R = 1$ g. The volumetric flow rate υ through the reactor is in liters per minute. $\alpha_c = 10^6 \,\mu\text{mol/mol}$ is a conversion factor, and the kinetic constants are $V_m = 200 \,\mu\text{mol/(min mg}_E)$, $K_m = 0.201$ M, $K_{si} = 0.5616$ M. Plot the inlet substrate concentration c_{S0} that maximizes the outlet molar flow rate of product, as a function of υ.

5.B.2. We wish to optimize the performance of a process with two adjustable parameters. Table 5.2 contains measured data of the cost function that we wish to minimize. Fit to this data some general (e.g. polynomial) model and use this model to propose a design that you think might have an even lower cost function value. To avoid extrapolating outside of the region of data, use a set of constraints that will limit the amount to which we can search far away from the existing data points.

5.B.3. We wish to determine the best path for a road connecting two points in hilly terrain. Let $r \in \Re^2$ be the coordinates of a point in kilometers and let the elevation at that point, also in kilometers, be $z(r)$. We represent the measured ground elevation data as a sum of contributions from individual hills, each hill being represented by a Gaussian function,

$$z(r) = \sum_{k=1}^{N_h} z_{max}^{[k]} \exp\left\{ -\tfrac{1}{2}(r - r_c^{[k]}) \cdot \left(\Sigma^{[k]}\right)^{-1} (r - r_c^{[k]}) \right\} \tag{5.177}$$

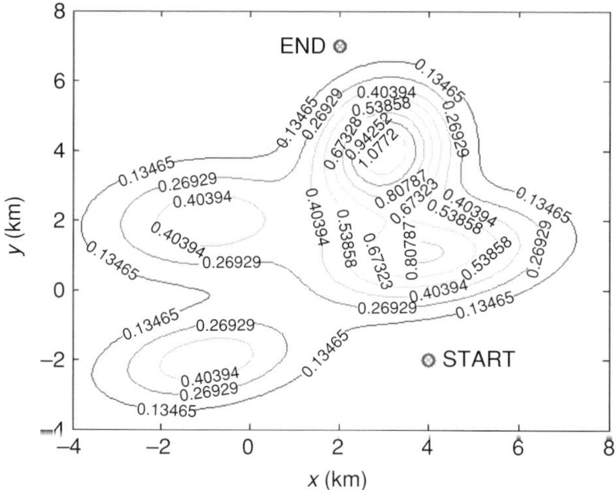

Figure 5.16 Contour map of region, showing lines at constant elevation. Start and end positions of the planned road are shown.

In the region of interest, we use a representation with four hills,

$$z_{\text{max}}^{[1]} = 1.2 \quad z_{\text{max}}^{[2]} = 0.8 \quad z_{\text{max}}^{[3]} = 0.5 \quad z_{\text{max}}^{[4]} = 0.5$$

$$r_{\text{c}}^{[1]} = \begin{bmatrix} 3 \\ 4 \end{bmatrix} \quad r_{\text{c}}^{[2]} = \begin{bmatrix} 4 \\ 1 \end{bmatrix} \quad r_{\text{c}}^{[3]} = \begin{bmatrix} -1 \\ -2 \end{bmatrix} \quad r_{\text{c}}^{[4]} = \begin{bmatrix} -1 \\ 2 \end{bmatrix}$$

$$\Sigma^{[1]} = \begin{bmatrix} 1.0 & 0.1 \\ 0.1 & 1.5 \end{bmatrix} \quad \Sigma^{[2]} = \begin{bmatrix} 3.0 & 0.5 \\ 0.5 & 1.0 \end{bmatrix} \quad \Sigma^{[3]} = \begin{bmatrix} 2.5 & 0.4 \\ 0.4 & 0.8 \end{bmatrix}$$

$$\Sigma^{[4]} = \begin{bmatrix} 3.0 & 0.2 \\ 0.2 & 1.2 \end{bmatrix}$$

(5.178)

Figure 5.16 shows the elevation contours along with the start $(4, -2)$ and end $(2, 7)$ positions of the planned road.

All land is available to build upon. Our task is to find the shortest path between the two end points subject to the constraint that the grade cannot be greater than 8%, i.e., the slope cannot be larger in magnitude than 0.08.

Let $0 \leq s \leq 1$ be a contour variable and $r(s)$ be the path of the road, subject to

$$r(0) = r_{\text{start}} = [4 \quad -2]^T \quad r(1) = r_{\text{end}} = [2 \quad 7]^T \tag{5.179}$$

We discretize the path by setting N contour positions $s_k = k(N+1)^{-1}$ and the coordinates $r^{[k]} = r(s_k) = [x^{[k]} \quad y^{[k]}]^T$. We wish to minimize

$$F_{\text{C}} = \left(r^{[1]} - r_{\text{start}}\right)^2 + \sum_{k=1}^{N-1} \left(r^{[k+1]} - r^{[k]}\right)^2 + \left(r_{\text{end}} - r^{[N]}\right)^2 \tag{5.180}$$

subject to the constraints that for each road segment,

$$\frac{\left|z\left(\mathbf{r}^{[k+1]}\right) - z\left(\mathbf{r}^{[k]}\right)\right|}{\sqrt{\left(x^{[k+1]} - x^{[k]}\right)^2 + \left(y^{[k+1]} - y^{[k]}\right)^2}} \leq \Gamma_{\max} \qquad \Gamma_{\max} = 0.08 \tag{5.181}$$

Using this approach, propose a path for the road to follow.

5.B.4. We wish to produce C from A and B by the reaction network

$$\begin{aligned}
A + B &\rightarrow C + D & r_{R1} &= k_1(T)c_A c_B \\
C + B &\rightarrow S_1 + D & r_{R2} &= k_2(T)c_C c_B \\
A + D &\rightarrow S_2 & r_{R3} &= k_3(T)c_A c_D \\
A + B &\rightarrow S_3 & r_{R4} &= k_4(T)c_A c_B \\
C + B &\rightarrow S_4 & r_{R5} &= k_5(T)c_C c_B
\end{aligned} \tag{5.182}$$

A CSTR has an input stream with a velocity of 1 l/s containing species A and B in a carrier solvent, such that

$$c_{A0} + c_{B0} < 2\,\text{M} \tag{5.183}$$

We have the following temperature-dependent rate constant data,

$$k_1(298\,\text{K}) = 0.01 \;\frac{1}{\text{mol s}} \quad k_1(310\,\text{K}) = 0.02 \;\frac{1}{\text{mol s}} \quad k_2(T) = k_1(T)$$

$$k_3(298\,\text{K}) = 0.001 \;\frac{1}{\text{mol s}} \quad k_3(310\,\text{K}) = 0.005 \;\frac{1}{\text{mol s}} \tag{5.184}$$

$$k_4(298\,\text{K}) = 0.001 \;\frac{1}{\text{mol s}} \quad k_4(310\,\text{K}) = 0.005 \;\frac{1}{\text{mol s}} \quad k_5(T) = k_4(T)$$

We wish to design the reactor (assumed operated isothermally) to maximize the concentration of C in the output stream. We vary the inlet concentrations c_{A0} and c_{B0}, the volume of the reactor V, within the range

$$10\,\text{l} \leq V \leq 10\,000\,\text{l} \tag{5.185}$$

and the temperature T within the range

$$298\,\text{K} \leq T \leq 335\,\text{K} \tag{5.186}$$

Propose an optimal steady-state CSTR design.

5.C.1. You wish to control the height $h(t)$ of water in a cylindrical tank of diameter 50 cm by varying the inlet volumetric flow rate $v_0(t)$ in liters per second. There is an outlet hole at the bottom of the tank of diameter 1 cm. Use Bernoulli's equation to propose an ODE model for $h(t)$. Then, compute the optimal feed control law $v_0(h)$, based on minimizing the cost functional

$$F\left[v_0(t); h^{[0]}\right] = \int_0^{t_H} \left\{ \frac{C_U}{2}[v_0(s) - v_{0,\,\text{set}}]^2 + [h(s) - h_{\text{set}}]^2 \right\} ds + C_H[h(t_H) - h_{\text{set}}]^2 \tag{5.187}$$

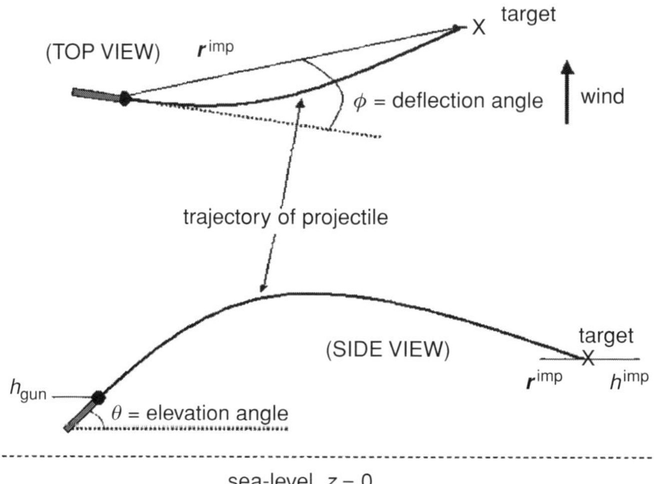

Figure 5.17 Shooting angles for the targeting of a projectile.

$h_{\mathrm{set}} = 1$ m is the set-point. $v_{0,\mathrm{set}}$ is the flow rate that maintains the height at the set-point at steady state. Use $t_{\mathrm{H}} = 600$ s, $C_{\mathrm{U}} = 0.1(h_{\mathrm{set}}/v_{0,\mathrm{set}})^2$, and $C_{\mathrm{H}} = 10^3 h_{\mathrm{set}}^{-2}$. Enforce the control input constraints that $0 \leq v_0(t) \leq 10v_{0,\mathrm{set}}$.

5.C.2. An early application of computing was the tabulation of accurate ballistic tables for artillery to account for wind and drag. Consider the case shown in (Figure 5.17) in which a projectile leaves a gun at an elevation of h_{gun} and is intended to hit a stationary target at an elevation h_{tar} and relative coordinates $(x_{\mathrm{tar}}, y_{\mathrm{tar}})$. For specified values of the elevation and deflection angles (θ, ϕ), we can integrate Newton's equation of motion to predict the impact location $(x_{\mathrm{imp}}, y_{\mathrm{imp}})$, as the position where the projectile passes through the target's elevation on the way down. Then, to aim the projectile we minimize the cost function

$$F^{(\mathrm{drag})}(\theta, \phi) = [x_{\mathrm{imp}}(\theta, \phi) - x_{\mathrm{tar}}]^2 + [y_{\mathrm{imp}}(\theta, \phi) - y_{\mathrm{tar}}]^2 \tag{5.188}$$

The equation of motion for the projectile is

$$m_{\mathrm{p}} \frac{d}{dt} v = m_{\mathrm{p}} g - \left(\tfrac{1}{2}\rho_{\mathrm{air}} A_{\mathrm{p}}\right) C_{\mathrm{D}} U u \tag{5.189}$$

ρ_{air} is the density of air, A_{p} is the cross-sectional area of the projectile, C_{D} is an empirical drag coefficient, and u is the relative velocity of the projectile with respect to that of the wind w,

$$u = v - w \quad U = |u| \tag{5.190}$$

For projectile velocities less than about a third of the speed of sound in air, compressibility effects may be neglected and the drag coefficient is a function of Reynolds' number alone, defined as

$$Re = \frac{\rho_{\mathrm{air}} U(2R_{\mathrm{p}})}{\mu_{\mathrm{air}}} \tag{5.191}$$

The viscosity of air is μ_{air} and the projectile radius is R_{p}. Drag coefficient data may be found in Table 5-22 of Perry & Green (1984) or the empirical correlations in (2.163) and the data of Table 2.2 may be used instead. We assume that the atmospheric conditions take their standard sea-level values (25 °C, 1 atm). For simplicity we neglect any variation of wind speed with altitude.

Write a program that computes the optimal angles for a specified compass bearing Φ and distance D to the target, the gun elevation h_{gun} and that of the target h_{tar}, the density ρ_s and mass m_p of the projectile, the wind speed W and the compass bearing of the wind Φ_W (e.g. 360° if the wind is from the north), and the speed v_{gun} at which the projectile leaves the gun.

Using this program, compute the optimal angles for the case where the gun and target are both at the same elevation, the projectile is made of steel and weighs 10 Kg, and the gun barrel velocity is 100 m/s (chosen to be quite low so that compressibility effects are negligible, for a more realistic case the gun velocity would be higher). The target is at a distance of 500 m to the east, and the wind is at 15 mph from the north.

How much of a lateral distance error would you have made if you had neglected wind and drag?

5.C.3. Modify your program from Problem 5.C.2 to account for a moving target.

6 Boundary value problems

Boundary value problems (BVPs) involve the solution of ODEs or partial differential equations (PDEs) on a spatial domain, subject to boundary conditions that hold on the domain boundary. Many problems from solid and fluid mechanics, electromagnetics, and heat and mass transfer are expressed naturally as BVPs. The forms of these differential equations often resemble each other because they arise from similar conservation principles. Here the emphasis is upon BVPs that arise from problems in transport phenomena.

This chapter focuses upon real-space methods, in which a computational grid is overlaid upon the domain. The BVP is then converted into a set of ODEs for a time-dependent problem or a set of algebraic equations for a steady problem. This technique can be used even when no analytical solution exists, and can be extended to BVPs with multiple equations or complex domain geometries. Here, the focus is upon the methods of finite differences, finite volumes, and finite elements. These methods have many characteristics in common; therefore, particular attention is paid to the finite difference method, as it is the easiest to code. The finite volume and finite element methods also are discussed; however, as the reader is most likely to use these in the context of prewritten software, the emphasis is upon conceptual understanding as opposed to implementation.

BVPs from conservation principles

Let $\varphi(\mathbf{r}, t)$ be some time-varying *field*, i.e., a function that assigns to each position $\mathbf{r}$ and time t a unique value $\varphi(\mathbf{r}, t)$. Common examples of fields in chemical engineering include

$$
\begin{aligned}
\varphi &= \rho & &\text{mass density} \\
\varphi &= \rho \mathbf{v} & &\text{linear momentum density} \\
\varphi &= \tfrac{1}{2}\rho|\mathbf{v}|^2 + \rho\hat{u} & &\text{total kinetic and internal energy density} \\
\varphi &= c_i & &\text{concentration of species } i
\end{aligned}
\tag{6.1}
$$

Each of these fields represents the density of some quantity Φ. Let us consider a closed domain Ω, a *control volume (CV)*, with boundary $\partial\Omega$ (Figure 6.1), and write a balance for the total amount of Φ within Ω,

$$
\frac{d}{dt}\left[\int_{\Omega}\varphi(\mathbf{r}, t)\,d\mathbf{r}\right] = \int_{\partial\Omega}\varphi[\mathbf{v}\cdot(-\mathbf{n})]\,dS + \int_{\partial\Omega}[\mathbf{J}_{\mathrm{D}}\cdot(-\mathbf{n})]\,dS + \int_{\Omega}s(\mathbf{r}, t, \varphi)\,d\mathbf{r}
\tag{6.2}
$$

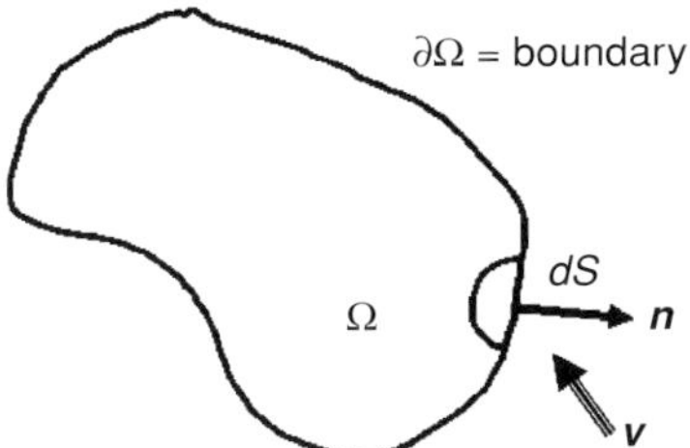

Figure 6.1 Fixed control volume (CV) in space.

The left-hand side is the rate of change of the total amount of Φ within Ω. The first term on the right-hand side is the net *convective transport* of Φ across $\partial\Omega$ into Ω by the velocity field of the medium $v(r, t)$, n being the outward normal vector at the boundary. The second term is the net *diffusive transport* of Φ across $\partial\Omega$ into Ω, with the flux vector J_D often being related to the local field gradient by a *constitutive equation* of the form of Fick's law,

$$
\begin{aligned}
\text{anisotropic diffusion} \qquad & J_D = -\Gamma \cdot \nabla\varphi \qquad & \Gamma = \Sigma_m \Sigma_n \Gamma_{mn} e^{[m]} e^{[n]} \\
\text{isotropic diffusion} \qquad & J_D = -\Gamma \nabla\varphi \qquad & \Gamma \in \mathfrak{R}
\end{aligned}
\tag{6.3}
$$

The third term on the right-hand side is a *source term*, $s(r, t, \varphi)$ being the amount of Φ generated per unit volume per unit time at (r, t). Equation (6.2) is known as a *macroscopic balance*, as it applies to a control volume of finite size. A corresponding *microscopic balance* is obtained through use of the *divergence theorem*,

$$
\int_{\partial\Omega} n \cdot w \, dS = \int_\Omega \nabla \cdot w \, dV
\tag{6.4}
$$

to convert the surface integrals into volume integrals,

$$
\int_\Omega \frac{\partial\varphi}{\partial t} dr = -\int_\Omega [\nabla \cdot (\varphi v)] dr - \int_\Omega [\nabla \cdot J_D] dr + \int_\Omega s(r, t, \varphi) dr
\tag{6.5}
$$

As all integrals are over the same domain, we can combine them,

$$
\int_\Omega \left[\frac{\partial\varphi}{\partial t} + \nabla \cdot (\varphi v) + \nabla \cdot J_D - s(r, t, \varphi) \right] dr = 0
\tag{6.6}
$$

This balance must hold for any arbitrary fixed control volume, requiring the field to satisfy everywhere the partial differential equation

$$
\frac{\partial\varphi}{\partial t} = -\nabla \cdot (\varphi v) - \nabla \cdot J_D + s(r, t, \varphi)
\tag{6.7}
$$

Substituting (6.3) for the diffusive flux yields

$$
\frac{\partial\varphi}{\partial t} = -\nabla \cdot (\varphi v) + \nabla \cdot [\Gamma \nabla\varphi] + s(r, t, \varphi)
\tag{6.8}
$$

For isotropic diffusion with a constant Γ, this takes the form of a classic convection/diffusion equation with a source term,

$$
\frac{\partial\varphi}{\partial t} = -\nabla \cdot (\varphi v) + \Gamma \nabla^2 \varphi + s(r, t, \varphi)
\tag{6.9}
$$

As source terms often arise from chemical reaction, (6.9) is also known as a *convection/diffusion/reaction equation.*

In this chapter, we consider the numerical solution of PDEs of the form (6.9). We focus first upon stationary problems, for which the steady-state field $\varphi(r)$ satisfies the PDE

$$0 = -\nabla \cdot (\varphi v) + \Gamma \nabla^2 \varphi + s(r, t, \varphi) \tag{6.10}$$

and then treat time-dependent problems. Here, we consider general transport-type PDEs, but do not address the particular numerical issues that arise in computational fluid dynamics (CFD). While the methods of finite differences, finite volumes, and finite elements, described here, are used in CFD, a few additional concerns arise, particularly involving the coupling of the pressure and velocity fields. To do justice to this subject would require a dedicated text; therefore, for further details the reader is referred to Ferziger & Peric (2001).

Real-space vs. function-space BVP methods

There are two general approaches to solving BVPs. In a *function-space method*, we write the solution as a linear combination of basis functions $\{\chi_m(r)\}$, each satisfying all boundary conditions,

$$\varphi(r, t) = \sum_m c_m(t) \chi_m(r) \tag{6.11}$$

$\varphi(r, t)$ then automatically satisfies all boundary conditions (assumed to be linear in φ), and we have merely to find the $\{c_m(r)\}$ that best satisfy (6.9).

Alternatively, in a *real-space method*, we specify a number of grid points $r^{[j]} \in \Omega$ and compute numerically the field values $\varphi(r^{[j]}, t)$ at these points. While function-space methods can yield analytical solutions for some linear PDEs in simple domains, real-space methods require numerical solution yet are more generally applicable, especially for problems with nonlinear source terms or complex domain geometries. We thus restrict our focus to real-space methods, although the finite element method will be seen to mix both approaches. For further discussion of function-space approaches, consult Bird *et al.* (2002), Deen (1998), and Stakgold (1979).

The finite difference method applied to a 2-D BVP

Let us consider a steady 2-D BVP on a rectangular domain involving only diffusion and a position-dependent source term, $-\nabla^2 \varphi = f(r)$, the *Poisson equation.* We require the solution to be zero on all boundaries, a *Dirichlet-type boundary condition.* Thus, the BVP is

$$-\nabla^2 \varphi = -\frac{\partial^2 \varphi}{\partial x^2} - \frac{\partial^2 \varphi}{\partial y^2} = f(x, y) \qquad 0 \le x \le L \qquad 0 \le y \le H$$

$$\begin{array}{llll} \text{BC 1} & \varphi(0, y) = 0 & 0 \le y \le H & \\ \text{BC 2} & \varphi(L, y) = 0 & 0 \le y \le H & \\ \text{BC 3} & \varphi(x, 0) = 0 & 0 \le x \le L & \\ \text{BC 4} & \varphi(x, H) = 0 & 0 \le x \le L & \end{array} \tag{6.12}$$

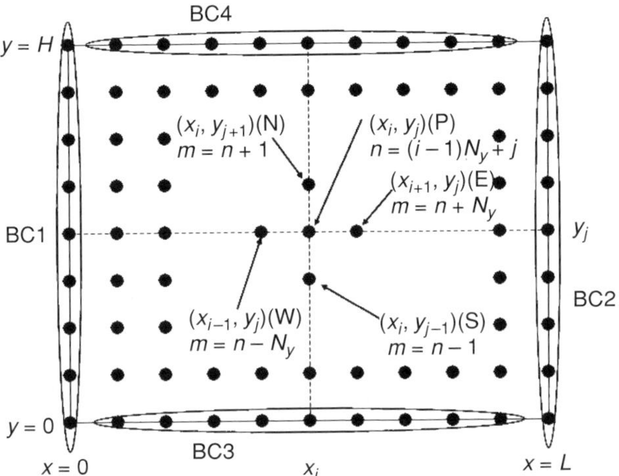

Figure 6.2 Regular 2-D grid with the labeling scheme for the finite difference method.

This BVP arises in fluid mechanics for the case of laminar flow of a Newtonian fluid through a rectangular duct, where

$$\varphi(x, y) = v_z(x, y) \qquad f(x, y) = \frac{1}{\mu}\left[-\left(\frac{\Delta P}{\Delta z}\right) + \rho g_z\right] \tag{6.13}$$

We solve (6.12) numerically using the *method of finite differences*, encountered earlier in Chapters 1 and 2. We place a regular grid of points as shown in Figure 6.2. To the point at (x_i, y_j), we assign a unique integer label $n = (i - 1)N_y + j$. The neighboring points, and their labels, are

$$
\begin{array}{llll}
\text{north (N)} & (x_i, y_{j+1}) & m = n + 1 \\
\text{east (E)} & (x_{i+1}, y_j) & m = n + N_y \\
\text{south (S)} & (x_i, y_{j-1}) & m = n - 1 \\
\text{west (W)} & (x_{i-1}, y_j) & m = n - N_y
\end{array}
\tag{6.14}
$$

We wish to compute the vector of grid point values,

$$\varphi = [\varphi_n] \in \Re^N \quad N = N_x N_y \quad \varphi_n = \varphi(x_i, y_j) \quad n = (i - 1)N_y + j \tag{6.15}$$

For each grid point on a boundary (circled in Figure 6.2), we obtain from the boundary condition a corresponding linear algebraic equation

$$
\begin{array}{lll}
\varphi_n = \varphi(x_i, y_j) = 0 & n = (i - 1)N_y + j \\
\text{BC 1} & i = 1 & 1 \le j \le N_y \\
\text{BC 2} & i = N_x & 1 \le j \le N_y \\
\text{BC 3} & 1 < i < N_x & j = 1 \\
\text{BC 4} & 1 < i < N_x & j = N_y
\end{array}
\tag{6.16}
$$

We obtain an algebraic equation for each grid point in the interior by requiring the PDE to be satisfied locally:

$$-\frac{\partial^2 \varphi}{\partial x^2}\bigg|_{(x_i,y_j)} - \frac{\partial^2 \varphi}{\partial y^2}\bigg|_{(x_i,y_j)} = f(x_i, y_j) \tag{6.17}$$

The key idea of finite differences is to approximate these local derivatives by algebraic differences of the field values at neighboring grid points.

Finite difference approximations

In the finite difference method, we approximate the value of the derivative of $f(x)$ at x_0 using a truncated Taylor series approximation for small Δx,

$$f(x_0 + \Delta x) = f(x_0) + (\Delta x)\frac{df}{dx}\bigg|_{x_0} + O[(\Delta x)^2] \tag{6.18}$$

to yield

$$\frac{df}{dx}\bigg|_{x_0} = \frac{f(x_0 + \Delta x) - f(x_0)}{\Delta x} + O[\Delta x] \tag{6.19}$$

Similarly, from the expansion in the opposite direction,

$$f(x_0 - \Delta x) = f(x_0) + (-\Delta x)\frac{df}{dx}\bigg|_{x_0} + O[(\Delta x)^2] \tag{6.20}$$

we obtain

$$\frac{df}{dx}\bigg|_{x_0} = \frac{f(x_0) - f(x_0 - \Delta x)}{\Delta x} + O[\Delta x] \tag{6.21}$$

Both of these *one-sided difference* approximations have errors proportional to Δx, the spacing between successive grid points. Subtracting the two,

$$\left\{ f(x_0 + \Delta x) = f(x_0) + (\Delta x)\frac{df}{dx}\bigg|_{x_0} + \frac{(\Delta x)^2}{2}\frac{d^2 f}{dx^2}\bigg|_{x_0} + O[(\Delta x)^3] \right\}$$

$$- \left\{ f(x_0 - \Delta x) = f(x_0) - (\Delta x)\frac{df}{dx}\bigg|_{x_0} + \frac{(\Delta x)^2}{2}\frac{d^2 f}{dx^2}\bigg|_{x_0} + O[(\Delta x)^3] \right\} \tag{6.22}$$

the zeroth-and second-order terms cancel out, and the resulting *central difference approximation* is second-order accurate,

$$\frac{df}{dx}\bigg|_{x_0} = \frac{f(x_0 + \Delta x) - f(x_0 - \Delta x)}{2(\Delta x)} + O[(\Delta x)^2] \tag{6.23}$$

The error of (6.23) is proportional to the square of the grid point spacing, thus it decays to zero as $\Delta x \to 0$ more rapidly than the errors of one-sided formulas. Equation (6.23) is more accurate than (6.19) or (6.21).

We can approximate second derivatives similarly by adding the expansions,

$$
\left\{
\begin{aligned}
f(x_0 + \Delta x) &= f(x_0) + (\Delta x)\left.\frac{df}{dx}\right|_{x_0} + \frac{(\Delta x)^2}{2}\left.\frac{d^2 f}{dx^2}\right|_{x_0} \\
&\quad + \frac{(\Delta x)^3}{3}\left.\frac{d^3 f}{dx^3}\right|_{x_0} + O[(\Delta x)^4]
\end{aligned}
\right\}
$$

$$
+ \left\{
\begin{aligned}
f(x_0 - \Delta x) &= f(x_0) - (\Delta x)\left.\frac{df}{dx}\right|_{x_0} + \frac{(\Delta x)^2}{2}\left.\frac{d^2 f}{dx^2}\right|_{x_0} \\
&\quad - \frac{(\Delta x)^3}{3!}\left.\frac{d^3 f}{dx^3}\right|_{x_0} + O[(\Delta x)^4]
\end{aligned}
\right\}
\tag{6.24}
$$

The first-and third-order terms cancel, yielding

$$
f(x_0 + \Delta x) + f(x_0 - \Delta x) = 2f(x_0) + [(\Delta x)^2]\left.\frac{d^2 f}{dx^2}\right|_{x_0} + O[(\Delta x)^4]
\tag{6.25}
$$

and an approximation for the second derivative,

$$
\left.\frac{d^2 f}{dx^2}\right|_{x_0} = \frac{f(x_0 - \Delta x) - 2f(x_0) + f(x_0 + \Delta x)}{(\Delta x)^2} + O[(\Delta x)^2]
\tag{6.26}
$$

that has the same order of accuracy as (6.23). For partial derivatives, similar finite difference approximations exist:

$$
\left.\frac{\partial f}{\partial x}\right|_{(x_0,y_0)} \approx \frac{f(x_0 + \Delta x, y_0) - f(x_0 - \Delta x, y_0)}{2(\Delta x)}
$$

$$
\approx \frac{f(x_0 + \Delta x, y_0) - f(x_0, y_0)}{(\Delta x)} \approx \frac{f(x_0, y_0) - (x_0 - \Delta x, y_0)}{(\Delta x)}
\tag{6.27}
$$

$$
\left.\frac{\partial^2 f}{\partial x^2}\right|_{(x_0,y_0)} \approx \frac{f(x_0 - \Delta x, y_0) - 2f(x_0, y_0) + f(x_0 + \Delta x, y_0)}{(\Delta x)^2}
\tag{6.28}
$$

Finite difference solution of a Poisson BVP

We apply (6.28) to approximate the local partial derivatives in (6.17),

$$
\left.\frac{\partial^2 \varphi}{\partial x^2}\right|_{(x_i,y_j)} \approx \frac{\varphi(x_{i-1}, y_j) - 2\varphi(x_i, y_j) + \varphi(x_{i+1}, y_j)}{(\Delta x)^2}
$$

$$
\left.\frac{\partial^2 \varphi}{\partial y^2}\right|_{(x_i,y_j)} \approx \frac{\varphi(x_i, y_{j-1}) - 2\varphi(x_i, y_j) + \varphi(x_i, y_{j+1})}{(\Delta y)^2}
\tag{6.29}
$$

and obtain a linear equation for each (x_i, y_j) (a *discretization* of the PDE),

$$
\frac{-\varphi(x_{i-1}, y_j) + 2\varphi(x_i, y_j) - \varphi(x_{i+1}, y_j)}{(\Delta x)^2} + \frac{-\varphi(x_i, y_{j-1}) + 2\varphi(x_i, y_j) - \varphi(x_i, y_{j+1})}{(\Delta y)^2}
$$

$$
= f(x_i, y_j)
\tag{6.30}
$$

We write this equation in standard matrix-vector form by employing the labeling

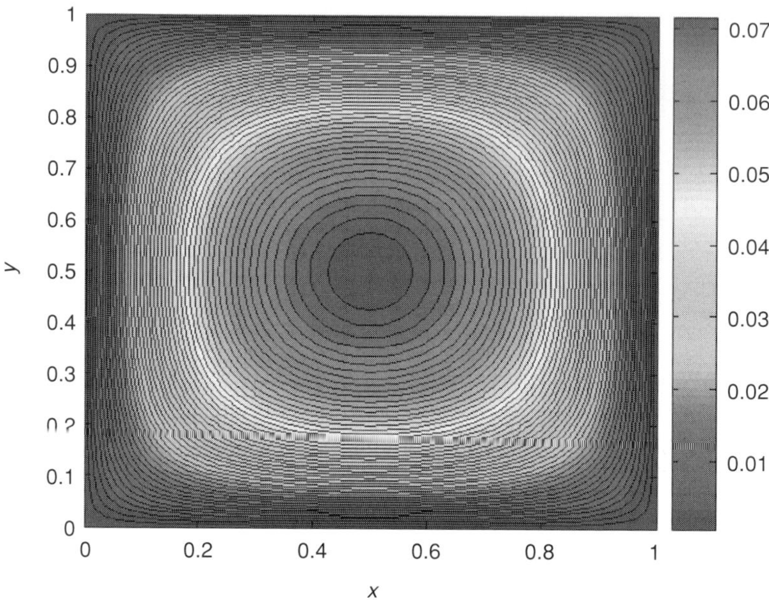

Figure 6.3 Solution by finite differences of the Poisson BVP with a source term $f = 1$.

scheme, $\varphi_n = \varphi(x_i, y_j)$, $n = (i-1)N_y + j$,

$$A_{n,n-N_y}\varphi_{n-N_y} + A_{n,n-1}\varphi_{n-1} + A_{nn}\varphi_n + A_{n,n+1}\varphi_{n+1} + A_{n,n+N_y}\varphi_{n+N_y} = b_n$$

$$A_{n,n-N_y} = A_{n,n+N_y} = \left[\frac{-1}{(\Delta x)^2}\right] \quad A_{n,n-1} = A_{n,n+1} = \left[\frac{-1}{(\Delta y)^2}\right] \tag{6.31}$$

$$A_{nn} = \left[\frac{2}{(\Delta x)^2} + \frac{2}{(\Delta y)^2}\right] \quad b_n = f(x_i, y_j)$$

For each row corresponding to a boundary point, we merely set $A_{nn} = 1, b_n = 0$ to enforce the boundary condition. BVP_2D_Poisson_FD.m solves this BVP (solution shown in Figure 6.3) and is invoked with the code

```
L = 1; H = 1; fun_name = 'f_rD_uniform'; num_pts = 51;
BVP_2D_Poisson_FD(fun_name,L,H,num_pts);
```

Extending the finite difference method

We next extend the finite difference method to treat BVPs of greater complexity, with non-Cartesian coordinates and nonuniform grids, von Neumann-type boundary conditions, multiple fields, time dependence, and PDEs in more than two spatial dimensions. We do so through the examples in the following sections.

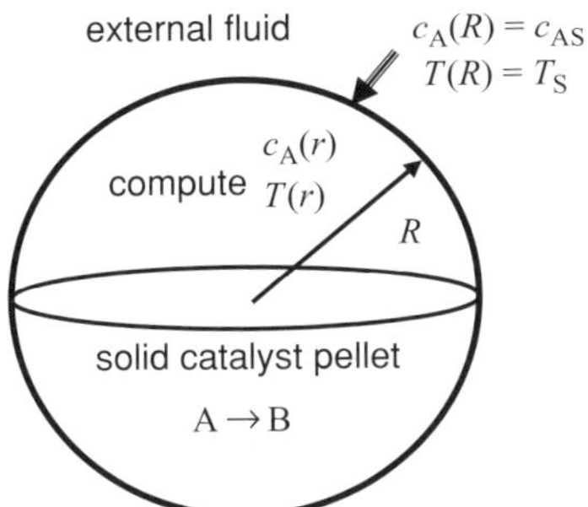

Figure 6.4 Nonisothermal reaction within a catalyst pellet with internal diffusion and thermal conduction.

Chemical reaction and diffusion in a spherical catalyst pellet

Consider the case of a nonisothermal reaction $A \rightarrow B$ occurring in the interior of a spherical catalyst pellet of radius R (Figure 6.4). We wish to compute the effect of internal heat and mass transfer resistance upon the reaction rate and the concentration and temperature profiles within the pellet. If D_A is the effective binary diffusivity of A within the pellet, and we have first-order kinetics, the concentration profile $c_A(r)$ is governed by the mole balance

$$\frac{d}{dr}\left[r^2 D_A \frac{dc_A}{dr}\right] - r^2[k(T)c_A] = 0 \tag{6.32}$$

Similarly, if λ is the effective thermal conductivity of the pellet, the temperature profile $T(r)$ is governed by the enthalpy balance

$$\frac{d}{dr}\left[r^2 \lambda \frac{dT}{dr}\right] + r^2(-\Delta H)[k(T)c_A] = 0 \tag{6.33}$$

Neglecting external heat or mass transfer resistance, we have known values of the concentration and temperature at the surface, $r = R$. At the pellet center, we use the symmetry conditions $dc_A/dr = dT/dr = 0$. Thus, we solve (6.32) and (6.33) subject to the boundary conditions

$$\text{BC 1} \quad c_A(R) = c_{AS} \quad T(R) = T_S \tag{6.34}$$

$$\text{BC 2} \quad \left.\frac{dc_A}{dr}\right|_{r=0} = 0 \quad \left.\frac{dT}{dr}\right|_{r=0} = 0 \tag{6.35}$$

The temperature dependence of the rate constant is

$$k(T) = k(T_S) \exp\left[-\frac{E_a}{RT_S}\left(\frac{T_S}{T} - 1\right)\right] \tag{6.36}$$

This BVP introduces several new issues: (1) nonCartesian (spherical) coordinates, (2) more than one coupled PDE, and (3) a BC at $r = 0$ that specifies the local value of the gradient (a *von Neumann-type boundary condition*). Also, experience tells us that when internal mass transfer resistance is strong, reaction only occurs within a thin layer near the surface over which the local concentration of A drops rapidly to zero. Thus, we use a computational grid that is nonuniform, with closer spacing between the points near the surface than deep within the interior of the pellet.

Actually, for this problem if we assume constant D_A, λ, and ΔH, we can reduce the BVP to a single equation; however, we return to the question of multiple fields later. Let us divide (6.33) by $-\Delta H$,

$$\frac{d}{dr}\left[r^2 \frac{\lambda}{(-\Delta H)}\frac{dT}{dr}\right] + r^2(kc_A) = 0 \tag{6.37}$$

and add (6.37) to (6.32),

$$\frac{d}{dr}\left[r^2 D_A \frac{dc_A}{dr}\right] + \frac{d}{dr}\left[r^2 \frac{\lambda}{(-\Delta H)}\frac{dT}{dr}\right] = 0 \tag{6.38}$$

Integrating this equation once yields

$$r^2\left[D_A \frac{dc_A}{dr} + \frac{\lambda}{(-\Delta H)}\frac{dT}{dr}\right] = C_1 \tag{6.39}$$

Dividing by r^2 gives

$$\frac{d}{dr}\left[D_A c_A + \frac{\lambda}{(-\Delta H)}T\right] = \frac{C_1}{r^2} \tag{6.40}$$

From the symmetry boundary condition (6.35), $C_1 = 0$, and a second integration yields

$$D_A c_A + \frac{\lambda}{(-\Delta H)}T = C_2 = D_A c_{AS} + \frac{\lambda}{(-\Delta H)}T_S \tag{6.41}$$

Thus, there exists a linear relation between $c_A(r)$ and $T(r)$,

$$T(r) - T_S = \frac{D_A(\Delta H)}{\lambda}[c_A(r) - c_{AS}] \tag{6.42}$$

that allows us to solve only a single equation, (6.32).

Dimensionless formulation

We next reduce the number of independent parameters by rephrasing the problem in terms of the dimensionless quantities

$$\xi = \frac{r}{R} \quad \varphi_A(\xi) = \frac{c_A(r = \xi R)}{c_{AS}} \quad \theta(\xi) = \frac{T(r = \xi R)}{T_S} \tag{6.43}$$

After some manipulation, the BVP takes the dimensionless form

$$\frac{d}{d\xi}\left[\xi^2 \frac{d\varphi_A}{d\xi}\right] - \xi^2 \Phi^2 \exp\left[\frac{\gamma \beta(1 - \varphi_A)}{1 + \beta(1 - \varphi_A)}\right]\varphi_A = 0 \tag{6.44}$$

$$\varphi_A(1) = 1 \quad \left.\frac{d\varphi_A}{d\xi}\right|_{\xi=0} = 0 \tag{6.45}$$

There are now only three dimensionless parameters

$$\Phi = R\sqrt{\frac{k(T_S)}{D_A}} \quad \beta = \frac{D_A(-\Delta H)c_{AS}}{\lambda T_S} \quad \gamma = \frac{E_a}{RT_S} \tag{6.46}$$

Φ is the *Thiele modulus*, and represents the strength of internal mass transfer resistance. When $\Phi \ll 1$, diffusion is so fast compared to reaction that mass transfer resistance is negligible. When $\Phi \geq 1$, the opposite is true and mass transfer resistance becomes rate-dominant. β is a measure of the relative importance of the heat of reaction, so that for $\beta > 1$ there is significant internal heating, $T(r) > T_S$, and when $\beta < -1$, significant internal cooling. γ is the dimensionless activation energy, and a large γ means that the reaction rate is very sensitive to the local temperature.

Finite differences on a nonCartesian, nonuniform grid

To solve (6.44), we use finite differences on a grid $0 < \xi_1 < \xi_2 < \cdots < \xi_N < 1$ and require that at each ξ_j, (6.44) be satisfied locally:

$$\frac{d}{d\xi}\left[\xi^2 \frac{d\varphi_A}{d\xi}\right]\bigg|_{\xi_j} - \xi_j^2 \Phi^2 \exp\left[\frac{\gamma\beta(1-\varphi_j)}{1+\beta(1-\varphi_j)}\right]\varphi_j = 0 \qquad (6.47)$$

where $\varphi_j = \varphi_A(\xi_j)$. As we expect strong gradients near $\xi = 1$ when $\Phi \geq 1$, we use a nonuniform grid with closer grid points near the surface. Defining the mid-points in the intervals between grid points,

$$\xi_{j+1/2} = \tfrac{1}{2}(\xi_j + \xi_{j+1}) \qquad \xi_{j-1/2} = \tfrac{1}{2}(\xi_j + \xi_{j-1}) \qquad (6.48)$$

we use a central difference approximation for the second derivative,

$$\frac{d}{d\xi}\left[\xi^2 \frac{d\varphi_A}{d\xi}\right]\bigg|_{\xi_j} \approx \frac{\xi_{j+1/2}^2 \left(\dfrac{d\varphi_A}{d\xi}\bigg|_{\xi_{j+1/2}}\right) - \xi_{j-1/2}^2 \left(\dfrac{d\varphi_A}{d\xi}\bigg|_{\xi_{j-1/2}}\right)}{(\xi_{j+1/2} - \xi_{j-1/2})} \qquad (6.49)$$

For the first derivatives, we use similar approximations,

$$\frac{d\varphi_A}{d\xi}\bigg|_{\xi_{j+1/2}} \approx \frac{\varphi_{j+1} - \varphi_j}{\xi_{j+1} - \xi_j} \qquad \frac{d\varphi_A}{d\xi}\bigg|_{\xi_{j-1/2}} \approx \frac{\varphi_j - \varphi_{j-1}}{\xi_j - \xi_{j-1}} \qquad (6.50)$$

to obtain from (6.49) and (6.50) the finite difference approximation

$$\frac{d}{d\xi}\left[\xi^2 \frac{d\varphi_A}{d\xi}\right]\bigg|_{\xi_j} \approx A_{j,j-1}\varphi_{j-1} + A_{jj}\varphi_j + A_{j,j+1}\varphi_{j+1} \qquad (6.51)$$

where

$$A_{j,j-1} = \alpha_j^{(lo)} \qquad A_{j,j} = -\left[\alpha_j^{(lo)} + \alpha_j^{(hi)}\right] \qquad A_{j,j+1} = \alpha_j^{(hi)} \qquad (6.52)$$

$$\alpha_j^{(lo)} = \frac{\xi_{j-1/2}^2}{(\xi_j - \xi_{j-1})(\xi_{j+1/2} - \xi_{j-1/2})} \qquad \alpha_j^{(hi)} = \frac{\xi_{j+1/2}^2}{(\xi_{j+1} - \xi_j)(\xi_{j+1/2} - \xi_{j-1/2})}$$

For each interior point $j = 2, 3, \ldots, N-1$ that does *not* neighbor a grid point at the boundary, the nonlinear algebraic equation obtained from (6.44) by finite differences is

$$f_j = A_{j,j-1}\varphi_{j-1} + A_{jj}\varphi_j + A_{j,j+1}\varphi_{j+1} - \xi_j^2 \Phi^2 \exp\left[\frac{\gamma\beta(1-\varphi_j)}{1+\beta(1-\varphi_j)}\right]\varphi_j = 0 \qquad (6.53)$$

Treatment of Dirichlet and von Neumann boundary conditions

The boundary conditions (6.45) are of Dirichlet-type (specified φ) at the surface and of von Neumann-type (specified $d\varphi/d\xi$) at the center. At the last grid point $\xi_N < 1$, we enforce

$\varphi_A(1) = 1$ by placing a hypothetical (nonexistent) grid point at $\xi_{N+1} = 1$ for which we set $\varphi_{N+1} = 1$. We then modify (6.53) for $j = N$ to use this value for the nonexistent grid point,

$$f_N = A_{N,N-1}\varphi_{N-1} + A_{NN}\varphi_N + A_{N,N+1} - \xi_N^2 \Phi^2 \exp\left[\frac{\gamma\beta(1-\varphi_N)}{1+\beta(1-\varphi_N)}\right]\varphi_N = 0 \tag{6.54}$$

We treat the von Neumann boundary condition similarly, applying (6.53) to $j = 1$, referring to a nonexistent point at $\xi_0 = 0$,

$$f_1 = A_{10}\varphi_0 + A_{11}\varphi_1 + A_{12}\varphi_2 - \xi_1^2 \Phi^2 \exp\left[\frac{\gamma\beta(1-\varphi_1)}{1+\beta(1-\varphi_1)}\right]\varphi_1 = 0 \tag{6.55}$$

To enforce $d\varphi/d\xi|_0 = 0$ we might think to set $\varphi_0 = \varphi_1$. However, this approximation is based upon the first-order finite difference

$$\left.\frac{d\varphi_A}{d\xi}\right|_0 = \frac{\varphi_1 - \varphi_0}{\xi_1} + O(\xi_1) \tag{6.56}$$

When solving diffusion equations it is common to use second-order accurate approximations, so that simply setting $\varphi_0 = \varphi_1$ is not the preferred way to treat a von Neumann boundary condition. Rather, we obtain second-order accuracy by fitting a quadratic polynomial to $\varphi(\xi)$ near $\xi = 0$,

$$\varphi(\xi) \approx \varphi_0 L_0(\xi) + \varphi_1 L_1(\xi) + \varphi_2 L_2(\xi) \quad L_j(\xi) = \prod_{\substack{k=0 \\ k\neq j}}^{2}\left[\frac{\xi - \xi_k}{\xi_j - \xi_k}\right] \tag{6.57}$$

The discretized form of the von Neumann boundary condition is then

$$\left.\frac{d\varphi_A}{d\xi}\right|_0 = 0 = \varphi_0 L_0'(0) + \varphi_1 L_1'(0) + \varphi_2 L_2'(0) \tag{6.58}$$

where

$$L_0'(0) = \frac{-(\xi_1 + \xi_2)}{\xi_1 \xi_2} \quad L_1'(0) = \frac{\xi_2}{\xi_1(\xi_2 - \xi_1)} \quad L_2'(0) = \frac{-\xi_1}{\xi_2(\xi_2 - \xi_1)} \tag{6.59}$$

For a locally uniform grid with $\xi_1 = \Delta\xi$, $\xi_2 = 2(\Delta\xi)$, (6.58) becomes

$$\left.\frac{d\varphi_A}{d\xi}\right|_0 = 0 = \frac{-3\varphi_0 + 4\varphi_1 - \varphi_2}{2(\Delta\xi)} \tag{6.60}$$

From this discretized boundary condition, we write the nonexistent grid value as

$$\varphi_0 = a_1\varphi_1 + a_2\varphi_2 \qquad a_j = -[L_j'(0)]/[L_0'(0)] \tag{6.61}$$

and substitute for φ_0 in (6.55),

$$f_1 = (A_{11} + a_1 A_{10})\varphi_1 + (A_{12} + a_2 A_{10})\varphi_2 - \xi_1^2 \Phi^2 \exp\left[\frac{\gamma\beta(1-\varphi_1)}{1+\beta(1-\varphi_1)}\right]\varphi_1 = 0 \tag{6.62}$$

Together, (6.53), (6.54), and (6.62) provide a set of N nonlinear algebraic equations for the N unknowns $\{\varphi_1, \varphi_2, \ldots, \varphi_N\}$ that can be solved numerically (e.g. by **fsolve**). The nonzero elements of the Jacobian in each interior row $j = 2, 3, \ldots, N - 1$ are

$$J_{j,j-1} = A_{j,j-1} \qquad J_{j,j+1} = A_{j,j+1}$$

$$J_{jj} = A_{jj} - \xi_j^2 \Phi^2 g(\varphi_j) \qquad g(\varphi) = \frac{d}{d\varphi} \exp\left[\frac{\gamma\beta(1-\varphi)}{1+\beta(1-\varphi)}\right] \tag{6.63}$$

In the first and last rows, the nonzero elements are

$$J_{11} = A_{11} + a_1 A_{10} - \xi_1^2 \Phi^2 g(\varphi_1) \qquad J_{12} = A_{12} + a_2 A_{10}$$

$$J_{N,N-1} = A_{N,N-1} \qquad J_{NN} = A_{NN} - \xi_N^2 \Phi^2 g(\varphi_N) \tag{6.64}$$

The linear system comprising (6.63) and (6.64) is thus tridiagonal, and elimination is performed rapidly, even for large N.

Definition of the effectiveness factor

From the solution of the BVP (6.44) and (6.45), we use the resulting concentration and temperature fields to compute the total rate of reaction within the pellet,

$$R_{\text{tot}} = \int_0^R k(T)c_A(r)(4\pi r^2)dr \tag{6.65}$$

Rewriting this integral in terms of the dimensionless quantities yields

$$R_{\text{tot}} = \left[\left(\frac{4}{3}\pi R^3\right)(k_s c_{A,s})\right]\left[3\int_0^1 \varphi_A(\xi)\exp\left[\frac{\gamma\beta(1-\varphi_A(\xi))}{1+\beta(1-\varphi_A(\xi))}\right]\xi^2 d\xi\right] \tag{6.66}$$

The product in the first set of square brackets is the total reaction rate if there were no concentration (or temperature) gradients within the catalyst particle. We define the *effectiveness factor* η from the relation

$$R_{\text{tot}} = \left(\frac{4}{3}\pi R^3\right)(k_s c_{A,s})\eta \tag{6.67}$$

such that

$$\eta = 3\int_0^1 \varphi_A(\xi)\exp\left[\frac{\gamma\beta(1-\varphi_A(\xi))}{1+\beta(1-\varphi_A(\xi))}\right]\xi^2 d\xi \tag{6.68}$$

Numerical solution in MATLAB

catalyst_nonisothermal_scan.m plots η vs. Φ for fixed γ and various values of β. For $\gamma = 1$, and $\Delta\xi = 0.01$ in the interior and $\Delta\xi = 0.001$ near the surface, the results are shown in Figure 6.5.

For zero or moderate reaction heating, $\beta \approx 0$, increasing Φ reduces η. This is easily understood, as slow diffusion causes a depletion zone of low A concentration to form in the center of the pellet, where the reaction rate is consequently low. By contrast, if there were significant heating relative to conduction, $\beta \geq 1$, for $\Phi \approx 1$, η would be greater than 1; i.e., the rates of reaction are *higher* than in the absence of internal transport resistance. This

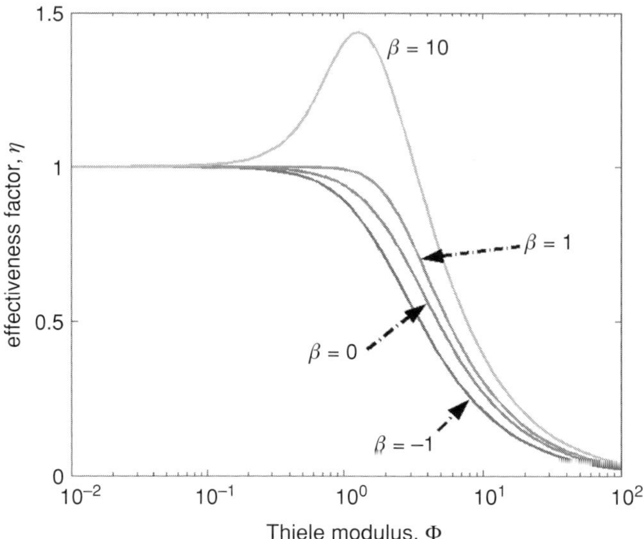

Figure 6.5 Effectiveness factor vs. Thiele modulus for nonisothermal first-order chemical reaction within a spherical catalyst pellet.

effect arises due to an increase in $k(T)$ by (6.36). In practice, this behavior is not observed, as solids are better transmitters of heat than mass. Therefore, $\eta < 1$ when internal mass transfer resistance is strong.

Finite differences for a convection/diffusion equation

Above we have considered problems in which transport is solely diffusive. We now consider the treatment of convection terms that introduce new numerical difficulties. We consider the simple PDE

$$\frac{\partial \varphi}{\partial t} = -v\frac{\partial \varphi}{\partial z} + \Gamma\frac{\partial^2 \varphi}{\partial z^2} + s(z, t, \varphi) \tag{6.69}$$

At steady state and with no source term, this equation becomes

$$-v\frac{d\varphi}{dz} + \Gamma\frac{d^2\varphi}{dz^2} = 0 \tag{6.70}$$

With the Dirichlet boundary conditions

$$\varphi(0) = \varphi_0 \quad \varphi(L) = \varphi_L \tag{6.71}$$

the solution for $0 \leq z \leq L$ is

$$\varphi(z) = \varphi_0 + \frac{e^{z(Pe)/L} - 1}{e^{(Pe)} - 1}(\varphi_L - \varphi_0) \tag{6.72}$$

where the dimensionless *Peclet number* is defined as

$$Pe = \frac{vL}{\Gamma} = \frac{\text{strength of convection}}{\text{strength of diffusion}} \tag{6.73}$$

This solution is plotted in Figure 6.6 for various Pe values. When $Pe \ll 1$, diffusion is dominant and we observe a linear increase from $\varphi_0 = 0$ to $\varphi_L = 1$. When $Pe \gg 1$, convection

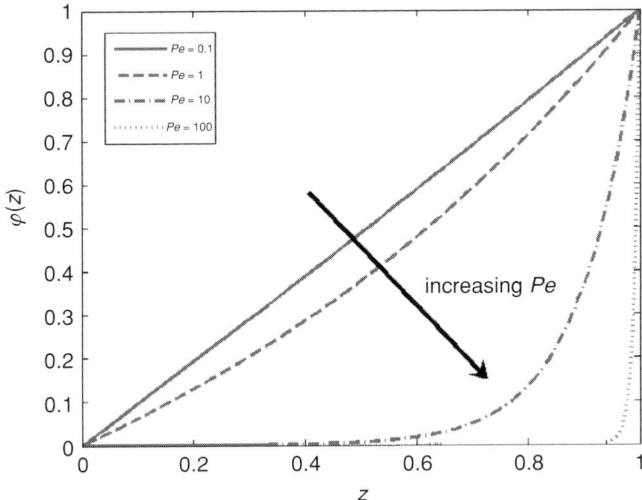

Figure 6.6 Analytical solution of a 1-D convection/diffusion equation at various Peclet numbers.

is dominant, and the flow (which is from left to right) "pushes" the incoming fluid at $z = 0$ downwind to the right and causes a sharp increase in φ near the exit outflow at $z = 1$.

Central difference scheme (CDS)

To solve this problem numerically, we apply the finite difference method for a grid of points, $0 < z_1 < z_2 < \cdots < z_N < L$, and require the differential equation to be satisfied locally at each point,

$$-v\left.\frac{d\varphi}{dz}\right|_{z_j} + \Gamma \left.\frac{d^2\varphi}{dz^2}\right|_{z_j} = 0 \tag{6.74}$$

For simplicity, we place the grid points uniformly with $\Delta z = L/(N+1)$, $z_j = j(\Delta z)$. For the second derivative, we have the approximation

$$\left.\frac{d^2\varphi}{dz^2}\right|_{z_j} \approx \frac{\varphi_{j-1} - 2\varphi_j + \varphi_{j+1}}{(\Delta z)^2} \tag{6.75}$$

For the first derivative, we consider two alternatives. In the *CDS*, we choose the more accurate approximation

$$\left.\frac{d\varphi}{dz}\right|_{z_j}^{(CDS)} = \frac{\varphi_{j+1} - \varphi_{j-1}}{2(\Delta z)} + O[(\Delta z)^2] \tag{6.76}$$

which yields the linear equation for grid point j,

$$-v\left[\frac{\varphi_{j+1} - \varphi_{j-1}}{2(\Delta z)}\right] + \Gamma \left[\frac{\varphi_{j-1} - 2\varphi_j + \varphi_{j+1}}{(\Delta z)^2}\right] = 0 \tag{6.77}$$

We multiply by $-2(\Delta z)^2/\Gamma$ to obtain

$$\frac{v(\Delta z)}{\Gamma}[\varphi_{j+1} - \varphi_{j-1}] - 2\varphi_{j-1} + 4\varphi_j - 2\varphi_{j+1} = 0 \tag{6.78}$$

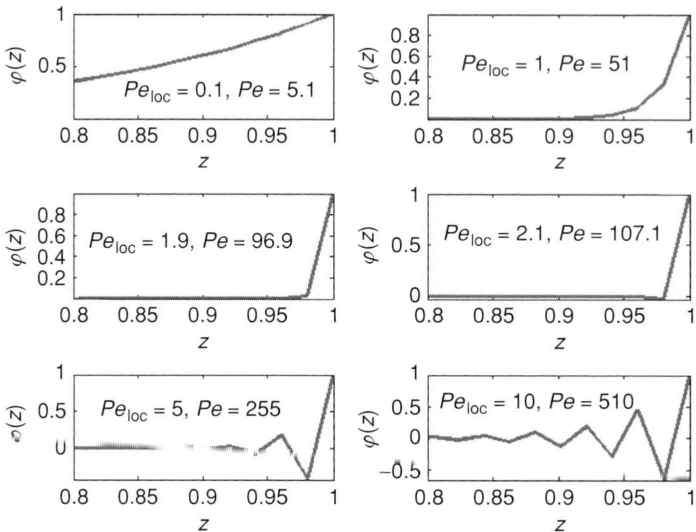

Figure 6.7 CDS solution of a 1-D convection/diffusion equation for various values of the local Peclet number below and above the critical value of 2 ($N = 50$).

We next define the *local Peclet number*

$$Pe_{\text{loc}} \equiv \frac{v(\Delta z)}{\Gamma} = (Pe)\left(\frac{(\Delta z)}{L}\right) = \frac{Pe}{N + 1} \tag{6.79}$$

to write the CDS equation for grid point j as

$$-(Pe_{\text{loc}} + 2)\varphi_{j-1} + 4\varphi_j + (Pe_{\text{loc}} - 2)\varphi_{j+1} = 0 \tag{6.80}$$

With $\alpha \equiv Pe_{\text{loc}} - 2$, $\beta \equiv Pe_{\text{loc}} + 2$, the CDS linear system is

$$\begin{bmatrix} 4 & \alpha & & & \\ -\beta & 4 & \alpha & & \\ & -\beta & 4 & \ddots & \\ & & \ddots & \ddots & \alpha \\ & & & -\beta & 4 \end{bmatrix} \begin{bmatrix} \varphi_1 \\ \varphi_2 \\ \vdots \\ \varphi_{N-1} \\ \varphi_N \end{bmatrix} = \begin{bmatrix} \beta\varphi_0 \\ 0 \\ \vdots \\ 0 \\ -\alpha\varphi_1 \end{bmatrix} \tag{6.81}$$

While for all $Pe_{\text{loc}} \geq 0$ we have $\beta > 0$, α changes sign at $Pe_{\text{loc}} = 2$. We expect some qualitative change in the solution when this occurs. The CDS solution is plotted in Figure 6.7 for various values of Pe_{loc} for a grid of 50 points. When $Pe_{\text{loc}} > 2$, the numerical solution exhibits oscillations that are not present in the true solution. Also, as this equation models the convection and diffusion of a density field, it is physically unrealistic for φ to take negative values, but it does so in the numerical solution when $Pe_{\text{loc}} > 2$. To remove such spurious oscillations, we can make the grid finer, thus decreasing Pe_{loc} for fixed Pe. However, often we do not know a priori how small to make the grid to avoid the oscillations, whose effect on the convergence of numerical algorithms may be severe.

Upwind difference scheme (UDS)

We now show that it is possible to remove the oscillations by using a different approximation for the first derivative in the convection term, an *upwind difference*, defined in this case for $v > 0$ (flow from left to right) as

$$\left. \frac{d\varphi}{dz} \right|_{z_j}^{(\text{UDS})} = \frac{\varphi_j - \varphi_{j-1}}{\Delta z} + O[(\Delta z)] \tag{6.82}$$

We take here a one-sided difference, and thus this formula is not as accurate, for finite Δz, as the central difference approximation. The *UDS* equation for grid point j is

$$- v \left[\frac{\varphi_j - \varphi_{j-1}}{\Delta z} \right] + \Gamma \left[\frac{\varphi_{j-1} - 2\varphi_j + \varphi_{j+1}}{(\Delta z)^2} \right] = 0 \tag{6.83}$$

Multiplying by $-(\Delta z)^2/L$ and defining Pe_{loc} as before yields

$$-(Pe_{\text{loc}} + 1)\varphi_{j-1} + (2 + Pe_{\text{loc}})\varphi_j - \varphi_{j+1} = 0 \tag{6.84}$$

Defining $\beta \equiv Pe_{\text{loc}} + 2$ and $\gamma \equiv Pe_{\text{loc}} + 1$, the UDS linear system is

$$\begin{bmatrix} \beta & -1 & & & & & \\ -\gamma & \beta & -1 & & & & \\ & -\gamma & \beta & -1 & & & \\ & & -\gamma & \ddots & \ddots & & \\ & & & \ddots & \beta & -1 \\ & & & & -\gamma & \beta \end{bmatrix} \begin{bmatrix} \varphi_1 \\ \varphi_2 \\ \varphi_3 \\ \vdots \\ \varphi_{N-1} \\ \varphi_N \end{bmatrix} = \begin{bmatrix} \gamma \varphi_0 \\ 0 \\ 0 \\ \vdots \\ 0 \\ \varphi_L \end{bmatrix} \tag{6.85}$$

Note that for all $Pe_{\text{loc}} \geq 0$, $\beta > 0$, $\gamma > 0$, and that the matrix elements have the same signs as they do in the CDS system when $Pe_{loc} < 2$. As Figure 6.8 shows, the UDS solution remains well-behaved for all values of Pe_{loc}.

Even though the UDS equations use a less accurate approximation for the convection derivative than the CDS equations, they are safer and better behaved. If too few grid points are used with the CDS method, large oscillations appear that can be disastrous with nonlinear source terms, but with the UDS method, the effect of too large a Δz is merely a loss in accuracy.

Why does upwind differencing work?

To see why upwind differencing works, we relate the upwind and central difference first-derivative approximations through the identity

$$\frac{\varphi_j - \varphi_{j-1}}{\Delta z} = \frac{\varphi_{j+1} - \varphi_{j-1}}{2(\Delta z)} + \left[\frac{\varphi_j - \varphi_{j-1}}{\Delta z} - \frac{\varphi_{j+1} - \varphi_{j-1}}{2(\Delta z)} \right]$$

$$\left. \frac{d\varphi}{dz} \right|_{z_j}^{(\text{UDS})} = \left. \frac{d\varphi}{dz} \right|_{z_j}^{(\text{CDS})} + \left[\frac{2\varphi_j - 2\varphi_{j-1}}{2(\Delta z)} - \frac{\varphi_{j+1} - \varphi_{j-1}}{2(\Delta z)} \right] \tag{6.86}$$

$$\left. \frac{d\varphi}{dz} \right|_{z_j}^{(\text{UDS})} = \left. \frac{d\varphi}{dz} \right|_{z_j}^{(\text{CDS})} + \left(\frac{\Delta z}{2} \right) \left[\frac{-\varphi_{j-1} + 2\varphi_j - \varphi_{j+1}}{(\Delta z)^2} \right]$$

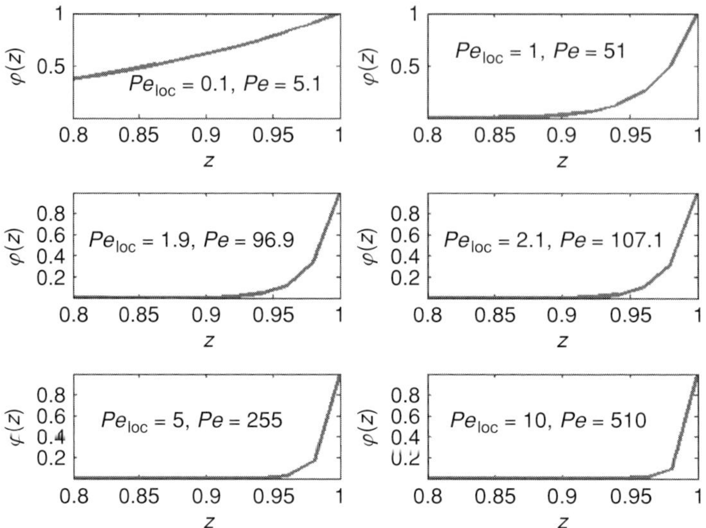

Figure 6.8 UDS solution of 1-D convection/diffusion equation at various Peclet numbers ($N = 50$).

Recognizing the factor within the square brackets as an approximation of $-d^2\varphi/d^2z|_{z_j}$, we have the relation

$$\left.\frac{d\varphi}{dz}\right|_{z_j}^{(UDS)} = \left.\frac{d\varphi}{dz}\right|_{z_j}^{(CDS)} - \left(\frac{\Delta z}{2}\right)\left.\frac{d^2\varphi}{dz^2}\right|_{z_j} \tag{6.87}$$

Thus, the UDS discretization of the equation

$$v\left.\frac{d\varphi}{dz}\right|_{z_j}^{(UDS)} - \Gamma\left.\frac{d^2\varphi}{dz^2}\right|_{z_j} = 0 \tag{6.88}$$

is equivalent to a CDS discretization with "extra" diffusion,

$$v\left\{\left.\frac{d\varphi}{dz}\right|_{z_j}^{(CDS)} - \left(\frac{\Delta z}{2}\right)\left.\frac{d^2\varphi}{dz^2}\right|_{z_j}\right\} - \Gamma\left.\frac{d^2\varphi}{dz^2}\right|_{z_j} = 0$$

$$v\left.\frac{d\varphi}{dz}\right|_{z_j}^{(CDS)} - \left[\Gamma + \frac{v(\Delta z)}{2}\right]\left.\frac{d^2\varphi}{dz^2}\right|_{z_j} = 0 \tag{6.89}$$

Upwind differencing is equivalent to adding additional *numerical diffusion*, or *artificial diffusion*, to obtain an effective diffusion constant

$$\Gamma_{eff} = \Gamma + \frac{v(\Delta z)}{2} \tag{6.90}$$

such that the effective local Peclet number is always less than 2:

$$Pe_{loc,eff} = \frac{v(\Delta z)}{\Gamma_{eff}} = \frac{v(\Delta z)}{\Gamma + v(\Delta z)/2} = \frac{2}{1 + 2/Pe_{loc}} \tag{6.91}$$

We could achieve the same effect if we used the CDS equations and increased the effective diffusion constant by a sufficient amount to avoid oscillations. In multiple spatial dimensions, it is common to add numerical diffusion only in the streamline direction, to obtain an effective anisotropic diffusion tensor,

$$\Gamma = \Gamma I + \Gamma_{\text{num}} \frac{\boldsymbol{v}\boldsymbol{v}^T}{|\boldsymbol{v}|^2} \tag{6.92}$$

For an in-depth treatment of methods for BVPs with strong convection, see Finlayson (1992).

Numerical solution of the HJB equation of optimal control

We now are in a position to analyze the numerical solution of the HJB equation from optimal control (Chapter 5) to minimize the cost functional

$$F\left[\boldsymbol{u}(t); \boldsymbol{x}^{[0]}\right] = \int_{t_0}^{t_{\text{H}}} \sigma(s, \boldsymbol{x}(s), \boldsymbol{u}(s))ds + \pi(\boldsymbol{x}(t_{\text{H}})) \tag{6.93}$$

for a system governed by $\dot{\boldsymbol{x}} = \boldsymbol{f}(t, \boldsymbol{x}, \boldsymbol{u})$. In terms of the backward time $\tau = t_{\text{H}} - t$, we solve the HJB equation

$$\frac{\partial \varphi}{\partial \tau} = \min_{\boldsymbol{u}(\tau, \boldsymbol{x})}[\sigma(t_{\text{H}} - \tau, \boldsymbol{x}, \boldsymbol{u}) + \nabla\varphi \cdot \boldsymbol{f}(t_{\text{H}} - \tau, \boldsymbol{x}, \boldsymbol{u})] \tag{6.94}$$

with the initial condition $\varphi(0, \boldsymbol{x}) = \pi(\boldsymbol{x})$. The optimal value of the cost functional $F[\boldsymbol{u}(t); \boldsymbol{x}^{[0]}]$ is $\varphi\left(\tau = t_{\text{H}} - t_0, \boldsymbol{x}^{[0]}\right)$ and the "best" feedback control law for the system is $\boldsymbol{u}_{\text{con}}(\boldsymbol{x}) = \boldsymbol{u}(\tau = t_{\text{H}} - t_0, \boldsymbol{x})$. In Chapter 5 we solved the HJB equation for a simple 1-D control problem, and are now in a position to explain the choice of discretization used there. We now recognize that (6.94) is a reaction/convection equation with σ taking the role of a source term and $\boldsymbol{f}$ taking the role of $-\boldsymbol{v}$. Thus, at each point in the calculation, we should examine the local sign of $\boldsymbol{f}$ and choose the appropriate upwind one-sided difference. This is in fact what was done in the example of Chapter 5, but without explanation. In practice, it is also common to add an artificial diffusion term $\varepsilon\nabla^2\varphi$ to the HJB equation to obtain a "viscosity solution." Diffusion makes dynamic programming more robust as it smooths out any discontinuities in the dynamics, cost functional, or control input.

When solving (6.94), the spatial domain should be extended to very large positive and negative values of x to limit any corruption of the solution in the region of interest by the boundary condition employed at the limits. To see how one might treat the boundaries, refer again to the example in Chapter 5.

Characteristics and types of PDEs

We see from our discussion of the 1-D convection/diffusion equation that the qualitative nature of an equation, and of its numerical solution, can change as we vary the parameters. These changes are related to the nature of how information about the field is propagated by the differential equation in space and time. The importance of information flow is reflected in a common naming convention for second-order PDEs. As this nomenclature is employed often in the literature, we briefly review it here.

Consider the general form of a second-order differential equation,

$$a\frac{\partial^2 \varphi}{\partial t^2} + b\frac{\partial^2 \varphi}{\partial t \partial z} + c\frac{\partial^2 \varphi}{\partial z^2} + f\left(t, z, \varphi, \frac{\partial \varphi}{\partial t}, \frac{\partial \varphi}{\partial z}\right) = 0 \tag{6.95}$$

For example, the time-varying 1-D convection/diffusion equation

$$\frac{\partial \varphi}{\partial t} - \Gamma\frac{\partial^2 \varphi}{\partial z^2} + \left[v\frac{\partial \varphi}{\partial z}\right] = 0 \tag{6.96}$$

in the limit of $Pe \ll 1$ takes the form above with

$$a = 1 \quad b = 0 \quad c = -\Gamma \tag{6.97}$$

In the limit $Pe \ll 1$, $b^2 - 4ac < 0$, and the PDE is said to be *parabolic*. By contrast, in the limit $Pe \gg 1$, we have a PDE dominated by convection,

$$\frac{\partial \varphi}{\partial t} + \left[v\frac{\partial \varphi}{\partial z}\right] - 0 \tag{6.98}$$

Differentiating once with respect to time yields

$$\frac{\partial^2 \varphi}{\partial t^2} + v\frac{\partial^2 \varphi}{\partial t \partial z} = 0 \quad \Rightarrow \quad \begin{aligned} a &= 1 \\ b &= v \\ c &= 0 \end{aligned} \tag{6.99}$$

Now, $b^2 - 4ac > 0$, and the equation is said to be *hyperbolic*. Thus, by changing Pe we alter the type of the PDE, and as we see below, this changes significantly the way the field is propagated.

The steady-state diffusion equation

$$-\Gamma\frac{\partial^2 \varphi}{\partial z^2} = 0 \tag{6.100}$$

has the general form of a second-order PDE with

$$a = 0 \quad\quad b = 0 \quad\quad c = -\Gamma \tag{6.101}$$

Here, $b^2 - 4ac = 0$ and the equation is said to be *elliptic*.

Consider some point P at position z_p and time t_p. What are the set of points in space-time whose field values influence the field value at (P) and the set of points in space-time whose field values are influenced in turn by the value at (P)? As a concrete example, consider the 1-D convection equation

$$\frac{\partial \varphi}{\partial t} + \left[v\frac{\partial \varphi}{\partial z}\right] = 0 \tag{6.102}$$

which describes purely convective transport of φ in the direction of increasing z for $v > 0$. At future times $t > t_p$, P only influences points (t, z) that are downstream with $z_p < z < z_p + v(t - t_p)$. Similarly, only past points (t, z) that are upstream with $z_p - v(t_p - t) < z < z_p$ influence P. The space-time diagram is shown in Figure 6.9. The two lines that separate the regions of influence from those of no influence are the *characteristic lines* for P.

We now relate the characteristic lines to the coefficients a, b, c in (6.95). Let us consider the point P and a point Q on one of its characteristic lines (Figure 6.9). If P and Q are

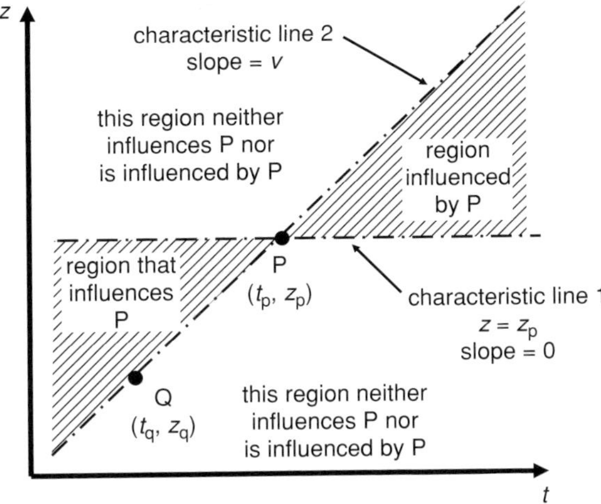

Figure 6.9 Space-time diagram of the 1-D convection equation showing characteristic lines.

infinitesimally close together,

$$z_q = z_p + dz \qquad t_q = t_p + dt \tag{6.103}$$

we use a truncated Taylor expansion to relate the field values at P and Q,

$$\varphi_q - \varphi_p = dt \left.\frac{\partial \varphi}{\partial t}\right|_{(P)} + dz \left.\frac{\partial \varphi}{\partial z}\right|_{(P)} \tag{6.104}$$

We apply similar expansions to the first derivatives,

$$\left.\frac{\partial \varphi}{\partial t}\right|_{(Q)} - \left.\frac{\partial \varphi}{\partial t}\right|_{(P)} = dt \left.\frac{\partial^2 \varphi}{\partial t^2}\right|_{(P)} + dz \left.\frac{\partial^2 \varphi}{\partial z \partial t}\right|_{(P)}$$

$$\left.\frac{\partial \varphi}{\partial z}\right|_{(Q)} - \left.\frac{\partial \varphi}{\partial z}\right|_{(P)} = dt \left.\frac{\partial^2 \varphi}{\partial t \partial z}\right|_{(P)} + dz \left.\frac{\partial^2 \varphi}{\partial z^2}\right|_{(P)} \tag{6.105}$$

We now combine these two expansions of the first derivatives with the condition that the differential equation be satisfied at P,

$$a \left.\frac{\partial^2 \varphi}{\partial t^2}\right|_{(P)} + b \left.\frac{\partial^2 \varphi}{\partial t \partial z}\right|_{(P)} + c \left.\frac{\partial^2 \varphi}{\partial z^2}\right|_{(P)} + f_{(P)} = 0 \tag{6.106}$$

to obtain the linear system

$$\begin{bmatrix} dt & dz & \\ & dt & dz \\ a & b & c \end{bmatrix} \begin{bmatrix} \left.\dfrac{\partial^2 \varphi}{\partial t^2}\right|_{(P)} \\[6pt] \left.\dfrac{\partial^2 \varphi}{\partial t \partial z}\right|_{(P)} \\[6pt] \left.\dfrac{\partial^2 \varphi}{\partial z^2}\right|_{(P)} \end{bmatrix} = \begin{bmatrix} \left.\dfrac{\partial \varphi}{\partial t}\right|_{(Q)} - \left.\dfrac{\partial \varphi}{\partial t}\right|_{(P)} \\[6pt] \left.\dfrac{\partial \varphi}{\partial z}\right|_{(Q)} - \left.\dfrac{\partial \varphi}{\partial z}\right|_{(P)} \\[6pt] -f_{(P)} \end{bmatrix} \tag{6.107}$$

Now, if Q were either within the interior of the region of influence or somewhere outside of it, then small changes in the first derivatives would correspond to small changes in the second derivatives. If Q is on a characteristic line, however, even an infinitesimal change in the first derivatives may have a finite effect upon the second derivatives, requiring the matrix of (6.107) to be singular,

$$\det \begin{bmatrix} dt & dz & \\ & dt & dz \\ a & b & c \end{bmatrix} = 0 = c(dt)^2 + b(dz)(dt) + a(dz)^2 \tag{6.108}$$

Dividing by $(dt)^2$, we obtain the slope of a characteristic line,

$$\frac{dz}{dt} = \frac{b \pm \sqrt{b^2 - 4ac}}{2a} \tag{6.109}$$

The sign of $b^2 - 4ac$ controls the number of real characteristic lines. For the 1-D time-varying convection equation, we find as expected,

$$\frac{dz}{dt} = \frac{v \pm \sqrt{v^2 - 4(1)(0)}}{2(1)} = \frac{v \pm v}{2} = v, 0 \tag{6.110}$$

Thus, we distinguish the three characteristic types of second-order PDEs.

Hyperbolic equations, $b^2 - 4ac > 0$

Two real characteristic lines exist; therefore, there exist distinct regions of space-time that are influenced, or not influenced, by each point P. Examples of hyperbolic equations are, of course, the 1-D convection equation, and also the wave equation

$$\frac{\partial^2 \varphi}{\partial t^2} - \frac{\partial^2 \varphi}{\partial z^2} = 0 \tag{6.111}$$

which has eigenfunctions of the form of traveling waves,

$$\left[\frac{\partial^2}{\partial t^2} - \frac{\partial^2}{\partial z^2} \right] e^{i(kz - \omega t)} = (k^2 - \omega^2) e^{i(kz - \omega t)} \tag{6.112}$$

In hyperbolic equations, information about the field is transmitted as traveling waves, and the effects of numerical error usually appear as oscillations.

Elliptic equations, $b^2 - 4ac < 0$

Here, there exist no real characteristic lines and no distinction can be made between regions of influence and no influence. In such problems, the value of the field at each point affects the value of the field at all other points. As an example, consider the 2-D steady-state diffusion equation

$$\frac{\partial^2 \varphi}{\partial x^2} + \frac{\partial^2 \varphi}{\partial y^2} = 0 \tag{6.113}$$

that, with $a = 0, b = 0, c = 1$ has $b^2 - 4ac = -4$.

In elliptic problems, information about the field is transmitted diffusively in all directions, and thus numerical error is "smoothed out."

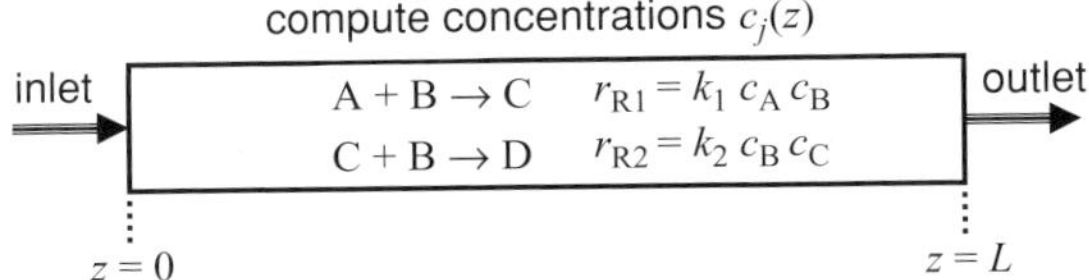

Figure 6.10 Tubular chemical reactor with dispersion.

Parabolic equations, $b^2 - 4ac = 0$

Here, there exists only one real characteristic line. Consider the time-varying 1-D diffusion equation

$$\frac{\partial \varphi}{\partial t} = \Gamma \frac{\partial^2 \varphi}{\partial z^2} \tag{6.114}$$

that, with $a = 0, b = 0, c = \Gamma$ has $b^2 - 4ac = 0$. The slope of the single characteristic line is indeterminate as $a = 0$, and thus corresponds to a vertical line running through P in the space-time diagram. Therefore, we find that each point P influences all points in space at all future times, and is influenced in turn by the field values at all spatial positions at all past times. Past numerical errors therefore are smoothed out with increasing time.

Modeling a tubular chemical reactor with dispersion; treating multiple fields

Many problems of interest involve multiple fields, each with its own governing equation. Consider a tubular chemical reactor of length L (Figure 6.10) with the reactions A + B → C and B + C → D. A fluid medium comprising initially only A and B flows through the reactor with a mean axial velocity v_z. In addition to convection, we also have a diffusive-like mixing, known as *dispersion*, due to the coupling of diffusion to inhomogeneities in the velocity field. We employ a common dispersion coefficient D as an effective axial diffusivity for each species. Dispersion is usually of minor importance compared with convection, with high values observed for the *Peclet number*

$$Pe = \frac{v_z L}{D} \tag{6.115}$$

The concentration fields of A, B, C, and D at steady state are governed by the coupled set of PDEs

$$\frac{\partial c_A}{\partial t} = -v_z \frac{\partial c_A}{\partial z} + D \frac{\partial^2 c_A}{\partial z^2} - k_1 c_A c_B = 0$$

$$\frac{\partial c_B}{\partial t} = -v_z \frac{\partial c_B}{\partial z} + D \frac{\partial^2 c_B}{\partial z^2} - k_1 c_A c_B - k_2 c_B c_C = 0$$

$$\frac{\partial c_C}{\partial t} = -v_z \frac{\partial c_C}{\partial z} + D \frac{\partial^2 c_C}{\partial z^2} + k_1 c_A c_B - k_2 c_B c_C = 0 \tag{6.116}$$

$$\frac{\partial c_D}{\partial t} = -v_z \frac{\partial c_D}{\partial z} + D \frac{\partial^2 c_D}{\partial z^2} + k_2 c_B c_C = 0$$

At the reactor inlet and outlet we use *Danckwerts' boundary conditions*. At the inlet, the flux of species $j = A, B, C, D$ entering the reactor is $v_z c_{j0}$, but once inside the reactor and in the presence of dispersion, the flux is $v_z c_j - D(dc_j/dz)$. Balancing these two fluxes at $z = 0$ yields the inlet boundary condition

$$v_z[c_j(0) - c_{j0}] - D\frac{dc_j}{dz}\bigg|_0 = 0 \tag{6.117}$$

When $D \to 0$, $c_j(0) = c_{j0}$, but when $D \to \infty$, this boundary condition correctly enforces $dc_j/dz|_0 = 0$. If the reaction stops once the stream leaves the reactor, the concentration profile becomes uniform, and we use the outlet boundary condition

$$\frac{dc_j}{dz}\bigg|_L = 0 \tag{6.118}$$

Solution by upwind finite differences

We solve the coupled set of PDEs (6.116) with the inlet boundary conditions (6.117) and outlet boundary conditions (6.118) using upwind finite differencing. We place a grid of N uniformly-spaced points $0 < z_1 < z_2 < \cdots < z_N < L$, $z_k = k(\Delta z)$, $\Delta z = L(N+1)^{-1}$ and write each PDE of (6.116) as

$$0 = -v_z\frac{dc_j}{dz} + D\frac{d^2c_j}{dz^2} + s_j[\{c_m(z)\}] \tag{6.119}$$

where $\{c_m(z)\}$ denotes the set of local concentrations, and the source terms for each field are

$$\begin{aligned}
s_A[\{c_m(z)\}] = -k_1 c_A c_B \quad & s_B[\{c_m(z)\}] = -k_1 c_A c_B - k_2 c_B c_C \\
s_C[\{c_m(z)\}] = k_1 c_A c_B - k_2 c_B c_C \quad & s_D[\{c_m(z)\}] = k_2 c_B c_C
\end{aligned} \tag{6.120}$$

Applying upwind finite differences, we obtain for each grid point z_k and each species $j = A, B, C, D$ a nonlinear algebraic equation

$$0 = -v_z\left[\frac{c_j(z_k) - c_j(z_{k-1})}{\Delta z}\right] + D\left[\frac{c_j(z_{k-1}) - 2c_j(z_k) + c_j(z_{k+1})}{(\Delta z)^2}\right] + s_j[\{c_m(z_k)\}] \tag{6.121}$$

Collecting terms, we have

$$0 = \alpha_{lo}c_j(z_{k-1}) + \alpha_{mid}c_j(z_k) + \alpha_{hi}c_j(z_{k+1}) + s_j[\{c_m(z_k)\}] \tag{6.122}$$

with the coefficients

$$\alpha_{lo} = \frac{v_z}{\Delta z} + \frac{D}{(\Delta z)^2} \quad \alpha_{mid} = -\frac{v_z}{\Delta z} - \frac{2D}{(\Delta z)^2} \quad \alpha_{hi} = \frac{D}{(\Delta z)^2} \tag{6.123}$$

We enforce the boundary conditions by removing the nonexistent unknowns $c_j(z_0)$ and $c_j(z_{N+1})$ through the discretizations of (6.117) and (6.118),

$$0 = v_z[c_j(z_0) - c_{j0}^{(in)}] - D\left[\frac{c_j(z_1) - c_j(z_0)}{\Delta z}\right] \qquad c_j(z_{N+1}) = c_j(z_N) \tag{6.124}$$

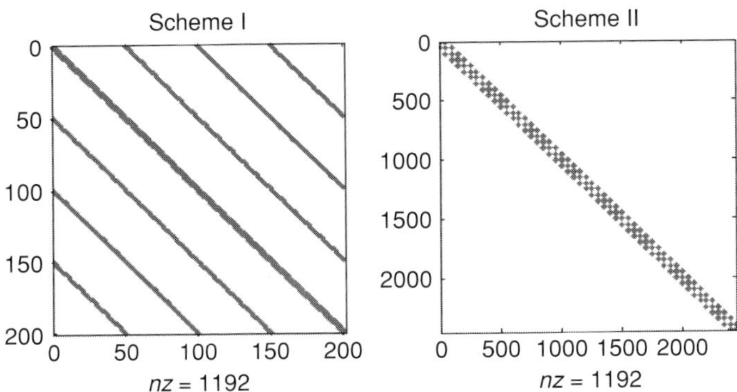

Figure 6.11 Effect of stacking on the sparsity pattern of the Jacobian for a BVP involving the 1-D transport of four fields, with local coupling of each field through the source terms. A grid of 50 points is used to discretize the PDEs. Plot generated by **BVP_field_JPattern.m**.

From the inlet boundary condition, we obtain

$$c_j(z_0) = \frac{Pe_{\mathrm{loc}}c_{j0} + c_j(z_1)}{1 + Pe_{\mathrm{loc}}} \qquad Pe_{\mathrm{loc}} = \frac{v_z z_1}{D} \tag{6.125}$$

Thus, for $k = 1$ we have the modified version of (6.122),

$$0 = \left[\alpha_{\mathrm{mid}} + \frac{\alpha_{\mathrm{lo}}}{1 + Pe_{\mathrm{loc}}}\right] c_j(z_1) + \frac{\alpha_{\mathrm{lo}} Pe_{\mathrm{loc}} c_{j0}}{1 + Pe_{\mathrm{loc}}} + \alpha_{\mathrm{hi}} c_j(z_2) + s_j[\{c_m(z_1)\}] \tag{6.126}$$

Similarly, for $k = N$ we have the modified version of (6.122),

$$0 = \alpha_{\mathrm{lo}} c_j(z_{N-1}) + [\alpha_{\mathrm{mid}} + \alpha_{\mathrm{hi}}] c_j(z_N) + s_j[\{c_m(z_N)\}] \tag{6.127}$$

We now stack the set of all unknowns into a single *state vector* using a labeling system that assigns to each unknown a unique integer. Equation (6.122) forms a set of nonlinear algebraic equations with a sparse Jacobian, and thus to avoid the fill-in problems discussed in Chapter 1, we make the bandwidth as small as possible (i.e., cluster the nonzero values around the principal diagonal). We see that (6.122) relates $c_j(z_k)$, the values of the same field c_j at the neighboring points, $c_j(z_{k-1})$ and $c_j(z_{k+1})$, and the values of all other fields at the same point, $c_{m \neq j}(z_k)$. Thus, from the two ways to stack the state vector,

$$\begin{aligned}
&\text{Scheme I } \mathbf{x}^{\mathrm{T}} = [c_A(z_1) c_A(z_2) \cdots c_A(z_N) c_B(z_1) \cdots c_B(z_N) \cdots c_D(z_N)] \\
&\text{Scheme II } \mathbf{x}^{\mathrm{T}} = [c_A(z_1) c_B(z_1) c_C(z_1) c_D(z_1) c_A(z_2) c_B(z_2) \cdots c_C(z_N) c_D(z_N)]
\end{aligned} \tag{6.128}$$

we choose scheme II, as it yields a Jacobian with a smaller bandwidth (Figure 6.11). We thus assign to $c_j(z_k)$ the label $n = N_{\mathrm{f}}(k-1) + j$, $N_{\mathrm{f}} = 4$ being the number of fields.

When coding a BVP with multiple fields, it is easiest to write the routines to use, as much as is possible, natural names and indices, e.g. $c_j(z_k)$, and to rely upon routines to stack and unstack the state vector and to place the matrix and vector elements in the appropriate positions. This approach is used in **tubular_reactor_2rxn_SS.m**. Sample results are shown

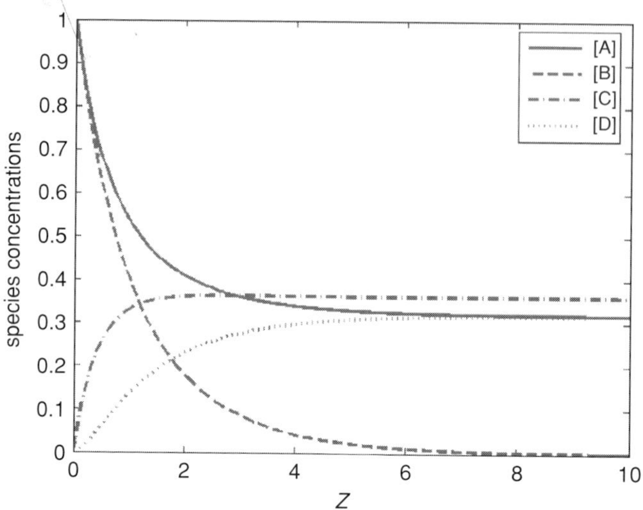

Figure 6.12 Steady-state concentration profiles in a tubular reactor (with reactions – $(A + B \rightarrow C, B + C \rightarrow D)$).

in Figure 6.12 for a simulation with the parameters

$$L = 10 \quad v_z = 1 \quad D = 10^{-4} \quad k_1 = k_2 = 1$$
$$c_{A0} = c_{B0} = 1 \quad c_{C0} = c_{D0} = 0 \tag{6.129}$$

Time-dependent simulation

We now simulate the dynamics of this tubular reactor by retaining the time derivative in each PDE of (6.116). Then, instead of a nonlinear algebraic equation (6.122), we obtain for each $c_j(z_k)$ an ODE

$$\frac{dc_j(z_k)}{dt} = \alpha_{\text{lo}} c_j(z_{k-1}) + \alpha_{\text{mid}} c_j(z_k) + \alpha_{\text{hi}} c_j(z_{k+1}) + s_j[\{c_m(z_k)\}] \tag{6.130}$$

As discussed in Chapter 4, discretized PDEs yield ODE systems that are very stiff; therefore, to avoid a very small time step, an implicit method such as the Crank–Nicholson method, **ode23s**, or **ode15s** should be used. Using **ode15s**, tubular_reactor_2rxn_dyn_sim.m simulates the reactor start-up dynamics. Initially the reactor is at steady state with an input stream containing only A, and then B is introduced to start the reaction (Figure 6.13).

Numerical issues for discretized PDEs with more than two spatial dimensions

Consider a BVP involving the Poisson equation in three dimensions

$$-\nabla^2 \varphi = -\frac{\partial^2 \varphi}{\partial x^2} - \frac{\partial^2 \varphi}{\partial y^2} - \frac{\partial^2 \varphi}{\partial z^2} = f(x, y, z) \tag{6.131}$$

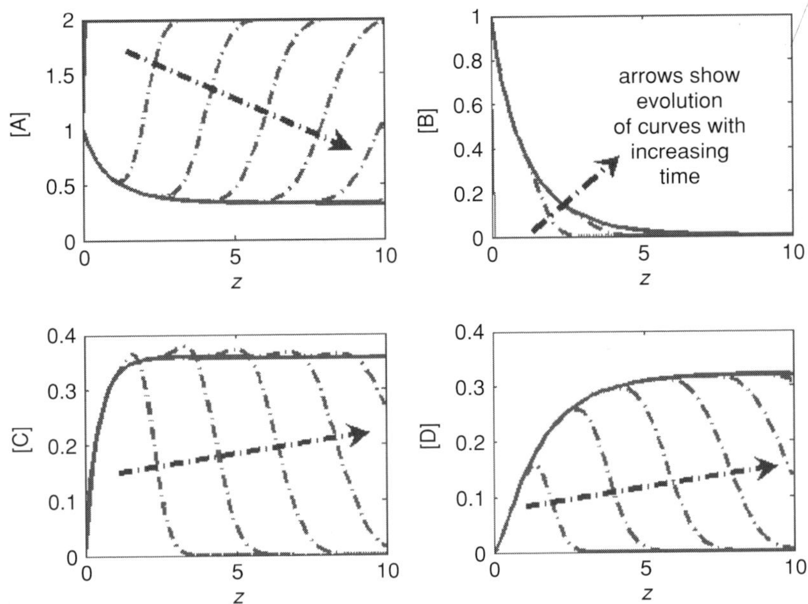

Figure 6.13 Dynamic concentration profiles during reactor start-up. Solid lines show the initial profile and the final profile at a time twice that of the mean residence time. Dashed lines show profiles at every 0.2 of the mean residence time.

on $0 \leq x \leq L, 0 \leq y \leq W, 0 \leq z \leq H$ subject to the Dirichlet boundary conditions

$$
\begin{array}{llll}
\text{BC 1} & \varphi(0, y, z) = 0 & 0 \leq y \leq W & 0 \leq z \leq H \\
\text{BC 2} & \varphi(L, y, z) = 0 & 0 \leq y \leq W & 0 \leq z \leq H \\
\text{BC 3} & \varphi(x, 0, z) = 0 & 0 \leq x \leq L & 0 \leq z \leq H \\
\text{BC 4} & \varphi(x, W, z) = 0 & 0 \leq x \leq L & 0 \leq z \leq H \\
\text{BC 5} & \varphi(x, y, 0) = 0 & 0 \leq x \leq L & 0 \leq y \leq W \\
\text{BC 6} & \varphi(x, y, H) = 0 & 0 \leq x \leq L & 0 \leq y \leq W
\end{array}
\tag{6.132}
$$

We set a uniform grid of points over the domain and substitute finite difference approximations for each second derivative to obtain for each interior point (x_i, y_j, z_k) a linear equation

$$
\begin{aligned}
&\frac{-\varphi(x_{i-1}, y_j, z_k) + 2\varphi(x_i, y_j, z_k) - \varphi(x_{i+1}, y_j, z_k)}{(\Delta x)^2} \\
&+ \frac{-\varphi(x_i, y_{j-1}, z_k) + 2\varphi(x_i, y_j, z_k) - \varphi(x_i, y_{j+1}, z_k)}{(\Delta y)^2} \\
&+ \frac{-\varphi(x_i, y_j, z_{k-1}) + 2\varphi(x_i, y_j, z_k) - \varphi(x_i, y_j, z_{k+1})}{(\Delta z)^2} = f(x_i, y_j, z_k)
\end{aligned}
\tag{6.133}
$$

N_x, N_y, N_z are the numbers of grid points in the x, y, and z directions respectively. We stack all unknowns in a single vector of dimension $N_x N_y N_z$ by assigning to each grid point (x_i, y_j, z_k) the unique label

$$
n = (k - 1)N_x N_y + (i - 1)N_y + j
\tag{6.134}
$$

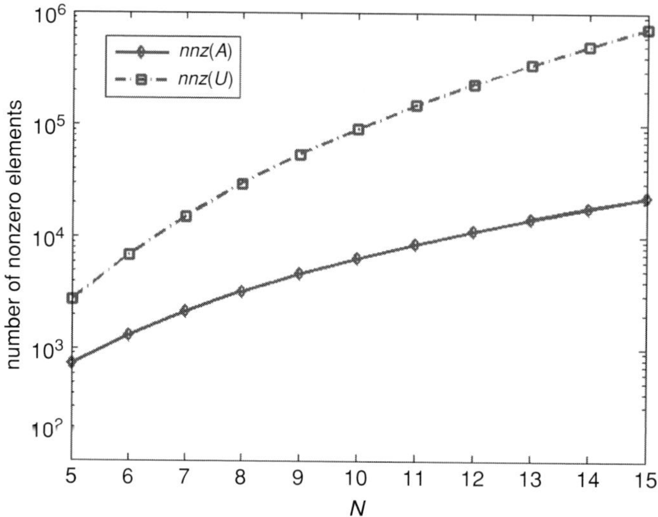

Figure 6.14 Numbers of nonzero elements in the 3-D diffusion matrix A before elimination and in the upper triangular matrix U obtained by LU decomposition as functions of N.

such that

$$\varphi(x_i, y_j, z_k) = \varphi_n \qquad \varphi(x_i, y_{j\pm1}, z_k) = \varphi_{n\pm1}$$
$$\varphi(x_{i\pm1}, y_j, z_k) = \varphi_{n\pm N_y} \qquad \varphi(x_i, y_j, z_{k\pm1}) = \varphi_{n\pm N_x N_y} \tag{6.135}$$

We then have the linear equation for each interior grid point

$$A_{n,n-N_x N_y}\varphi_{n-N_x N_y} + A_{n,n-N_y}\varphi_{n-N_y} + A_{n,n-1}\varphi_{n-1} + A_{nn}\varphi_n$$
$$+ A_{n,n+1}\varphi_{n+1} + A_{n,n+N_y}\varphi_{n+N_y} + A_{n,n+N_x N_y}\varphi_{n+N_x N_y} = f_n \tag{6.136}$$

with nonzero elements

$$A_{n,n-N_x N_y} = A_{n,n+N_x N_y} = -(\Delta z)^{-2}$$
$$A_{n,n-N_y} = A_{n,n+N_y} = -(\Delta x)^{-2}$$
$$A_{n,n-1} = A_{n,n+1} = -(\Delta y)^{-2}$$
$$A_{nn} = 2[(\Delta x)^{-2} + (\Delta y)^{-2} + (\Delta z)^{-2}] \tag{6.137}$$

This yields a positive-definite system $A\mathbf{x} = \mathbf{f}$, with a matrix of bandwidth $N_x N_y$, containing seven nonzero elements per row ($7N_x N_y N_z$ in total).

In Figure 6.14, we plot as a function $N_x = N_y = N_z = N$ the numbers of nonzero elements in the original matrix A and the upper triangular matrix U obtained by Gaussian elimination. Even at very small N, the number of nonzero elements increases significantly during Gaussian elimination, due to the problem of fill-in discussed in Chapter 1 (Figure 6.15). For all but very small 3-D grids, it is impossible to perform Gaussian elimination because we would run out of available memory. Even without fill-in, the number of nonzero elements in A, $7N^3$, makes it difficult to store A, even in sparse-matrix format, when N is large. Thus, although solving a BVP in three dimensions is not any different conceptually than solving are in one or two dimensions, we must solve the resulting linear systems with alternative methods.

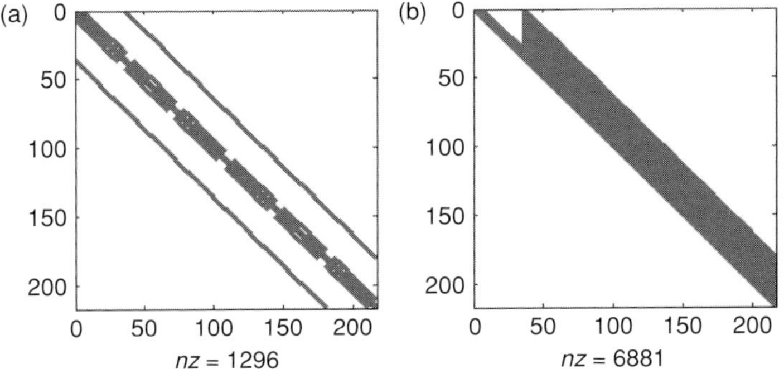

Figure 6.15 Sparsity patterns of a 3-D diffusion operator matrix (a) before and (b) after Gaussian elimination for $N = 6$.

We consider here iterative methods to solve $Ax = b$ that begin at an initial guess $x^{[0]}$ and generate a sequence $x^{[1]}$, $x^{[2]}$, . . . that (hopefully) converges to a solution. We have encountered a few such methods earlier in this text.

The Jacobi, Gauss–Seidel, and successive over-relaxation (SOR) methods

In Chapter 3, we examined the Jacobi method for diagonally-dominant A,

$$Bx^{[k+1]} = b + (B - A)x^{[k]} \tag{6.138}$$

where B contains only the diagonal values of A. Thus, solving (6.138) requires no elimination. Some improvement in the convergence rate is obtained in the *Gauss–Seidel method*, where again we apply (6.138) but now with B being either the upper-triangular, B = triu(A), or lower-triangular, B = tril(A), part of A. The *SOR* method is a more efficient modification of the Gauss–Seidel method, in which we partition A as

$$
A =
\underbrace{\begin{bmatrix}
D_{11} & & & & \\
& D_{22} & & & \\
& & D_{33} & & \\
& & & \ddots & \\
& & & & D_{NN}
\end{bmatrix}}_{D_A}
+
\underbrace{\begin{bmatrix}
0 & & & & \\
L_{21} & 0 & & & \\
L_{31} & L_{32} & 0 & & \\
\vdots & \vdots & \ddots & \ddots & \\
L_{N1} & L_{N2} & L_{N3} & \cdots & 0
\end{bmatrix}}_{L_A}
$$

$$
+
\underbrace{\begin{bmatrix}
0 & U_{12} & U_{13} & \cdots & U_{1N} \\
& 0 & U_{23} & \cdots & U_{2N} \\
& & 0 & \cdots & U_{3N} \\
& & & \ddots & \vdots \\
& & & & 0
\end{bmatrix}}_{U_A}
\tag{6.139}
$$

In the *lower-SOR method*, we extract the lower-triangular part of A and divide the diagonal values by an *over-relaxation parameter* $\omega > 1$,

$$B = \frac{1}{\omega}D_A + L_A \quad 1 < \omega < 2 \tag{6.140}$$

to obtain an update system

$$\left(\omega^{-1}D_A + L_A\right)x^{[k+1]} = b + \left[(\omega^{-1} - 1)D_A - U_A\right]x^{[k]} \tag{6.141}$$

that is solved by forward substitution. In the *upper-SOR method*, we select

$$B = \frac{1}{\omega}D_A + U_A \quad 1 < \omega < 2 \tag{6.142}$$

to obtain the update system

$$\left(\omega^{-1}D_A + U_A\right)x^{[k+1]} = b + \left[(\omega^{-1} - 1)D_A - L_A\right]x^{[k]} \tag{6.143}$$

which is solved by backward substitution. For a positive-definite matrix, SOR converges for all $1 < \omega < 2$. Some tuning of ω is required to optimize the rate of convergence (1.9 being a good initial guess); however, the convergence rate can be made somewhat less sensitive to the choice of ω if we alternate lower and upper-SOR steps, yielding the *symmetric SOR (SSOR)* method. We do not consider these methods in further detail, because the methods described below are preferred and are implemented directly in MATLAB. For further discussion, consult Stoer & Bulirsch (1993) and Quateroni *et al.* (2000).

The conjugate gradient method for positive-definite matrices

In Chapter 5, we considered the conjugate gradient method which solves $Ax = b$ by minimizing the quadratic cost function

$$F(x) = \frac{1}{2}x^{\mathrm{T}}Ax - b^{\mathrm{T}}x \tag{6.144}$$

in no more than N steps. At each iteration, we need only compute a single matrix-vector product; therefore, no elimination occurs and we take full advantage of sparsity. The method is implemented in MATLAB as **pcg**. It can be used for A stored in either full- or sparse-matrix format,

```
A = [2 -1 0 0; -1 2 -1 0; 0 -1 2 -1; 0 0 -1 2];
b = [1;1;1;1];
x = pcg(A,b),
pcg converged at iteration 2 to a solution with relative residual 0
x =
    2
    3
    3
    2
```

It can also be used without even having to store A in memory if we supply a routine that returns the vector $A\boldsymbol{v}$ for an input vector $\boldsymbol{v}$. Such use of **pcg** is demonstrated in BVP_2D_Poisson_FD_cg.m.

The generalized minimum residual (GMRES) Krylov subspace method

The efficiency of the conjugate gradient method leads us to consider the existence of methods that solve linear systems in a similar manner but without requiring positive-definiteness. Such methods exist and are invoked in MATLAB by the keywords **bicg**, **bicgstab**, and **gmres**. We present here a brief discussion of **gmres**.

We wish to solve $A\boldsymbol{x} = \boldsymbol{b}$, but, as A is no longer required to be positive-definite, we introduce the term *residual* for the vector that we have previously called the steepest descent direction, $\boldsymbol{r} = A\boldsymbol{x} - \boldsymbol{b}$. Given an initial guess $\boldsymbol{x}^{[0]}$, the initial residual is

$$\boldsymbol{r}^{[0]} = \boldsymbol{b} - A\boldsymbol{x}^{[0]} \tag{6.145}$$

We wish to generate a sequence $\boldsymbol{x}^{[1]}, \boldsymbol{x}^{[2]}, \ldots$ such that the sequence of residual norms $\|\boldsymbol{r}^{[1]}\|, \|\boldsymbol{r}^{[2]}\|, \ldots$ converges to zero. The GMRES method looks for new solution estimates that minimize the residual among the members of a *Krylov subspace* of $\boldsymbol{r}^{[0]}$, that at order $k+1$ is

$$K_{k+1}\!\left(\boldsymbol{r}^{[0]}; A\right) \equiv \text{span}\left\{\boldsymbol{r}^{[0]}, A\boldsymbol{r}^{[0]}, A^2\boldsymbol{r}^{[0]}, \ldots, A^k\boldsymbol{r}^{[0]}\right\} \tag{6.146}$$

This subspace is sufficiently large to include all searches based on taking some step $a^{[k]}$ in the residual direction at each iteration,

$$\boldsymbol{x}^{[k+1]} = \boldsymbol{x}^{[k]} + a^{[k]}\boldsymbol{r}^{[k]} \tag{6.147}$$

This follows from the relation among residuals at successive iterations,

$$\boldsymbol{b} - A\boldsymbol{x}^{[k+1]} = \boldsymbol{b} - A\boldsymbol{x}^{[k]} - a^{[k]}A\boldsymbol{r}^{[k]}$$

$$\boldsymbol{r}^{[k+1]} = \left(I - a^{[k]}A\right)\boldsymbol{r}^{[k]} \tag{6.148}$$

from which we obtain

$$\boldsymbol{r}^{[k]} = \sum_{j=0}^{k-1}\left(I - a^{[j]}A\right)\boldsymbol{r}^{[j]} \tag{6.149}$$

We thus find that $\boldsymbol{r}^{[k]} \in K_{k+1}(\boldsymbol{r}^{[0]}; A)$. It also follows that

$$\boldsymbol{x}^{[k]} = \boldsymbol{x}^{[0]} + \boldsymbol{p}^{[k]} \qquad \boldsymbol{p}^{[k]} \in K_k\!\left(\boldsymbol{r}^{[0]}; A\right) \tag{6.150}$$

In GMRES, we compute successive estimates

$$\begin{aligned}
\boldsymbol{x}^{[1]} &= \boldsymbol{x}^{[0]} + \boldsymbol{p}^{[1]} & \boldsymbol{p}^{[1]} &= \alpha_{10}\boldsymbol{r}^{[0]} \\
\boldsymbol{x}^{[2]} &= \boldsymbol{x}^{[0]} + \boldsymbol{p}^{[2]} & \boldsymbol{p}^{[2]} &= \alpha_{20}\boldsymbol{r}^{[0]} + \alpha_{21}A\boldsymbol{r}^{[0]} \\
\boldsymbol{x}^{[3]} &= \boldsymbol{x}^{[0]} + \boldsymbol{p}^{[3]} & \boldsymbol{p}^{[3]} &= \alpha_{30}\boldsymbol{r}^{[0]} + \alpha_{31}A\boldsymbol{r}^{[0]} + \alpha_{32}A^2\boldsymbol{r}^{[0]}
\end{aligned} \tag{6.151}$$

by finding the $\boldsymbol{p}^{[k]} \in K_k(\boldsymbol{r}^{[0]}; A)$ that minimizes the 2-norm of the residual,

$$\left\|\boldsymbol{r}^{[k]}\right\|_2^2 = \left\|\boldsymbol{b} - A(\boldsymbol{x}^{[0]} + \boldsymbol{p}^{[k]})\right\|_2^2 \leq \left\|\boldsymbol{b} - A(\boldsymbol{x}^{[0]} + \boldsymbol{p})\right\|_2^2 \quad \forall \boldsymbol{p} \in K_k\!\left(\boldsymbol{r}^{[0]}; A\right) \tag{6.152}$$

This is done by storing at iteration k an $N \times k$ matrix $V^{[k]}$ whose column vectors are orthonormal basis vectors of $K_k(r^{[0]}; A)$. We then can write any $p \in K_k(r^{[0]}; A)$ as the linear combination

$$p = V^{[k]}c \quad c \in \Re^k \tag{6.153}$$

We find $p^{[k]} = V^{[k]}c^{[k]}$ by minimizing the quadratic cost function

$$F(c) = \left\| b - A\left(x^{[0]} + V^{[k]}c\right) \right\|_2^2 = \left\| r^{[0]} - AV^{[k]}c \right\|_2^2 \tag{6.154}$$

It may be shown that the GMRES method converges in at most N iterations; however, the memory required to store $V^{[k]}$ for large k is considerable, and the GMRES method becomes unwieldy after a few iterations. In practice, we restart the method every $m < N$ steps, so that $V^{[k]}$ never becomes larger than an $N \times m$ matrix.

To demonstrate **gmres**, we start with the matrix A used to demonstrate **pcg** above, and add 1 to the $(1, 4)$ element, destroying its symmetry. While this is a problem for **pcg**, it is not for **gmres**.

A2 = A; A2(1,4) - A(1,4) + 1;
x = pcg(A2,b), % this should not work
pcg stopped at iteration 4 without converging to the desired tolerance
1e-006 because the maximum number of iterations was reached.
The iterate returned (number 4) has relative residual 0.16
x =
* 0.6356*
* 2.1460*
* 2.6616*
* 1.8970*
x = gmres(A2,b), % but this should work
gmres converged at iteration 4 to a solution with relative residual 1.2e-015
* x =*
* 0.6667*
* 2.0000*
* 2.3333*
* 1.6667*

To restart **gmres** every restart iterations, use **x = gmres(A,b,restart)**.

The use of preconditioners

The conjugate gradient and GMRES methods converge in at most N iterations, but is it possible for them to converge in significantly less than N iterations? Consider for the linear system $Ax = b$ the first conjugate gradient or GMRES update

$$x^{[1]} = x^{[0]} + a^{[0]}r^{[0]} \quad r^{[0]} = b - Ax^{[0]} \tag{6.155}$$

Let W be a matrix whose columns are eigenvectors of A and Λ be the diagonal matrix containing the eigenvalues of A, $AW = \Lambda W$. Let us assume that A is diagonalizable, such that

$$A = W \Lambda W^{-1} \quad b = Ax = W \Lambda W^{-1} x \tag{6.156}$$

Then, the first update is

$$x^{[1]} = x^{[0]} + a^{[0]}\left[-Ax^{[0]} + b \right] = \left(I - a^{[0]} A \right) x^{[0]} + a^{[0]} b$$
$$x^{[1]} = W\left(I - a^{[0]} \Lambda \right) W^{-1} x^{[0]} + a^{[0]} W \Lambda W^{-1} x \tag{6.157}$$

Consider what happens when all eigenvalues of A are equal; i.e., A has a condition number of 1 and Λ is isotropic, $\Lambda = \lambda I$. Then,

$$x^{[1]} = W\left(I - a^{[0]} \lambda I \right) W^{-1} x^{[0]} + a^{[0]} W(\lambda I) W^{-1} x \tag{6.158}$$

and as $WW^{-1} = I$, we have

$$x^{[1]} = \left(1 - a^{[0]} \lambda \right) x^{[0]} + a^{[0]} \lambda x \tag{6.159}$$

Therefore, when $\Lambda = \lambda I$, the conjugate gradient and GMRES methods converge in a single iteration to the solution x with a step-size $a^{[0]} = \lambda^{-1}$, *from any initial guess.*

This result is the motivation behind the use of a *preconditioner*, a transformation of the linear system into a related one whose condition number is closer to 1. This reduces the number of iterations necessary for iterative methods to converge to a solution, and is common practice in the numerical solution of BVPs. Let us choose some nonsingular upper-triangular matrix M_2 defining a coordinate transformation,

$$\hat{x} = M_2 x \quad x = M_2^{-1} \hat{x} \tag{6.160}$$

We then write $Ax = b$ as

$$A M_2^{-1} \hat{x} = b \tag{6.161}$$

We next define a lower-triangular matrix M_1 and a perturbation matrix P such that $PA \approx M$, $M = M_1 M_2$. That is, if the exact LU factorization is $PA = LU$, $M_1 \approx L$, $M_2 \approx U$. Multiplying (6.161) by $M_1^{-1} P$, we have

$$\left[M_1^{-1} P A M_2^{-1} \right] \hat{x} = M_1^{-1} P b \tag{6.162}$$

Substituting for PA, this becomes

$$\left[M_1^{-1} P A M_2^{-1} \right] \hat{x} = \left[M_1^{-1} L U M_2^{-1} \right] \hat{x} = \left[\left(M_1^{-1} L \right)\left(U M_2^{-1} \right) \right] \hat{x} = c \tag{6.163}$$

where we obtain c by solving

$$c = M_1^{-1} P b \quad \Leftrightarrow \quad M_1 c = P b \tag{6.164}$$

by forward substitution.

Let the current estimate of the solution to (6.163) be $\hat{x}^{[k]}$, and let the conjugate gradient or GMRES update be

$$\hat{x}^{[k+1]} = \hat{x}^{[k]} + \hat{p}^{[k]} \tag{6.165}$$

Now, if we indeed use the exact LU factors $M_1 = L$ and $M_2 = U$, (6.163) becomes $\hat{x} = c$ and (6.165) converges in a single iteration. Of course, we cannot use the exact factors because, being generated by elimination, fill-in causes them to have many more nonzero elements than the original matrix A. However, it still may be possible to find approximate factors $PA \approx M_1 M_2$ such that $M_1^{-1} P A M_2^{-1}$ has a condition number closer to 1 than the original system PA. Then, the conjugate gradient or GMRES search directions point more closely towards the solution and require fewer iterations to reduce the residual norm to near zero. $PA \approx M$, $M = M_1 M_2$ is then said to serve as a *preconditioner* for A. The update to the original system at each iteration is obtained by backward substitution of

$$M_2 x^{[k+1]} = \hat{x}^{[k]} + \hat{p}^{[k]} \tag{6.166}$$

For a positive-definite matrix A, the existence of the Cholesky factorization $A = R^T R$, allows us to use a preconditioner $A \approx M_2^T M_2$, such that the transformed system

$$\left[M_2^{T(-1)} A M_2^{-1} \right] \hat{x} = c \quad M_2^T c = b \tag{6.167}$$

is also positive-definite, and thus can be solved by conjugate gradients.

How should we choose the preconditioner?

There are many possible approaches, and sometimes one uses a preconditioner specifically tailored to the PDE being solved. Here, we consider only general approaches. The simplest preconditioner is the *Jacobi preconditioner*, for which M is merely the diagonal part of A, **M = diag(diag(A))**. More effective preconditioners are obtained from incomplete factorizations of A, using an *incomplete Cholesky factorization* if A is positive-definite (and we are using **pcg**) or an *incomplete LU factorization* if A is not (and we are using **gmres** or **bicgstab**).

What do we mean by an incomplete factorization?

Consider the Cholesky factorization of a positive-definite matrix, $A = R^T R$, where R is upper-triangular. While this decomposition is exact, it is inefficient to use because it fills in zero elements of A with nonzero values. However, we can generate an incomplete Cholesky factorization $A \approx M$, $M = M_2^T M_2$ if we execute the Cholesky algorithm, but throw out some fraction of the new nonzero values introduced by fill-in to control the number of nonzero values to a manageable amount. Alternatively, in a *modified incomplete Cholesky factorization*, we subsume the discarded nonzero values into the diagonal elements. A similar procedure for Gaussian elimination yields an incomplete LU factorization, $A \approx M$, $M = LU$, that can be applied if A is not positive-definite. Using either of these factorizations, $A \approx M$, $M = M_1 M_2$ we supply the preconditioner with the syntax

```
x = pcg(A,b,tol,maxit,M); x = pcg(A,b,tol,maxit,M1,M2);
x = gmres (A,b,restart,tol,maxit,M1,M2);
```

tol sets the desired level of accuracy, maxit is the allowable number of iterations, and restart asks GMRES to rebuild the Krylov subspace every restart iterations. The preconditioner is supplied as M or as M1, M2. To specify the initial guess of the solution,

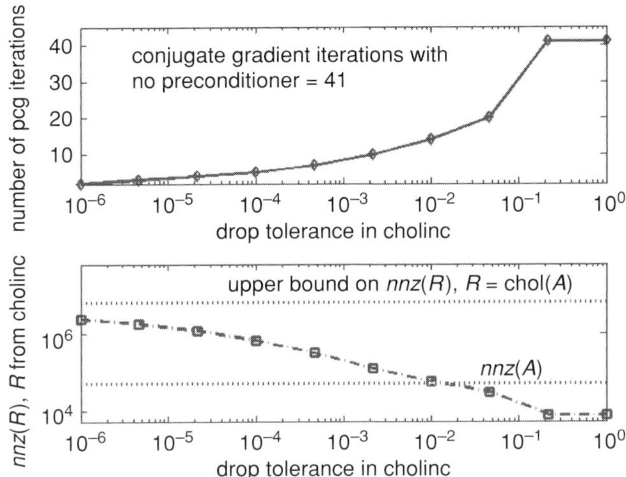

Figure 6.16 Effect of an incomplete Cholesky preconditioner on the convergence of the conjugate gradient method ($N = 20$).

use an additional argument x0 after these. Incomplete factorizations with no fill-in are done by

R = cholinc(A,'0'); [L,U,P] = luinc(A,'0');

The "0" argument demands that no fill-in be allowed. In practice, better performance is obtained if we allow at least some fill-in, as controlled by a drop tolerance **droptol**, passed with the syntax

R = cholinc(A,droptol);

cholinc_pcg_test.m examines the performance of **pcg** with an incomplete Cholesky preconditioner to solve $-\nabla^2\varphi = 1$ in three dimensions for a grid of $N \times N \times N$ points (Figure 6.16). As **droptol** is increased, more nonzero values are discarded. While the memory usage decreases, the incomplete Cholesky factorization becomes less effective at preconditioning the system and more conjugate gradient iterations are necessary. Here, **droptol** values of $10^{-3} - 10^{-2}$ result in moderate fill-in, but are effective at reducing the number of conjugate gradient iterations necessary to find the solution. Sparsity patterns of A and of the incomplete Cholesky factors for various values of **droptol** are shown in Figure 6.17. Such tuning of the preconditioner (here by varying **droptol**) to improve its performance is a standard part of the practice of numerically solving BVPs. cholinc_pcg_test.m further demonstrates how an options structure can be substituted for **droptol** to control further details of the algorithm.

It should be noted that the factors M_1 and M_2 may be singular if too much discarding is done (from too high a drop tolerance). This can be avoided by adding **droptol** to any zero diagonal elements of the incomplete U matrix to avoid zero values there, by calling **[L,U,P] = luinc(A, OPTS);** with **OPTS.udiag = 1**. **OPTS.droptol** stores the drop tolerance. Type **doc** luinc and doc cholinc for further details.

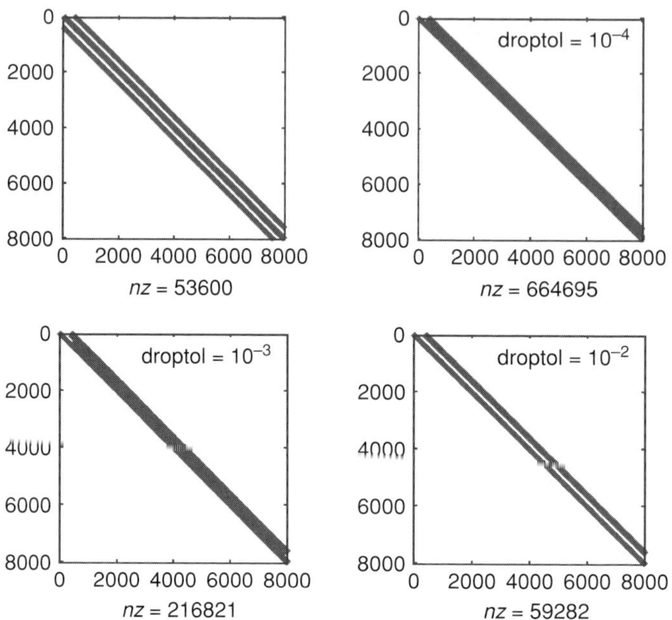

Figure 6.17 Effect of drop tolerance on the sparsity of incomplete Cholesky factors.

Example. 3-D heat transfer in a stove top element

Consider the following "kitchen" transport problem. We have an electric stove top comprising several heating elements (Figure 6.18). Each element contains within a ceramic matrix two annular regions in which heat is generated by electrical resistance at a volumetric rate S. The thermal conductivities of the ceramic matrix and heat generation material have a constant value λ. The dimensions of an individual element are defined in Figure 6.18.

We wish to calculate the temperature profile within a single element, assuming that it is part of a periodic array (the same boundary conditions apply if it is insulated on all sides). We assume that a metal pot containing boiling water has been placed on top of the heating element. If the thermal conductivity of the metal pot is much higher than the thermal conductivity of the ceramic matrix, we expect that at steady state, the temperature of the upper surface will be equal uniformly to that of boiling water, T_b. On the bottom surface, we assume that there is an underlying insulator layer, so that the heat flux out of the bottom is zero, implying a zero gradient there.

We define a dimensionless temperature and dimensionless coordinates,

$$\theta = \frac{T - T_b}{T_b} \qquad \chi = \frac{x}{L} - \frac{1}{2} \qquad \eta = \frac{y}{L} - \frac{1}{2} \qquad \zeta = \frac{z}{L} \tag{6.168}$$

and convert the governing heat conduction equation to dimensionless form

$$-\frac{\partial^2 \theta}{\partial \chi^2} - \frac{\partial^2 \theta}{\partial \eta^2} - \frac{\partial^2 \theta}{\partial \zeta^2} = \sigma H(\chi, \eta, \zeta) \tag{6.169}$$

The following function "switches on" the heat generation only within the specified annular

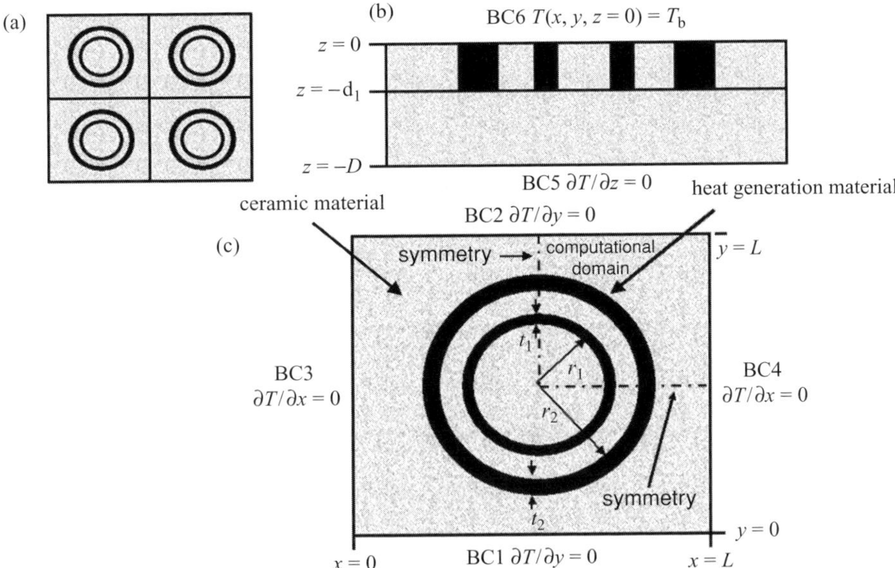

Figure 6.18 Stove top geometry: (a) 2×2 grid of heating elements; (b) side view of an individual element in a periodic array; (c) top view of an individual element.

regions,

$$H(\chi, \eta, \zeta) = \begin{cases} 1, & \text{if } (\chi, \eta, \zeta) \text{ is within an annular region} \\ 0, & \text{otherwise} \end{cases} \tag{6.170}$$

This reduces the number of independent parameters to seven,

$$r_1/L \quad t_1/L \quad r_2/L \quad t_2/L \quad a \equiv L/D \quad b \equiv d_1/L \quad \sigma \equiv \frac{SL^2}{T_b \lambda} \tag{6.171}$$

As we expect the solution to possess the symmetry

$$\theta(\chi, \eta, \zeta) = \theta(-\chi, \eta, \zeta) \quad \theta(\chi, \eta, \zeta) = \theta(\chi, -\eta, \zeta) \tag{6.172}$$

we restrict the domain to $0 \leq \chi \leq 1/2, 0 \leq \eta \leq 1/2$ (dashed lines in Figure 6.18), and enforce symmetry at $\chi = 0, \eta = 0$, to yield the boundary conditions:

$$\begin{array}{lllll} \text{BC 1} & \eta = 0 & 0 \leq \chi \leq \dfrac{1}{2} & -\dfrac{1}{a} \leq \zeta \leq 0 & \dfrac{\partial \theta}{\partial \eta} = 0 \\[2mm] \text{BC 2} & \eta = \dfrac{1}{2} & 0 \leq \chi \leq \dfrac{1}{2} & -\dfrac{1}{a} \leq \zeta \leq 0 & \dfrac{\partial \theta}{\partial \eta} = 0 \\[2mm] \text{BC 3} & \chi = 0 & 0 \leq \eta \leq \dfrac{1}{2} & -\dfrac{1}{a} \leq \zeta \leq 0 & \dfrac{\partial \theta}{\partial \chi} = 0 \\[2mm] \text{BC 4} & \chi = \dfrac{1}{2} & 0 \leq \eta \leq \dfrac{1}{2} & -\dfrac{1}{a} \leq \zeta \leq 0 & \dfrac{\partial \theta}{\partial \chi} = 0 \\[2mm] \text{BC 5} & \zeta = -\dfrac{1}{a} & 0 \leq \chi \leq \dfrac{1}{2} & 0 \leq \eta \leq \dfrac{1}{2} & \dfrac{\partial \theta}{\partial \zeta} = 0 \\[2mm] \text{BC 6} & \zeta = 0 & 0 \leq \chi \leq \dfrac{1}{2} & 0 \leq \eta \leq \dfrac{1}{2} & \theta = 0 \end{array} \tag{6.173}$$

We now apply the finite difference method, to obtain a linear equation for each interior point $(\chi_i, \eta_j, \zeta_k)$ with the label $n = (k-1)N_x N_y + (i-1)N_y + j$,

$$A_{n,n-N_x N_y}\theta_{n-N_x N_y} + A_{n,n-N_y}\theta_{n-N_y} + A_{n,n-1}\theta_{n-1} + A_{nn}\theta_n + A_{n,n+1}\theta_{n+1}$$
$$+ A_{n,n+N_y}\theta_{n+N_y} + A_{n,n+N_x N_y}\theta_{n+N_x N_y} = \sigma H(\chi_i, \eta_j, \zeta_k) \tag{6.174}$$

Previously, when solving the Poisson equation with Dirichlet boundary conditions, we obtained a matrix that was positive-definite and could be solved with the conjugate gradient method. For this problem, however, we have a number of von Neumann boundary conditions, e.g. at the grid points, $(\chi_i, \eta_1 = 0, \zeta_k)$, for which an approximation of the boundary derivative,

$$\frac{\partial \theta}{\partial \eta} = 0 \approx \frac{-\theta(\chi_i, \eta_3, \zeta_k) + 4\theta(\chi_i, \eta_2, \zeta_k) - 3\theta(\chi_i, \eta_1, \zeta_k)}{2(\Delta\eta)} \tag{6.175}$$

yields a linear equation for row n of the system,

$$3\theta_n - 4\theta_{n+1} + \theta_{n+2} = 0 \tag{6.176}$$

This row destroys the symmetry of the matrix (and its diagonal dominance as well). Therefore, we use a method that does not require A to be positive-definite, e.g. **bicgstab** or **gmres**. stove_top_3D_FD.m performs this calculation for the parameters

$$\sigma = 1 \qquad a = 0.67 \quad b = 0.5$$
$$r_1/L = 0.1 \quad t_1/L = 0.05 \quad r_2/L = 0.25 \quad t_2/L = 0.05 \tag{6.177}$$

Options exist for solving the linear system by Gaussian elimination (impractical except for very small grids), or by the **bicgstab** and **gmres** iterative methods. The program also compares the total heat generation rate within the element to the net heat flux across the upper surface. As these numbers must agree for the exact solution, comparing their values provides a measure of accuracy. The results for a grid of $51 \times 51 \times 25$ points are shown in Figure 6.19 and Figure 6.20.

The MATLAB 1-D parabolic and elliptic solver pdepe

Above, we have focused on solving BVPs by implementing the finite difference method directly. MATLAB also has a dedicated 1-D BVP solver, **pdepe**, for systems of equations of parabolic and elliptic type. Its use is rather straightforward, and for further details type doc pdepe.

Finite differences in complex geometries

We have used the finite difference method to solve BVPs in rather simple geometries. The finite difference method can be extended to domains that are composites of these basic shapes (Figure 6.21) or that can be "stretched" into one of them (Figure 6.22).

For the system in Figure 6.22, we define the transformed coordinates

$$\xi = \frac{x}{L} \quad \eta = \frac{y}{h(x)} \tag{6.178}$$

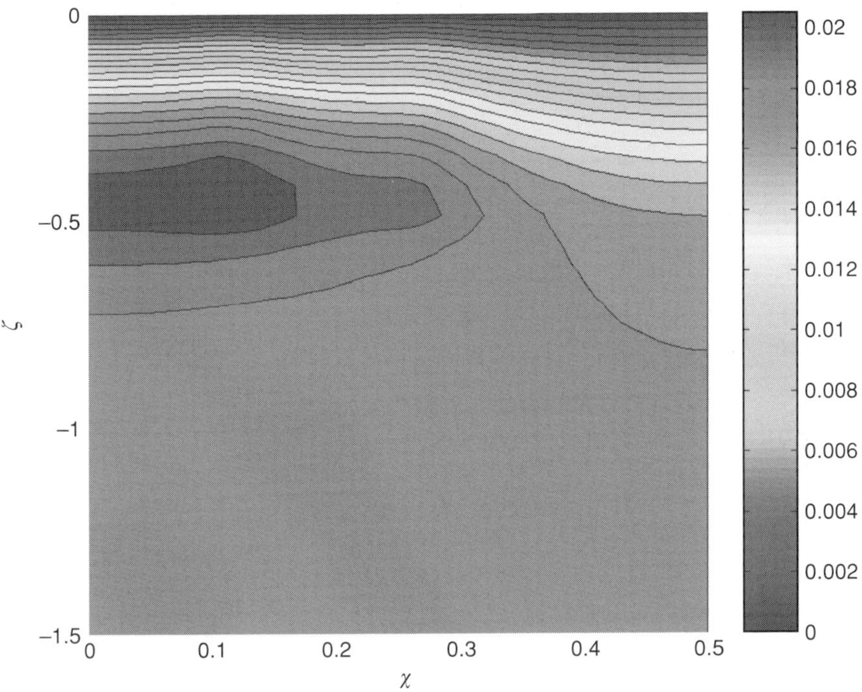

Figure 6.19 Side view of temperature field within stove top element. The highest temperature occurs in the center of the element, near the bottom of the heat generating regions. ($\sigma = 1$, $N_{xy} = 51$, $N_z = 25$.)

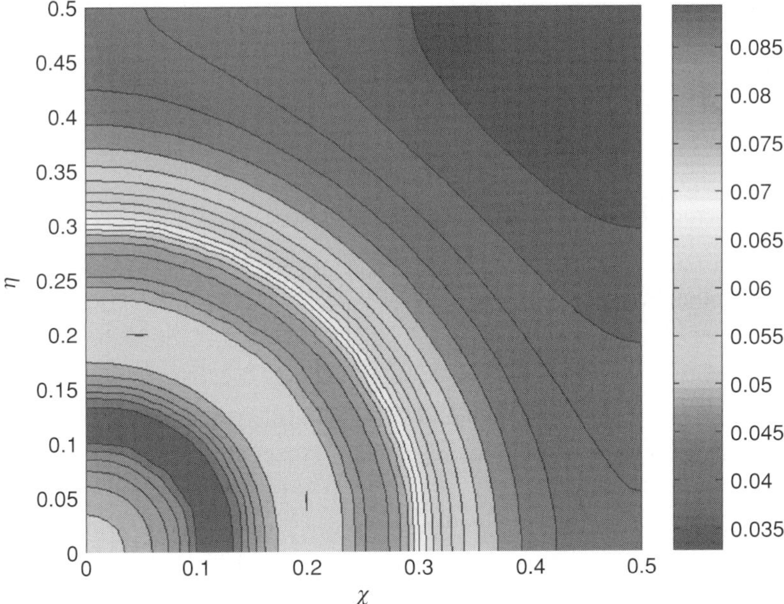

Figure 6.20 Top view of heat flux at upper surface. The center of the element is at the lower left. (Total heat flux out = 0.054, total heat generation rate = 0.063.)

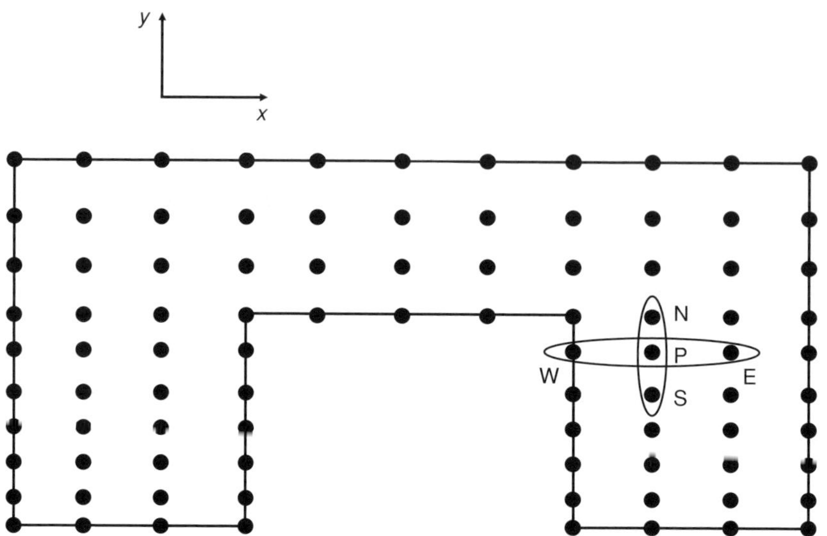

Figure 6.21 Grid placement in a domain comprising fused rectangles.

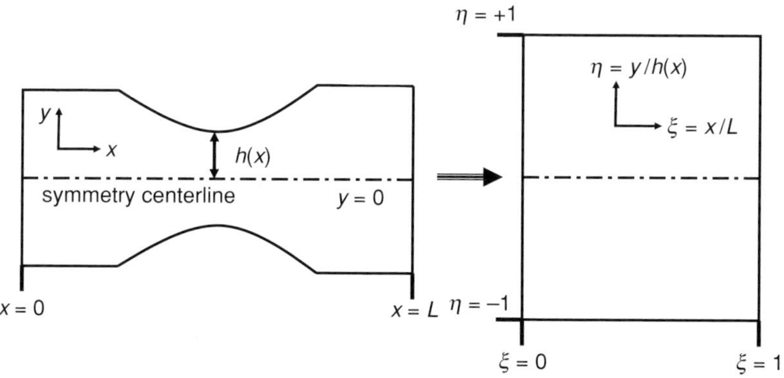

Figure 6.22 Coordinate transformation of a domain into a rectangle.

so that the domain becomes a simple rectangle,

$$0 \leq \xi \leq 1 \quad -1 \leq \eta \leq 1 \tag{6.179}$$

We then place a grid in this rectangular domain, and solve the BVP in transformed space. To do so, we need to express the derivatives with respect to x and y in terms of (ξ, η). For (6.178), the chain rule yields

$$\frac{\partial \varphi}{\partial x} = \frac{\partial \varphi}{\partial \xi}\frac{\partial \xi}{\partial x} + \frac{\partial \varphi}{\partial \eta}\frac{\partial \eta}{\partial x} = \frac{1}{L}\frac{\partial \varphi}{\partial \xi} + \frac{\partial \varphi}{\partial \eta}\left[-\frac{y}{[h(x)]^2}\frac{dh}{dx}\right] = \frac{1}{L}\frac{\partial \varphi}{\partial \xi} - \left[\frac{\eta h'(\xi L)}{h(\xi L)}\right]\frac{\partial \varphi}{\partial \eta}$$

$$\frac{\partial \varphi}{\partial y} = \frac{\partial \varphi}{\partial \xi}\frac{\partial \xi}{\partial y} + \frac{\partial \varphi}{\partial \eta}\frac{\partial \eta}{\partial y} = \frac{\partial \varphi}{\partial \xi}(0) + \frac{\partial \varphi}{\partial \eta}\left[\frac{1}{h(x)}\right] = \left[\frac{1}{h(\xi L)}\right]\frac{\partial \varphi}{\partial \eta} \tag{6.180}$$

The second derivative with respect to y then takes a simple form

$$\frac{\partial^2 \varphi}{\partial y^2} = \frac{\partial}{\partial y}\left(\frac{\partial \varphi}{\partial y}\right) = \frac{1}{[h(\xi L)]^2}\frac{\partial^2 \varphi}{\partial \eta^2} \tag{6.181}$$

By contrast, the second derivative with respect to x is much more complex,

$$\frac{\partial^2 \varphi}{\partial x^2} = \frac{\partial}{\partial x}\left(\frac{\partial \varphi}{\partial x}\right) = \left\{\frac{1}{L}\frac{\partial}{\partial \xi} - \left[\frac{\eta h'(\xi L)}{h(\xi L)}\right]\frac{\partial}{\partial \eta}\right\}\left\{\frac{1}{L}\frac{\partial \varphi}{\partial \xi} - \left[\frac{\eta h'(\xi L)}{h(\xi L)}\right]\frac{\partial \varphi}{\partial \eta}\right\} \tag{6.182}$$

Through this approach, we can employ a finite difference discretization on a regular grid in (ξ, η) space; however, the differential equation now involves more complex derivatives. The finite element method, described below, allows us to solve BVPs in complex geometries without performing such coordinate transformations (which are not always possible anyway).

The finite volume method

Above, our focus has been on the finite difference method, which is easy to implement in domains of rather simple geometry. In complex domains, it is difficult to place a grid and keep track of neighbors when the grid points are required to lie along the coordinate axes. Here, we discuss another method that is not subject to this condition. We again consider the 2-D Poisson equation but now instead of the microscopic equation

$$-\nabla^2 \varphi = -\frac{\partial^2 \varphi}{\partial x^2} - \frac{\partial^2 \varphi}{\partial y^2} = f(x, y) \tag{6.183}$$

we consider the corresponding macroscopic balance

$$0 = \int_{\partial \Omega} [(-\boldsymbol{n}) \cdot (-\nabla \varphi)]dS + \int_{\Omega} f(x, y)dV \tag{6.184}$$

where we have used the equivalence between (6.2) and (6.7). In the finite volume method, we apply this macroscopic balance to each of a number of control volumes, or cells, that partition the computational domain (Figure 6.23). In the center of each cell, we place a computational node, at which we would like to compute the solution. To compare the resulting *finite volume method* to that of finite differences, let us consider a regular grid of uniformly-spaced points, with each cell having a "2-D volume" $V_c = (\Delta x)(\Delta y)$. The nodal coordinates are (x_i, y_j), $x_i = i(\Delta x)$, $y_j = j(\Delta y)$ and each cell is surrounded by neighbors to the "north," "south," "east," and "west."

To obtain a set of algebraic equations for the node field values, we apply the macroscopic balance to each cell, denoted as the control volume $\Omega_{(i,j)}$, that is centered on the nodal position (x_i, y_j),

$$0 = \int_{\partial \Omega_{(i,j)}} [\boldsymbol{n} \cdot \nabla \varphi]dS + \int_{\Omega_{(i,j)}} f(x, y)dV \tag{6.185}$$

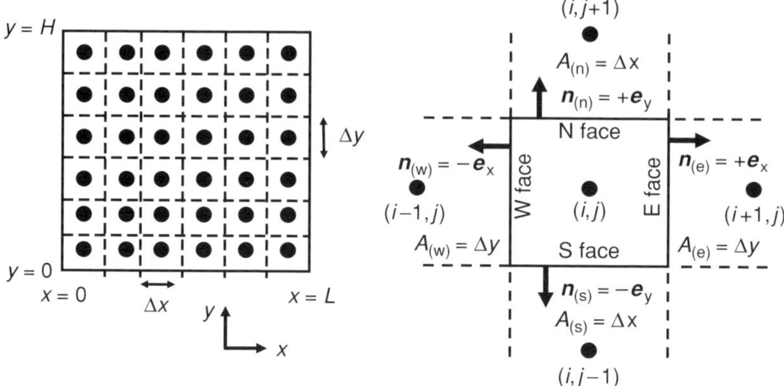

Figure 6.23 Regular control volume/node pattern in the finite volume method applied to a regular 2-D grid.

For the volume integral, we use for each cell the approximation

$$\int_{\Omega_{(i,j)}} f(x, y)\,dV \approx f(x_i, y_j)[(\Delta x)(\Delta y)] \tag{6.186}$$

We partition the surface integral into contributions from each face,

$$\int_{\partial\Omega_{(i,j)}} [\boldsymbol{n} \cdot \nabla\varphi]\,dS \approx A_{(n)}\big[\boldsymbol{n}_{(n)} \cdot \nabla\varphi|_{(n)}\big] + A_{(e)}\big[\boldsymbol{n}_{(e)} \cdot \nabla\varphi|_{(e)}\big]$$
$$+ A_{(s)}\big[\boldsymbol{n}_{(s)} \cdot \nabla\varphi|_{(s)}\big] + A_{(w)}\big[\boldsymbol{n}_{(w)} \cdot \nabla\varphi|_{(w)}\big] \tag{6.187}$$

$\nabla\varphi|_{(n)}$ is an averaged gradient value over the north face. Substituting for the "areas" and unit normals of each face,

$$\int_{\partial\Omega_{(i,j)}} [\boldsymbol{n} \cdot \nabla\varphi]\,dS \approx (\Delta x)\big[\boldsymbol{e}_y \cdot \nabla\varphi|_{(n)}\big] + (\Delta y)\big[\boldsymbol{e}_x \cdot \nabla\varphi|_{(e)}\big]$$
$$+ (\Delta x)\big[-\boldsymbol{e}_y \cdot \nabla\varphi|_{(s)}\big] + (\Delta y)\big[-\boldsymbol{e}_x \cdot \nabla\varphi|_{(w)}\big] \tag{6.188}$$

and using the fact that the normals point in the coordinate axes, we have

$$\int_{\partial\Omega_{(i,j)}} [\boldsymbol{n} \cdot \nabla\varphi]\,dS \approx (\Delta x)\,\frac{\partial\varphi}{\partial y}\bigg|_{(n)} + (\Delta y)\,\frac{\partial\varphi}{\partial x}\bigg|_{(e)} - (\Delta x)\,\frac{\partial\varphi}{\partial y}\bigg|_{(s)} - (\Delta y)\,\frac{\partial\varphi}{\partial x}\bigg|_{(w)}$$
$$\approx (\Delta x)\left[\frac{\partial\varphi}{\partial y}\bigg|_{(n)} - \frac{\partial\varphi}{\partial y}\bigg|_{(s)}\right] + (\Delta y)\left[\frac{\partial\varphi}{\partial x}\bigg|_{(e)} - \frac{\partial\varphi}{\partial x}\bigg|_{(w)}\right] \tag{6.189}$$

For the averaged partial derivative on each face, we use the value at the face center, which is simply approximated from the difference between the nodal values on each side:

$$\frac{\partial\varphi}{\partial y}\bigg|_{(n)} \approx \frac{\varphi_{(i,j+1)} - \varphi_{(i,j)}}{\Delta y} \qquad \frac{\partial\varphi}{\partial y}\bigg|_{(s)} \approx \frac{\varphi_{(i,j)} - \varphi_{(i,j-1)}}{\Delta y}$$
$$\frac{\partial\varphi}{\partial x}\bigg|_{(e)} \approx \frac{\varphi_{(i+1,j)} - \varphi_{(i,j)}}{\Delta x} \qquad \frac{\partial\varphi}{\partial x}\bigg|_{(w)} \approx \frac{\varphi_{(i,j)} - \varphi_{(i-1,j)}}{\Delta x} \tag{6.190}$$

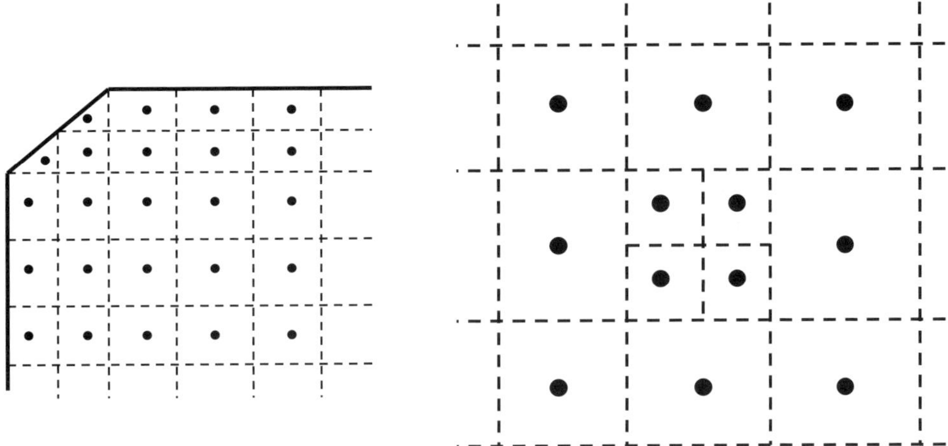

Figure 6.24 Irregular cell geometries that may be treated using the finite volume method.

The macroscopic balance for each cell is then

$$0 = (\Delta x)\left[\left.\frac{\partial \varphi}{\partial y}\right|_{(n)} - \left.\frac{\partial \varphi}{\partial y}\right|_{(s)}\right] + (\Delta y)\left[\left.\frac{\partial \varphi}{\partial x}\right|_{(e)} - \left.\frac{\partial \varphi}{\partial x}\right|_{(w)}\right] + f(x_i, y_j)[(\Delta x)(\Delta y)]$$

(6.191)

If we now divide by the cell "volume" $(\Delta x)(\Delta y)$, we have

$$0 = (\Delta y)^{-1}\left[\left.\frac{\partial \varphi}{\partial y}\right|_{(n)} - \left.\frac{\partial \varphi}{\partial y}\right|_{(s)}\right] + (\Delta x)^{-1}\left[\left.\frac{\partial \varphi}{\partial x}\right|_{(e)} - \left.\frac{\partial \varphi}{\partial x}\right|_{(w)}\right] + f(x_i, y_j) \qquad (6.192)$$

Substituting in the approximations for the derivatives, we obtain exactly the same algebraic equations as we had for the finite difference method using the central difference approximation.

Why introduce the finite volume method if it gives the same algebraic system as the finite difference method in this example? One reason is that we can extend the finite volume method to complex geometries. With finite differences, the grid points must lie along the coordinate axes, a restriction that is inconvenient in complex geometries or when we wish to subdivide only particular cells (Figure 6.24). Also, if we use the same approximation for the integrals over each face for the cells on both sides, the approximation errors cancel out when we apply a macroscopic balance to the total system. This property is particularly convenient in computational fluid dynamics, and thus the popular CFD package FLUENT™ uses a finite volume approach.

The finite element method (FEM)

Use of a regular grid, with the points lined up along the coordinate axes, is difficult in a complex geometry as labeling the points and identifying the neighbors is tedious. It is much easier to generate an irregular, or unstructured grid. The *FEM* is popular as unlike the finite difference method it does not require the grid to be regular.

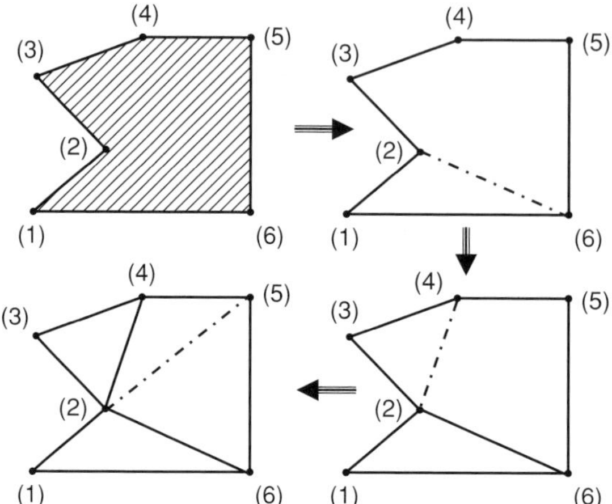

Figure 6.25 Automatic partitioning of a 2-D region into nonoverlapping triangles, subsequent steps in the clockwise direction from upper left.

Automatic mesh generation

Here we consider a simple algorithm to form a nonstructured grid on a closed region in two dimensions. First, we parameterize the outer boundary as a number of line segments, and number the vertices that connect the line segments (upper left in Figure 6.25). Subsequent steps are shown moving clockwise. Next, we identify (1) as the vertex with the smallest interior angle, and draw a new line segment between its neighbors (2) and (6). Then, in the remaining polygon connecting nodes (2)–(3)–(4)–(5)–(6)–(2), we identify (3) as the vertex with the smallest interior angle and connect its neighbors (2) and (4). Finally, we identify in the remaining polygon (6) as the vertex with the smallest interior angle and connect its neighbors (2) and (5) to finish the partition of the domain into nonoverlapping triangles. Sometimes, the algorithm calls for a line segment to be drawn that is unacceptable as it lies partially outside of the domain (Figure 6.26). The remedy is to identify a node, here (4), such that (1)–(4) does lie within the domain, and then to proceed independently for the two polygons on either side of (1)–(4).

Once the domain has been partitioned into nonoverlapping triangles to form an initial *mesh* (i.e., node positions with the associated topology of connections that describe the partition), it can be refined for more accurate calculations. In *global mesh refinement* (Figure 6.27), we add new nodes at the mid-points of the segments, and connect them so that each old triangle now contains four smaller ones. Alternatively, if there is some region where the solution changes rapidly, say near (2), we can perform a *local mesh refinement* only within this region.

Here, we have described only a simple algorithm for partitioning a 2-D domain, but more efficient alternatives (also in three dimensions) exist and are described in O'Rourke (1993). FEM is often performed with rectangular elements rather than triangular (or in three dimensions, tetrahedral) elements, but here for brevity we restrict our discussion to triangular elements in two dimensions.

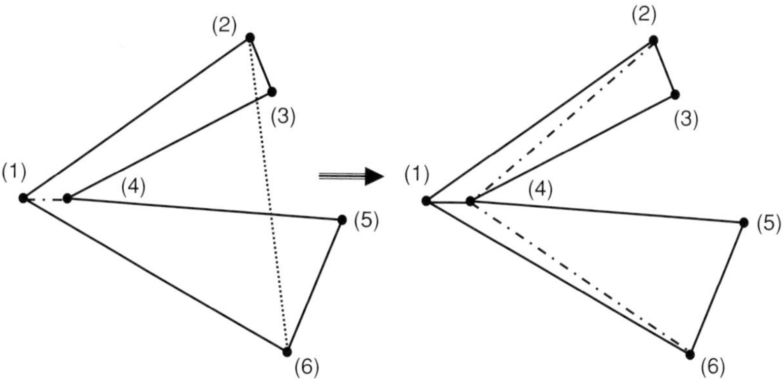

Figure 6.26 Corrective action to avoid a vertex connection that lies partially outside of the domain.

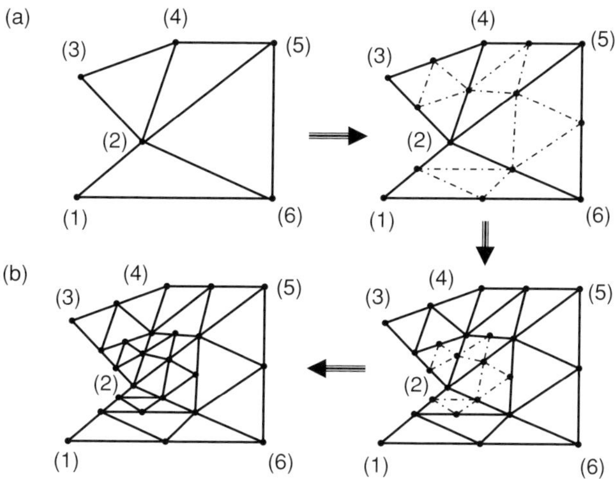

Figure 6.27 (a) Global and (b) local refinement of a mesh, starting from upper left and moving clockwise.

The optional MATLAB PDE toolkit (doc pdetool), created by the developers of FEMLAB^{TM} (www.comsol.com), has tools for forming meshes and solving simple PDEs in two dimensions. **pdetool** opens a graphical user interface (GUI), in which we can draw the domain, mesh it, specify boundary conditions and PDE parameters, solve, and plot the solution. As tutorials are provided on the use of the GUI, here our focus is upon use of the command-line interface to access the functions of the **PDE toolkit** directly.

First, we demonstrate specifying the domain geometry, using a polygon of the shape shown in Figure 6.25 with the vertex positions

$$r^{[1]} = \begin{bmatrix} 0 \\ 0 \end{bmatrix} \quad r^{[2]} = \begin{bmatrix} 1 \\ 1 \end{bmatrix} \quad r^{[3]} = \begin{bmatrix} 0 \\ 2 \end{bmatrix} \quad r^{[4]} = \begin{bmatrix} 1.5 \\ 3 \end{bmatrix}$$

$$r^{[5]} = \begin{bmatrix} 3 \\ 3 \end{bmatrix} \quad r^{[6]} = \begin{bmatrix} 3 \\ 0 \end{bmatrix} \tag{6.193}$$

We substitute into (6.218) the expansion (6.200) that we now write as

$$\varphi(\mathbf{r}) = \sum_{q \notin \partial\Omega^{[d]}} \varphi_q \chi_q(\mathbf{r}) + \sum_{q \in \partial\Omega^{[d]}} \varphi_B(\mathbf{r}^{[q]}) \chi_q(\mathbf{r}) \tag{6.219}$$

so that for each $p \notin \partial\Omega^{[d]}$, (6.218) becomes

$$0 = \int_\Omega \left\{ (\nabla\chi_p) \cdot \left[\sum_{q \notin \partial\Omega^{[d]}} \varphi_q \nabla\chi_q + \sum_{q \in \partial\Omega^{[d]}} \varphi_B(\mathbf{r}^{[q]}) \nabla\chi_q \right] \right\} d\mathbf{r} - \int_\Omega \chi_p f \, d\mathbf{r} \tag{6.220}$$

Upon rearrangement, this becomes

$$0 = \sum_{q \notin \partial\Omega^{[d]}} A_{pq} \varphi_q + \sum_{q \in \partial\Omega^{[d]}} A_{pq} \varphi_B(\mathbf{r}^{[q]}) - \int_\Omega \chi_p(\mathbf{r}) f(\mathbf{r}, \varphi) d\mathbf{r} \tag{6.221}$$

where

$$A_{pq} = \int_\Omega [(\nabla\chi_p) \cdot (\nabla\chi_q)] d\mathbf{r} \tag{6.222}$$

The elements of A are easy to compute once we know the node positions and mesh topology. Let $\Omega^{[k]}$ denote the region occupied by triangular element k, and let the volume of this element be $V^{[k]}$. Then, since the elements are nonoverlapping, we can partition the integral in (6.222) into contributions from each element,

$$A_{pq} = \sum_k \int_{\Omega^{[k]}} [(\nabla\chi_p) \cdot (\nabla\chi_q)] d\mathbf{r} \tag{6.223}$$

Now, each element integral is zero unless *both* p and q form a vertex of the element; therefore, A is a very sparse matrix. For those elements k that do have p and q as vertices, the gradients of the shape functions are locally constant since we use linear interpolation (Figure 6.29). Thus, the elements of A are simply

$$A_{pq} = \sum_{k \in S_{\mathrm{E}}^{[p,q]}} \left[\nabla\chi_p \big|_{\Omega^{[k]}} \cdot \nabla\chi_q \big|_{\Omega^{[k]}} \right] V^{[k]} \tag{6.224}$$

$S_{\mathrm{E}}^{[p,q]}$ is the set of elements that have both p and q as vertices.

To evaluate the final integral of (6.221), we interpolate the source term from its values at the nodes,

$$f(\mathbf{r}, \varphi) = \sum_{q \notin \partial\Omega^{[d]}} f(\mathbf{r}^{[q]}, \varphi_q) \chi_q(\mathbf{r}) + \sum_{q \in \partial\Omega^{[d]}} f(\mathbf{r}^{[q]}, \varphi_B(\mathbf{r}^{[q]})) \chi_q(\mathbf{r}) \tag{6.225}$$

to obtain

$$\int_\Omega \chi_p(\mathbf{r}) f(\mathbf{r}, \varphi) d\mathbf{r} = \sum_{q \notin \partial\Omega^{[d]}} S_{pq} f(\mathbf{r}^{[q]}, \varphi_q) + \sum_{q \in \partial\Omega^{[d]}} S_{pq} f(\mathbf{r}^{[q]}, \varphi_B(\mathbf{r}^{[q]})) \tag{6.226}$$

where we have another sparse matrix S, with elements

$$S_{pq} = \int_\Omega \chi_p(\mathbf{r}) \chi_q(\mathbf{r}) d\mathbf{r} = \sum_{k \in S_{\mathrm{E}}^{[p,q]}} \int_{\Omega^{[k]}} \chi_p(\mathbf{r}) \chi_q(\mathbf{r}) d\mathbf{r} \tag{6.227}$$

Therefore for each node $p \notin \partial\Omega^{[d]}$, we have an algebraic equation

$$0 = \sum_{q \notin \partial\Omega^{[d]}} A_{pq}\varphi_q - b_p(\varphi) \tag{6.228}$$

where

$$b_p(\varphi) = \sum_{q \notin \partial\Omega^{[d]}} S_{pq} f\left(\mathbf{r}^{[q]}, \varphi_q\right) + \sum_{q \in \partial\Omega^{[d]}} S_{pq} f\left(\mathbf{r}^{[q]}, \varphi_{\mathrm{B}}\left(\mathbf{r}^{[q]}\right)\right) - \sum_{q \in \partial\Omega^{[d]}} A_{pq}\varphi_{\mathrm{B}}\left(\mathbf{r}^{[q]}\right)$$
$$\tag{6.229}$$

For a source term with no field-dependence, (6.228) is a sparse linear system. If not, it must be solved with Newton's method; however, the Jacobian matrix is also sparse.

Convection terms in FEM

Above, we have considered only diffusive transport; however, it is not difficult to extend the method to include convective transport as well. The choice of linear shape functions used above results in a system of algebraic equations with similar accuracy to central finite difference approximations and shares the tendency of CDS to exhibit unphysical oscillations when convection is strong. The oscillations are reduced by mesh refinement (to reduce the local Peclet number) or by adding additional numerical diffusion, especially in the streamline direction. Upwind versions of FEM are obtained by biasing the shape functions to weight more heavily the upstream direction. Such subjects are beyond the scope of this chapter, and the reader is referred to a dedicated text on FEM such as Akin (1994). The relationship between FEM with linear shape functions and CDS finite differences is examined in the supplemental material in the accompanying website for the case of the 1-D convection/diffusion equation.

FEM in MATLAB

In MATLAB, an optional **pdetool** toolbox allows one to solve 2-D BVPs and to perform various low-level mesh generation and assembly operations. There exists also a more advanced software package, FEMLAB™, (www.comsol.com), from the developers of **pdetool**, that can solve multiple PDEs of various types in two and three dimensions. The supplemental material in the accompanying website contains an example of the use of FEMLAB™ to model a microfluidic H-filer, requiring the simultaneous solution of the Navier–Stokes equations and a mass balance. A second FEMLAB™ example in the supplemental material models natural convection between two flat, vertical plates at different temperatures (Figure 6.30).

In general, if you have a simple geometry and are solving a system of transport-type PDEs of the form of (6.9), it is not too difficult to write your own finite difference program. For more complicated systems, it is not worth writing an FEM program from scratch, as the FEM is more tedious to implement than finite differences, and canned software libraries and packages already exist that are suitable for these problems (unless your system has PDEs that are not of standard type, e.g. when solving a fluid mechanics problem with a

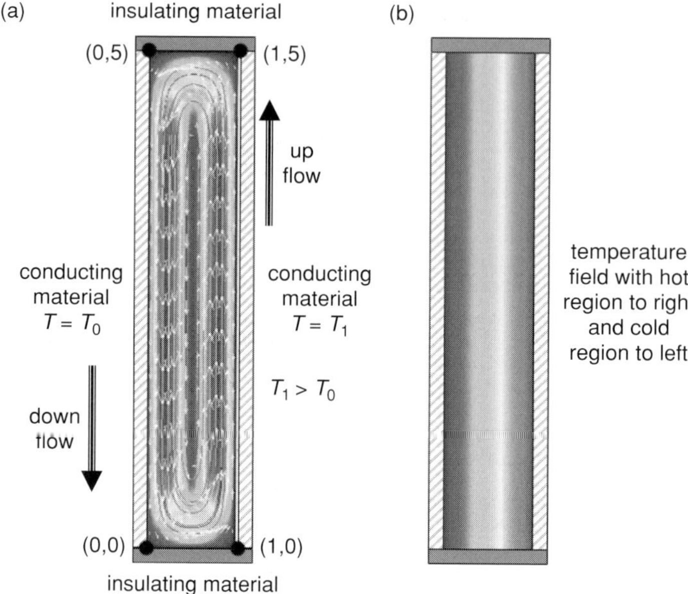

Figure 6.30 (a) Velocity and (b) temperature profiles for natural convection of a fluid between two vertical plates maintained at different temperatures. Results from FEMLAB™ (www.comsol.com).

novel non Newtonian constitutive equation). Below, we demonstrate use of the MATLAB PDE toolkit.

Numerical solution of a 2-D BVP using the MATLAB PDE toolkit

pde_ex1.m demonstrates the use of the PDE toolkit functions to solve a BVP with Poisson's equation

$$-\nabla^2 u = f(x, y) = x^2 + y^2 + 1 \tag{6.230}$$

on the domain described above, whose geometry is defined by polygon1_geom.m. On the boundary sections on the left-hand side, a Dirichlet condition $u = 1$ is enforced, and on the right-hand boundary section, we again enforce $u = 1$. At the top and bottom boundaries, we use zero-flux von Neumann boundary conditions. These boundary conditions are defined in a boundary m-file pde_ex1_bound.m. Finally, pde_ex1.m calls the adaptive mesh solver **adaptmesh**, which computes the solution to a system of elliptic PDEs on the domain. Plots of the mesh and solution are shown in Figure 6.31. During the solution process, the routine estimates where the discretization errors are highest and adds new nodes.

Here, we have only a single field, but the solver can treat multiple fields and field-dependent source terms (if we set the "nlin" flag to "on" or use **pdenonlin**). Type doc adaptmesh for further details. There are also lower-level commands available that perform isolated tasks such as assembling the various matrices, and interpolating fields from node values; however, the use of such routines is beyond the scope of this text. Finally, routines

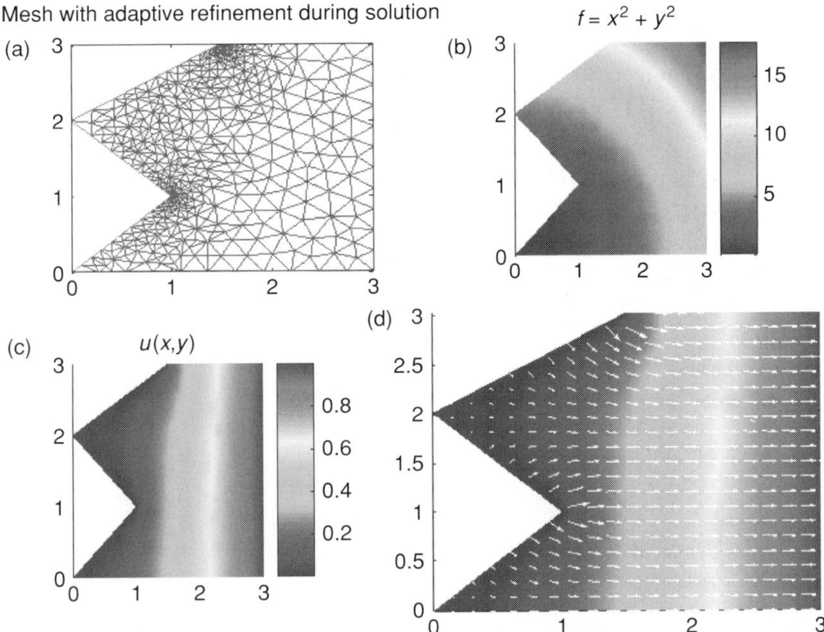

Figure 6.31 Solution of Poisson's equation on an irregular domain in two dimensions; (a) mesh showing adaptive refinement; (b) source term; (c) contour plot of solution; (d) solution with arrows showing local gradient vector of solution.

also exist for solving systems of parabolic and hyperbolic equations. The GUI also has specialized modes for heat and mass transfer, solid mechanics, and electromagnetics. Here, we have demonstrated use of the command-line interface, but the GUI often makes solving problems easier.

Further study in the numerical solution of BVPs

This chapter has introduced the major real-space numerical methods to solving BVPs; however, this subject is far more vast than could be covered even in a dedicated text. For further reading on the subject of CFD, consult Ferziger & Peric (2001). An in-depth discussion of simulations involving coupled transport and chemical reaction is provided by Oran & Boris (2001). Much current research in numerical methods for BVPs involves multigrid methods in which the computation cycles through coarse to fine grids and back again to improve the performance of iterative methods (Trottenberg *et al.*, 2000).

MATLAB summary

For a BVP in a simple geometry, it is fairly straightforward to discretize the system oneself using finite differences. For 1-D BVPs with equations of parabolic and elliptic type, **pdepe** can be used instead. For BVPs in a complex domain in two dimensions, the PDE toolkit can solve systems of equations of elliptic, hyperbolic, and parabolic type using FEM. The software package **FEMLAB**^{TM}, built upon **MATLAB** by the developers of the PDE toolkit (www.comsol.com) can solve more general and complex BVPs in two and three dimensions. Three dimensional BVPs do not pose any new conceptual issues, but elimination can no longer be used to solve the resulting linear systems. Iterative methods are necessary, such as **pcg** if the system is positive-definite and **bicgstab** or **gmres** if it is not. Preconditioners improve significantly the efficiency of these methods, and one may either use **cholinc** or **luinc** to perform an incomplete Cholesky or LU factorization respectively.

Problems

6.A.1. Use finite differences to discretize the following BVP in three dimensions:

$$-\nabla^2\varphi = \exp[-(x^2 + y^2 + z^2)/2] \quad -1 \le x, y, z \le 1$$
$$\text{Dirichlet condition} \quad \varphi = 0 \quad \text{on all boundaries} \tag{6.231}$$

Solve the linear system with **pcg**, for a grid of $50 \times 50 \times 50$ points. How many iterations are necessary with no preconditioner? Next, use an incomplete Cholesky preconditioner with no fill-in, and report the number of iterations required for convergence. Finally, as a function of **droptol**, plot the number of iterations and the number of nonzero elements in the Cholesky factor. What value of drop tolerance do you recommend using? For this optimal value of the drop tolerance, change the number of grid points, and report how the number of iterations and CPU time (**doc cputime**) varies with the size of the grid.

6.A.2. Solve the system $Ax = b$ that discretizes the BVP of problem 6.A.1 without storing the matrix A in memory. Supply a routine that returns Av for input v.

6.A.3. Solve the 2-D version of problem 6.A.1 with $z = 0$ by the FEM using the **PDE toolkit** or **FEMLAB**^{TM}.

6.B.1. Consider the following heat transfer problem. A fluid with thermal conductivity λ, density ρ, and specific heat capacity $\hat{C}_p$ flows at a Reynolds number $Re < 100$ through a cylindrical pipe of radius R. The fluid viscosity is μ, and you may neglect viscous heating. If z is the axial position, for $z < 0$ the wall temperature is T_0 and at $z = 0$ the wall temperature jumps abruptly to T_1 for $z > 0$. This is known as the Graetz problem, and an analytical solution exists if conduction in the axial direction is neglected. Transform this problem into dimensionless variables and write a program to compute numerically the steady-state temperature profile without neglecting axial conduction. Have your program take as input the values of the dimensionless parameters that characterize the system. Report your results for values of the parameters in the range $[10^{-2}, 10^2]$. Try reducing the number of independent parameters through clever rescaling.

6.B.2. You are conducting the enzymatic conversion of a substrate S into a product P, with the micromoles of S converted per minute per milligram of enzyme being described by Michaelis Menten kinetics,

$$-\hat{r}_S = \frac{V_m S}{K_m + S} \quad V_m = 200 \, \frac{\mu\text{mol}}{\text{min mg}_E} \quad K_m = 0.2 \, \text{M} \tag{6.232}$$

You use an immobilized enzyme system in which the enzyme is encapsulated in beads of radius R of a polymer hydrogel at a mass loading density $\rho_E = 10^{-2}$ mg$_E$/ml of gel. The substrate diffusivity in the hydrogel is 10^{-6} cm^2/s. The bulk substrate concentration is 1 M. Neglect external mass transfer resistance. Define an internal effectiveness factor, and compute its value as a function of R in the range $10^{-4} - 10^{-2}$ m. Since the enzyme is expensive, what radius would you use, and why?

6.B.3. Consider a system in which a charged surface, maintained at a constant electric potential Φ_0, is in contact with an aqueous solution of NaCl, with a bulk salt concentration c_{NaCl}. If $\Phi_0 > 0$, the surface selectively attracts the Cl$^-$ anions, else if $\Phi_0 < 0$ it attracts the Na$^+$ cations. In either case, near the surface there is a region of charge imbalance in the salt solution that counteracts the surface potential. As the distance z from the surface increases, the electric potential decays to zero, as the surface electric charge is screened by the salt solution.

We present here a model for the electric potential near a charged planar surface known as *Gouy–Chapman theory*. Let $c_+(z)$ and $c_-(z)$ be the ion molar concentrations as functions of the distance z from the surface. The charge density, also a function of z, is then

$$\rho(z) = q_e N_{av}[c_+(z) - c_-(z)] \tag{6.233}$$

$q_e = 1.602 \times 10^{-19}$ C is the charge of an electron and $N_{av} = 6.022 \times 10^{23}$. The electric potential is governed by the Poisson equation

$$-\frac{d^2\Phi}{dz^2} = \frac{\rho(z)}{\varepsilon_0 \varepsilon_r} \tag{6.234}$$

$\varepsilon_0 = 8.854 \times 10^{-12}$ C^2/(J m) is the permittivity of free space and $\varepsilon_r = 78$ is the dielectric constant of water.

We obtain a closed-form equation for $\Phi(z)$ at equilibrium using statistical mechanics. The potential energy of a cation is $E_+(z) = q_e\Phi(z)$, and that of an anion is $E_-(z) = -q_e\Phi(z)$. Then, with the far-field condition $\Phi(z) \to 0$ as $z \to \infty$, the *Boltzmann distribution* predicts the ion concentration fields

$$c_+(z) = c_{NaCl} \exp\left[-\frac{q_e\Phi(z)}{k_b T}\right] \quad c_-(z) = c_{NaCl} \exp\left[+\frac{q_e\Phi(z)}{k_b T}\right] \tag{6.235}$$

The charge density is then

$$\rho(z) = q_e N_{av} c_{NaCl} \left\{\exp\left[-\frac{q_e\Phi(z)}{k_b T}\right] - \exp\left[+\frac{q_e\Phi(z)}{k_b T}\right]\right\} \tag{6.236}$$

The electric potential field at equilibrium is obtained from the following BVP, involving the nonlinear *Poisson–Boltzmann equation*,

$$-\frac{d^2\Phi}{dz^2} = \frac{\rho(z)}{\varepsilon_0 \varepsilon_r} = \frac{q_e N_{av} c_{NaCl}}{\varepsilon_0 \varepsilon_r} \left\{\exp\left[-\frac{q_e\Phi(z)}{k_b T}\right] - \exp\left[+\frac{q_e\Phi(z)}{k_b T}\right]\right\}$$

$$\text{BC 1} \qquad \Phi(0) = \Phi_0$$
$$\text{BC 2} \qquad \Phi(\infty) = 0 \tag{6.237}$$

The quantity $q_e N_{av} c_{NaCl}/\varepsilon_0 \varepsilon_r$ on the right-hand side must have the same units as the left-hand side (electric potential divided by length squared). Likewise, from the argument of the exponential functions, $k_b T/q_e$ must have the units of electric potential. Therefore, from dimensional analysis, the characteristic decay length of $\varphi(z)$ satisfies

$$\frac{q_e N_{av}(2c_{NaCl})}{\varepsilon_0 \varepsilon_r} = \frac{k_b T/q_e}{\lambda^2} \quad \Rightarrow \lambda = \left[\frac{\varepsilon_0 \varepsilon_r k_b T}{q_e^2 N_{av}(2c_{NaCl})}\right]^{1/2} \tag{6.238}$$

We have placed an additional factor of 2 here to agree with the usual definition of λ as the *Debye screening length*. Note that in this equation, c_{NaCl} is in units of moles per cubic meter. We next define the dimensionless quantities

$$\varphi = \frac{q_e \Phi}{k_b T} \qquad \xi = \frac{z}{\lambda} \tag{6.239}$$

so that the BVP in dimensionless form is

$$-\frac{d^2\varphi}{d\xi^2} = \frac{1}{2}\left[e^{-\varphi(\xi)} - e^{+\varphi(\xi)}\right]$$

$$\text{BC 1 (surface)} \qquad \varphi(0) = q_e \Phi_0/(k_b T)$$
$$\text{BC 2 (far-field)} \qquad \varphi(\infty) = 0 \tag{6.240}$$

Make a plot of Debye length (in meters) vs. [NaCl] in moles at room temperature in the range of 10^{-6}–1 M. Does it make sense to have a Debye length less than 10^{-10} m? Solve this BVP numerically using finite differences, and plot $\varphi(\xi)$ for various values of $\varphi(0)$. Show that when $|\Phi_0| \ll (k_b T/q_e)$, the solution is a simple exponential decay. For a further discussion of screening in ionic solutions, consult Stokes & Evans (1997).

6.B.4. In problem 6.B.3, we assumed a constant surface potential; however, we can modify the problem to impose a known surface charge density σ_0. Electroneutrality of the surface and the neighboring salt solution requires

$$\sigma_0 = -\int_0^\infty \rho(z)dz = \varepsilon_0 \varepsilon_r \int_0^\infty \frac{d^2\Phi}{dz^2}dz = \varepsilon_0 \varepsilon_r \frac{d\Phi}{dz}\bigg|_0^\infty \tag{6.241}$$

In the second equality we have used the Poisson equation. Since as $z \to \infty$, $\Phi(z)$ becomes uniformly zero, the derivative at $z = \infty$ is zero and the surface charge density is simply related to the derivative value at the surface:

$$\sigma_0 = -\varepsilon_0 \varepsilon_r \frac{d\Phi}{dz}\bigg|_0 \tag{6.242}$$

Thus, we need only modify the boundary condition at the surface to move from a specified surface potential value to a specified charge density. Modify your program from problem 6.B.3 to plot the dimensionless solution as a function of dimensionless charge density.

6.B.5. We focus in this text on problems of direct interest to chemical engineers; however, the solution of BVPs is important in other fields as well. In finance, a common means to assign a value to a derivative is to solve the *Black–Scholes equation*. Let $S(t)$ be the spot price of some asset (e.g. a stock) at time t. We are considering purchasing a European call option that gives us the opportunity to buy the asset at specified future time $T > t$ for an exercise price E, yielding a payoff if the future price $S(T)$ is above E of

$$\text{payoff} = \max\{S(T) - E, 0\} \tag{6.243}$$

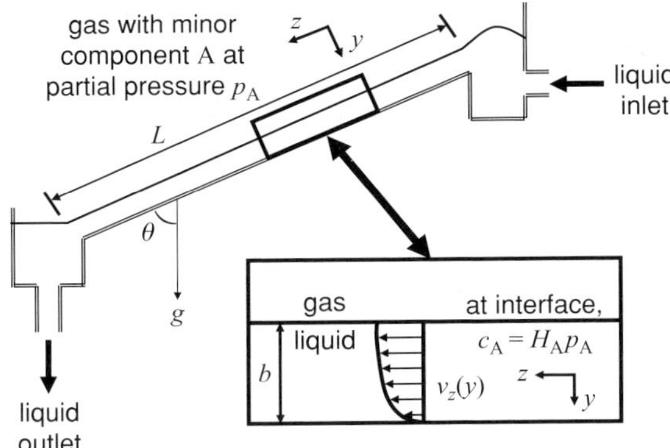

Figure 6.32 Reaction-enhanced diffusion into a falling film.

The current value (fair price) V of the option should depend upon the current asset price S and the time remaining to expiry, $T - t$. The Black-Scholes equation (derived in Chapter 7) states that $V(S, t)$ satisfies

$$\frac{\partial V}{\partial t} + \frac{1}{2}\sigma^2 S^2 \frac{\partial^2 V}{\partial S^2} + rS\frac{\partial V}{\partial S} - rV = 0 \tag{6.244}$$

with the final and boundary conditions

$$\text{final condition} \qquad V(S, T) = \max\{S - E, 0\}$$
$$\text{spatial boundary conditions} \qquad 0 = \left.\frac{\partial^2 V}{\partial S^2}\right|_0 = \left.\frac{\partial^2 V}{\partial S^2}\right|_{S_{\max} \gg \max(S(t), E)} \tag{6.245}$$

r is the (risk-free) interest rate used to define the time value of money. σ is the volatility of the underlying asset, and measures how vigorously the price varies with time,

$$\sigma^2 = \frac{\sum_{k=1}^{N_s}(R_k - \langle R \rangle)^2}{(N_s - 1)(\Delta t)} \qquad R_k = \frac{S(t_k + \Delta t) - S(t_k)}{S(t_k)} \qquad \langle R \rangle = \frac{\sum_{k=1}^{N_s} R_k}{N_s} \tag{6.246}$$

This model is based upon a treatment of the asset price as a random walk, the mathematics of which will be discussed in Chapter 7. While an analytical solution exists for this "plain vanilla option," more realistic cases generally require numerical solution. Write a program that computes $V(S, t)$ for a European call option. For more on this subject, consult Wilmott (2000).

6.C.1. Consider a system with a reversible reaction $A + B \Leftrightarrow C + D$, with kinetics $r_R = k(c_A c_B - K_{eq}^{-1} c_C c_D)$ occurring inside a catalyst pellet that is the shape of a long, narrow cylinder of radius R. Given R, k, K_{eq}, the effective binary diffusivities D_j within the catalyst pellet, and the surface concentrations c_{jS} of each species $j = A, B, C, D$, write a program that computes the net reaction rate per unit volume of catalyst. Report your results for the following parameter values (make sure to use proper concentration units that agree with the other terms in your equations),

$$R = 0.1 \text{ cm} \quad D_j = 10^{-7} \text{ cm}^2/\text{s} \quad k = 10^{-3}\frac{1}{\text{mol s}} \quad K_{eq} = 10^3$$
$$c_{AS} = 1 \text{ M} \quad c_{BS} = 1 \text{ M} \quad c_{CS} = c_{DS} = 0 \tag{6.247}$$

6.C.2. Modify the program of problem 6.C.1 to account for external mass transfer resistance, given the effective binary diffusivities D_{jf} of each species in the external fluid, the Sherwood number Sh, and the bulk concentrations c_{jb}. Report your results for the parameter values of problem 6.C.1, replacing the known surface concentrations with those computed by external mass transfer with

$$D_{jf} = 10^{-5} \text{ cm}^2/\text{s} \qquad Sh = 10$$
$$c_{Ab} = 1 \text{ M} \qquad c_{Bb} = 1 \text{ M} \qquad c_{Cb} = c_{Db} = 0 \tag{6.248}$$

6.C.3. Solve problem 6.C.1. by FEM using the **PDE toolkit** or **FEMLAB**^{TM}.

6.C.4. Instead of assuming a given Sherwood number for the external mass transfer coefficients, use **FEMLAB**^{TM} to compute the laminar velocity profile around the cylinder to directly model the effect of forced convection on the external mass transfer rate. Report your results for Reynolds' numbers of 10^{-1}, 10^{-2}, 10^{-1}, 1, 10, using a viscosity $\mu = 10^{-3}$ Pa s and density $\rho = 10^3 \text{ kg/m}^3$.

6.C.5. A common technique to remove a minor component from a gas stream is reactive absorption (Cussler & Varma, 1997). A gas phase containing a minor component A is passed through a column where it contacts a liquid phase containing a species B that reacts with A. The depletion of A through reaction with B allows the removal of more A from the gas stream than would be possible from considerations of solubility alone.

Let us consider the problem of reactive absorption into a falling film (Figure 6.32) from a large volume of gas with a constant partial pressure of A, $p_A = 10^{-4}$ atm. At the interface, the concentration of A is in equilibrium with the gas according to Henry's law, $H_A = 10^3$ M/atm. We assume that the amount of A entering the film has a negligible effect on the viscosity μ and density ρ (use the values for water) and that the film travels so slowly that the flow is laminar. Within the liquid, A reacts with B according to a reversible first-order reaction $A + B \Leftrightarrow AB$ at the rate $r_R = k(c_A c_B - K^{-1} c_{AB})$, $K = 10^3$ and $k = 10^{-2}$ L/(mol s). The feed stream has $c_{B0} = 1M$, $c_{A0} = c_{ABO} = 0$, and model B and AB as being nonvolatile by using no-flux BCs at the liquid–gas interface. For a dilute solution of A and B, we neglect the heat of absorption and assume a uniform temperature.

Propose a model for this system in the form of a BVP, specifying both the set of PDEs and the appropriate boundary conditions. Assume a film of thickness $b = 1$ mm, length $L = 50$ cm, inclined at $\theta = 80°$. Use effective binary diffusivities in the film of $D_j = 10^{-5} \text{ cm}^2/\text{s}$, $j = A, B, AB$. Compute the steady-state concentration profile of each species within the film and the average absorption rate per unit area. Then, decrease the rate constant to zero to see what the mass transfer rate would be without reaction.

7 Probability theory and stochastic simulation

We now consider probability theory, and its applications in stochastic simulation. First, we define some basic probabilistic concepts, and demonstrate how they may be used to model physical phenomena. Next, we derive some important probability distributions, in particular, the Gaussian (normal) and Poisson distributions. Following this is a treatment of stochastic calculus, with a particular focus upon Brownian dynamics. Monte Carlo methods are then presented, with applications in statistical physics, integration, and global minimization (simulated annealing). Finally, genetic optimization is discussed. This chapter serves as a prelude to the discussion of statistics and parameter estimation, in which the Monte Carlo method will prove highly useful in Bayesian analysis.

The theory of probability

A *stochastic system* is one whose behavior is not purely deterministic and predictable, but rather has (assumed inherent) randomness. The theory of probability provides a mathematical framework for understanding and modeling such systems. Rather than provide abstract definitions, we introduce the subject through an example: modeling the distribution of polymer chain lengths in condensation polymerization.

Condensation polymers

Let us consider a reacting system comprising two chemical species. The first has a number α_1 of acid groups, e.g. $-COOH$, and the second has a number β_2 of base end groups, e.g. $-OH$ or $-NH_2$. These acid and base end groups react with each other to form linkages, $-CONH-$ or $-COO-$ and a condensate molecule (e.g. water) (Figure 7.1). An example is terephthalic acid and ethylene glycol (Figure 7.2), which react to form poly(ethylene terephathalate), PET, a common material found in soda bottles and clothing. As each molecule contains two functional groups, the reaction produces linear polymer chains with no branching.

If one of the species contains more than two functional groups, the polymer chains are not linear, but rather contain many branches, and at sufficiently high conversions (above the gel point) a cross-linked network is formed (Figure 7.3). We wish to apply probability theory to understand how the distribution of chain lengths depends upon the conversion of

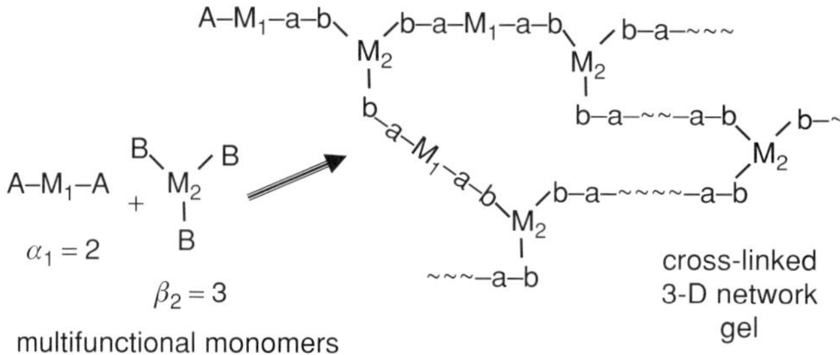

Figure 7.1 Condensation reactions among acid and base groups.

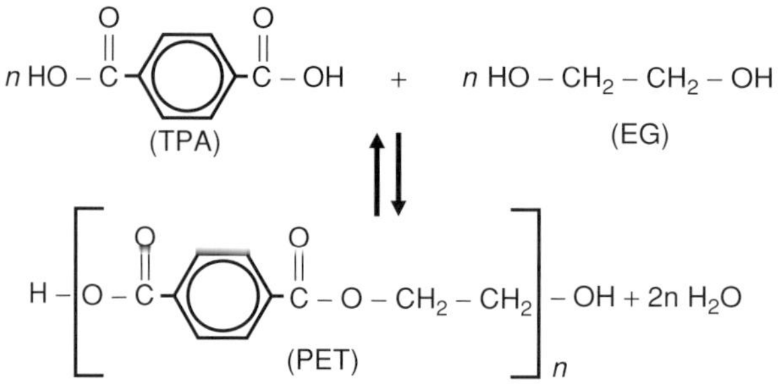

Figure 7.2 Condensation polymerization to produce linear PET chains.

Figure 7.3 Cross-linked network forms for multifunctional monomers when the conversion exceeds
the gel point.

the end groups, in both the linear and nonlinear (multifunctional) cases. By doing so, we
encounter many basic definitions in probability theory.

Chain length distribution in linear condensation polymers; joint and conditional probabilities

First, we consider the linear case with $\alpha_1 = \beta_2 = 2$. Initially, we have equal concentrations
of the two monomers, $[M_1]_0 = [M_2]_0$. As each monomer has two end groups, the initial
acid and base concentrations are also equal, $[A]_0 = [B]_0 = 2[M_1]_0$. As the acid and base
groups react, they form longer and longer chains. How does the distribution of chain lengths
vary as a function of conversion?

We define the conversions of the acid and base groups respectively as

$$p_A = 1 - \frac{[A]}{[A]_0} \qquad p_B = 1 - \frac{[B]}{[B]_0} \tag{7.1}$$

Because $[A]_0 = [B]_0$ and the reaction consumes equal numbers of acid and base end groups, $p_A = p_B = p$. At $0 \leq p \leq 1$, let $[P_n]$ be the concentration of chains that comprise n monomer units, i.e., that have a *chain length* or *degree of polymerization* equal to n. Each of the monomers has a chain length of 1; thus, at $p = 0$, $[P_1] = 2[M_1]_0 = [A]_0$ and all other $[P_{n \neq 1}] = 0$.

We wish to compute as functions of $0 \leq p \leq 1$ the *chain length distribution* $[P_n]$, the *number (DP_n) and weight (DP_w) averaged chain lengths*, and their ratio (the *polydispersity* P_{disp}), which is a measure of the breadth of the distribution. If $P_{\mathrm{disp}} = 1$, all chains are of the same length, and if $P_{\mathrm{disp}} \gg 1$ there is considerable variation in chain length.

$$DP_n \equiv \frac{\sum_{n=1}^{\infty} n[P_n]}{\sum_{n=1}^{\infty} [P_n]} \qquad DP_w \equiv \frac{\sum_{n=1}^{\infty} n^2[P_n]}{\sum_{n=1}^{\infty} n[P_n]} \qquad P_{\mathrm{disp}} \equiv \frac{DP_w}{DP_n} \tag{7.2}$$

We compute these quantities using statistical techniques developed by Paul Flory (Flory, 1953, Odian, 1991, Dotson *et al.*, 1996).

First, we make the assumption (the *equal reactivity hypothesis*) that the reactivity of an end group does not depend upon the length of the chain to which it is attached. As defined above, $p_A = p_B = p$ is the fraction of all acid or base groups that have reacted, and the fraction that remain unreacted is $(1 - p)$. If all groups are equally likely to have reacted or not, and we randomly select a group, the probability that it will have reacted is p and the probability that it will not have reacted is $(1 - p)$.

Defining the probability of an event

Above, we have invoked an argument of *frequentist* statistics to define a probability p from the conversion p. That is, the probability of observing an event E is determined by the expected number of observations of E, N_E, in a very large number N of independent random trials,

$$p(E) \approx \frac{N_E}{N} \tag{7.3}$$

It is not necessary to invoke such a large-population argument to define a probability. We can say that the probability of observing an event E is $p(E)$ if we have no preference between making the following two bets:

In a random trial, event E is observed.

or

A perfectly uniform random number generator in $[0, 1]$ returns a value u
less than or equal to $p(E)$.

The latter definition of a probability is less restrictive, but it also appears to be open to subjective interpretation. That is, we are making a personal value (belief) judgement about which bet to accept. This subject will be revisited in our discussion of statistics and parameter estimation. For now, we accept the latter definition as being more general, but note that the frequentist approach provides a convenient choice of belief when the necessary frequency information is available. It is always incumbent upon us to assign probabilities in such a way as to avoid inconsistencies. For example, the sum of the probabilities that E occurs and that E does not occur must always equal 1. Similarly, we define probabilities such that they satisfy the rules of joint and conditional probabilities outlined below.

We now use this probabilistic (frequentist) interpretation to compute $[P_n]$. Each chain in the system, being linear, has on average one unreacted acid group and one unreacted base group. As the acid and base concentrations are equal, the total number of chains is

$$\sum_{n=1}^{\infty}[P_n] = [A] = (1 - p)[A]_0 \tag{7.4}$$

What fraction of this total number of chains has a length equal to n? Or, equivalently, what is the probability that a randomly selected unreacted acid group is attached to a chain with exactly n monomer units? If this probability is $P(A$ is attached to chain of n units), then

$$[P_n] = (1 - p)[A]_0 \times P(A \text{ is attached to chain of } n \text{ units}) \tag{7.5}$$

To compute this quantity, we use the rules of probability theory, for which we need a few basic definitions.

Probability theory deals with *events*. Here, the event of interest is the observation that a randomly-selected chain has exactly n monomer units. Let us say that we have selected at random an unreacted acid end group that must lie at a chain end. If we march down the chain, we find that to achieve a length of exactly n units, the first $(n - 1)$ encountered acid groups must have reacted and the nth group must *not* have reacted. Thus, the event that our chain has exactly n units can be related to a sequence of simpler events – whether each of the acid groups encountered along the chain has reacted or not. We now compute the probability of the composite event (the chain contains n units) from the probabilities of these simpler events.

For each $j = 1, 2, \ldots, N$, let R_j be the event that the jth acid group along the chain has reacted, and U_j be the event that it is unreacted. Since these are the only two possibilities, their probabilities must sum to 1,

$$P(R_j) + P(U_j) = 1 \tag{7.6}$$

Our composite event, (A is attached to chain of n units), is equivalent to saying that the following sequence of events occurs, $R_1, R_2, \ldots, R_{n-1}, U_n$ so that $P(A$ is attached to chain of n units) is equal to the joint probability

$$P(A \text{ is attached to chain of } n \text{ units}) = P(R_1 \cap R_2 \cap \cdots \cap R_{n-1} \cap U_n) \tag{7.7}$$

$\cap$ is the symbol for intersection, and here signifies that the events on both sides of the sign occur.

Defining joint and conditional probabilities

The *joint probability* of two events E_1 and E_2 is the probability $P(E_1 \cap E_2)$, also written $P(E_1, E_2)$, that both occur. If $P(E_1)$ is the probability that E_1 occurs,

$$P(E_1 \cap E_2) = P(E_1)P(E_2 \mid E_1) \tag{7.8}$$

where $P(E_2|E_1)$ is the *conditional probability* that if E_1 occurs, so does E_2. If the probability of observing event E_2 does not depend upon whether E_1 occurs or not, the two events are said to be *independent*, and

$$P(E_2|E_1) = P(E_2) \quad P(E_1 \cap E_2) = P(E_1)P(E_2) \tag{7.9}$$

The conditional and joint probabilities are related by *Bayes' theorem*, whose application to statistics forms the basis of the next chapter,

$$P(E_1 \cap E_2) = P(E_1)P(E_2|E_1) = P(E_2)P(E_1|E_2) \tag{7.10}$$

This structure is quite flexible and is easily extended to multiple events. Using this framework, we write the probability that a randomly-selected chain has exactly n units as a product of conditional probabilities as

$$P(\text{A is attached to chain of } n \text{ units})$$
$$= P(R_1) \times P(R_2|R_1) \times P(R_3|R_1 \cap R_2) \times \cdots \times P(R_{n-1}|R_1 \cap R_2 \cap \cdots \cap R_{n-2})$$
$$\times P(U_n|R_1 \cap R_2 \cap \cdots \cap R_{n-1}) \tag{7.11}$$

If we assume that the probabilities of every group having reacted or not are equal and independent, the events in this sequence are independent, and we have

$$P(\text{A is attached to chain of } n \text{ units}) = P(R_1) \times P(R_2) \times \cdots \times P(R_{n-1}) \times P(U_n) \tag{7.12}$$

The probabilities of observing a reacted or unreacted group are

$$P(R_j) = p \quad P(U_j) = (1 - p) \tag{7.13}$$

therefore,

$$P(\text{A is attached to chain of } n \text{ units}) = p^{n-1}(1 - p) \tag{7.14}$$

The concentrations of each chain length polymer at conversion p thus are predicted to follow the *Flory most-probable chain length distribution*:

$$[P_n] = (1 - p)[A]_0 \times p^{n-1}(1 - p) = [A]_0(1 - p)^2 p^{n-1} \tag{7.15}$$

The average chain lengths and polydispersity are computed using the formulas of geometric progression

$$DP_n \equiv \frac{\sum_{n=1}^{\infty} n[P_n]}{\sum_{n=1}^{\infty} [P_n]} = \frac{1}{1-p} \qquad DP_w \equiv \frac{\sum_{n=1}^{\infty} n^2[P_n]}{\sum_{n=1}^{\infty} n[P_n]} = \frac{1+p}{1-p}$$
$$P_{\text{disp}} \equiv \frac{DP_w}{DP_n} = 1 + p \tag{7.16}$$

We only achieve high chain lengths at conversions near 1, and in this limit $P_{\text{disp}} \approx 2$. Figure 7.4 shows the distribution (7.15) at 99% conversion.

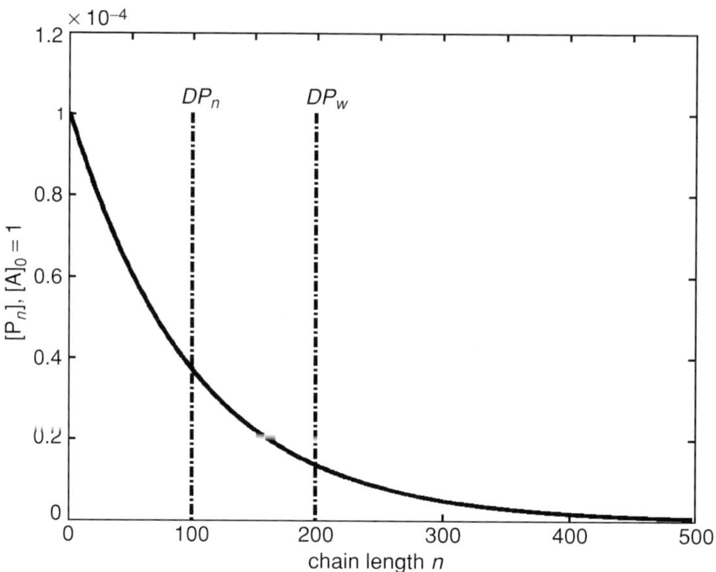

Figure 7.4 Flory most-probable chain length distribution at a conversion of 0.99. The locations of number and weight averaged chain lengths are noted. Polydispersity is 1.99.

Gelation of multifunctional monomers (more on conditional probabilities and mathematical expectation)

We now extend this statistical approach to the case of multifunctional monomers in which either α_1 or β_2 exceeds 2 so that above a critical *gel point* conversion, the system forms a cross-linked network that extends throughout space (Figure 7.3). To do so, we use the *Macosko–Miller method* (Macosko & Miller, 1976), in which we compute at a given conversion the average mass W of a chain that is attached to a randomly-selected monomer unit. The gel point is the conversion at which W diverges to infinity.

First, we must be more precise about what we mean by "average." Let us say that we perform a number of trial experiments $v = 1, 2, 3, \ldots, N_{\text{exp}}$ in which we select at random a monomer unit in the system, and for each trial we measure the mass W_v of the chain that contains the randomly-selected unit. As W_v varies in a stochastic manner from one experiment to the next, it is a *random variable*. If M_1 is the mass of a single monomer unit of type 1 and M_2 is the mass of a single monomer unit of type 2, W_v takes the value for a chain of m_1 type-1 units and m_2 type-2 units,

$$W_{m_1,m_2} = m_1 M_1 + m_2 M_2 \tag{7.17}$$

We obtain DP_w by setting $M_1 = M_2 = 1$. If $P(W_{m_1,m_2})$ is the probability of observing in any single trial the value W_{m_1,m_2}, we define the *expectation* of the random variable W as

$$E(W) = \sum_{m_1,m_2} W_{m_1,m_2} P(W_{m_1,m_2}) \tag{7.18}$$

If the number of trials N_{exp} that we perform is very large, we expect

$$E(W) = \langle W \rangle \approx \frac{1}{N_{\text{exp}}} \sum_{v=1}^{N_{\text{exp}}} W_v \tag{7.19}$$

To predict the gel point, we compute $E(W)$ at the specified conversions

$$p_A = 1 - \frac{[A]}{[A]_0} \quad p_B = 1 - \frac{[B]}{[B]_0} \tag{7.20}$$

Unlike our previous discussion, we do *not* assume that $p_A = p_B$. We start with N_1 monomers of type 1 per unit volume, each with α_1 acid groups, and N_2 monomers per unit volume of type 2, each with β_2 base groups. The initial acid and base end group concentrations are

$$[A]_0 = \alpha_1 N_1 \quad [B]_0 = \beta_2 N_2 \tag{7.21}$$

As the numbers of acid and base groups consumed by reaction are equal,

$$[A]_0 - [A] = [B]_0 - [B] \quad \Rightarrow \quad [A]_0 p_A = [B]_0 p_B \tag{7.22}$$

Hence, the conversions of the acid and base groups are related,

$$p_B = \frac{[A]_0}{[B]_0} p_A = \left(\frac{\alpha_1 N_1}{\beta_2 N_2}\right) p_A \tag{7.23}$$

The conditional probability that if we select an acid group, it is unreacted, is

$$P(A|a) = 1 - p_A \tag{7.24}$$

and the conditional probability that it has reacted to form a linkage is

$$P(L|a) = p_A \tag{7.25}$$

As any randomly-selected acid group must be either reacted or unreacted, these two conditional probabilities must sum to 1:

$$P(L|a) + P(A|a) = 1 \tag{7.26}$$

Similarly for the base group, we have the conditional probabilities

$$P(B|b) = 1 - p_B \quad P(L|b) = p_B \tag{7.27}$$

From these conditional probabilities, we compute $E(W)$. First, we note that if we randomly select a monomer unit at random, the probabilities that it is of type 1 or 2 are

$$P(M_1) = \frac{N_1}{N_1 + N_2} \quad P(M_2) = \frac{N_2}{N_1 + N_2} \tag{7.28}$$

Here M_1 and M_2 denote the events that the randomly-selected monomer unit is type 1 or type 2 respectively.

If $E(W|M_1)$ is the *conditional expectation* of W when we have selected a type 1 monomer unit, and $E(W|M_2)$ is the corresponding value for a type-2 monomer, we expand $E(W)$ in terms of the mutually exclusive events M_1 and M_2:

$$E(W) = E(W|M_1)P(M_1) + E(W|M_2)P(M_2) \tag{7.29}$$

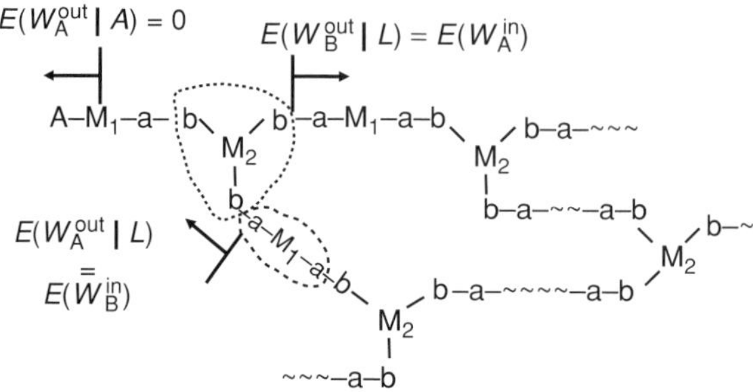

Figure 7.5 Expected weights looking "out" or "in" from functional groups.

We first consider $E(W|M_1)$. If we select a type-1 unit, the mass W of the chain must at least equal the mass M_1 of a single type-1 unit, as well as the contributions from the masses observed by looking "outwards" from each of the α_1 acid groups (Figure 7.5). $E(W_A^{out})$ is the expected weight of the section of chain attached to a type-1 unit through one of its acid groups. Thus,

$$E(W|M_1) = M_1 + \alpha_1 E\left(W_A^{out}\right) \tag{7.30}$$

Similarly, the expected weight observed if we select a type-2 monomer equals the weight of a single type-2 monomer unit plus the expected weights $E(W_B^{out})$ looking outwards across each of the β_2 base groups:

$$E(W|M_2) = M_2 + \beta_2 E\left(W_B^{out}\right) \tag{7.31}$$

To compute $E(W_A^{out})$, we note that a randomly-selected acid group must be either reacted or unreacted, so that we expand in the possible outcomes:

$$E\left(W_A^{out}\right) = E\left(W_A^{out}|L\right)P(L|a) + E\left(W_A^{out}|A\right)P(A|a) \tag{7.32}$$

If the selected acid group is unreacted, there is no chain attached to this group and $E(W_A^{out}|A) = 0$. Also, as $P(L|a) = p_A$, we have

$$E\left(W_A^{out}\right) = E\left(W_A^{out}|L\right) \times p_A \tag{7.33}$$

Next, from Figure 7.5, we note that the expected weight looking "out" from an acid group on monomer 1 equals the expected weight observed looking "in" from a base group on a monomer of type 2, and hence

$$E\left(W_A^{out}|L\right) = E\left(W_B^{in}\right) = M_2 + (\beta_2 - 1)E\left(W_B^{out}\right) \tag{7.34}$$

Here, we have used the fact that if we come "in" across one of the base groups, there are only $(\beta_2 - 1)$ other possible ways to go "out." Combining (7.33) and (7.34) yields

$$E\left(W_A^{out}\right) = E\left(W_B^{in}\right) \times p_A = p_A\left[M_2 + (\beta_2 - 1)E\left(W_B^{out}\right)\right] \tag{7.35}$$

Applying the same logic to $E(W_B^{out})$,

$$E\left(W_B^{out}\right) = E\left(W_A^{in}\right) \times p_B = p_B\left[M_1 + (\alpha_1 - 1)E\left(W_A^{out}\right)\right] \tag{7.36}$$

Equations (7.35) and (7.36) provide two equations for the two unknowns $E(W_A^{out})$ and $E(W_B^{out})$, yielding

$$E\left(W_A^{out}\right) = \frac{p_A[M_2 + (\beta_2 - 1)p_B M_1]}{1 - p_A(\alpha_1 - 1)p_B(\beta_2 - 1)} \tag{7.37}$$

from which $E(W_B^{out})$ is computed by (7.36). The expected weight attached to a randomly selected group is then

$$E(W) = \left[M_1 + \alpha_1 E\left(W_A^{out}\right)\right]P(M_1) + \left[M_2 + \beta_2 E\left(W_B^{out}\right)\right]P(M_2) \tag{7.38}$$

Gelation occurs when $E(W) \to \infty$, which happens as $1 - p_A(\alpha_1 - 1)p_B(\beta_2 - 1)$, the denominator of $E(W_A^{out})$, goes to zero, yielding the gel point condition

$$p_A(\alpha_1 - 1)p_B(\beta_2 - 1) = 1 \tag{7.39}$$

Using the relation between the two conversions posed by the reaction stoichiometry (7.23), the conversion of A at the gel point is

$$p_{A,c} = \sqrt{\left(\frac{\beta_2 N_2}{\alpha_1 N_1}\right) \frac{1}{(\alpha_1 - 1)(\beta_2 - 1)}} \tag{7.40}$$

For balanced end group concentrations, with $\alpha_1 = 2, \beta_2 = 3$,

$$p_{A,c} = p_{B,c} = p_c = \sqrt{\frac{1}{(2 - 1)(3 - 1)}} = \sqrt{\frac{1}{2}} = 0.7071 \tag{7.41}$$

Figure 7.6 shows the predicted DP_w vs. p, up to the point where it diverges to infinity as the gel forms (denoted by vertical dashed line).

Above, we have computed the gel point only for a mixture of two reactants. This theory is extended to the case of multiple monomers (with a more general treatment of the conditional probabilities) in Beers & Ray (2001). This probabilistic approach can also be used to compute the gel and sol mass fractions following gelation.

Important probability distributions

We now consider some important, and common, forms of probability distributions for a random variable. First, we state some basic definitions.

Definition Probability distribution of a discrete random variable
Let X be a random variable that may take one of a countable number M of discrete values X_1, $X_2, \ldots, X_M$. Let us conduct some very large number T of trials in which we measure the value of X. Let $N(X_j)$ be the number of times that we observe the value X_j, $\Sigma_{j=1}^{M} N(X_j) = T$. Then, the *probability distribution* of X is defined in the frequentist manner for very large T as

$$P(X_j) = \frac{N(X_j)}{T} \qquad \sum_{j=1}^{M} P(X_j) = 1 \tag{7.42}$$

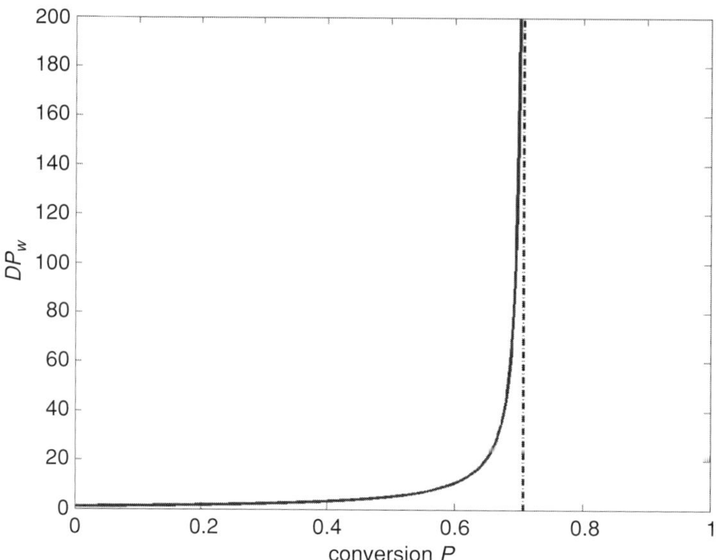

Figure 7.6 DP_w vs. p for gel formation with a bifunctional acid and a trifunctional base, with balanced end group concentrations. Gelation occurs at a conversion of 70.7%. ($[A]_0 = [B]_0$, $\alpha_1 = 2$, $\beta_2 = 3$.)

Definition Probability distribution of a continuous random variable

Let x be a random variable that may take any value between x_{lo} and x_{hi}. We define the *continuous probability distribution* of x to be the function $p(x)$, such that the probability of observing a value between x and $x + dx$ is $p(x)dx$. This probability distribution is normalized to 1:

$$\int_{x_{\text{lo}}}^{x_{\text{hi}}} p(x)dx = 1 \qquad (7.43)$$

and the expectation, or average, value of x is

$$E(x) = \langle x \rangle = \int_{x_{\text{lo}}}^{x_{\text{hi}}} x p(x)dx \qquad (7.44)$$

To generate the continuous probability distribution from a number of trial measurements, we subdivide the region $x_{\text{lo}} \leq x \leq x_{\text{hi}}$ into B nonoverlapping bins, each of width $\Delta x = (x_{\text{hi}} - x_{\text{lo}})/B$. Bin j contains the subdomain $x_j - (\Delta x)/2 \leq x \leq x_j + (\Delta x)/2$. Again we perform a very large number T of trials, in which we count the number of times $N(x_j)$ that we observe a value of x in bin j. Then, the value of $p(x_j)$ is approximately

$$p(x_j) \approx \frac{N(x_j)}{(\Delta x)T} \qquad (7.45)$$

and we approximate the distribution using piecewise-constant interpolation,

$$p(x) \approx \sum_{j=1}^{B} \left[\frac{N(x_j)}{(\Delta x)T} \right] \Omega_j(x)$$

$$\Omega_j(x) = \begin{cases} 1, & \text{if}\,[x_j - (\Delta x)/2] \leq x < [x_j + (\Delta x)/2] \\ 0, & \text{otherwise} \end{cases} \qquad (7.46)$$

Definition Cumulative probability distribution

We define from either a discrete or a continuous probability distribution the *cumulative probability distribution*,

$$F(X_k) = \sum_{j=1}^{k} P(X_j) \quad \text{or} \quad F(x) = \int_{x_{\text{lo}}}^{x} p(x')dx' \tag{7.47}$$

that satisfies the normalization condition

$$F(X_M) = 1 \quad \text{or} \quad F(x_{\text{hi}}) = 1 \tag{7.48}$$

From the cumulative probability distribution, we can generate values of X or x at random according to the specified probability distribution. For example, in the case of a continuous variable, we generate a uniformly distributed number $0 \leq u \leq 1$ using **rand**. The corresponding random value of x, generated according to the probability distribution $p(x)$ satisfies

$$F(x) = \int_{x_{\text{lo}}}^{x} p(x')dx' = u \tag{7.49}$$

Definition Variance and standard deviation

We have defined the expectation for both discrete and continuous probability distributions that gives the average, or mean value, of a random variable. To obtain a measure of the breadth of the distribution of X, we define the *variance* to be the expected quadratic variation from the mean,

$$\text{var}(X) = E[(X - E(X))^2] = E(X^2) - [E(X)]^2 \tag{7.50}$$

The square root of the variance is known as the *standard deviation*,

$$\sigma = \sqrt{\text{var}(X)} \tag{7.51}$$

If X and Y are independent, the variance of their sum is simply

$$\text{var}(X + Y) = \text{var}(X) + \text{var}(Y) \tag{7.52}$$

Later we extend these definitions to multiple, interacting random variables, but first consider some common forms of discrete and continuous probability distributions.

Bernoulli trials

Let us say that we are tossing a coin for which we have a probability p_{H} of observing heads and a probability p_{T} of observing tails. Each coin toss, assumed independent, is an example of a *Bernoulli trial*. If the coin is fair, $p_{\text{H}} = p_{\text{T}} = 1/2$. We now ask the following question:

If we make n independent coin tosses, what is the probability that we will observe heads n_{H} number of times, i.e. that our sequence of tosses will return heads n_{H} times and tails $n_{\text{T}} = n - n_{\text{HT}}$ times?

While this may seem like a question that is only important for those planning a trip to Las Vegas, it is directly relevant to a wide variety of physical phenomena. Let us say that

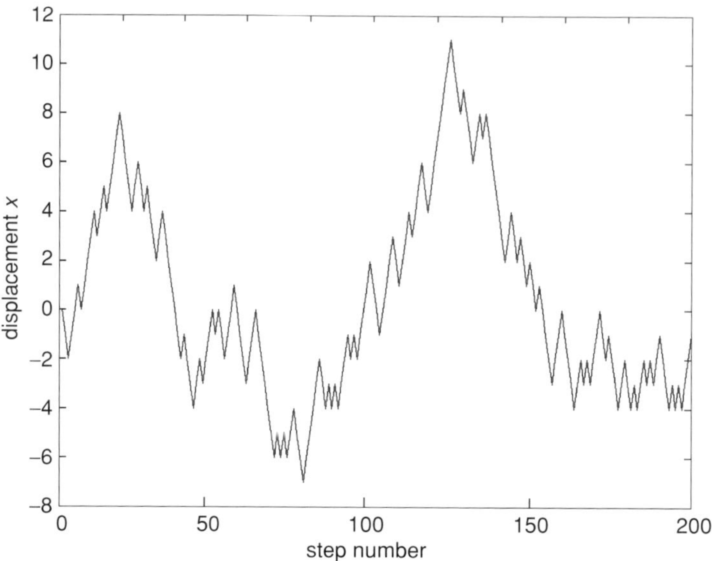

Figure 7.7 One-dimensional random walk showing the net displacement x vs. the number of random steps of length 1.

we are performing an experiment, and observe that even if we hold all of the parameters constant (insofar as we can identify, measure, and control them), we still do not measure exactly the same result from one experiment to the next. The result of any single experiment contains some random effect of noise. Often, we do not know exactly from where this noise originates, yet we do suspect that it comes not just from one source. Rather, the observed error is the net sum of many (say N_e) small random errors. A reasonable model for ε, the random error in an experiment, is then

$$\varepsilon = c(2\zeta_1 - 1) + c(2\zeta_2 - 1) + \cdots + c(2\zeta_{N_e} - 1) \tag{7.53}$$

ζ_j is a random variable whose value is determined by a coin toss,

$$\zeta_j = \begin{cases} 1, & \text{if heads} \\ 0, & \text{if tails} \end{cases} \tag{7.54}$$

and is independent of the values of the other error contributions. Thus, $(2\zeta_j - 1) = \pm 1$ With this model, the error in any single experiment is determined by the total number of heads $\sum_{j=1}^{N_e} \zeta_j$ in a set of N_e coin tosses.

The random walk problem

Models of Brownian motion and of the geometry of ideal polymer chains are based on the concept of a *random walk*. Let us start at time $t = 0$ and flip a coin. If heads, we take a step to the right of length l. If tails, we take a step to the left. We continue to do this, taking n steps, and measure the net displacement x (our final position) (Figure 7.7).

After N_s steps (coin tosses) the net displacement that we have traveled is

$$x = l \sum_{j=1}^{n} (2\zeta_j - 1) \quad \zeta_j = \begin{cases} 1, & \text{if heads} \\ 0, & \text{if tails} \end{cases} \tag{7.55}$$

The average (and most likely) displacement is zero, as we double back upon our path at random,

$$\langle x \rangle = l \left\langle \sum_{j=1}^{n} (2\zeta_j - 1) \right\rangle = l \sum_{j=1}^{n} (2\langle \zeta_j \rangle - 1) = l \sum_{j=1}^{n} \left(2 \left(\frac{1}{2} \right) - 1 \right) = 0 \tag{7.56}$$

but, the mean-squared displacement is *not* zero,

$$\langle x^2 \rangle = \left\langle \left[l \sum_{j=1}^{n} (2\zeta_j - 1) \right] \left[l \sum_{k=1}^{n} (2\zeta_k - 1) \right] \right\rangle = l^2 \sum_{j=1}^{n} \sum_{k=1}^{n} \langle (2\zeta_j - 1)(2\zeta_k - 1) \rangle$$

$$\langle x^2 \rangle = l^2 \sum_{j=1}^{n} \sum_{k=1}^{n} \langle \zeta_j' \zeta_k' \rangle \quad \zeta_j' = 2\zeta_j - 1 = \begin{cases} 1, & \text{if heads} \\ -1, & \text{if tails} \end{cases} \tag{7.57}$$

Using the independence of the coin tosses,

$$\langle \zeta_j' \zeta_k' \rangle = \delta_{jk} = \begin{cases} 1, & \text{if } j = k \\ 0, & \text{if } j \neq k \end{cases} \tag{7.58}$$

we find that the mean-squared displacement is linear in the number of steps,

$$\langle x^2 \rangle = l^2 \sum_{j=1}^{n} \sum_{k=1}^{n} \langle \zeta_j' \zeta_k' \rangle = l^2 \sum_{j=1}^{n} \sum_{k=1}^{n} \delta_{jk} = l^2 \sum_{j=1}^{n} (1) = l^2 n \tag{7.59}$$

The binomial distribution

Now that we have seen some possible applications of this coin toss question, let us derive the answer. First, let us say that we make ten coin tosses. We wish to compute the probability of observing an exact ordered sequence of heads and tails, e.g.

H   T   T   H   H   T   H   H   T   H

If the outcomes of each coin toss are independent, and if p_H is the probability of a coin toss returning heads and $p_T = 1 - p_H$ is the probability that it returns tails, the probability of observing the exact sequence above is

$$p_H \times p_T \times p_T \times p_H \times p_H \times p_T \times p_H \times p_H \times p_T \times p_H = p_H^6 p_T^4 \tag{7.60}$$

In general, the probability of observing an exact ordered sequence of n_H heads and n_T tails is $p_H^{n_H} p_T^{n_T}$. Note that the sequence above is only one of the possible sequences containing n_H heads and n_T tails. Others include

H   H   T   T   H   H   T   H   T   H
H   H   T   H   H   T   T   H   H   T
H   T   H   H   T   H   T   H   H   T

Since every sequence with the same number of heads and tails has the same probability of being observed, the probability that we observe *any* sequence of n_H heads and $n_T = n - n_H$

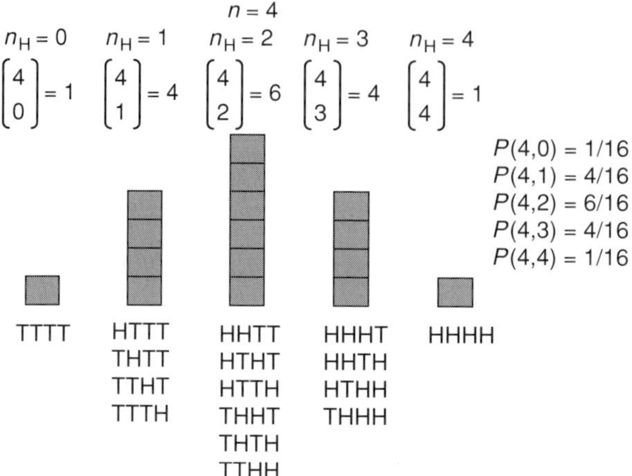

Figure 7.8 The binomial distribution for four fair coin tosses.

tails (without regard to the order in which they appear) follows the *binomial distribution*

$$P(n, n_\mathrm{H}) = \binom{n}{n_\mathrm{H}} p_\mathrm{H}^{n_\mathrm{H}} p_\mathrm{T}^{n_\mathrm{T}} = \binom{n}{n_\mathrm{H}} p_\mathrm{H}^{n_\mathrm{H}} (1 - p_\mathrm{H})^{(n - n_\mathrm{H})} \tag{7.61}$$

$\binom{n}{n_\mathrm{H}}$ is the number of possible sequences of n tosses with n_H heads, and is known as a *binomial coefficient*,

$$\binom{n}{n_\mathrm{H}} = \frac{n!}{n_\mathrm{H}!(n - n_\mathrm{H})!} \quad n! = n \times (n - 1) \times \cdots \times 3 \times 2 \times 1 \tag{7.62}$$

As a check, consider the case of $n = 4$, $n_\mathrm{H} = 2$, for which

$$\binom{4}{2} = \frac{4!}{2!(4 - 2)!} = \frac{4 \times 3 \times 2 \times 1}{(2 \times 1)(2 \times 1)} = \frac{24}{4} = 6 \tag{7.63}$$

The six possible sequences are

$$\mathrm{H\,H\,T\,T} \qquad \mathrm{T\,H\,H\,T} \qquad \mathrm{T\,T\,H\,H} \qquad \mathrm{H\,T\,H\,T} \qquad \mathrm{T\,H\,T\,H} \qquad \mathrm{H\,T\,T\,H}$$

For a fair coin,

$$P(n, n_\mathrm{H}) = \binom{n}{n_\mathrm{H}} \left(\frac{1}{2}\right)^{n_\mathrm{H}} \left(\frac{1}{2}\right)^{n_\mathrm{T}} = \binom{n}{n_\mathrm{H}} \left(\frac{1}{2}\right)^{n} \tag{7.64}$$

hence, we have the distribution of sequences shown in Figure 7.8.

The optional MATLAB statistics toolkit contains several functions for evaluating the binomial distribution. **binornd** generates random numbers according to the binomial distribution, **binofit** fits a binomial distribution to a data set, **binostat** computes the mean and variance, **binopdf** returns the probability distribution, and the cumulative distribution and its inverse are returned by **binocdf** and **binoinv**. Similar routines are available for a host of other common probability distributions (see the toolkit documentation for a list).

The Gaussian (normal) distribution

Let us return to the example of the random walk, in which the net displacement after n steps of length l is

$$x = l \sum_{j=1}^{n} (2\zeta_j - 1) \quad \zeta_j = \begin{cases} 1, & \text{if heads} \\ 0, & \text{if tails} \end{cases} \tag{7.65}$$

If we perform n coin tosses, and n_H are heads, the net displacement is

$$\frac{x}{l} = n_\mathrm{H} - n_\mathrm{T} = n_\mathrm{H} - (n - n_\mathrm{H}) = 2n_\mathrm{H} - n \tag{7.66}$$

Hence for a given displacement, the numbers of heads and tails are

$$n_\mathrm{H} = \frac{1}{2}\left(n + \frac{x}{l}\right) \quad n_\mathrm{T} = n - n_\mathrm{H} = \frac{1}{2}\left(n - \frac{x}{l}\right) \tag{7.67}$$

For a fair coin, we obtain from the binomial distribution,

$$P(n, n_\mathrm{H}) = \binom{n}{n_\mathrm{H}}\left(\frac{1}{2}\right)^n = \left[\frac{n!}{n_\mathrm{H}!(n - n_\mathrm{H})!}\right]\left(\frac{1}{2}\right)^n \tag{7.68}$$

the probability of observing a net displacement x after n steps of length l,

$$P(x; n, l) = \frac{n!}{[(n + x/l)/2]![(n - x/l)/2]!}\left(\frac{1}{2}\right)^n \tag{7.69}$$

We evaluate this equation in the limit $n \to \infty$, taking the natural logarithm,

$$\ln[P(x; n, l)] = \ln(n!) - \ln\{[(n + x/l)/2]!\} - \ln\{[(n - x/l)/2]!\} - n \ln 2 \tag{7.70}$$

We remove the factorials by using *Stirling's approximation*

$$\ln(N!) = \ln\left(\prod_{m=1}^{N} m\right) = \sum_{m=1}^{N} \ln(m) \approx \int_{1}^{N} \ln x \, dx = N \ln N - N \tag{7.71}$$

Therefore, for large n, we have

$$\ln[P(x; n, l)] \approx n \ln n - n - \left[\frac{n + x/l}{2}\right]\ln\left[\frac{n + x/l}{2}\right] + \left[\frac{n + x/l}{2}\right]$$
$$- \left[\frac{n - x/l}{2}\right]\ln\left[\frac{n - x/l}{2}\right] + \left[\frac{n - x/l}{2}\right] - n \ln 2 \tag{7.72}$$

After cancelling out terms, this simplifies to

$$\ln[P(x; n, l)] \approx -\left[\frac{n + x/l}{2}\right]\ln\left[\frac{n + x/l}{2}\right]$$
$$- \left[\frac{n - x/l}{2}\right]\ln\left[\frac{n - x/l}{2}\right] - n \ln\left(\frac{n}{2}\right) \tag{7.73}$$

We further simplify the expression by noting

$$n \ln\left(\frac{n}{2}\right) = \left\{\left[\frac{n + x/l}{2}\right] + \left[\frac{n - x/l}{2}\right]\right\}\ln\left(\frac{n}{2}\right) \tag{7.74}$$

hence,

$$\ln[P(x;n,l)] \approx -\left[\frac{n+x/l}{2}\right]\left\{\ln\left[\frac{n+x/l}{2}\right] + \ln\left(\frac{n}{2}\right)\right\}$$
$$-\left[\frac{n-x/l}{2}\right]\left\{\ln\left[\frac{n-x/l}{2}\right] + \ln\left(\frac{n}{2}\right)\right\} \tag{7.75}$$

The rule $\ln(ab) = \ln(a) + \ln(b)$ yields

$$\ln[P(x;n,l)] \approx -\left[\frac{n+x/l}{2}\right]\left\{\ln\left[\frac{n+x/l}{2}\left(\frac{2}{n}\right)\right]\right\}$$
$$-\left[\frac{n-x/l}{2}\right]\left\{\ln\left[\frac{n-x/l}{2}\left(\frac{2}{n}\right)\right]\right\}$$
$$\approx -\left[\frac{n+x/l}{2}\right]\left\{\ln\left[1 + \frac{x}{nl}\right]\right\} - \left[\frac{n-x/l}{2}\right]\left\{\ln\left[1 - \frac{x}{nl}\right]\right\} \tag{7.76}$$

In the limit of large n, the high degree of backtracking in the random walk means that $|x|$ $\ll nl$, hence we use the Taylor expansions around $x/(nl) = 0$,

$$\ln\left[1 + \frac{x}{nl}\right] \approx \frac{x}{nl} - \frac{1}{2}\left(\frac{x}{nl}\right)^2 \qquad \ln\left[1 - \frac{x}{nl}\right] \approx -\frac{x}{nl} - \frac{1}{2}\left(\frac{x}{nl}\right)^2 \tag{7.77}$$

to obtain

$$\ln[P(x;n,l)] \approx -\left[\frac{n+x/l}{2}\right]\left\{\frac{x}{nl} - \frac{1}{2}\left(\frac{x}{nl}\right)^2\right\} - \left[\frac{n-x/l}{2}\right]\left\{-\frac{x}{nl} - \frac{1}{2}\left(\frac{x}{nl}\right)^2\right\} \tag{7.78}$$

Collecting terms yields the particularly simple result

$$\ln[P(x;n,l)] \approx -\frac{x^2}{2nl^2} \tag{7.79}$$

Taking the exponential and renormalizing to $\int_{-\infty}^{+\infty} P(x;n,l)dx = 1$ yields

$$P(x;n,l) = \frac{1}{\sqrt{2\pi nl^2}}\exp\left[-\frac{x^2}{2nl^2}\right] \tag{7.80}$$

Defining $\sigma^2 = nl^2$ produces the *Gaussian (normal) distribution*

$$P(x;\sigma) = \frac{1}{\sigma\sqrt{2\pi}}\exp\left[-\frac{x^2}{2\sigma^2}\right] \tag{7.81}$$

We note by symmetry of the exponential argument that the average is zero:

$$\langle x \rangle = E(x) = \int_{-\infty}^{+\infty} x P(x;\sigma)dx = \int_{-\infty}^{+\infty} \frac{x}{\sigma\sqrt{2\pi}}\exp\left[-\frac{x^2}{2\sigma^2}\right]dx = 0 \tag{7.82}$$

The *variance* is

$$\mathrm{var}(x) = E[(x - E(x))^2] = \int_{-\infty}^{+\infty} x^2 P(x;\sigma)dx = \int_{-\infty}^{+\infty} \frac{x^2}{\sigma\sqrt{2\pi}}\exp\left[-\frac{x^2}{2\sigma^2}\right]dx = \sigma^2 \tag{7.83}$$

Therefore, σ is the *standard deviation* of the Gaussian distribution. The Gaussian distribution is plotted in Figure 7.9 for various values of σ.

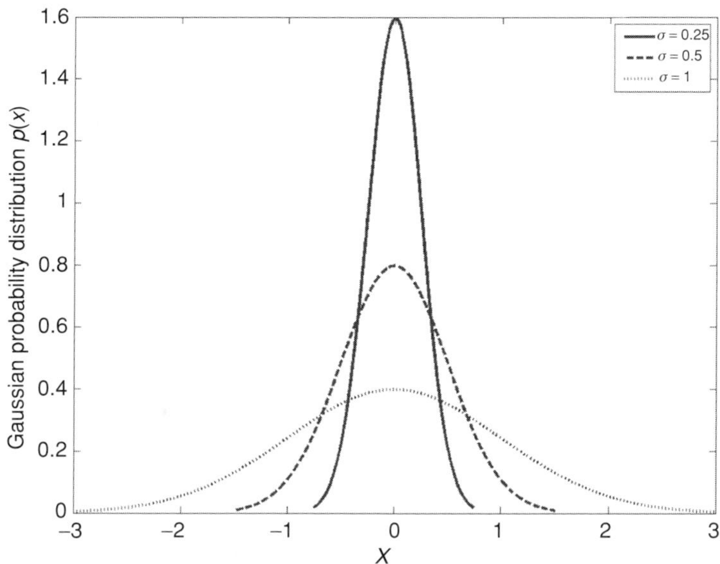

Figure 7.9 Gaussian (normal) probability distributions with means of zero and various values of the standard deviation.

The central limit theorem of statistics

We have seen that the binomial distribution of a random walk reduces to the Gaussian distribution as $n \to \infty$. We now make this result more general.

Let $\zeta_1, \zeta_2, \ldots, \zeta_n$ be a set of independent random variables with means $\mu_j = E(\zeta_j)$ and variances $\sigma_j^2 = \mathrm{var}(\zeta_j)$. The distributions of these random variables need *not* be Gaussian. According to the *central limit theorem of statistics*, the statistic

$$S_n = \frac{1}{\sqrt{n}} \sum_{j=1}^{n} \frac{\zeta_j - \mu_j}{\sigma_j} \tag{7.84}$$

itself a random variable, is normally distributed in the limit $n \to \infty$ with a variance of 1:

$$P(S_n) = \frac{1}{\sqrt{2\pi}} \exp\left[-\frac{S_n^2}{2}\right] \quad \text{as } n \to \infty \tag{7.85}$$

This theorem is what makes the normal distribution "normal."

The Gaussian distribution with nonzero mean

We define the probability distribution $N(\mu, \sigma^2)$ to be the normal distribution with a mean μ and variance σ^2,

$$N(\mu, \sigma^2) = P(x; \mu, \sigma) = \frac{1}{\sigma\sqrt{2\pi}} \exp\left[-\frac{(x - \mu)^2}{2\sigma^2}\right]$$

$$E(x) = \langle x \rangle = \mu \qquad \mathrm{var}(x) = \sigma^2 \tag{7.86}$$

In MATLAB, we generate a random number distributed according to $N(0, 1)$ using **randn**. From such a random number r, we obtain a random number r' distributed according to $N(\mu, \sigma^2)$ from the rule

$$r' = \mu + \sigma r \tag{7.87}$$

or we can use the **statistics toolkit** function **normrnd** to directly generate random numbers from $N(\mu, \sigma^2)$. Both **randn** and **normrnd** can also generate matrices of independent random numbers.

The Poisson distribution

We next consider another limiting case of the binomial distribution. Let us perform a number n of Bernoulli trials in which we have for each a probability p of "success" and a probability $(1 - p)$ of "failure". For each trial, we define the random variable

$$\zeta_j = \begin{cases} 1, & \text{if trial is a success} \\ 0, & \text{if trial is a failure} \end{cases} \tag{7.88}$$

The expectation is $E(\zeta_j) = p$ and the variance is

$$\text{var}(\zeta_j) = E\left(\zeta_j^2\right) - [E(\zeta_j)]^2 = p - [p]^2 = p(1 - p) \tag{7.89}$$

The total number of successes in the trial, itself a random variable, is

$$\zeta = \sum_{j=1}^{n} \zeta_j \qquad E(\zeta) = \sum_{j=1}^{n} E(\zeta_j) = \sum_{j=1}^{n} (p) = pn \tag{7.90}$$

As each trial is independent,

$$\text{var}(\zeta) = \sum_{j=1}^{n} \text{var}(\zeta_j) = \sum_{j=1}^{n} p(1 - p) = np(1 - p) \tag{7.91}$$

This sum is distributed according to the binomial distribution,

$$P(\zeta; n, p) = \binom{n}{\zeta} p^\zeta (1 - p)^{n-\zeta} \tag{7.92}$$

We now derive a limiting form of this probability distribution that is valid in the limit $n \to \infty$ when the probability of success in any single trial is very small, $p \ll 1$. First, from the series expansion

$$e^{-p} \approx 1 - p + \frac{p^2}{2} - \cdots \tag{7.93}$$

we obtain for $p \ll 1$, and the corresponding condition $\zeta \ll n$,

$$(1 - p)^{n-\zeta} \approx (e^{-p})^{n-\zeta} \approx (e^{-p})^n = e^{-pn} \tag{7.94}$$

Writing the binomial coefficient explicitly yields

$$P(\zeta; n, p) \approx \frac{p^\zeta}{\zeta!} e^{-pn} \left[\frac{n!}{(n-\zeta)!} \right] \tag{7.95}$$

Next, we take the natural logarithm and apply Stirling's approximation,

$$\ln\left[\frac{n!}{(n-\zeta)!}\right] = \ln(n!) - \ln[(n-\zeta)!] \approx n\ln n - n - (n-\zeta)\ln(n-\zeta) + n - \zeta$$
$$\approx n\ln n - n - (n-\zeta)\ln(n-\zeta) + n - \zeta \approx \zeta\ln n \qquad (7.96)$$

Hence,

$$\left[\frac{n!}{(n-\zeta)!}\right] \approx n^{\zeta} \qquad (7.97)$$

and the binomial distribution reduces to the *Poisson distribution*

$$P(\zeta; n, p) = \frac{(pn)^{\zeta}}{\zeta!} e^{-pn} \qquad (7.98)$$

with the normalization

$$\sum_{\zeta=0}^{n} P(\zeta; n, p) = e^{-pn} \sum_{\zeta=0}^{n} \frac{(pn)^{\zeta}}{\zeta!} = e^{-pn} e^{+pn} = 1 \qquad (7.99)$$

and the same expectation and variance as the binomial distribution,

$$E(\zeta) = pn \quad \mathrm{var}(\zeta) = np(1-p) \qquad (7.100)$$

The **statistics toolkit** offers several functions for the Poisson distribution, whose value is returned by **poisspdf**. The cumulative distribution and its inverse are returned by **poisscdf** and **poissinv**. To fit a Poisson distribution to a data set use **poissfit**; the mean and standard deviation are returned by **poissstat**. Random numbers are returned by **poissrnd**.

A classic application of the Poisson distribution is the question

If we buy a very large number n of lottery tickets, each with a very small probability p of winning, what is the probability that we will have bought at least one winner?

Applying the Poisson distribution, this probability is

$$P(\zeta \geq 1; n, p) = \sum_{\zeta=1}^{n} P(\zeta; n, p) = 1 - P(0; n, p) = 1 - e^{-pn} \qquad (7.101)$$

The Poisson distribution finds common use in the study of many physical phenomena. For example, consider the case of anionic living polymerization. Using an initiator such as n-butyl lithium that forms a carbanion, we can polymerize vinyl monomers (Figure 7.10). Initiation is rapid, so that we start growing each chain at the same time. In a very small time interval Δt, the probability that we add a monomer unit to a specific chain during this interval is $k_p[M](\Delta t)$, where k_p is a propagation rate constant and $[M]$ is the monomer concentration. To account for the changing monomer concentration, we define a scaled time

$$\tau = k_p \int_0^t [M](t')dt' \qquad (7.102)$$

and take n steps forward in τ, each of duration $\Delta\tau$. The probability of adding a monomer

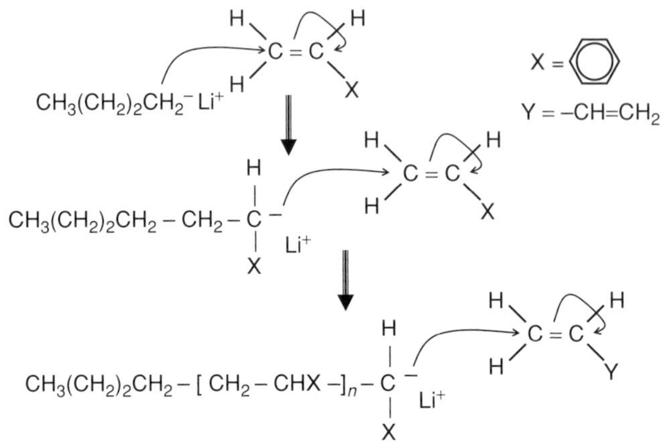

Figure 7.10 Anionic polymerization of a block copolymer.

unit during this time interval is

$$p = k_{\mathrm{p}}[\mathrm{M}](\Delta t) = \Delta \tau \tag{7.103}$$

Taking this event to be a success, and taking the limit $\Delta \tau \to 0$, $n \to \infty$ so that the Poisson distribution applies, we find that the probability that a particular chain has grown to a length x, starting from a length $x = 1$ at $\tau = 0$ is

$$P(x; n, \Delta \tau) = \frac{[(\Delta \tau)n]^{(x-1)}}{(x-1)!} e^{-[(\Delta \tau)n]} \tag{7.104}$$

The distribution of chain lengths at a scaled time $\tau = n(\Delta \tau)$ is

$$P(x; \tau) = \frac{\tau^{x-1}}{(x-1)!} e^{-\tau} \tag{7.105}$$

and the average chain lengths and polydispersity are

$$DP_n = 1 + \tau \quad DP_w = 1 + \tau + \frac{\tau}{1 + \tau} \quad P_{\mathrm{disp}} = 1 + \frac{\tau}{(1 + \tau)^2} \tag{7.106}$$

In the limit of very long chain lengths, $P_{\mathrm{disp}} \to 1$; i.e., the chains are of uniform length. This ability to generate chains of precise, uniform lengths, combined with the ability to switch monomers in the middle of the synthesis to produce block copolymers, makes anionic living polymerization an important tool in polymer science, particularly in the formation of nanoscale-ordered materials through microphase separation.

Random vectors and multivariate distributions

We now extend the concept of random variables to treat random vectors, for which we need a number of additional definitions.

Definition Covariance and correlation of two random variables
Let X and Y be two random variables. The *covariance* of X and Y is

$$\mathrm{cov}(X, Y) = E\{[X - E(X)][Y - E(Y)]\} \tag{7.107}$$

A related concept is the correlation of X and Y, $\mathrm{corr}(X, Y) = \mathrm{cov}(X, Y)/\sqrt{\mathrm{var}(X) + \mathrm{var}(Y)}$. If X and Y are independent, $\mathrm{cov}(X, Y) = 0$; however, a covariance of zero does not necessarily imply that the two variables must be independent (although it suggests that they are). If $\mathrm{cov}(X, Y) > 0$, then when X is greater than its mean $E(X)$, Y tends also to be greater than its mean $E(Y)$. Conversely, if $\mathrm{cov}(X, Y) < 0$, then if $X > E(X)$, it is more probable that Y is less than $E(Y)$. A nonzero covariance means only that the two variables tend to behave in a related manner, it does not mean that there is a cause and effect relationship among them. Asserting the latter is a common fallacy.

Definition Covariance matrix of a random vector
Let v be a vector whose components are random variables, not necessarily independent. Then, the *covariance matrix* of v, $\mathrm{cov}(v)$, has elements

$$[cov(v)]_{ij} = E\{[v_i - E(v_i)][v_j - E(v_j)]\} \tag{7.108}$$

If each component of v is independent of all others, $\mathrm{cov}(v)$ is diagonal,

$$\mathrm{cov}(v) = \begin{bmatrix} \mathrm{var}(v_1) & & & \\ & \mathrm{var}(v_2) & & \\ & & \ddots & \\ & & & \mathrm{var}(v_N) \end{bmatrix} \tag{7.109}$$

If in addition, each component of v has the same variance σ^2, then

$$\mathrm{cov}(v) = \sigma^2 I \tag{7.110}$$

If $v = Ax$, where A is a constant matrix and x is another random vector, then

$$\mathrm{cov}(v) = \mathrm{cov}(Ax) = A[\mathrm{cov}(x)]A^{\mathrm{T}} \tag{7.111}$$

If v is a random vector and c is a constant vector,

$$\mathrm{var}(c \cdot v) = \mathrm{var}(c^{\mathrm{T}}v) = c^{\mathrm{T}}[\mathrm{cov}(v)]c = c \cdot [\mathrm{cov}(v)]c \tag{7.112}$$

The covariance matrix is always symmetric and positive-definite.

Definition Multivariate Gaussian (normal) distribution
Let v be a random N-dimensional vector with a mean $\mu = E(v)$ and a covariance matrix Σ. Since Σ is symmetric, positive-definite, Σ^{-1} always exists. The Gaussian (normal) distribution of v is

$$P(v; \mu, \Sigma) = \frac{1}{(2\pi)^{N/2}\sqrt{|\Sigma|}} \exp\left\{-\frac{1}{2}(v - \mu)^{\mathrm{T}}\Sigma^{-1}(v - \mu)\right\} \tag{7.113}$$

The Boltzmann and Maxwell distributions

Many applications of probability theory to chemical engineering arise in statistical mechanics, the microscopic theory that underpins thermodynamics. Consider a system whose state is described by the state vector q, such that the energy in this microstate is $E(q)$. A key result of statistical mechanics is the *Boltzmann distribution*. For a system closed to its surroundings with respect to the exchange of mass, held at a constant temperature T and volume V, the probability of observing the system in microstate q is

$$P(q) = \frac{1}{Q} \exp\left[-\frac{E(q)}{k_b T}\right] \qquad Q = \sum_q \exp\left[-\frac{E(q)}{k_b T}\right] \qquad (7.114)$$

k_b is Boltzmann's constant, the ideal gas constant R divided by Avagadro's number. T is the absolute temperature, in kelvin. Applying this formula to the kinetic energy distribution of a moving particle of mass m at thermal equilibrium, we obtain the *Maxwell velocity distribution*:

$$P(v) \propto \exp\left[-\frac{m|v|^2}{2k_b T}\right] \qquad (7.115)$$

Brownian dynamics and stochastic differential equations (SDEs)

We next consider an important application of probability theory to physical science, the *theory of Brownian motion*, and introduce the subject of stochastic calculus. Let us consider the x-direction motion of a small spherical particle immersed in a Newtonian fluid. As observed by the botanist Robert Brown in the early 1800s, the motion of the particle is very irregular, and apparently random. Let $V_x(t)$ be the x-direction velocity as a function of time. For a particle of mass m and radius R in a fluid of viscosity μ, the equation of motion is

$$m\frac{dV_x}{dt} = -\zeta V_x + F_R(t) \qquad (7.116)$$

where $\zeta = 6\pi\mu R$ is the drag constant (predicted for very small particles by Stokes' law) and $F_R(t)$ is a random, fluctuating force due to collisions between the particle and the fluid molecules. Even though $V_x(t)$ fluctuates randomly, we can characterize deterministically the *velocity autocorrelation function* $C_{V_x}(t)$. Let $V_x(t_1)$ be the velocity at time t_1 and $V_x(t_2)$ be the velocity at time t_2. If we measure the product of these two values and take the average, we obtain a function that should depend only upon the time elapsed between the two measurements,

$$\langle V_x(t_1)V_x(t_2)\rangle = C_{V_x}(t_2 - t_1) = \langle V_x(t_1 - t_2)V_x(0)\rangle \qquad (7.117)$$

This function should have the property $C_{V_x}(-t) = C_{V_x}(t)$, and at $t = 0$ should agree with the average $\langle V_x^2 \rangle$ predicted by the Maxwell velocity distribution,

$$C_{V_x}(0) = \langle V_x^2 \rangle = \frac{k_b T}{m} \qquad (7.118)$$

Also, since the velocities measured at very different times should be uncorrelated, we expect $\lim_{t\to\infty} C_{V_x}(t) = 0$. In general, we expect this correlation function to take the approximate form of an exponential decay,

$$C_{V_x}(t \geq 0) \approx C_{V_x}(0)e^{-t/\tau_{V_x}} \qquad (7.119)$$

where τ_{V_x} is a velocity correlation time.

How is τ_{V_x} related to the properties of the particle and fluid? Let us say that the particle is moving through the fluid at some velocity, and then at time $t = 0$, we turn off the random force and allow the drag force to dissipate the kinetic energy. The particle motion then follows

$$m\frac{dV_x}{dt} = -\zeta V_x \implies V_x(t) = V_x(0)\exp\left[-\frac{\zeta t}{m}\right] \tag{7.120}$$

Therefore, the velocity correlation time is

$$\tau_{V_x} = \frac{m}{\zeta} = \left(\frac{4}{3}\pi\rho R^3\right)\left(\frac{1}{6\pi\mu R}\right) = \frac{2\rho R^2}{9\mu} \tag{7.121}$$

For the example case of a neutrally buoyant particle in water, $\rho = 10^3\,\text{kg/m}^3$, $\mu = 10^{-3}\,\text{Pa s}$. For a very small particle with a diameter of 10 nm, on the size of macromolecules, the velocity autocorrelation time is

$$\tau_{V_x} \sim \frac{(10^3\,\text{kg/m}^3)(10^{-8}\,\text{m})^2}{(10^{-3}\,\text{Pa s})} = 10^{3-16+3}\,\text{s} = 10^{-10}\,\text{s} \tag{7.122}$$

Even for a much larger particle of $100\,\mu\text{m} = 10^{-4}\,\text{m}$, the correlation time is still quite short,

$$\tau_{V_x} \sim \frac{(10^3\,\text{kg/m}^3)(10^{-4}\,\text{m})^2}{(10^{-3}\,\text{Pa s})} = 10^{3-8+3}\,\text{s} = 10^{-2}\,\text{s} \tag{7.123}$$

Now, if we take velocity measurements at times much further apart than τ_{V_x}, the results will be independent and uncorrelated from each other. From the above analysis, we see that the correlation times of a particle in water are quite short. Thus, in many applications where we are only concerned with the dynamics of the particle on time scales larger than τ_{V_x}, we can neglect the effect of velocity correlation and derive our governing equation in the limit $\tau_{V_x} \to 0$. Even in this limit, however, we must have a nonzero value of $\langle V_x(0)V_x(0)\rangle = \langle [V_x(0)]^2\rangle$; therefore, we write our approximate velocity autocorrelation as

$$\langle V_x(t)V_x(0)\rangle = 2D\delta(t) \tag{7.124}$$

where $\delta(t)$ is the *Dirac delta function*:

$$\delta(t) = \lim_{\sigma \to 0}\frac{1}{\sigma\sqrt{2\pi}}\exp\left[-\frac{x^2}{2\sigma^2}\right] \qquad \int_{-\infty}^{+\infty}f(t)\delta(t)dt = f(0) \tag{7.125}$$

Strictly speaking, the Dirac delta function is not a function at all, but is rather defined solely through the integral relation in (7.125). See the discussion of the theory of distributions in Stakgold (1979).

To see that D is consistent with the common definition of the diffusivity, we compute the average displacement over the time period $[0, t]$,

$$\Delta x(t) = \int_0^t V_x(t_1)dt_1 \tag{7.126}$$

with

$$\langle [\Delta x(t)]^2\rangle = \left\langle\left[\int_0^t V_x(t_1)dt_1\right]\left[\int_0^t V_x(t_2)dt_2\right]\right\rangle = \int_0^t\int_0^t \langle V_x(t_1)V_x(t_2)\rangle dt_1 dt_2 \tag{7.127}$$

Using the fact that the statistical properties of the velocity are independent of absolute time,

$$\langle V_x(t_1)V_x(t_2)\rangle = \langle V_x(t_1 - t_2)V_x(0)\rangle \tag{7.128}$$

and taking the limit $\tau_{V_x} \to 0$, we have

$$\langle V_x(t_1)V_x(t_2)\rangle = 2D\delta(t_1 - t_2) \tag{7.129}$$

Therefore,

$$\langle [\Delta x(t)]^2\rangle = \int_0^t \int_0^t [2D\delta(t_1 - t_2)]dt_1 dt_2 = 2D \int_0^t \int_0^t \delta(t_1 - t_2)dt_1 dt_2$$

$$= 2D \int_0^t (1)dt_2 = 2Dt \tag{7.130}$$

The mean-squared displacement varies linearly with time, and D is indeed the diffusivity, as commonly defined.

The Langevin equation

We again consider the 1-D motion of a spherical particle, and now include a conservative force arising from an external potential energy field $U(x)$. The equation of motion is then

$$m\frac{dV_x}{dt} = -\zeta V_x - \frac{dU}{dx} + F_R(t) \tag{7.131}$$

We want to take the limit $\tau_{V_x} \to 0$. As $\tau_{V_x} = m/\zeta$, we achieve this limit by letting $m \to 0$, but retain the constant value of the drag constant ζ (i.e., we neglect inertial effects). The equation of motion then becomes

$$V_x = \frac{dx}{dt} = -\frac{1}{\zeta}\frac{dU}{dx} + \frac{1}{\zeta}F_R(t) \tag{7.132}$$

This is known as the *Langevin equation*. Assuming that the statistical properties of the random force are independent of $U(x)$, we analyze $F_R(t)$ in the absence of an external potential, where

$$V_x(t) = \frac{dx}{dt} = \frac{1}{\zeta}F_R(t) \tag{7.133}$$

The autocorrelation of this equation yields the statistical properties of $F_R(t)$,

$$\langle F_R(t)F_R(0)\rangle = \zeta^2\langle V_x(t)V_x(0)\rangle = 2D\zeta^2\delta(t)$$
$$\langle F_R(t)\rangle = 0 \tag{7.134}$$

We next reintroduce the potential $U(x)$ and integrate the Langevin equation over an interval from time t to time $t + \Delta t$,

$$x(t + \Delta t) - x(t) = -\frac{1}{\zeta}\int_t^{t+\Delta t} \frac{dU}{dx}\bigg|_{x(t')} dt' + \frac{1}{\zeta}\int_t^{t+\Delta t} F_R(t')dt' \tag{7.135}$$

Assuming the gradient of the potential to be constant over the interval Δt,

$$x(t + \Delta t) - x(t) = -\frac{1}{\zeta}\left(\frac{dU}{dx}\right)(\Delta t) + X_{\mathrm{R}}(t, t + \Delta t) \tag{7.136}$$

where we have defined the random displacement due to the random force,

$$X_{\mathrm{R}}(t, t + \Delta t) = \frac{1}{\zeta}\int_t^{t+\Delta t} F_{\mathrm{R}}(t')dt' \tag{7.137}$$

The Wiener process

Let us consider the statistical properties of the random displacement (7.137). First, we see that the average displacement is zero:

$$\langle X_{\mathrm{R}}(t, t + \Delta t)\rangle = \frac{1}{\zeta}\int_t^{t+\Delta t} \langle F_{\mathrm{R}}(t')\rangle dt' = 0 \tag{7.138}$$

Next, we compute the autocorrelation function of X_{R}:

$$\langle X_{\mathrm{R}}(t, t + \Delta t)X_{\mathrm{R}}(t', t' + \Delta t)\rangle = \left\langle \left[\frac{1}{\zeta}\int_t^{t+\Delta t} F_{\mathrm{R}}(t_1)dt_1\right]\left[\frac{1}{\zeta}\int_{t'}^{t'+\Delta t'} F_{\mathrm{R}}(t_2)dt_2\right]\right\rangle$$

$$= \frac{1}{\zeta^2}\int_t^{t+\Delta t}\left\{\int_{t'}^{t'+\Delta t} \langle F_{\mathrm{R}}(t_1)F_{\mathrm{R}}(t_2)\rangle dt_2\right\}dt_1 \tag{7.139}$$

Substituting $\langle F_{\mathrm{R}}(t_1)F_{\mathrm{R}}(t_2)\rangle = 2D\zeta^2\delta(t_1 - t_2)$ yields

$$\langle X_{\mathrm{R}}(t, t + \Delta t)X_{\mathrm{R}}(t', t' + \Delta t)\rangle = 2D\int_t^{t+\Delta t}\left\{\int_{t'}^{t'+\Delta t}\delta(t_1 - t_2)dt_2\right\}dt_1 \tag{7.140}$$

If $t' \neq t$, as $\Delta t \to 0$, the right-hand side of (7.140) goes to zero as there is no $t_2 \in [t', t' + \Delta t]$ that equals any $t_1 \in [t, t + \Delta t]$. But, if $t' = t$, then

$$\langle X_{\mathrm{R}}(t, t + \Delta t)X_{\mathrm{R}}(t, t + \Delta t)\rangle = 2D\int_t^{t+\Delta t}\left\{\int_t^{t+\Delta t}\delta(t_1 - t_2)dt_2\right\}dt_1$$

$$= 2D\int_t^{t+\Delta t}\{1\}\,dt_1 = 2D(\Delta t) \tag{7.141}$$

Therefore, the correlation function of the displacement due to the random force during a time interval Δt is

$$\langle X_{\mathrm{R}}(t, t + \Delta t)X_{\mathrm{R}}(t', t' + \Delta t)\rangle = 2D(\Delta t)\delta(t - t') \tag{7.142}$$

We see that the statistical properties of this random displacement depend upon the value of the diffusivity and upon the time step Δt. We separate these two dependences by defining the random variable ΔW_t with the statistical properties,

$$\langle \Delta W_t\rangle = 0 \quad \langle \Delta W_t \Delta W_{t'}\rangle = (\Delta t)\delta(t - t') \tag{7.143}$$

such that

$$X_{\mathrm{R}}(t, t + \Delta t) = (2D)^{1/2}\Delta W_t \tag{7.144}$$

ΔW_t is said to be the finite increment of a *Wiener process*.

Let us take a random walk in one dimension, stepping a distance l every δt time interval. After n such steps, the elapsed time is $t = n(\delta t)$. The mean-squared displacement during the random walk is

$$\langle [\Delta x(t)]^2 \rangle = l^2 n = l^2 \left(\frac{t}{\delta t} \right) \tag{7.145}$$

Now, to be a model of diffusive motion, we must have the mean-squared displacement growing linearly with elapsed time; therefore, we set

$$l^2 = \delta t \quad l = \sqrt{\delta t} \tag{7.146}$$

so that

$$\langle [\Delta x(t)]^2 \rangle = t \tag{7.147}$$

In the limit $\delta t \to 0$, this random walk becomes a Wiener process. A Wiener process has the same long-time behavior as a random walk with steps of $\sqrt{\delta t}$ each δt time period, but the steps are taken infinitely close together. This is unphysical; however, when modeling Brownian diffusion, we are really only interested in behavior on time scales longer than the velocity autocorrelation time.

Note that for $\delta t \ll 1$, $\sqrt{\delta t} \gg \delta t$. The "derivative" of this process for small, but finite δt, is approximately

$$\frac{\Delta x}{\Delta t} \sim \frac{l}{\delta t} = \frac{\sqrt{\delta t}}{\delta t} = \frac{1}{\sqrt{\delta t}} \tag{7.148}$$

Thus, as $\delta t \to 0$, the "derivative" of $x(t)$ diverges. In fact, it diverges so badly that the integral

$$x(t + \delta t) - x(t) = \int_t^{t+\delta t} \left. \frac{dx}{dt} \right|_{t'} dt' \tag{7.149}$$

is *not* defined according to the rules of deterministic calculus.

Stochastic Differential Equations (SDEs)

The lack of a proper definition for (7.149) means that we *cannot* apply the traditional rules of calculus to Brownian motion; rather, we must use the special rules of *stochastic calculus*. Thus, integrals of the form of (7.137) and ODEs of the form of (7.132) are not to be defined using deterministic calculus as we have done above. Let us now write (7.132) in a form that *is* well defined by multiplying it by dt,

$$dx = -\frac{1}{\zeta} \left(\frac{dU}{dx} \right) dt + \frac{1}{\zeta} F_R(t) dt \tag{7.150}$$

For a small time interval, (7.144) yields

$$\frac{1}{\zeta} F_R dt = dX_R = (2D)^{1/2} dW_t \tag{7.151}$$

dW_t is a differential update of a Wiener process. We are now faced with choosing the time in $[t, t + dt]$ at which we evaluate dU/dx; here, we do so at the beginning of the time step, to obtain an *SDE* that is an *Itô-type SDE*,

$$dx = -\frac{1}{\zeta} \left(\left. \frac{dU}{dx} \right|_{x(t)} \right) dt + [2D]^{1/2} dW_t \tag{7.152}$$

Equation (7.152) is also called the *Langevin equation* for the particle.

For a description of SDEs, see Kloeden & Platen (2000). An abbreviated discussion is found in Öttinger (1996), with a focus on polymer science.

The simplest rule to integrate (7.152) is the *explicit Euler SDE method*:

$$x(t + \Delta t) - x(t) = -\frac{1}{\zeta} \left(\left. \frac{dU}{dx} \right|_{x(t)} \right) (\Delta t) + [2D]^{1/2} (\Delta W_t) \tag{7.153}$$

ΔW_t is an approximation to dW_t for a finite Δt and is drawn at random from a Gaussian distribution with mean $\mu = 0$ and variance $\sigma^2 = \Delta t$. In **MATLAB**, we generate ΔW_t by the code,

dW_t = sqrt(dt) * randn;

BD_1D.m uses the explicit Euler method to simulate the Brownian dynamics of a spherical particle in a quadratic energy well, $U(x) = k_{sp} x^2 / 2$ with $k_{sp} = 1$. Trajectories are plotted in Figure 7.11 for $\zeta = 1$, $D = 1$ and various Δt. As Δt decreases, the path fluctuates more wildly for short times, but has the same long-time properties.

Itô's stochastic calculus

While the explicit Euler method is simple, it is not very accurate. For a deterministic differential equation, we build higher-order methods through Taylor series expansions; however, the rules of stochastic calculus are different. Consider the SDE

$$dX = a(t, X)dt + b(t, X)dW_t \tag{7.154}$$

Using $t_{k+1} - t_k = \Delta t$, $X_{t_k} = X(t_k)$, we have an exact update

$$X_{t_{k+1}} - X_{t_k} = \int_{t_k}^{t_{k+1}} a(t, X(t))dt + \int_{t_k}^{t_{k+1}} b(t, X(t))dW_t \tag{7.155}$$

In deterministic calculus (no dW_t term), we expand $a(t, X)$ in time using the differential $da = (\partial a/\partial t)dt + (\partial a/\partial X)(dX/dt)dt$; however, here we have a stochastic integral involving a Wiener process. How do we interpret this integral, and what is the proper form of differential for a stochastic process?

We use here *Itô's stochastic calculus*, in which we approximate the stochastic integral by quadrature using the values at the beginning of each subinterval:

$$\int_0^{t_F} b(t)dW_t \approx \sum_{j=1}^{N} b(t_{j-1})\left[W_{t_j} - W_{t_{j-1}} \right] \qquad t_j = \frac{j t_F}{N} \tag{7.156}$$

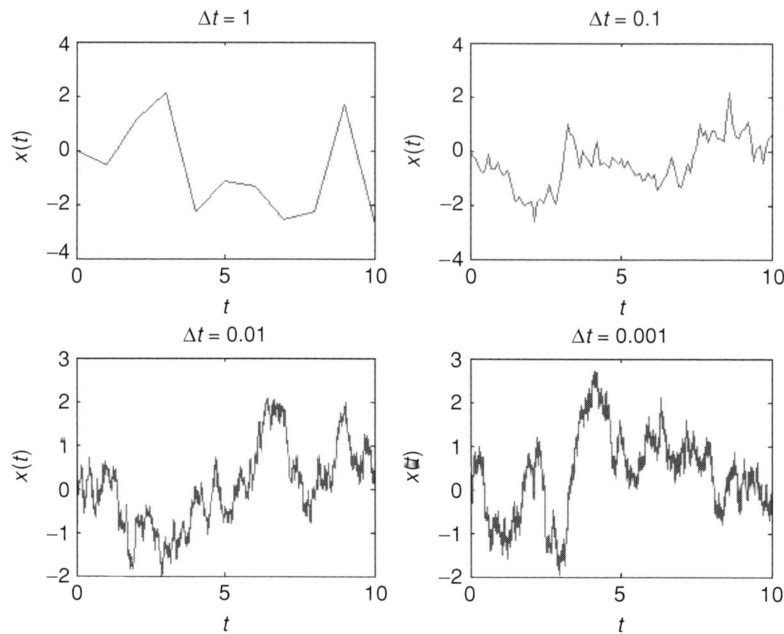

Figure 7.11 Brownian dynamics trajectories of the 1-D motion of a particle in a quadratic potential. As the time step is decreased, the path becomes more irregular at short times, but the long-time properties remain similar.

Next, we write the differential of any $F(t, X)$, with X being governed by the SDE (7.154). Assuming F depends continuously upon t and X, we use an expansion to second order:

$$F(t + dt, X(t + dt)) - F(t, X) \approx \left.\frac{\partial F}{\partial t}\right|_{(t,X)} dt + \left.\frac{\partial F}{\partial X}\right|_{(t,X)} dX + \frac{1}{2}\left.\frac{\partial^2 F}{\partial X^2}\right|_{(t,X)} dX^2$$

$$+ \frac{1}{2}\left.\frac{\partial^2 F}{\partial X \partial t}\right|_{(X,t)} dX dt + \frac{1}{2}\left.\frac{\partial^2 F}{\partial t^2}\right|_{(t,X)} dt^2 \quad (7.157)$$

For the first-order term in X, we substitute the SDE (7.154),

$$\left.\frac{\partial F}{\partial X}\right|_{(t,X)} dX = \left.\frac{\partial F}{\partial X}\right|_{(t,X)} [a(t, X)dt + b(t, X)dW_t] \quad (7.158)$$

Substituting similarly for the mixed partial derivative term,

$$\frac{1}{2}\left.\frac{\partial^2 F}{\partial X \partial t}\right|_{(X,t)} dX dt \approx \frac{1}{2}\left.\frac{\partial^2 F}{\partial X \partial t}\right|_{(X,t)} [a(t, X)dt + b(t, X)dW_t]dt \quad (7.159)$$

As dW_t is a random number of magnitude $\sqrt{dt}$, this term is of higher overall order in time than (7.158) and thus is dropped. Similarly, the last term of (7.157) that is second order in time is dropped. We then have

$$dF \approx \left.\frac{\partial F}{\partial t}\right|_{(t,X)} dt + \left.\frac{\partial F}{\partial X}\right|_{(t,X)} [a(t, X)dt + b(t, X)dW_t] + \frac{1}{2}\left.\frac{\partial^2 F}{\partial X^2}\right|_{(t,X)} dX^2 \quad (7.160)$$

Substituting for dX^2,

$$dX^2 = [a\,dt + b\,dW_t]^2 = a^2(dt)^2 + 2ab(dt)(dW_t) + b^2(dW_t)^2 \tag{7.161}$$

While the first two terms are of higher order than 1 in time, the last term is *not*, as $dW_t \sim \sqrt{dt}$. Thus, we must retain it, and replace it with its "average" value $(dW_t)^2 \to dt$.

We provide here a heuristic argument for accepting this replacement. Consider approximating the Wiener process as a random walk in which in one unit of time we make n steps each of length l. After one unit time interval, the mean-squared displacement is $nl^2 = 1$. Thus, we obtain the same mean-square displacement if we replace the Wiener process by a random walk with $l^2 = n^{-1} = dt$. Thus, the replacement $(dW_t)^2 \to dt$ is valid, in the "mean-square sense".

Using $dX^2 \approx b^2(dW_t)^2 \to b^2 dt$ in (7.160), we have *Itô's lemma*

$$dF = \left[\left.\frac{\partial F}{\partial t}\right|_{(t,X)} + \left.\frac{\partial F}{\partial X}\right|_{(t,X)} a(t,X) + \frac{1}{2}\left.\frac{\partial^2 F}{\partial X^2}\right|_{(t,X)} [b(t,X)]^2 \right] dt + \left.\frac{\partial F}{\partial X}\right|_{(t,X)} b(t,X)dW_t \tag{7.162}$$

This demonstrates the major difference between the stochastic and deterministic forms of calculus. In stochastic calculus, we expand functions to higher orders, replace $(dW_t)^2 \to dt$, and then keep all terms that contribute up to the desired order.

We now use this formalism to demonstrate the derivation of a higher order integration method than the explicit Euler one. In the explicit Euler method we neglect the time-variation of a and b over the time step. This is particularly bad for the second integral as dW_t is of order $t^{1/2}$, and thus the explicit Euler method is only 1/2-order accurate for predicting the actual trajectory. Thus, let us increase the order of accuracy of this term by using a $t^{1/2}$ accurate expansion of b in $t_k \le t \le t_{k+1}$:

$$b(t,X_t) \approx b(t_k,X_{t_k}) + \left.\frac{\partial b}{\partial X}\right|_{(t_k,X_{t_k})} \left.\frac{\partial X}{\partial W}\right|_{(t_k,X_{t_k})} [W_t - W_{t_k}] \tag{7.163}$$

Using $\partial X/\partial W = b$, we have for the second integral of (7.155),

$$\int_{t_k}^{t_{k-1}} b(t,X(t))dW_t \approx \int_{t_k}^{t_{k+1}} \left\{ b(t_k,X_{t_k}) + \left.\frac{\partial b}{\partial X}\right|_{(t_k,X_{t_k})} b(t_k,X_{t_k})[W_t - W_{t_k}] \right\} dW_t \tag{7.164}$$

This yields

$$X_{t_{k+1}} \approx X_{t_k} + a\big(t_k,X_{t_k}\big)(t_{k+1} - t_k) + b\big(t_k,X_{t_k}\big)\big[W_{t_{k+1}} - W_{t_k}\big]$$
$$+ \left.\frac{\partial b}{\partial X}\right|_{(t_k,X_{t_k})} b\big(t_k,X_{t_k}\big) \int_{t_k}^{t_{k+1}} [W_t - W_{t_k}]dW_t \tag{7.165}$$

The last term is the leading-order correction to the explicit Euler method to raise the order of accuracy to 1. We next evaluate the stochastic integral

$$I_{t_k,t_{k+1}} = \int_{t_k}^{t_{k+1}} [W_t - W_{t_k}]dW_t = \tfrac{1}{2}\big\{[W_{t_{k+1}} - W_{t_k}]^2 - (t_{k+1} - t_k)\big\} \tag{7.166}$$

To show that (7.166) is valid, we define

$$G(t,Y) = 2I_{t,t_{k+1}} = [Y - W_{t_k}]^2 - (t - t_k) \quad dY = dW_t \tag{7.167}$$

and apply (7.162),

$$dG = \left[-1 + \tfrac{1}{2}(2)[1]^2 \right]dt + 2\left[Y - W_{t_k} \right](1)dW_t = \left[Y - W_{t_k} \right]dW_t \tag{7.168}$$

to yield the integrand of (7.166). Using (7.166) in (7.165), we obtain the *Mil'shtein rule*

$$X_{t_{k+1}} \approx X_{t_k} + a(t_k, X_{t_k})(\Delta t) + b(t_k, X_{t_k})(\Delta W_t) + \left. \frac{\partial b}{\partial X} \right|_{(t_k, X_{t_k})} b(t_k, X_{t_k})\tfrac{1}{2}[(\Delta W_t)^2 - \Delta t]$$

$$\tag{7.169}$$

Further higher-order methods are discussed in Kloeden & Platen (2000).

Example. Stochastic calculus in quantitative finance

Stochastic calculus is used heavily in quantitative finance, a significant employer of numerate engineers. In Problem 6.B.5, we solved the Black–Scholes equation for the fair value of an option. Here, we show how this equation is obtained, through stochastic calculus.

Consider the spot (market) price $S(t)$ of some financial asset as a function of time. We sample the price at uniform time periods $t_k = k(\Delta t)$ and let $S_k = S(t_k)$. A reasonable model of the behavior of many assets is the lognormal random walk, which assumes that the return between successive time periods

$$R_k = \frac{S_{k+1} - S_k}{S_k} \tag{7.170}$$

is normally distributed so that the spot price is governed by the SDE

$$dS = \mu S dt + \sigma S dW_t \tag{7.171}$$

μ is the drift rate, and is computed from a sequence of returns by

$$\mu = \frac{1}{N_s(\Delta t)} \sum_{k=1}^{N_s} R_k \tag{7.172}$$

σ is the volatility of the asset, and is estimated from

$$\sigma^2 = \frac{1}{(N_s - 1)(\Delta t)} \sum_{k=1}^{N_s}(R_k - \langle R \rangle)^2 \qquad \langle R \rangle = \mu(\Delta t) \tag{7.173}$$

In a simple type of derivative, a European option, we purchase at time t the right to either buy (a call option) or sell (a put option) the underlying asset at some time $T > t$ in the future at an exercise price E. Thus, at time T, if we purchase the option, we will have a payoff

$$\text{payoff}(S(T)) = \begin{cases} \max(S(T) - E, 0), & \text{call option} \\ \max(E - S(T), 0), & \text{put option} \end{cases} \tag{7.174}$$

What is the fair price $V(S, t)$ of this option at $t < T$ when the spot price of the underlying asset is S?

To compute this value, assume that we purchase an option and at the same time short (i.e., sell assets we don't actually have – this is often possible and legal) a quantity Δ of the underlying asset. The value of this portfolio is $\Pi = V(S, t) - \Delta S$. Applying (7.162), the

SDE for the portfolio value is

$$d\Pi = \left\{ \frac{\partial V}{\partial t} + \frac{\partial V}{\partial S}(\mu S) + \frac{1}{2}\frac{\partial^2 V}{\partial S^2}[\sigma^2 S^2] \right\} dt + \frac{\partial V}{\partial S}(\sigma S)dW_t - \Delta[\mu S dt + \sigma S dW_t] \tag{7.175}$$

Collecting terms, we have

$$d\Pi = \left\{ \frac{\partial V}{\partial t} + \mu S\left(\frac{\partial V}{\partial S} - \Delta\right) + \frac{1}{2}\sigma^2 S^2 \frac{\partial^2 V}{\partial S^2} \right\} dt + \sigma S\left(\frac{\partial V}{\partial S} - \Delta\right) dW_t \tag{7.176}$$

Now, if we make the special choice $\Delta = \partial V/\partial S$, then the random nature of the portfolio disappears and we have the purely deterministic result

$$d\Pi = \left\{ \frac{\partial V}{\partial t} + \frac{1}{2}\sigma^2 S^2 \frac{\partial^2 V}{\partial S^2} \right\} dt \tag{7.177}$$

Such reduction of risk is known as hedging, and the strategy above is called delta-hedging. Of course, in practice not all risk is removed as the model is not completely accurate. Also, this approach involves modifying Δ continually as $\partial V/\partial S$ changes, which exposes the holder to transaction costs that should be modeled as well to design an optional hedging strategy.

If we follow this strategy, our model predicts that all risk will have been removed. It would not be fair if this strategy were to yield higher or lower returns than the alternative risk-free strategy of taking the initial value of our portfolio $\Pi(S, t)$ and putting it in a bank account to earn interest at a rate r. Using this "no free lunch," or "no arbitrage," argument, we should have

$$d\Pi = r\Pi dt = r(V - \Delta S)dt$$
$$\left\{ \frac{\partial V}{\partial t} + \frac{1}{2}\sigma^2 S^2 \frac{\partial^2 V}{\partial S^2} \right\} dt = r\left(V - \frac{\partial V}{\partial S}S \right) dt \tag{7.178}$$

This yields the famous *Black–Scholes equation*

$$\frac{\partial V}{\partial t} + rS\frac{\partial V}{\partial S} + \frac{1}{2}\sigma^2 S^2 \frac{\partial^2 V}{\partial S^2} - rV = 0 \tag{7.179}$$

which is solved backwards in time starting at the final condition (7.174). For more on the modeling of derivatives, consult Wilmott (2000).

The Fokker–Planck equation

We have proposed an SDE model (7.152) for the 1-D Brownian motion of a particle, and now relate this SDE to the time-evolution of the probability distribution $p(t, x)$, where the probability of finding the particle at time t in $[x, x + dx]$ is dx. We know that at equilibrium, this probability density should converge to the Boltzmann distribution

$$p_{eq}(x) = Z^{-1} \exp\left[-\frac{U(x)}{k_b T} \right] \qquad Z = \int_{-\infty}^{+\infty} \exp\left[-\frac{U(x)}{k_b T} \right] dx \tag{7.180}$$

Figure 7.12 shows the plot produced by BD_1D.m that contains the measured probability distribution of x generated by a histogram of the Brownian dynamics trajectory along with

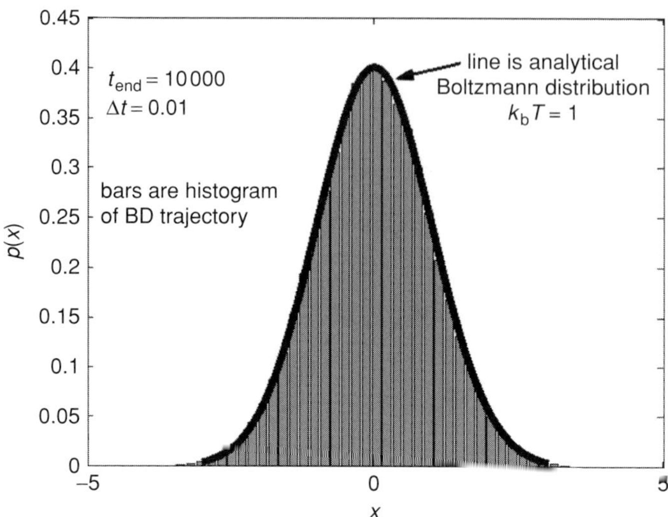

Figure 7.12 Comparison between probability distribution generated by Brownian dynamics (BD) trajectory and that from Boltzmann distribution.

the expected result at equilibrium with $k_b T = 1$. We see from their agreement that somehow setting $\zeta = 1$ and $D = 1$ results in the correct sampling at $k_b T = 1$.

What is the fundamental relationship between ζ and D, and how can we obtain an equation for the probability distribution from the SDE? Let us frame our discussion somewhat more generally for the SDE

$$dX = a(t, X)dt + b(t, X)dW_t \tag{7.181}$$

The update in a time interval δt is

$$X_{t+\delta t} - X_t = \int_t^{(t+\delta t)} a(t', X_{t'})dt' + \int_t^{(t+\delta t)} b(t', X_{t'})dW_{t'} \tag{7.182}$$

Consider some $F(x)$, for which (7.162) yields the differential at t',

$$dF = \left[\left.\frac{\partial F}{\partial X}\right|_{(t',X_{t'})} a(t', X_{t'}) + \frac{1}{2}\left.\frac{\partial^2 F}{\partial X^2}\right|_{(t',X_{t'})} [b(t', X_{t'})]^2 \right] dt + \left.\frac{\partial F}{\partial X}\right|_{(t',X_{t'})} b(t', X_{t'})dW_{t'} \tag{7.183}$$

Integrating this differential over the path $X_t \rightarrow X_{t+\delta t}$ yields

$$F(X_{t+\delta t}) - F(X_t) = \int_t^{(t+\delta t)} \left[\left.\frac{\partial F}{\partial X}\right|_{(t',X_{t'})} a(t', X_{t'}) + \frac{1}{2}\left.\frac{\partial^2 F}{\partial X^2}\right|_{(t',X_{t'})} [b(t', X_{t'})]^2 \right] dt'$$

$$+ \int_t^{(t+\delta t)} \left.\frac{\partial F}{\partial X}\right|_{(t',X_{t'})} b(t', X_{t'})dW_{t'} \tag{7.184}$$

Let us define an operator L_t that generates an "expected time derivative,"

$$L_t F(x) = \lim_{\delta t \to 0} \frac{1}{\delta t} \{ E[F(X_{t+\delta t})|X_t = x] - F(x) \} \tag{7.185}$$

Taking the expectation of (7.184) over a very small time step, the stochastic contribution cancels out as $\langle dW_{t'} \rangle = 0$. Thus,

$$L_t F(x) = \left(\frac{\partial F}{\partial x} \right) a(t, x) + \frac{1}{2} \left(\frac{\partial^2 F}{\partial x^2} \right) [b(t, x)]^2 \tag{7.186}$$

Let $p(t, x|t', x')$ be the *transition probability* that if the particle is at x' at time t', then at time t, it is at x. Then, we could write (7.185) as

$$L_t F(x) = \lim_{\delta t \to 0} \frac{1}{\delta t} \left\{ \int_{-\infty}^{+\infty} F(x'')p(t + \delta t, x''|t, x)dx'' - F(x) \right\} \tag{7.187}$$

Let us multiply (7.187) by $p(t, x|t', x')$ with $t' < t$ and integrate over x:

$$\int_{-\infty}^{+\infty} [L_t F(x)]p(t, x|t', x')dx$$

$$= \lim_{\delta t \to 0} \frac{1}{\delta t} \left\{ \int_{-\infty}^{+\infty} \left[\int_{-\infty}^{+\infty} F(x'')p(t + \delta t, x''|t, x)dx'' \right] p(t, x|t', x')dx \right.$$

$$\left. - \int_{-\infty}^{+\infty} F(x)p(t, x|t', x')dx \right\} \tag{7.188}$$

We now use the *Chapman–Kolmogorov equation*,

$$\int_{-\infty}^{+\infty} p(t + \delta t, x''|t, x)p(t, x|t', x')dx = p(t + \delta t, x''|t', x') \tag{7.189}$$

to write (7.188) as

$$\int_{-\infty}^{+\infty} [L_t F(x)]p(t, x|t', x')dx$$

$$= \lim_{\delta t \to 0} \frac{1}{\delta t} \left\{ \int_{-\infty}^{+\infty} F(x'')p(t + \delta t, x''|t', x')dx'' - \int_{-\infty}^{+\infty} F(x)p(t, x|t', x')dx \right\} \tag{7.190}$$

In the first integral on the right-hand side x'' is only a dummy variable of integration, and we are free to replace it with x,

$$\int_{-\infty}^{+\infty} [L_t F(x)]p(t, x|t', x')dx$$

$$= \lim_{\delta t \to 0} \frac{1}{\delta t} \left\{ \int_{-\infty}^{+\infty} F(x)p(t + \delta t, x|t', x')dx - \int_{-\infty}^{+\infty} F(x)p(t, x|t', x')dx \right\} \tag{7.191}$$

Collecting the integrals on the right-hand side, taking the limit $\delta t \to 0$, and using the finite difference approximation

$$\frac{\partial}{\partial t} p(t, x|t', x') \approx \frac{1}{\delta t} [p(t + \delta t, x|t', x') - p(t, x|t', x')] \tag{7.192}$$

we have

$$\int_{-\infty}^{+\infty} [L_t F(x)]p(t, x|t', x')dx = \int_{-\infty}^{+\infty} F(x) \left[\frac{\partial}{\partial t} p(t, x|t', x') \right] dx \tag{7.193}$$

If we define the adjoint operator $L_t^\dagger$ such that

$$\int_{-\infty}^{+\infty} [L_t F(x)] p(t, x | t', x') dx = \int_{-\infty}^{+\infty} F(x)[L_t^\dagger p(t, x | t', x')] dx \tag{7.194}$$

then (7.193) becomes

$$\int_{-\infty}^{+\infty} F(x)[L_t^\dagger p(t, x | t', x')] dx = \int_{-\infty}^{+\infty} F(x)\left[\frac{\partial}{\partial t} p(t, x | t', x')\right] dx \tag{7.195}$$

and hence the transition probability distribution is governed by the *forward Kolmogorov equation*

$$\frac{\partial}{\partial t} p(t, x | t', x') = L_t^\dagger p(t, x | t', x') \tag{7.196}$$

For the operator (7.186), one can show by integration by parts that

$$L_t^\dagger = -\frac{\partial}{\partial x} a(t, x) + \frac{1}{2} \frac{\partial^2}{\partial x^2} [b(t, x)]^2 \tag{7.197}$$

and thus

$$\frac{\partial}{\partial t} p(t, x | t', x') = \left[-\frac{\partial}{\partial x} a(t, x) + \frac{1}{2} \frac{\partial^2}{\partial x^2} [b(t, x)]^2\right] p(t, x | t', x') \tag{7.198}$$

Using the relation

$$p(t, x) = \int_{-\infty}^{+\infty} p(t, x | t', x') p(t', x') dx' \tag{7.199}$$

we multiply (7.198) by $p(t', x')$ and integrate over all x' to obtain the *Fokker–Planck equation* for $p(t, x)$,

$$\frac{\partial p}{\partial t} = -\frac{\partial}{\partial x}[a(t, x) p(t, x)] + \frac{1}{2} \frac{\partial^2}{\partial x^2}\{[b(t, x)]^2 p(t, x)\} \tag{7.200}$$

For 1-D Brownian motion with the SDE (7.152),

$$a(t, x) = J_c(x) = -\frac{1}{\zeta}\left(\frac{dU}{dx}\right) \qquad [b(t, x)]^2 = 2D \tag{7.201}$$

the Fokker–Planck equation is

$$\frac{\partial p}{\partial t} = -\frac{\partial}{\partial x}[J_c(x) p(t, x)] + \frac{\partial^2}{\partial x^2}\{Dp(t, x)\} \tag{7.202}$$

This looks somewhat like a convection/diffusion equation, where the first term is the convective flux due to the presence of the external potential and the second term is the diffusion caused by the random Brownian motion. If we have a system of N noninteracting particles, each governed by an independent SDE (7.152), the concentration field of the particles is $c(t, x) = Np(t, x)$ and is governed by

$$\frac{\partial c}{\partial t} = -\frac{\partial}{\partial x}[J_c(x) c(t, x)] + \frac{\partial^2}{\partial x^2}\{Dc(t, x)\} \tag{7.203}$$

Thus, we see that there is a strong relationship between macroscopic diffusion and microscopic Brownian motion.

Some care must be taken if the diffusivity is itself a function of position, as the correct microscopic balance for the concentration field is

$$\frac{\partial c}{\partial t} = -\frac{\partial}{\partial x}[J_c c] + \frac{\partial}{\partial x}\left[D(x)\frac{\partial c}{\partial x}\right] \tag{7.204}$$

Applying the chain rule,

$$\frac{\partial}{\partial x}\frac{\partial}{\partial x}[Dc] = \frac{\partial}{\partial x}\left[D\frac{\partial c}{\partial x}\right] + \frac{\partial}{\partial x}\left[c\frac{dD}{dx}\right] \tag{7.205}$$

we rewrite (7.204) as the correct form of the Fokker–Planck equation for a position-dependent diffusivity $D(x)$:

$$\frac{\partial c}{\partial t} = -\frac{\partial}{\partial x}\left[\left(J_c + \frac{dD}{dx}\right)c\right] + \frac{\partial^2}{\partial x^2}[D(x)c(t, x)] \tag{7.206}$$

The corresponding Langevin equation is

$$dx = \left\{J_c(x) + \frac{dD}{dx}\right\}dt + [2D(x)]^{1/2}dW_t \tag{7.207}$$

The additional deterministic contribution from the position-dependent diffusivity is known as *spurious drift*.

The Einstein relation

We are now ready to derive the proper relationship between the drag constant ζ and the diffusivity D. If the diffusion is constant, the Fokker–Planck equation (7.202) takes the form

$$\frac{\partial p}{\partial t} = \frac{\partial}{\partial x}\left[\frac{1}{\zeta}\left(\frac{dU}{dx}\right)p(t, x) + D\frac{\partial p}{\partial x}\right] \tag{7.208}$$

As $t \to \infty$, the system approaches a stable steady state for $D > 0$ at which

$$\frac{d}{dx}\left[\frac{1}{\zeta}\left(\frac{dU}{dx}\right)p(x) + D\frac{dp}{dx}\right] = 0 \tag{7.209}$$

Integrating yields

$$\frac{1}{\zeta}\left(\frac{dU}{dx}\right)p(x) + D\frac{dp}{dx} = \text{constant} = 0 \tag{7.210}$$

In the latter equality we use the fact that since the probability field is conserved (there is no source term), the net flux (convective and diffusive) is zero everywhere at the steady state. We want this condition to be satisfied by the equilibrium Boltzmann distribution (7.180). Taking the derivative of $p_{\mathrm{eq}}(x)$ yields

$$\frac{dp_{\mathrm{eq}}}{dx} = \frac{d}{dx}\left\{Z^{-1}\exp\left[-\frac{U(x)}{k_b T}\right]\right\} = Z^{-1}\left[-\frac{1}{k_b T}\frac{dU}{dx}\right]\exp\left[-\frac{U(x)}{k_b T}\right] = -\frac{1}{k_b T}\frac{dU}{dx}p_{\mathrm{eq}} \tag{7.211}$$

Substituting this into the no-flux condition (7.210) we obtain

$$0 = \frac{1}{\zeta}\left(\frac{dU}{dx}\right)p_{\mathrm{eq}} + D\left\{-\frac{1}{k_b T}\frac{dU}{dx}p_{\mathrm{eq}}\right\} = \left(\frac{dU}{dx}\right)\left(\frac{1}{\zeta} - \frac{D}{k_b T}\right)p_{\mathrm{eq}} \tag{7.212}$$

This yields the famous *Einstein relation*

$$D = \frac{k_b T}{\zeta} \tag{7.213}$$

Combining this expression with Stokes' law for the drag constant yields the *Stokes–Einstein relation*, predicting the diffusivity of a sphere through a Newtonian fluid,

$$D = \frac{k_b T}{6\pi \mu R} \tag{7.214}$$

The Einstein relation is a special case of a more general result known as the *fluctuation-dissipation theorem* (*FDT*). The FDT relates the strength of the random thermal fluctuations (here D) to the corresponding susceptibility to external perturbations (here ζ^{-1}) in such a way that ensures that the probability distribution converges to the proper equilibrium result at steady state.

Using the Einstein relation, the random Brownian force on the particle has the statistical properties

$$\langle F_R(t)\rangle = 0 \quad \langle F_R(t)F_R(0)\rangle = [2\zeta k_b T]\delta(t) \tag{7.215}$$

General formulation of SDEs; Brownian motion in multiple dimensions

Above we have considered only a single SDE, but systems of coupled SDEs can also be solved. Consider the 3-D isotropic Brownian motion of a paticle, in which each component of the position vector is governed by a SDE:

$$dx = -\zeta^{-1}\frac{\partial U}{\partial x}dt + (2D)^{1/2}dW_t^{(x)}$$

$$dy = -\zeta^{-1}\frac{\partial U}{\partial y}dt + (2D)^{1/2}dW_t^{(y)} \tag{7.216}$$

$$dz = -\zeta^{-1}\frac{\partial U}{\partial z}dt + (2D)^{1/2}dW_t^{(z)}$$

$dW_t^{(x)}, dW_t^{(y)}$, and $dW_t^{(z)}$ are increments of independent Wiener processes and the Einstein relation requires $\zeta^{-1} = D/(k_b T)$. Defining the position, conservative force, and Wiener update vectors,

$$\mathbf{r} = \begin{bmatrix} x \\ y \\ z \end{bmatrix} \quad \mathbf{F}^{(c)} = -\nabla U = \begin{bmatrix} -\partial U/\partial x \\ -\partial U/\partial y \\ -\partial U/\partial z \end{bmatrix} \quad d\mathbf{W}_t = \begin{bmatrix} dW_t^{(x)} \\ dW_t^{(y)} \\ dW_t^{(z)} \end{bmatrix} \tag{7.217}$$

(7.216) is written more compactly as

$$d\mathbf{r} = \zeta^{-1}\mathbf{F}^{(c)}dt + (2D)^{1/2}d\mathbf{W}_t \tag{7.218}$$

For diffusion in three dimensions, in the absence of an external potential, we have

$$dx = (2D)^{1/2}dW_t^{(x)} \quad dy = (2D)^{1/2}dW_t^{(y)} \quad dz = (2D)^{1/2}dW_t^{(z)} \tag{7.219}$$

The mean-squared displacement in 3-D space is

$$\langle(\Delta r)^2\rangle = \langle(\Delta x)^2 + (\Delta y)^2 + (\Delta z)^2\rangle \tag{7.220}$$

But, since the random motion in each direction is independent,

$$\langle(\Delta r)^2\rangle = \langle(\Delta x)^2\rangle + \langle(\Delta y)^2\rangle + \langle(\Delta z)^2\rangle \tag{7.221}$$

and from the stochastic properties of the Wiener process, we have

$$\langle(\Delta x)^2\rangle = 2Dt = \langle(\Delta y)^2\rangle = \langle(\Delta z)^2\rangle \tag{7.222}$$

Therefore, in three dimensions,

$$\langle(\Delta r)^2\rangle = 2Dt + 2Dt + 2Dt = 6Dt \tag{7.223}$$

Similarly, for diffusion in two dimensions, $\langle(\Delta r)^2\rangle = 4Dt$.

In general, we relate the Langevin and Fokker–Planck equations as follows. Let us have an ensemble whose members propagate according to an SDE

$$dx = a(t, x)dt + B(t, x) \cdot dW_t \tag{7.224}$$

with a state-dependent drift $a(t, x)$ and a state-dependent tensor $B(t, x)$. Then, defining from $B(t, x)$ the positive-semidefinite diffusion tensor

$$D(t, x) = B(t, x) \cdot B^{\mathrm{T}}(t, x) \tag{7.225}$$

the probability distribution of the ensemble follows the Fokker–Planck equation

$$\frac{\partial p}{\partial t} = -\nabla \cdot [a(t, x)p(t, x)] + \tfrac{1}{2}\nabla\,\nabla : [D(t, x)p(t, x)] \tag{7.226}$$

Given $D(t, x)$, a corresponding $B(t, x)$ may be obtained by Cholesky factorization; however, this choice of B is not unique as BQ, for any orthogonal Q, also satisfies (7.225).

Markov chains and processes; Monte Carlo methods

In our discussion of polymerization and random walks, we have been using the concept of *Markov chains*. Many simulation and computational methods are based on the random generation of states (events) according to a defined probability distribution. Because these techniques involve random number generation, they are known generally as Monte Carlo methods, after the famous casino in Monaco.

Markov chains

Let us consider a stochastic system whose state is characterized by a vector q. Using the laws of conditional probability, we write the probability of observing a particular sequence of states $q^{[0]}, q^{[1]}, \ldots, q^{[N]}$ as

$$P\big(q^{[0]}, q^{[1]}, \ldots, q^{[N]}\big) = P\big(q^{[0]}\big) \times P\big(q^{[1]}\big|q^{[0]}\big) \times P\big(q^{[2]}\big|q^{[1]}, q^{[0]}\big)$$

$$\times \cdots \times P\big(q^{[N]}\big|q^{[N-1]}, \ldots, q^{[1]}, q^{[0]}\big)$$

$$\tag{7.227}$$

In a *Markov process of order m*, each conditional probability of adding a new value $q^{[k]}$

depends only upon the values of the previous m members of the sequence,

$$P\big(q^{[k]}\big|q^{[k-1]}, q^{[k-2]}, \ldots, q^{[1]}, q^{[0]}\big) = P\big(q^{[k]}\big|q^{[k-1]}, q^{[k-2]}, \ldots, q^{[k-m]}\big) \tag{7.228}$$

Of particular importance are *first-order Markov processes*,

$$P\big(q^{[0]}, q^{[1]}, \ldots, q^{[N]}\big) = P\big(q^{[0]}\big) \times P\big(q^{[1]}\big|q^{[0]}\big) \times P\big(q^{[2]}\big|q^{[1]}\big) \times P\big(q^{[3]}\big|q^{[2]}\big)$$
$$\times \cdots \times P\big(q^{[N]}\big|q^{[N-1]}\big) \tag{7.229}$$

Here, the conditional probabilities take the form of *transition probabilities*,

$$P\big(q^{[k]}\big|q^{[k-1]}\big) \equiv T\big(q^{[k-1]} \to q^{[k]}\big) \tag{7.230}$$

Each sequence $q^{[0]}$, $q^{[1]}$, $\ldots$,$q^{[N]}$ generated by a Markov process is called a *Markov chain. Markov chain Monte Carlo (MCMC) simulation* is the general term given to the use of computers (with hopefully good random number generators) to generate Markov chains. Here, we focus on the case where the transition probabilities are defined such that the members of the Markov chain are distributed according to some specified distribution $P(q)$.

Monte Carlo simulation in statistical mechanics

Consider a system at constant temperature and volume, with a state vector q and an energy function $U(q)$, such that at equilibrium, the probability of finding the system in state q is described by the Boltzmann distribution

$$P(q) = \frac{1}{Z} \exp\left[-\frac{U(q)}{k_{\mathrm{b}} T}\right] \qquad Z = \sum_q \exp\left[-\frac{U(q)}{k_{\mathrm{b}} T}\right] \tag{7.231}$$

As $T \to 0$, the system is confined more closely to its minimum potential energy state, until at $T = 0^+$, only the minimum energy state has a nonzero probability of being observed.

In Monte Carlo simulation, we generate a random sequence $q^{[1]}, q^{[2]}, q^{[3]}, \ldots$ of states that are sampled according to the probability distribution $P(q)$. That is, the number of members of such a sequence within some region of volume dq around q is proportional to $P(q)dq$. Let us assume that we have some way of generating a random sequence with the correct probability distribution, and that at iteration k we are at the state $q^{[k]}$. We then propose at random a move to a new state $q^{(\mathrm{new})}$, where the probability of proposing this move is $\gamma(q^{[k]} \to q^{(\mathrm{new})})$. We now either accept or reject this move in such a way as to sample correctly from $P(q)$.

To do so, we define an acceptance probability $\alpha(q^{[k]} \to q^{(\mathrm{new})})$, and if we generate a random number u, distributed uniformly on $[0, 1]$, that is less than or equal to this probability, we accept the move. That is, the new state is generated by the rule

$$q^{[k+1]} = \begin{cases} q^{(\mathrm{new})}, & \text{if } u \leq \alpha\big(q^{[k]} \to q^{(\mathrm{new})}\big) \\ q^{[k]}, & \text{if } u > \alpha\big(q^{[k]} \to q^{(\mathrm{new})}\big) \end{cases} \tag{7.232}$$

Note that if we reject the move, we then count the current state again.

The acceptance probability is related to the generating probability distribution $P(\boldsymbol{q})$ by the *condition of detailed balance,*

$$P\big(\boldsymbol{q}^{[k]}\big)\gamma\big(\boldsymbol{q}^{[k]} \to \boldsymbol{q}^{(\mathrm{new})}\big)\alpha\big(\boldsymbol{q}^{[k]} \to \boldsymbol{q}^{(\mathrm{new})}\big)$$
$$= P\big(\boldsymbol{q}^{(\mathrm{new})}\big)\gamma\big(\boldsymbol{q}^{(\mathrm{new})} \to \boldsymbol{q}^{[k]}\big)\alpha\big(\boldsymbol{q}^{(\mathrm{new})} \to \boldsymbol{q}^{[k]}\big) \tag{7.233}$$

To understand the origin and meaning of this relation, consider the following thought experiment. Let us say that we were generating not one sequence, but many sequences independently and in parallel. At each step, the number of sequences in our ensemble at or near $\boldsymbol{q}$ should be proportional to $P(\boldsymbol{q})$. Therefore, from one step to the next in simulating our ensemble of sequences, the number of sequences that move away from any point $\boldsymbol{q}$ to any other point should be balanced by the number moving to $\boldsymbol{q}$ from all other points. The condition of detailed balance goes yet further, and states that the number of sequences making the move $\boldsymbol{q}^{[k]} \to \boldsymbol{q}^{(\mathrm{new})}$ must be balanced by the number making the move $\boldsymbol{q}^{(\mathrm{new})} \to \boldsymbol{q}^{[k]}$. This is represented mathematically as the flux balance (7.233). From this condition, we obtain the following rule for the ratio of the acceptance probabilities,

$$\frac{\alpha\big(\boldsymbol{q}^{[k]} \to \boldsymbol{q}^{(\mathrm{new})}\big)}{\alpha\big(\boldsymbol{q}^{(\mathrm{new})} \to \boldsymbol{q}^{[k]}\big)} = \frac{P\big(\boldsymbol{q}^{(\mathrm{new})}\big)\gamma\big(\boldsymbol{q}^{(\mathrm{new})} \to \boldsymbol{q}^{[k]}\big)}{P\big(\boldsymbol{q}^{[k]}\big)\gamma\big(\boldsymbol{q}^{[k]} \to \boldsymbol{q}^{(\mathrm{new})}\big)} \tag{7.234}$$

We often pick a generating process that is symmetric,

$$\gamma\big(\boldsymbol{q}^{[k]} \to \boldsymbol{q}^{(\mathrm{new})}\big) = \gamma\big(\boldsymbol{q}^{(\mathrm{new})} \to \boldsymbol{q}^{[k]}\big) \tag{7.235}$$

For example, we can displace each component at random,

$$q_m^{(\mathrm{new})} \leftarrow q_m^{[k]} + \Delta_m(u_m - 0.5) \tag{7.236}$$

where u_m is a random variable (independent for each component) that is uniformly distributed between zero and one. Δ_m is some maximum allowable displacement that may be tuned dynamically to optimize the fraction of moves that are accepted to the range 0.1–0.5.

With a symmetric process for generating the moves, the ratio of acceptance probabilities becomes

$$\frac{\alpha\big(\boldsymbol{q}^{[k]} \to \boldsymbol{q}^{(\mathrm{new})}\big)}{\alpha\big(\boldsymbol{q}^{(\mathrm{new})} \to \boldsymbol{q}^{[k]}\big)} = \frac{P\big(\boldsymbol{q}^{(\mathrm{new})}\big)}{P\big(\boldsymbol{q}^{[k]}\big)} \tag{7.237}$$

A choice that satisfies this condition is

$$\alpha\big(\boldsymbol{q}^{[k]} \to \boldsymbol{q}^{(\mathrm{new})}\big) = \min\big\{1, P\big(\boldsymbol{q}^{(\mathrm{new})}\big)/P\big(\boldsymbol{q}^{[k]}\big)\big\} \tag{7.238}$$

which yields the *Metropolis Monte Carlo method,* based upon iterating the following procedure:

generate a proposed move $\boldsymbol{q}^{[k]} \to \boldsymbol{q}^{(\mathrm{new})}$ at random from $\gamma\big(\boldsymbol{q}^{[k]} \to \boldsymbol{q}^{(\mathrm{new})}\big)$
compute $P(\boldsymbol{q}^{(\mathrm{new})})/P(\boldsymbol{q}^{[k]})$
generate a random variable u, uniformly distributed on $[0, 1]$
if $u \leq \min\{1, P(\boldsymbol{q}^{(\mathrm{new})})/P(\boldsymbol{q}^{[k]})\}$, $\boldsymbol{q}^{[k+1]} = \boldsymbol{q}^{(\mathrm{new})}$; else, $\boldsymbol{q}^{[k+1]} = \boldsymbol{q}^{[k]}$

To simulate a physical system using this method, we start at some initial state, and perform some large number of Monte Carlo steps to equilibrate the system. Initially, we may start at

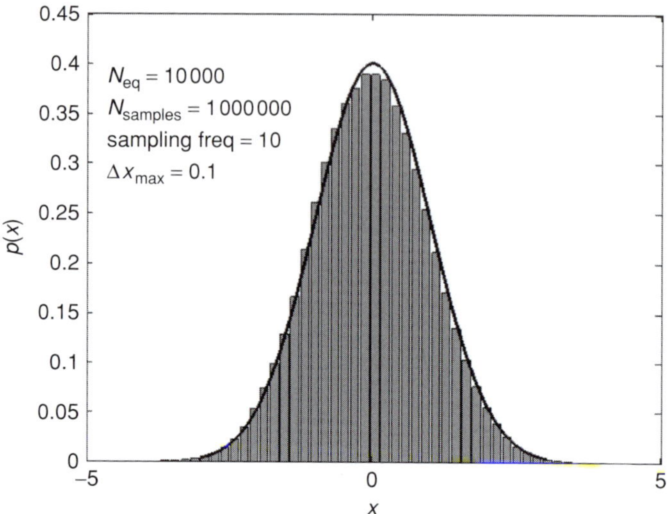

Figure 7.13 Exact and Monte Carlo sampled probability distributions of a particle in a 1-D quadratic energy well.

some state that is not very likely. After a large number of steps, the system will evolve to a more likely state at which point we can begin measuring the properties of the system. From our sequence of generated states, we measure any property $A(\boldsymbol{q})$ and compute the average,

$$\langle A \rangle = \frac{1}{N_s} \sum_{j=1}^{N_s} A\left(\boldsymbol{q}^{[j]}\right) \tag{7.239}$$

When simulating from the Boltzmann distribution, this should agree with the thermodynamic equilibrium value of A in the NVT (constant mole number, volume, and temperature) ensemble as $N_s \rightarrow \infty$. For more on Monte Carlo simulation, consult Frenkel & Smit (2002).

MC_NVT_sim1.m simulates a particle trapped in the same quadratic potential energy well as in the Brownian dynamics example of Figure 7.12. Figure 7.13 shows the probability distribution measured from the Monte Carlo simulation, compared to the exact result. For the large number of samples in this run, we see that the sampled distribution agrees quite well with the Boltzmann distribution.

Example. Monte Carlo simulation of a 2-D Ising lattice

Lattice models often are used to introduce statistical mechanics, because they are simple to understand and easy to simulate. A 2-D Ising lattice comprises $N \times N$ sites in a rectangular array, in which each state has a spin variable that takes on a value of $+1$ if the spin is "up" and -1 if the spin is "down." A state ν of the system assigns to each of the N^2 sites a spin

$$S_{ij}^{[\nu]} = \begin{cases} 1, & \text{if spin is up} \\ -1, & \text{if spin is down} \end{cases} \tag{7.240}$$

If two sites *(i, j)* and *(m, n)* are neighbors (to the "north," "south," "east," or "west"), they interact with a coupling strength J to favor parallel alignment (either both up or both down), with an energy contribution

$$E_{ij,mn} = -J S_{ij}^{[v]} S_{mn}^{[v]} = \begin{cases} -J, & \text{if spins are parallel} \\ J, & \text{if spins are anti-parallel} \end{cases} \tag{7.241}$$

In addition, each site interacts with an external magnetic field H according to the state of its spin, which has a magnetic moment μ. The total energy of the lattice in state v is then

$$E^{[v]} = H\mu \sum_{i=1}^{N} \sum_{j=1}^{N} S_{ij}^{[v]} - \frac{J}{2} \sum_{i=1}^{N} \sum_{j=1}^{N} S_{ij}^{[v]} \left[S_{i-1,j}^{[v]} + S_{i+1,j}^{[v]} + S_{i,j-1}^{[v]} + S_{i,j+1}^{[v]} \right] \tag{7.242}$$

We divide by 2 in the second sum in the energy expression to avoid overcounting each interacting pair. To mimic the behavior of an infinite lattice, we use periodic boundary conditions, in which for neighboring points outside the $N \times N$ simulation cell we assume

$$S_{-1,j}^{[v]} = S_{N,j}^{[v]} \quad S_{N+1,j}^{[v]} = S_{1,j}^{[v]} \quad S_{i,-1}^{[v]} = S_{i,N}^{[v]} \quad S_{i,N+1}^{[v]} = S_{i,1}^{[v]} \tag{7.243}$$

For each state v, we define the net magnetization and the order parameter

$$m^{[v]} = \mu \sum_{i=1}^{N} \sum_{j=1}^{N} S_{ij}^{[v]} \qquad \sigma^{[v]} = \frac{1}{N^2} \sum_{i=1}^{N} \sum_{j=1}^{N} S_{ij}^{[v]} \tag{7.244}$$

If all spins are "down," $\sigma^{[v]} = -1$, and if all spins are "up", $\sigma^{[v]} = 1$. If $\sigma^{[v]} = 0$ there is no net order of spins in the lattice. Ising_lattice_MC.m simulates a 2-D Ising lattice and is called by the following code:

```
MCOpts.N = 50; MCOpts.mu = 1; MCOpts.H = 0;
MCOpts.J = 1; MCOpts.kb_T = 5;
MCOpts.Nequil = 50*(MCOpts.N^2); MCOpts.Nsamples = 50000;
MCOpts.freq_sample = MCOpts.N; MCOpts.make_plots = 1;
Ising_lattice_MC(MCOpts);
```

make_Ising_lattice_MC_movie.m uses the results of Ising_lattice_MC.m to make a movie showing the spin fluctuations.

An infinite 2-D Ising lattice has a critical point at the *Curie temperature* T_c, $k_b T_c \approx 2.269\,J$, above which there is no net order in the absence of an external field, and below which the system has a net surplus of either "up" or "down" spins. A strong external field can induce spin order even above T_c, but higher fields are required at higher temperatures. Figure 7.14 shows two sample states from Monte Carlo simulations. Figure 7.14(a) shows the positions of spin-up sites in a disordered state for the base case simulated by the code above, while Figure 7.14(b) shows that imposing an external field $H < 0$ results in mostly spin-up sites. For further discussion of Ising lattices, see Chandler (1987).

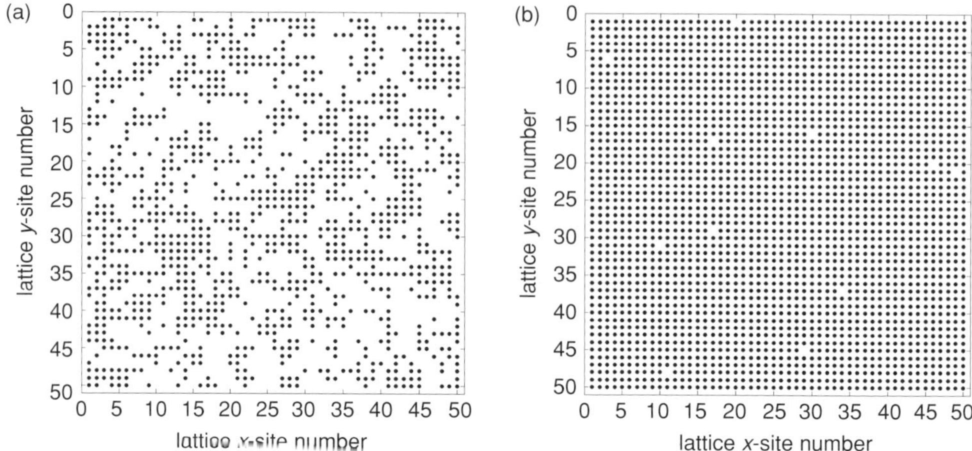

Figure 7.14 Sample states from Monte Carlo simulation of a 2-D Ising lattice at a temperature above the Curie point, with points showing spin-up sites: (a) in the absence of an external field, the lattice has a mixture of spin-up and spin-down sites; (b) with $H < 0$, lattice sites are mostly spin-up with a few spin-down sites due to thermal fluctuations ($N = 50$, $\mu = 1$, $H = 0$, $J = 1$, $K_b T = 5$.)

Field theory and stochastic PDEs

Above, we have considered only ordinary SDEs, but many models in statistical physics are of the form of stochastic PDEs. Let us consider a system such as the Ising lattice that at high temperatures is disordered, but in which as the temperature is lowered, some form of order emerges below a critical temperature T_c. We define an order parameter φ such that $\varphi = 0$ denotes a completely disordered state. We have defined such a spin order parameter for the lattice model (7.244); however, the concept is quite general. For example, when modeling the phase separation of two species A and B, we could take φ to be the difference in the local densities of each species, $\varphi = \rho_A - \rho_B$.

Commonly, near T_c, the system experiences significant long-range fluctuations in φ (you can observe this with the Ising MC program). Modeling the system on length scales large compared to the lattice size, we define a local order parameter field $\varphi(x)$ that characterizes the local degree of order at x. As φ is presumed to be small near T_c (the order is just beginning to emerge), we approximate the local free energy density $f(x)$ as a Taylor series in $\varphi(x)$, and write the total free energy Hamiltonian of the system as the *Landau phenomological free energy model*:

$$H[\varphi(x)] = \int \left\{ -w(x)\varphi(x) + \frac{1}{2}r[\varphi(x)]^2 + u[\varphi(x)]^4 + \frac{1}{2}c|\nabla\varphi|^2 \right\} dx \tag{7.245}$$

For an Ising lattice,

$$r = \frac{k_b}{a^d}(T - T^*) \quad T^* = \frac{J2^d}{k_b} \quad u = \frac{k_b T}{12a^d} \quad w = \frac{H\mu}{a^d} \quad c = Ja^{2-d} \tag{7.246}$$

where a is the lattice spacing and d is the dimension of the lattice. If we neglect any fluctuations in the order parameter, a *mean-field approximation*, we identify the thermodynamic prediction of the uniform order parameter $\langle \varphi \rangle$ as that minimizing the mean-field free energy

$$f_{\mathrm{MF}}(\langle \varphi \rangle) = -w \langle \varphi \rangle + \tfrac{1}{2} r \langle \varphi \rangle^2 + u \langle \varphi \rangle^4 \tag{7.247}$$

Taking $\partial f_{\mathrm{MF}}/\partial \langle \varphi \rangle = 0$, we have $-w + 2r \langle \varphi \rangle_{\mathrm{MF}} + 4u \langle \varphi \rangle_{\mathrm{MF}}^3 = 0$. In the absence of an external field, $w = 0$, this becomes

$$2\langle \varphi \rangle_{\mathrm{MF}}\left[r + 2u \langle \varphi \rangle_{\mathrm{MF}}^2 \right] = 0 \quad \Rightarrow \quad \langle \varphi \rangle_{\mathrm{MF}} = \begin{cases} 0, & r > 0, T > T^* \\ \pm\sqrt{\dfrac{-r}{2u}}, & r < 0, T < T^* \end{cases} \tag{7.248}$$

Thus T^* is the predicted transition temperature from mean-field theory. This mean-field picture of the phase transition is only valid when c is so large that spatial modulations in the local order parameter field are suppressed; however, we see from (7.246) that this is unlikely, especially when the dimension d is small.

One way to sample the fluctuations in the order parameter, and thus model their effect upon the phase transition, is to propose a stochastic model for the order parameter field such as the *time dependent Ginzburg–Landau model A (TDGL-A) dynamics*:

$$\frac{\partial \varphi}{\partial t} = -\Gamma \frac{\delta H}{\delta \varphi} + \eta(t, \boldsymbol{x}) \tag{7.249}$$

where $\Gamma > 0$. The *functional derivative* of $H\left[\varphi(\boldsymbol{x})\right]$ is

$$\left.\frac{\delta H}{\delta \varphi}\right|_{\boldsymbol{x}'} = \lim_{\varepsilon \to 0} \frac{H[\varphi(\boldsymbol{x}) + \varepsilon \delta(\boldsymbol{x} - \boldsymbol{x}')] - H[\varphi(\boldsymbol{x})]}{\varepsilon} = [-w + r\varphi + 4u\varphi^3 - c\nabla^2 \varphi]|_{\boldsymbol{x}'} \tag{7.250}$$

and $\eta(t, \boldsymbol{x})$ is a random noise field. Here, we have taken the "naive" approach of dividing the field differential by dt, as this is common in the physics community, even though the time derivative is not well defined. Equation (7.249) is a *stochastic PDE*, and to discretize it, we place a uniform grid of points $\boldsymbol{x}_m$, where $\boldsymbol{m}$ is a unique label. If φ_m is the field value at $\boldsymbol{m}$, we discretize (7.249) as the set of ordinary SDEs

$$d\varphi_m = -\Gamma \left[\left.\frac{\delta H}{\delta \varphi}\right|_{\boldsymbol{x}_m} \right] dt + \eta_m dt$$

$$\left.\frac{\delta H}{\delta \varphi}\right|_{\boldsymbol{x}_m} = -w(\boldsymbol{x}_m) + r\varphi_m + 4u\varphi_m^3 - c\nabla^2 \varphi \big|_{\boldsymbol{x}_m} \tag{7.251}$$

We use finite differences to relate $\nabla^2 \varphi|_{\boldsymbol{x}_m}$ to an algebraic difference of field values at neighboring grid points. Let us compare (7.251) to the set of SDEs for Brownian motion in multiple dimensions,

$$dx_m = -\frac{1}{\zeta} \frac{\partial U}{\partial x_m} dt + \frac{1}{\zeta} F_{\mathrm{R},m}(t) dt \tag{7.252}$$

where the random force vector has the statistical properties

$$\langle F_{\mathrm{R},m}(t) \rangle = 0 \qquad \langle F_{\mathrm{R},m}(t) F_{\mathrm{R},n}(t') \rangle = 2k_{\mathrm{b}} T \zeta \delta(t - t') \delta_{mn} \tag{7.253}$$

We see a strong resemblance between (7.251) and (7.252), except that (7.251) has the functional derivative evaluated at $\boldsymbol{x}_m$ rather than the "traditional" derivative of the energy with respect to the grid value. For a lattice of cells with a volume v_c around each point, if we use quadrature to approximate the functional $H[\varphi(x)]$ as a function $H(\varphi)$ of the grid values,

$$\left.\frac{\delta H}{\delta \varphi}\right|_{x_m} \approx v_c^{-1}\frac{\partial H}{\partial \varphi_m} \tag{7.254}$$

so that the analogous form to (7.252) is

$$d\varphi_m = -\Gamma v_c^{-1}\left[\frac{\partial H}{\partial \varphi_m}\right]dt + \eta_m dt \tag{7.255}$$

Thus, we associate $\zeta^{-1} \Leftrightarrow \Gamma v_c^{-1}$ and by analogy to (7.253) define the statistical properties of η_m to be

$$\langle \eta_m(t)\rangle = 0 \quad \langle \eta_m(t)\eta_n(t')\rangle = 2k_{\mathrm{b}}T\Gamma v_c^{-1}\delta_{m,n}\delta(t-t') \tag{7.256}$$

As $v_c \to 0$, $v_c^{-1}\delta_{m,n} \to \delta(x_m - x_n)$, so that the statistical properties of the random noise field $\eta(t, x)$ in (7.249) are

$$\langle \eta(t, x)\rangle = 0 \quad \langle \eta(t, x)\eta(t', x')\rangle = 2k_{\mathrm{b}}T\Gamma\delta(x-x')\delta(t-t') \tag{7.257}$$

The SDE for the field value at each grid point is then

$$d\varphi_m = -\Gamma\left[\left.\frac{\delta H}{\delta \varphi}\right|_{x_m}\right]dt + \left(2k_{\mathrm{b}}T\Gamma v_c^{-1}\right)^{1/2}dW_t^{(m)} \tag{7.258}$$

where the $dW_t^{(m)}$ are independent Wiener processes. TDGL_A_2D.m uses this method to sample the order-parameter field fluctuations at a specified temperature. For more on field theory applications in statistical physics, consult Chaikin & Lubensky (2000).

Monte Carlo integration

In Chapter 4, we considered a simple method to estimate by Monte Carlo simulation the value of a definite integral

$$\Phi = \int_\Omega f(x)dx \tag{7.259}$$

We consider here an alternative method employing importance sampling that may be used when we can compute the volume V_Ω of Ω easily. We start by writing (7.259) as the integral over all space

$$\Phi = \int_{R^N} H_\Omega(x)f(x)dx \quad H_\Omega(x) = \begin{cases} 1, & \text{if } x \in \Omega \\ 0, & \text{if } x \notin \Omega \end{cases} \tag{7.260}$$

Then, we write the integral in terms of the average of $f(x)$ over Ω, which we determine by Monte Carlo simulation,

$$\Phi = V_\Omega\langle f\rangle_\Omega \quad \langle f\rangle_\Omega = \frac{1}{N_S}\sum_{j=1}^{N_S} f(x^{[j]}) \tag{7.261}$$

The $\{x^{[j]}\}$, uniformly distributed in Ω, are obtained from Metropolis Monte Carlo sampling using the sampling distribution

$$P(x) = \frac{H_\Omega(x)}{V_\Omega} \tag{7.262}$$

Simulated annealing

Consider again the Boltzmann distribution for a system with a state vector x and an energy function $E(x)$, such that the probability of finding the system at state x is

$$P(x) \propto \exp\left[-\frac{E(x)}{k_b T}\right] \tag{7.263}$$

As $T \to 0$, the system is confined more closely to its minimum energy state, until at $T = 0^+$ only the minimum energy state is observed. By analogy to this result, we obtain a method for finding the global minimum of a cost function known as *simulated annealing*. Unlike the methods of Chapter 5, simulated annealing may be used for both continuously-varying and discretely-varying parameters. We merely replace the energy with the cost function and set k_b to 1:

$$P(x) \propto \exp\left[-\frac{F(x)}{T}\right] \tag{7.264}$$

We sample from this distribution using the Metropolis Monte Carlo method, starting initially at a very high temperature so that the system is able to move efficiently around parameter space. We then slowly decrease the temperature to zero, to allow the system sufficient time to escape from higher-lying local minima and find the global minimum (which is easier said than done).

This method is implemented by simulated_annealing.m, called by

[x,F,iflag,x_traj] = simulated_annealing(func_name,x0,OptParam,ModelParam);

func_name is the name of a routine that returns the cost function,

function f = fun_name(x,ModelParam);

ModelParam is a structure of fixed parameters passed to the cost function routine. x0 is the initial guess of the state vector. Here, all components of x are assumed to vary continuously, but the routine can be modified easily to treat discretely-varying parameters. OptParam is an optional structure that allows the user to modify the behavior of the algorithm. Type help simulated_annealing for further details. The output includes the estimated minimum x and its cost function value F, an integer iflag for which 1 denotes success, and x_traj, a trajectory of state values sampled during the annealing.

The performance of this simulated annealing routine is examined for the following cost function, which has two minima, one at $\tilde{x}^{(1)} = (-3, 3)$ and a lower one at $\tilde{x}^{(2)} = (3, -3)$:

$$F(x) = \begin{cases} 10 + 0.1\|x - \tilde{x}^{(1)}\|_2^2, & \text{if } \|x - \tilde{x}^{(1)}\|_2^2 \le \|x - \tilde{x}^{(2)}\|_2^2 \\ \|x - \tilde{x}^{(2)}\|_2^2, & \text{otherwise} \end{cases} \tag{7.265}$$

test_simulated_annealing.m performs this calculation, and calls global_min_cost_func.m, that returns the value of the cost function.

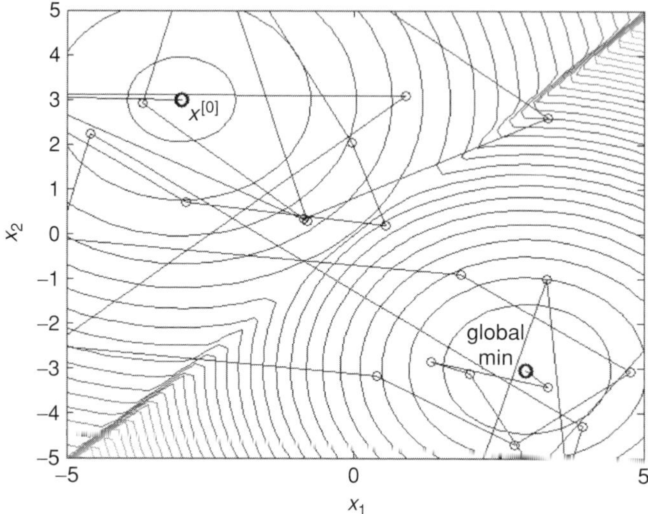

Figure 7.15 Partial trajectory of a simulated annealing run of a cost function with multiple minima.

Figure 7.15 shows the contour plot of this cost function, overlaid with the trajectory of the simulation (not all points are shown). The simulation was started at $\tilde{x}^{(1)}$, a local minimum but not the global minimum, at which any of the deterministic optimization methods of Chapter 5 terminate immediately. Eventually, the stochastic simulation is able to overcome the barrier to settle finally at the global minimum. We see also that the irregular shape of the cost function surface near the partition between points closest to $\tilde{x}^{(1)}$ and those nearest to $\tilde{x}^{(2)}$ does not pose a significant problem but would for the deterministic algorithms of Chapter 5. The program sim-anneal-ext.m generates a movie showing the progress of simulated annealing for a 1-D cost function with multiple local minima.

Genetic programming

We mention here another method, *genetic programming*, to identify global minima. While simulated annealing is based on an analogy from physics, genetic algorithms take their inspiration from biology. The general idea is to let an initial guess of the parameter vector evolve to improve the cost function in a manner that mimics natural selection. At each generation of the evolutionary process, there are n_p members of the population, but only the "best" $n_s < n_p$ survive to the next generation. In contrast to simulated annealing, at each iteration, the best population member is at least as good as the initial guess, so this algorithm is popular for those wishing merely to improve a design and not necessarily to obtain the optimum. Genetic algorithms are easily adapted to problems with discrete-valued parameters, although the method is presented here for a continuous parameter vector.

First, we start the calculation with some initial first member $x^{[1]}$. From this guess, the generation is filled by creating new members through a combined process of single-member mutation and dual-member crossing. For each $k = 2, 3, \ldots, n_p$, we add a new member

$x^{[k]}$ to the population. We define two probabilities: α_m, the probability that $x^{[k]}$ will be created by a mutation, and α_c, the probability that it will be created by crossing. These two probabilities must sum to 1. Here, we only present one type each of mutation and crossing, but more complicated combinations of random offspring production are used in practice.

To decide which process to use to add $x^{[k]}$, we generate a random number u_k, uniformly distributed on [0,1]. If $u_k \leq \alpha_m$, we generate $x^{[k]}$ by selecting at random some previous member of the population $x^{[j]}$, with $j < k$, and mutating it. A simple mutation is to displace each component randomly:

$$x_m^{[k]} \leftarrow x_m^{[j]} + \sigma_m g_m \tag{7.266}$$

g_m is a random number normally distributed with a standard deviation of 1 and σ_m is the standard deviation for the displacement of x_m. We choose a normally distributed random variable so that there is a finite probability of selecting even large displacements. If we have components that take on a discrete number of values, a common mutation is to select one discrete component randomly and to give it a new value at random from among the possibilities.

If $u_k > \alpha_m$ and $k > 2$, we generate $x^{[k]}$ by a random crossing of two previous members of the population. We select at random some $j < k$ and $l < k$ to serve as "parents." One simple way to perform the crossing is to select at random for each component the value of one parent or the other:

$$x_m^{[k]} = \begin{cases} x_m^{[j]}, & \text{if } u_m' \leq 0.5 \\ x_m^{[l]}, & \text{if } u_m' > 0.5 \end{cases} \tag{7.267}$$

Such a crossing generates a new offspring member that may be in a significantly different region of phase space than either of the parents. This allows the genetic algorithm to take large steps in parameter space, thus aiding the search for the global minimum.

Once a generation is filled completely, the selection of members that survive to the next generation begins. We order the population by increasing values of the cost function $F(x)$, or equivalently by decreasing order of the fitness, $-F(x)$. The best members are found at the beginning of this list. We want to select some sub population of $n_s < n_p$ survivors that will be passed to the next generation; however, it would be a mistake to take merely the first n_s members of the list. As in biology, in-breeding is to be avoided, so before we accept a new member as a survivor, we ensure that it is not too similar to one previously accepted. For some positive-definite metric matrix Γ, we define the squared distance between two members as

$$d_{kj}^2 \equiv \left(x^{[k]} - x^{[j]}\right) \cdot \Gamma \left(x^{[k]} - x^{[j]}\right) \tag{7.268}$$

When considering a member $x^{[k]}$ for admittance into the set of survivors, we check to see whether for all previously identified survivors $x^{[j]}$, $d_{kj}^2 \geq d_{min}^2$. Only if $x^{[k]}$ is sufficiently different from all of the previous survivors will it be accepted. If we cannot find n_s sufficiently diverse survivors, we move on to the next generation with less than n_s, as propagating two nearly identical members does little good.

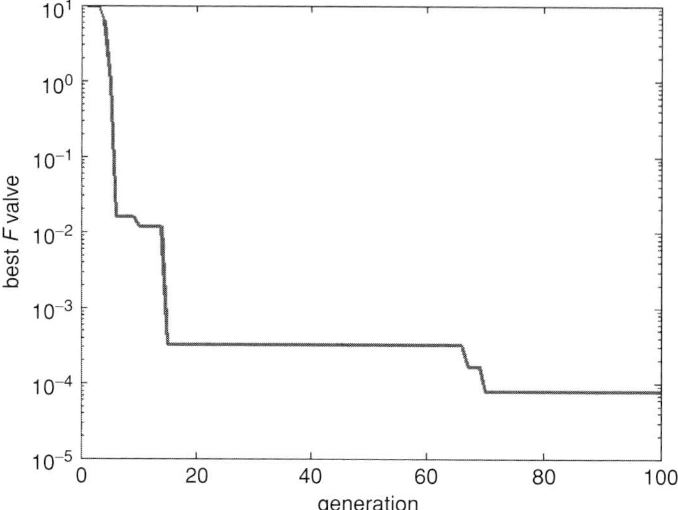

Figure 7.16 Cost function of best population member vs. generation number.

In Figure 7.16, we plot the best cost function value among members of the population vs. the generation number for a sample run of this genetic_minimizer.m routine with the cost function (7.265). Again, we start from the higher minimum at $(-3, 3)$, and simulate for 100 generations with $n_p = 100, n_s = 10, \Delta = 1$. After a few iterations, the algorithm has found its way to the vicinity of the global minimum, but more iterations are required to reduce the cost function value further.

The supplemental material at the accompanying website contains an additional example of a genetic algorithm. This example converts the process of solving a Sudoku puzzle into a discrete optimization that is then solved by a genetic algorithm.

MATLAB summary

The Statistics toolkit contains many useful functions for stochastic simulation. A uniform random number in $[0, 1]$ is returned by **rand**; **randn** returns a random number distributed by the normal distribution with a mean of zero and a variance of 1 (for more general, and multivariate, normal distributions, use **normrnd**). The normal probability distribution, cumulative distribution, and inverse cumulative probability distribution are returned by **normpdf**, **normcdf**, and **norminv** respectively. Similar routines are available for other distributions; for example, the Poisson probability density function is returned by **poisspdf**. A GUI tool, **dfittool**, is available to fit data to a probability distribution. The mean, standard variation, and variance of a data set are returned by **mean**, **std**, and **var** respectively. For a more comprehensive listing of the available functions, consult the documentation for the Statistics toolkit.

Problems

7.A.1. Consider a game involving two six-sided dice. Plot the probability of observing each total value 2–12 if the dice are fair. How many times, on average, must one roll the dice before a 7 is observed? How many times, on average, can one roll the dice before a 2, "snake eyes," is observed?

7.A.2. Consider again the log normal random walk model of the spot (market) price $S(t)$ of an asset,

$$dS = \mu S dt + \sigma S dW_t \tag{7.269}$$

Consider a stock with its price sampled each business day (assume 252 business days per year), with time measured in years. Let the (annual) drift rate be 8%, and let the volatility of the stock be 20%. Write a program that generates typical characteristic random trajectories over a year, starting from a price of 100. Make a histogram of the final stock price after one year.

7.A.3. The central limit theorem of statistics states that the sum of many independent random variables tends towards a normal distribution without regard to the shape of the underlying distribution. Consider the statistic X that is a sum of N random variables u_j, each uniformly distributed on $[0, 1]$,

$$X = \sum_{j=1}^{N} u_j \tag{7.270}$$

For large N, compute the distribution of X and fit it to a normal distribution. As N increases, does $P(X)$ indeed approach a normal distribution? Repeat the calculations for Y defined as

$$Y = \sum_{j=1}^{N} w_j u_j \tag{7.271}$$

where the w_j are random weights determined at the beginning of the calculation from a uniform distribution on $[0, 1]$. Does $P(Y)$ also approach a normal distribution?

7.A.4. Consider again the 2-D Ising lattice model, for which a Monte Carlo simulation program has been provided. Using this program, you are asked here to investigate the nature of the order–disorder transition as a function of the external field H and thermal energy $k_b T$. Simulate a 100×100 lattice, with $\mu = 1$, $J = 1$. In the absence of an external field, $H = 0$, the order–disorder transition occurs at the Curie point $k_b T_c \approx 2.269$ J (for an infinite lattice). Using the provided Monte Carlo program, compute the average magnetic order parameter as a function of H at various temperatures above and below the Curie point. Plot the order parameter for a very small field $H > 0$ as a function of temperature and describe the qualitative nature of the transition. Then, plot the order parameter vs. H at several temperatures. How does the transition from positive to negative H change above and below the Curie point?

7.A.5. A simple model for the geometry of a polymer chain is to treat it as a 3-D random walk of N steps, each of length l. The chain is described by a set of coordinates $r^{[0]}$, $r^{[1]}$, $r^{[2]}$, ..., $r^{[N]}$ where the vector $r^{[k+1]} - r^{[k]}$, between successive coordinates is of length l and is distributed isotropically in 3-D space. To generate a random vector $v \in \Re^3$ that is isotropically distributed, first generate a vector w with components $w_j = 2 \times (\text{rand} - 0.5) \in [-1, +1]$, where rand is uniformly distributed on $[0, 1]$. Then, if w lies within the unit sphere, rescale it to 1 and accept the scaled vector as v. Otherwise, if it lies outside of the unit sphere, generate a new w and try again.

Write a program that generates random conformations of a polymer chain with this model, and construct the resulting probability distribution of the end-to-end distance $|r_N - r_0|$. Show that this model results in a Gaussian distribution of the end-to-end distance,

$$P(|r_N - r_0|) \propto \exp\left\{-\frac{U(|r_N - r_0|)}{k_b T}\right\} \qquad U(|r_N - r_0|) = \frac{K}{2}|r_N - r_0|^2 \qquad (7.272)$$

K is an effective spring constant that describes the free energy required to stretch the polymer chain. Using your program, demonstrate that K is related to N and l by

$$K = \frac{3k_b T}{Nl^2} \qquad (7.273)$$

7.B.1. For the asset price model of Problem 7.A.2, compute the corresponding Fokker–Planck equation and solve it for the probability $P(S, t)$ that the asset has a price S at time t over the year-long simulation period. Use the parameters and the initial condition given in Problem 7.A.2.

7.B.2. Here, we demonstrate once more how Brownian dynamics relates to diffusive behavior, by simulating spherical particles of radius 1 mm in water at room temperature. At time $t = 0$, a particle is released at the origin and undergoes 3-D Brownian motion. Write a program that repeats this simulation many times and plots the radial concentration profile of particles as a function of time. It is easier to do the data analysis if you do the simulations concurrently. Then, solve the corresponding time-dependent diffusion equation in spherical coordinates and compare the results to that obtained from Brownian dynamics.

7.B.3. Diffusion limited aggregation is a process by which an agglomerate of small particles grows through the slow addition of particles to its surface through diffusion. Here, you are asked to simulate this process in two dimensions using a simple model of diffusion on a lattice, to observe the fractal shape that results from this mode of growth. We first define several radii $R_0 < R_{\text{end}} < R_1 < R_2$, and form a 2-D lattice of $N \times N$ unit cells with $N \geq 2R_2 + 3$. The center of this lattice is identified as the origin, and a matrix with elements φ_{mn} for each site (m, n) in the lattice is allocated to store a 0 if the cell is empty and 1 if it is filled with a small particle. Let r_{mn} be the distance from the cell to the origin. Initially, only cells within a "seed particle" region with $r_{mn} \leq R_0$ are filled and the others are empty.

Then, a simulation is begun that consists of a sequence of particle insertions in an empty cell (m, n) with $r_{mn} \approx R_1$, followed by random displacements of the particle by one cell to the top, right, bottom, or left. If at any time, the inserted particle neighbors a filled site, it is assumed to "stick" to the aggregate, the current site is filled, and a new particle insertion is performed. If at any time the particle diffuses outside of the escape radius R_2, the diffusion iterations are terminated, as it is assumed to be very unlikely that the particle will find its way back to stick to the aggregate. No site is then filled, but another particle insertion at R_1 is performed. After each insertion, compute the radius of gyration of the aggregate,

$$R_g^2 = \frac{\Sigma_{m,n}\varphi_{mn}\sqrt{(m - x_c)^2 + (n - y_c)^2}}{\Sigma_{m,n}\varphi_{mn}} \qquad \begin{aligned} x_c &= (\Sigma_{m,n}\varphi_{mn}m)/(\Sigma_{m,n}\varphi_{mn}) \\ y_c &= (\Sigma_{m,n}\varphi_{mn}n)/(\Sigma_{m,n}\varphi_{mn}) \end{aligned} \tag{7.274}$$

and stop the simulation when R_g exceeds R_{end}. Using **spy**, make a plot of the fractal geometry of the resulting aggregate for the parameters

$$R_0 = 1 \qquad R_1 = 30 \qquad R_2 = 60 \qquad R_{end} = 15 \tag{7.275}$$

7.B.4. Consider an ideal polymer chain of N freely-jointed segments of length l. We wish to simulate the evolution of the polymer chain under flow without accounting for any entanglements between chains (this is valid as long as the molecular weight is less than ~ 1000). We coarse-grain the geometry of the chain by treating it as a collection of N_b "beads," each subsequent pair of beads being connected by a "strand" of $Z = N/N_b$ segments. The set of SDEs that describe the motion of the chain under flow in a velocity field $v(x)$ is

$$d\boldsymbol{R}_v = \left[\boldsymbol{v}(0) + (\boldsymbol{\nabla v})^{\mathrm{T}} \cdot \boldsymbol{R}_v + \frac{1}{\zeta}\boldsymbol{F}_v^{(c)} \right] dt + \sqrt{\frac{2k_b T}{\zeta}} d\boldsymbol{W}_t^{(v)} \tag{7.276}$$

$d\boldsymbol{W}_t^{(v)}$ is an independent vector of three Wiener processes for each bead. ζ is the drag constant for each bead, and the total conservative spring force acting on bead $v \in [0, N]$ is

$$\boldsymbol{F}_v^{(c)} = \boldsymbol{F}(\boldsymbol{R}_v - \boldsymbol{R}_{v+1}) + \boldsymbol{F}(\boldsymbol{R}_v - \boldsymbol{R}_{v-1}) \tag{7.277}$$

where we use the finitely-extensible (FENE) spring law

$$\boldsymbol{F}(\boldsymbol{Q}) = \frac{-K\boldsymbol{Q}}{1 - |\boldsymbol{Q}|^2/Q_{max}^2} \qquad K = \frac{3k_b T}{Zl^2} \qquad Q_{max} = Zl \tag{7.278}$$

When the chain is stretched only a little, this force law agrees with the Gaussian model of Problem 7.A.5. As the spring is stretched more, it becomes stiffer so that the chain can never be extended beyond its contour length.

Write a **MATLAB** program that simulates and animates the motion of a polymer chain in a shear velocity field $v(x) = \dot{\gamma}x_2 e^{[1]}$. For clarity in the animation, shift the center of mass to the origin after each time step. Perform a number of simulations at varying shear rates $\dot{\gamma} > 0$ with the parameters

$$N = 100 \qquad l = 1 \qquad N_b = 10 \qquad k_b T = 1 \qquad \zeta = 1 \tag{7.279}$$

Describe the nature of the motion and how the shape of the chain is altered.

7.B.5. Another approach to global minimization is the *particle swarm optimization (PSO) method* (Kennedy & Eberhart, 1995). Let $F(x)$ be the cost function surface. We simulate a population of N particles using an algorithm that is meant to mimic the motion of a swarm of animals in a search for food (www.swarmintelligence.org). Each particle α has a position $x^{[\alpha]}$ and velocity $v^{[\alpha]}$ that are updated at each iteration. Let $p^{[\alpha]}$ be α's "personal best," i.e. the coordinates with the lowest $F(x)$ that have been visited to date by particle α. Let g be the position of lowest $F(\boldsymbol{x})$ found by *any* particle to date. The update in the velocity and position of each particle is then

$$v_k^{[\alpha]} \leftarrow v_k^{[\alpha]} + c_1 u_k \left(p_k^{[\alpha]} - x_k^{[\alpha]} \right) + c_2 u_k' \left(g_k - x_k^{[\alpha]} \right) \qquad k \in [1, \dim(\boldsymbol{x})]$$
$$\boldsymbol{x}^{[\alpha]} \leftarrow \boldsymbol{x}^{[\alpha]} + \boldsymbol{v}^{[\alpha]} \tag{7.280}$$

$\boldsymbol{u}$, $\boldsymbol{u}'$ are vectors of independent random numbers, uniformly distributed on $[0, 1]$ and c_1, c_2 are tunable learning parameters. A maximum allowable velocity may be specifed as well. Write a program that uses PSO with approximately 10–50 particles, and compare its performance to simulated annealing and genetic algorithms for the cost function (7.265).

7.C.1. Atoms of an ideal gas such as argon interact only through dispersion forces that are modeled by the Lennard–Jones pairwise interaction:

$$U_{\mathrm{LJ}}(R_{\alpha\beta}) = 4\varepsilon \left[\left(\frac{\sigma}{R_{\alpha\beta}} \right)^{12} - \left(\frac{\sigma}{R_{\alpha\beta}} \right)^{6} \right] \qquad r_{\alpha\beta} = \left| \boldsymbol{R}_\alpha - \boldsymbol{R}_\beta \right| \tag{7.281}$$

This interaction includes a strong short-range repulsion when $R_{\alpha\beta} \ll \sigma$ and a weaker long-range attraction when $R_{\alpha\beta} \gg \sigma$. As $U_{\mathrm{LJ}}(R_{\alpha\beta}) \to 0$ as $R_{\alpha\beta} \to \infty$, it is common to set the interaction energy to zero with $R_{\alpha\beta} > r_{\mathrm{cut}}$, where a common choice of cutoff radius is $r_{\mathrm{cut}} = 3.5\sigma$.

For a system of N Lennard–Jones atoms, the total potential energy of the system is the sum of the pairwise interactions from each unique pair of atoms:

$$U_{\mathrm{tot}}(\boldsymbol{q}) = \sum_{\alpha=1}^{N} \sum_{\beta=\alpha+1}^{N} U_{\mathrm{LJ}}(R_{\alpha\beta}) \tag{7.282}$$

$\boldsymbol{q}$ is the state vector of all atomic positions. Using a cutoff radius, when computing U_{tot} we need consider only those pairs of atoms with $R_{\alpha\beta} \leq r_{\mathrm{cut}}$.

If the molecular weight of each atom is m_{a}, we simulate the system at density ρ by using periodic boundary conditions on a cubic box of dimension $L \times L \times L$, such that $\rho L^3 = m_{\mathrm{a}} N$. We assume that this simulation cell is surrounded by identical copies, with images of atom α at

$$\boldsymbol{R}_\alpha^{[m_1, m_2, m_3]} = \boldsymbol{R}_\alpha + m_1 L \boldsymbol{e}^{[1]} + m_2 L \boldsymbol{e}^{[2]} + m_3 L \boldsymbol{e}^{[3]} \qquad m_j = 0, \pm 1, \pm 2, \ldots \tag{7.283}$$

If $L > r_{\mathrm{cut}}$, then for each pair (α, β) of atoms in (7.282), we need only consider the images of each atom that are closest to each other.

At constant N, constant volume $V = L^3$, and constant temperature T, we sample the system at equilibrium using the Metropolis Monte Carlo method for the Boltzmann distribution $P(\boldsymbol{q}) \propto \exp[-U_{\mathrm{tot}}(\boldsymbol{q})/k_{\mathrm{b}}T]$. The state is updated by selecting at random one or more atoms, and then randomly translating them. If only one atom is selected, then only a small fraction of the pairwise interactions in (7.282) change value and need to be recomputed. In practice, the CPU time necessary to compute the pairwise interactions is reduced by a strategy such as using a cell list, in which the simulation box is divided into nonoverlapping subcells of length just greater than r_{cut}. Then, each atom is assigned to one of these subcells, and when computing the total energy, only atoms in the same or neighboring subcells need to be considered.

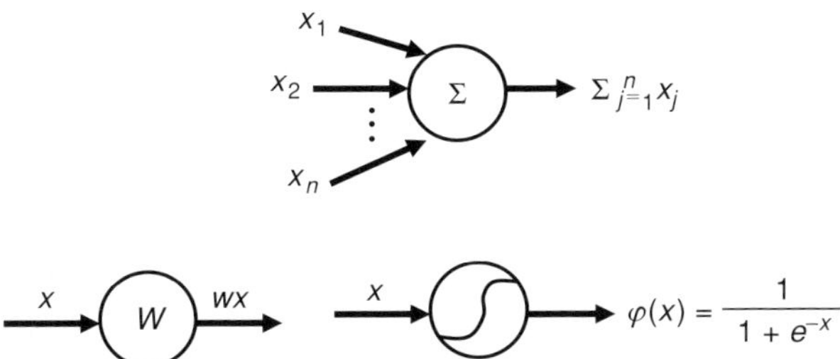

Figure 7.17 Summation, multiplication, and sigmoidal nodes used in neural networks.

For this problem, you are asked to write a program for Monte Carlo simulation of a system of Lennard–Jones atoms. For argon, the Lennard–Jones parameters are $\varepsilon/k_b = 119.8\,\mathrm{K}$, $\sigma = 3.405 \times 10^{-10}\,\mathrm{m}$. From the ideal gas law, compute the density of argon at standard atmospheric conditions, and then simulate a system of at least 100 atoms at these conditions. The pressure of the system can be computed from the formula

$$P = \rho k_b T + \frac{1}{3L^3}\left\langle \sum_{\alpha=1}^{N} \sum_{\beta=\alpha+1}^{N} \boldsymbol{F}_{\alpha\beta} \cdot \boldsymbol{R}_{\alpha\beta} \right\rangle \qquad \boldsymbol{F}_{\alpha\beta} = -\frac{\partial U}{\partial \boldsymbol{R}_{\alpha\beta}} \qquad (7.284)$$

Since evaluating the pressure requires computing the forces, to save CPU time it is common to only compute the average in (7.284) over some subset of states from the total trajectory, each separated by a large number of MC steps. Subsequent states in the Monte Carlo simulation are nearly identical, and thus it makes little sense, from the standpoint of sampling phase-space, to compute the pressure at each Monte Carlo step. From the simulation at room temperature, compute the compressibility factor $Z = PV/Nk_b T$, which is 1 for an ideal gas. Then, reduce the density, keeping the temperature constant at 298 K, and find where Z starts to diverge from the ideal gas result.

If the temperature were to be reduced and the density increased, this system would transition from a gas to a liquid and then to a solid. For more on atomistic simulation, how we extract material properties from such simulations, and how we simulate systems at constant pressures, constant chemical potentials, etc. consult Frenkel & Smit (2002).

7.C.2. Earlier, we have used statistical arguments to compute the average molecular weight as a function of conversion for condensation polymerization of multifunctional monomers. Here, you are asked to model the evolution of the chain length distribution by *kinetic Monte Carlo simulation*. Again, consider the case of a bifunctional acid type-1 monomer and a trifunctional base type-2 monomer, with the numbers N_1 and N_2 of each monomer chosen to balance the acid and base end group concentrations. We have estimated the gel point to occur at an acid conversion of $\sim70\%$. To simulate the evolution of the chain length distribution up to this point, rather than just compute DP_w, we generate a data structure that can represent the connectivity of the monomers at any time. Let this data structure be State, with the level-one members
.M1(N1), .M2(N2), .alpha1, .beta2, .molStartA, .molStartB.

State.alpha1 $= 2$ and State.beta2 $= 3$ for the case described above. State.M1(k) contains information about the state of the kth type-1 monomer in the array

Table 7.1 Measured data of system performance

θ_1	θ_2	$F(\boldsymbol{\theta})$
2.653	2.639	0.948
2.625	2.703	0.744
1.865	2.699	0.381
2.591	3.104	0.393
1.337	2.772	0.648
1.779	2.699	0.411
2.470	2.515	1.162
1.265	3.247	0.784

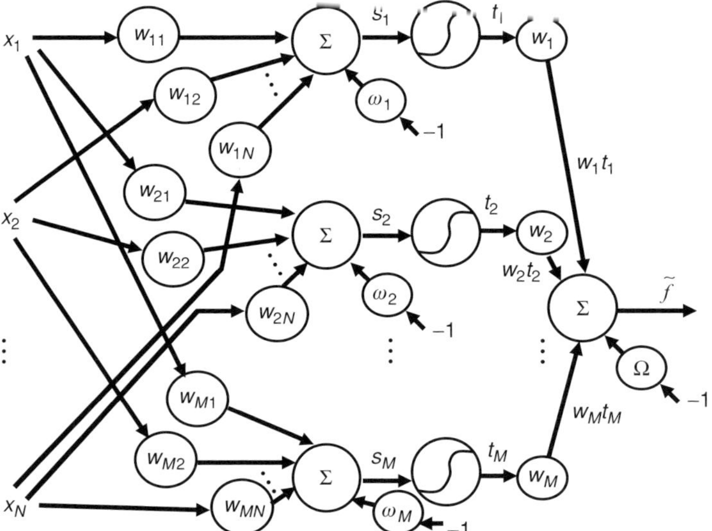

Figure 7.18 Three-layer neural network for representing a continuous function $f(x)$.

State.M1(k).A(State.alpha1,2). If acid group j of this monomer is unreacted, State.M1(k).A(j,1) = 0. Else, if it has reacted with base group m of type-2 monomer n, State.M1(k).A(j,1) = n and State.M1(k).A(j,2) = m. Similar state information is stored for each type-2 monomer k in State.M2(k).B(State.beta2,2).

State.molStartA (N1) and State.molStartB(N2) are vectors with components that take the value of 1 if the monomer is a unique "starting position" of a molecule and 0 if it is not. Using this approach, we can measure the chain length of the molecule attached to each "start" position by a simple recursive algorithm, and sample each unique molecule only once.

We start our simulation with all end groups unreacted, so that each monomer is a unique molecule and all "start" values are 1. Then, we conduct a simulation in which we select a pair of unreacted acid and base groups at random. We connect these by reaction, and set to

0 the "start" values of the participating monomer units. At each 0.05 increment of the acid conversion, plot the chain length distribution, up to near the gel point. Unless the number of monomers that you simulate is very large, the simulation will not be accurate near the gel point. Plot the measured DP_w as a function of conversion, and compare to the previous analytical result.

7.C.3. Often, we wish to represent some function f of $x \in \Re^N$ by an empirical approximation $\tilde{f}(x) \approx f(x)$. A common approach (especially in the IT community) is to construct a *neural network* from nodes such as those in Figure 7.17. In theory, a three-layer neural network (Figure 7.18) of such nodes can represent a continuous function, if the number M of nodes in the middle layer is sufficiently large (Dean *et al.* 1995).

To compute the value of $\tilde{f}(x)$, we work backwards, to obtain

$$\tilde{f}(x) = \sum_{j=1}^{M} w_j t_j - \Omega = \sum_{j=1}^{M} w_j \varphi(s_j) - \Omega \qquad \varphi(x) = \frac{1}{1 + e^{-x}} \tag{7.285}$$

where

$$s_j = \sum_{k=1}^{N} w_{jk} x_k - \omega_j \tag{7.286}$$

The parameters used to fit the network are stored in $\boldsymbol{\theta} \in \Re^P$, $P = M(N + 2) + 1$,

$$\boldsymbol{\theta} = [w_1 \cdots w_M w_{11} \cdots w_{1N} \, w_{21} \cdots w_{2N} \cdots w_{MN} \; \omega_1 \cdots \omega_M \; \Omega]^{\mathrm{T}} \tag{7.287}$$

We obtain $\boldsymbol{\theta}$ by minimizing the sum of squared errors between the predictions of the network and a set of training data, $\{f^{[p]} = f(x^{[p]})\}$, $p \in [1, N_d]$,

$$S(\boldsymbol{\theta}) = \frac{1}{2} \sum_{p=1}^{N_d} [\tilde{f}(x^{[p]}) - f^{[p]}]^2 \tag{7.288}$$

Write a program that uses a stochastic algorithm to fit the neural net. Demonstrate its use to fit the data of Table 7.1, which are measurements of a cost function representing how poorly a system performs as a function of two tunable parameters, θ_1 and θ_2. Using the fitted model, θ_1 and θ_2 could be varied automatically to improve the performance through learning from past experience.

8 Bayesian statistics and parameter estimation

Throughout this text, we have considered algorithms to perform simulations – given a model of a system, what is its behavior? We now consider the question of model development. Typically, to develop a model, we postulate a mathematical form, hopefully guided by physical insight, and then perform a number of experiments to determine the choice of parameters that best matches the model behavior to that observed in the set of experiments. This procedure of model proposition and comparison to experiment generally must be repeated iteratively until the model is deemed to be sufficiently reliable for the purpose at hand. The problem of drawing conclusions from data is known as *statistical inference*, and in particular, our focus here is upon *parameter estimation*. We use the powerful Bayesian framework for statistics, which provides a coherent approach to statistical inference and a procedure for making optimal decisions in the presence of uncertainty. We build upon the concepts of the last chapter and find, in particular, Monte Carlo simulation to be a powerful and general tool for Bayesian statistics.

General problem formulation

The basic *parameter estimation*, or *regression*, problem involves fitting the parameters of a proposed model to agree with the observed behavior of a system (Figure 8.1). We assume that, in any particular measurement of the system behavior, there is some set of *predictor variables* $x \in \Re^M$ that fully determines the behavior of the system (in the absence of any random noise or error). For each experiment, we measure some set of *response variables* $y^{(r)} \in \Re^L$. If $L = 1$, we have *single-response data* and if $L > 1$, *multiresponse data*. We write these predictor and response vectors in row form,

$$x = [x_1 \, x_2 \ldots x_M] \qquad y^{(r)} = [y_1 \, y_2 \ldots y_L] \tag{8.1}$$

We propose a mathematical relation, which maps the predictors x to the responses $y^{(r)}$, that involves a set of adjustable *model parameters* $\theta \in \Re^P$, whose values we wish to estimate from the measured response data. Let us say that we have a set of N experiments, in which for experiment $k = 1, 2, \ldots N$, $x^{[k]}$ is the row vector of predictor variables and the row vector of measured response data is $y^{[k]}$. For each experiment, we have a model prediction of the response

$$\hat{y}^{[k]}(\theta) = f(x^{[k]}; \theta) \tag{8.2}$$

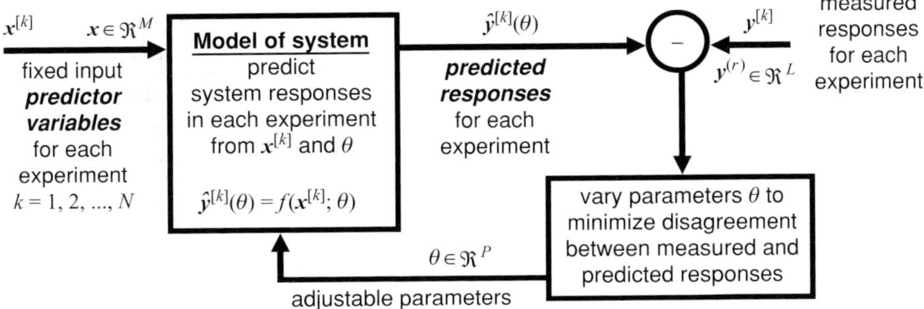

Figure 8.1 The parameter estimation problem.

The basic regression problem is: given a proposed model $f(x^{[k]}; \theta)$, how do we choose θ such that the model predictions $\hat{y}^{[k]}(\theta)$ agree most closely to the observations $y^{[k]}$? Of course, we need to ask – how do we define "close agreement," and how close is close enough to accept the model?

Given the data at hand, which generally include some uncontrolled random errors, how do we estimate the accuracy with which we have estimated θ? Here, we use Bayesian statistics, a framework for describing how our uncertainty in the values of the parameters changes by doing the experiments. Before doing the experiments, we characterize our knowledge about θ – which we treat as a random vector – by a prior probability density $p(\theta)$. If we have accurate prior knowledge, this distribution is sharply peaked; if not, it is diffuse. After obtaining new data $\{y^{[k]}\}$ from the experiments, we use the rules of Bayesian analysis to compute a posterior density $p(\theta|\{y^{[k]}\})$ that describes our new uncertainty after taking into account the additional data. With this posterior, we can test hypotheses, form confidence (credible) intervals, and make rational decisions based on uncertainty in θ using numerical simulation.

The Bayesian view is at odds with the traditional frequentist approach to statistics, which does not treat θ as itself being random. The formation of confidence intervals, selection of parameter estimation rules, etc. in the sampling approach are not as direct, and sometimes not as well behaved, as those of Bayesian statistics. A large fraction of the statistics community has been resistant to the Bayesian paradigm, but this situation is changing and the modern practice of statistics is increasingly Bayesian. Thus, we take this framework for this chapter, and provide an overview of its general and powerful tools for parameter estimation.

Example. Fitting kinetic parameters of a chemical reaction

Before proceeding, let us consider some simple examples of parameter estimation problems. We are studying the kinetics of the chemical reaction $A + B \rightarrow C$, which if assumed elementary, has the rate law

$$r_{R1} = k_1(T)c_A c_B \tag{8.3}$$

c_A and c_B are the concentrations of the two reactants in M, r_{R1} is the reaction rate per unit volume in M/s, and $k_1(T)$ is a temperature-dependent rate constant in $(Ms)^{-1}$. We determine the value of $k_1(T)$ by studying the behavior of well-characterized reactors such as a perfectly-mixed CSTR or a batch reactor (a well-mixed vessel with no inflow or outflow).

Fitting the rate law to steady-state measurements of a CSTR

For a CSTR, we measure $k_1(T)$ by operating the reactor at steady state at constant volume V, constant volumetric flow rate v, constant inlet concentrations $c_j^{(in)}$, and measuring the outlet concentrations c_j. We obtain a measurement of the rate though the balance on C,

$$\frac{dc_C}{dt} = 0 = v\left[c_C^{(in)} - c_C\right] + Vr_{R1} \Rightarrow r_{R1} = \frac{v}{V}\left[c_C - c_C^{(in)}\right] \tag{8.4}$$

Measurements of c_A and c_B are necessary to determine $k_1(T)$ from (8.3). For each CSTR experiment, the set of predictor variables is $x = [c_A\ c_B]$. The parameter that we wish to estimate is $\theta = [k_1]$. The response variable is $y^{(r)} = [r_{R1}]$.

In these equations, we have assumed that the reaction is elementary. To relax this assumption, we fit the data to the rate law $r_{R1} = k_1 c_A^{v_a} c_B^{v_b}$, in which case the model parameter vector is now $\theta = [k_1\ v_a\ v_b]^T$.

Alternatively, if we wish to study the temperature dependence of the rate constant, we repeat the experiments at different temperatures. The predictor vector is now $x = [c_A\ c_B\ T]$. Assuming an Arrenhius form of the rate law, $k_1(T) = A_1 \exp[-E_1/RT]$, the vector of model parameters is $\theta = [v_a\ v_b\ A_1\ E_1]^T$. In each case above, we have single-response data, with a nonlinear analytical model relating the predictors to the response.

Fitting the rate law to initial rate measurements in a batch reactor

We can also study the kinetics of the reaction in a batch reactor. We charge the reactor with known initial concentrations $c_A(0)$ and $c_B(0)$, and measure the concentrations of the species as functions of time, governed by the initial value problem:

$$\frac{dc_A}{dt} = -r_{R1} \qquad \frac{dc_B}{dt} = -r_{R1} \qquad \frac{dc_C}{dt} = r_{R1} \qquad r_{R1} = k_1 c_A^{v_a} c_B^{v_b}$$
$$c_A(t=0) = c_A(0) \qquad c_B(t=0) = c_B(0) \qquad c_C(t=0) = 0 \tag{8.5}$$

For each run, we keep the temperature constant at T. From the slope of $c_C(t)$ at $t = 0$, we obtain the initial reaction rate:

$$\left.\frac{dc_C}{dt}\right|_{t=0} = r_{R1} = k_1(T)[c_A(0)]^{v_a}[c_B(0)]^{v_b} \tag{8.6}$$

Let us say that the results of these experiments are those in Table 8.1. Now, the predictor variables are $x = [c_A(0)\ c_B(0)]$ and we have a single response, $y = r_{R1}$. The set of model parameters is $\theta = [k_1(T)\ v_a\ v_b]^T$. Again, we have single-response data and an analytical, nonlinear model relating the predictors to the response.

Table 8.1 Instantaneous rate data for chemical reaction $A + B \rightarrow C$

Experiment	c_A (M)	c_B (M)	Rate r_{R1} (M/s)
1	0.1	0.1	0.0246×10^{-3}
2	0.2	0.1	0.0483×10^{-3}
3	0.1	0.2	0.0501×10^{-3}
4	0.2	0.2	0.1003×10^{-3}
5	0.05	0.2	0.0239×10^{-3}
6	0.2	0.05	0.0262×10^{-3}

Table 8.2 Predictor and response variables in modified linear form

Experiment	$\log_{10} c_A = x_1$	$\log_{10} c_B = x_2$	$\log_{10} r_{R1} = y$
1	−1.0000	−1.0000	−4.6096
2	−0.6990	−1.0000	−4.3157
3	−1.0000	−0.6990	−4.2999
4	−0.6990	−0.6990	−3.9988
5	−1.3010	−0.6990	−4.6224
6	−0.6990	−1.3010	−4.5818

Transforming the batch reactor data to obtain a linear regression problem

As written, the model $r_{R1} = k_1(T) c_A^{v_a} c_B^{v_b}$ is nonlinear; however, if we take the base-10 logarithm,

$$\log_{10} r_{R1} = \log_{10} k_1(T) + v_a \log_{10} c_A + v_b \log_{10} c_B \tag{8.7}$$

define new predictor and response variables

$$y = \log_{10} r_{R1} \quad x_1 = \log_{10} c_A \quad x_2 = \log_{10} c_B \tag{8.8}$$

and define new model parameters

$$\beta_0 = \log_{10} k_1(T) \quad \beta_1 = v_a \quad \beta_2 = v_b \tag{8.9}$$

we obtain a model that depends linearly upon its parameters:

$$y = \beta_0 + \beta_1 x_1 + \beta_2 x_2 \tag{8.10}$$

The modified predictor and response variables are listed in Table 8.2. From this single-response linear regression, we obtain estimates for the reaction exponents v_a, v_b, and the rate constant $k_1(T)$. Such transformations are common in practice, but care must be taken in drawing the proper statistical conclusions as the transformations also act upon the random noise that is assumed present.

Table 8.3 Batch reaction kinetic data for reaction $A + B \rightarrow C$ with initial concentrations of 0.1 M for A and B

time (h)	c_A (M)	c_B (M)	c_C (M)
0.5	0.0985	0.0995	0.0010
1.0	0.0637	0.0651	0.0357
1.5	0.0500	0.0496	0.0501
2.0	0.0462	0.0453	0.0512
3.0	0.0363	0.0384	0.0682
4.0	0.0248	0.0247	0.0747
5.0	0.0171	0.0174	0.0809
6.0	0.0168	0.0203	0.0818
7.0	0.0131	0.0136	0.0858
8.0	0.0150	0.0121	0.0863
9.0	0.0140	0.0142	0.0872
10.0	0.0134	0.0134	0.0928

Fitting the rate law to the entire dynamic profile of a batch reactor run

Above, we have used only the initial slope of $c_C(t)$ to measure the reaction rate at the initial concentrations $c_A(0)$ and $c_B(0)$. We could fit the rate law to the complete data set of concentrations vs. time, such as is found in Table 8.3 for the experiment with $c_A(0) = 0.1$ M and $c_B(0) = 0.1$ M. If we use only the data for the concentration of C, we have a single-response data set. If we include all concentration values, we have a multiresponse data set with $L = 3$. For this regression problem, we obtain the model predictions by solving the IVP (8.5) numerically.

Fitting the rate law from multiple sources

Here, we have provided several data sets that each give information about the rate law, but each is obtained in different experiments, with perhaps different levels of experimental errors. When we have such a collection of data from multiple sources, we would like to consider all of it, even if some of the data may be more accurate than others. Below, we see how to treat such composite data sets, where some data sets may be single-response and others multiresponse.

These example data sets also provide a good overview of parameter estimation problems of varying complexity. Easiest is a single-response linear regression problem in which, as for the data of Table 8.2, we have a linear model that relates the predictors to the response. Next is the case of nonlinear single-response regression, such as for the data of Table 8.1, where now the response is an explicit nonlinear analytical function of the predictors. If we fit the rate law to the $c_C(t)$ data of Table 8.3, we still have a nonlinear single-response regression, but now the model predictions must be obtained by numerical simulation of (8.5). Finally, if we consider as well the $c_A(t)$ and $c_B(t)$ data of Table 8.3, we have a multiresponse nonlinear regression problem, for which the predictions again must be made by numerical simulation.

Single-response linear regression

We first discuss the regression, or estimation, of parameters in linear models from single-response data. Let us say that we have performed a set of N experiments in which for each experiment $k = 1, 2, \ldots, N$, the set of predictor variables $\{x_1^{[k]}, x_2^{[k]}, \ldots, x_M^{[k]}\}$ is known a priori, and a measurement is made of the single-response variable $y^{[k]}$. We assume that this single-response variable depends linearly upon the predictors,

$$y^{[k]} = \beta_0 + \beta_1 x_1^{[k]} + \beta_2 x_2^{[k]} + \cdots + \beta_M x_M^{[k]} + \varepsilon^{[k]} \tag{8.11}$$

$\varepsilon^{[k]}$ is some *random measurement error* for the kth experiment. Due to this error, the measured response is not equal to the "true" response of the system, and the statistical properties of the error are very important, although generally unknown at the time of measurement. The "true" parameters $\{\beta_0, \beta_1, \ldots, \beta_M\}$ are those that describe the system behavior perfectly in the absence of any random, model, or predictor error. That is, we hypothesize that the model is indeed a valid one. Thus, the set of predictor variables chosen indeed completely specifies the response of the system (in the absence of random error) and the relationship between the response and each predictor is truly linear.

In some instances, we wish to fit a model with a zero y-intercept $\beta_0 = 0$, such that

$$y^{[k]} = \beta_1 x_1^{[k]} + \beta_2 x_2^{[k]} + \cdots + \beta_M x_M^{[k]} + \varepsilon^{[k]} \tag{8.12}$$

We introduce a common notation for both cases by defining for each experiment the vectors of predictors and parameters,

$$\begin{aligned}
x^{[k]} &= \begin{bmatrix} 1 \; x_1^{[k]} \; x_2^{[k]} \; \ldots x_M^{[k]} \end{bmatrix} \quad \theta = [\beta_0 \, \beta_1 \, \beta_2 \ldots \beta_M]^{\mathrm{T}} \quad \text{with } y\text{-intercept} \\
x^{[k]} &= \begin{bmatrix} x_1^{[k]} \; x_2^{[k]} \; \ldots x_M^{[k]} \end{bmatrix} \quad \theta = [\beta_1 \, \beta_2 \ldots \beta_M]^{\mathrm{T}} \quad \text{without } y\text{-intercept}
\end{aligned} \tag{8.13}$$

Then, the response in the kth experiment is

$$y^{[k]} = x^{[k]} \cdot \theta^{(\text{true})} + \varepsilon^{[k]} \tag{8.14}$$

Because we do not know the "true" values $\theta^{(\text{true})}$ of the parameters, we must consider the parameter vector to be adjustable. The *predicted response* in the kth experiment, *as a function of* θ, is

$$\hat{y}^{[k]}(\theta) = x^{[k]} \cdot \theta \tag{8.15}$$

We further define for our set of experiments the *design matrix* X to contain for each experiment, the row vector of the predictor values,

$$X = \begin{bmatrix} - & x^{[1]} & - \\ - & x^{[2]} & - \\ & \vdots & \\ - & x^{[N]} & - \end{bmatrix} \tag{8.16}$$

For $\dim(\theta) = P$, X is an $N \times P$ matrix. X is specified when we design the set of experiments and is known prior to collecting the response data.

$$
X = \begin{bmatrix} 1 & x_1^{[1]} & x_2^{[1]} & \cdots & x_M^{[1]} \\ 1 & x_1^{[2]} & x_2^{[2]} & \cdots & x_M^{[2]} \\ \vdots & \vdots & \vdots & & \vdots \\ 1 & x_1^{[N]} & x_2^{[N]} & \cdots & x_M^{[N]} \end{bmatrix} \qquad X = \begin{bmatrix} x_1^{[1]} & x_2^{[1]} & \cdots & x_M^{[1]} \\ x_1^{[2]} & x_2^{[2]} & \cdots & x_M^{[2]} \\ \vdots & \vdots & & \vdots \\ x_1^{[N]} & x_2^{[N]} & \cdots & x_M^{[N]} \end{bmatrix} \tag{8.17}
$$
with y-intercept   without y-intercept

The vector of predicted responses in each experiment is then

$$
\hat{\boldsymbol{y}}(\boldsymbol{\theta}) = \begin{bmatrix} \hat{y}^{[1]}(\boldsymbol{\theta}) \\ \vdots \\ \hat{y}^{[N]}(\boldsymbol{\theta}) \end{bmatrix} = X\boldsymbol{\theta} \tag{8.18}
$$

Linear least-squares regression

We vary $\boldsymbol{\theta}$ until the model predictions $\hat{y}^{[k]}(\boldsymbol{\theta})$ agree most closely with the observed $y^{[k]}$. We must define what we mean by "close agreement," but a readily apparent choice of metric is that we select the value $\boldsymbol{\theta}_{\text{LS}}$ that minimizes the *sum of squared errors*

$$
S(\boldsymbol{\theta}) \equiv \sum_{k=1}^{N} \left[y^{[k]} - \hat{y}^{[k]}(\boldsymbol{\theta}) \right]^2 = |\boldsymbol{y} - \hat{\boldsymbol{y}}(\boldsymbol{\theta})|^2 \tag{8.19}
$$

That is,

$$
\left. \frac{\partial S}{\partial \boldsymbol{\theta}^{\mathrm{T}}} \right|_{\boldsymbol{\theta}_{LS}} = \boldsymbol{0} \quad \nabla^2 S(\boldsymbol{\theta})|_{\boldsymbol{\theta}_{\text{LS}}} > 0 \tag{8.20}
$$

Substituting $\hat{\boldsymbol{y}}(\boldsymbol{\theta}) = X\boldsymbol{\theta}$ yields

$$
S(\boldsymbol{\theta}) = [\boldsymbol{y} - X\boldsymbol{\theta}]^{\mathrm{T}}[\boldsymbol{y} - X\boldsymbol{\theta}] = \boldsymbol{y}^{\mathrm{T}}\boldsymbol{y} - (X\boldsymbol{\theta})^{\mathrm{T}}\boldsymbol{y} - \boldsymbol{y}^{\mathrm{T}}(X\boldsymbol{\theta}) + (X\boldsymbol{\theta})^{\mathrm{T}}(X\boldsymbol{\theta}) \tag{8.21}
$$

Taking the derivative with respect to $\boldsymbol{\theta}$,

$$
\frac{\partial S}{\partial \boldsymbol{\theta}^T} = \boldsymbol{0} - X^{\mathrm{T}}\boldsymbol{y} - X^{\mathrm{T}}\boldsymbol{y} + \left[X^{\mathrm{T}}X + (X^{\mathrm{T}}X)^{\mathrm{T}} \right]\boldsymbol{\theta} = -2X^{\mathrm{T}}\boldsymbol{y} + 2(X^{\mathrm{T}}X)\boldsymbol{\theta} \tag{8.22}
$$

and setting it equal to zero yields a linear system for $\boldsymbol{\theta}_{\text{LS}}$,

$$
(X^{\mathrm{T}}X)\boldsymbol{\theta}_{LS} = X^{\mathrm{T}}\boldsymbol{y} \quad \Rightarrow \quad \boldsymbol{\theta}_{LS} = \left[(X^{\mathrm{T}}X)^{-1}X^{\mathrm{T}} \right]\boldsymbol{y} \tag{8.23}
$$

$X^{\mathrm{T}}X$ is a $P \times P$ matrix; its size is governed by the number of fitted parameters. The (i, j) element of $X^{\mathrm{T}}X$ is

$$
(X^{\mathrm{T}}X)_{ij} = \sum_{k=1}^{N} X_{ki} X_{kj} \tag{8.24}
$$

As we increase the number of experiments N, the magnitudes of the elements of $X^{\mathrm{T}}X$ increase. $X^{\mathrm{T}}X$ contains information about the ability of the experimental design to probe the parameter values. $X^{\mathrm{T}}X$ is a real, symmetric matrix that is at least positive-semidefinite. For a well-designed set of experiments that provides sufficient data to estimate each parameter to at least some finite accuracy, $X^{\mathrm{T}}X$ is positive-definite.

Solving the least-squares linear system

Often, $X^{\mathrm{T}}X$ may be ill conditioned, i.e., it has one or more eigenvalues near zero, when the experimental design does not provide sufficient information to measure well one or more linear combinations of parameters. Therefore, QR decomposition is the favored solution method,

$$
\begin{aligned}
X^{\mathrm{T}}X &= QR \\
Q^{\mathrm{T}} = Q^{-1} \qquad &R \text{ is upper-triangular}
\end{aligned}
\tag{8.25}
$$

As Q is orthogonal, we write $(X^{\mathrm{T}}X)\boldsymbol{\theta}_{\mathrm{LS}} = X^{\mathrm{T}}\boldsymbol{y}$ as

$$
QR\boldsymbol{\theta}_{\mathrm{LS}} = X^{\mathrm{T}}\boldsymbol{y} \Rightarrow R\boldsymbol{\theta}_{\mathrm{LS}} = Q^{\mathrm{T}}X^{\mathrm{T}}\boldsymbol{y}
\tag{8.26}
$$

The solution of the latter system requires only backward substitution.

A more expensive approach, but one that provides further insight into experimental design, is to diagonalize the real, symmetric matrix $X^{\mathrm{T}}X$:

$$
X^{\mathrm{T}}X = V \Lambda V^{\mathrm{T}}
\tag{8.27}
$$

V is an orthogonal matrix, $V^{\mathrm{T}} = V^{-1}$, whose columns are the normalized eigenvectors of $X^{\mathrm{T}}X$ and Λ is a diagonal matrix whose principal diagonal contains the eigenvalues of $X^{\mathrm{T}}X$. The least-squares estimate $\boldsymbol{\theta}_{\mathrm{LS}}$ is

$$
\boldsymbol{\theta}_{\mathrm{LS}} = [V \Lambda^{-1} V^{\mathrm{T}}]X^{\mathrm{T}}\boldsymbol{y}
\tag{8.28}
$$

Note also that since $X^{\mathrm{T}}X\boldsymbol{\theta}_{\mathrm{LS}} = X^{\mathrm{T}}\boldsymbol{y}$, we have an overdetermined system,

$$
X\boldsymbol{\theta}_{\mathrm{LS}} = \boldsymbol{y}
\tag{8.29}
$$

that can be solved by SVD (Chapter 3). The SVD approach is popular; however, here we use the familiar eigenvalue decomposition (8.27) in lieu of SVD to analyze experimental designs.

Example. Least-squares fitting of rate law parameters to transformed batch data

We now apply the least-squares method to the transformed batch reactor data of Table 8.2 for the kinetics of $A + B \rightarrow C$, using the model

$$
\log_{10} r_{\mathrm{R1}} = \log_{10} k_1 + \nu_{\mathrm{a}} \log_{10} c_{\mathrm{A}} + \nu_{\mathrm{b}} \log_{10} c_{\mathrm{B}}
\tag{8.30}
$$

The data of Table 8.2 yield the design matrix and response vector

$$X = \begin{bmatrix} 1 & -1.0000 & -1.0000 \\ 1 & -0.6990 & -1.0000 \\ 1 & -1.0000 & -0.6990 \\ 1 & -0.6990 & -0.6990 \\ 1 & -1.3010 & -0.6990 \\ 1 & -0.6990 & -1.3010 \end{bmatrix} \quad y = \begin{bmatrix} -4.6096 \\ -4.3157 \\ -4.2999 \\ -3.9988 \\ -4.6224 \\ -4.5818 \end{bmatrix} \tag{8.31}$$

The vector of parameters that we fit is

$$\boldsymbol{\theta} = [\log_{10} k_1 \quad v_a \quad v_b]^{\mathrm{T}} \tag{8.32}$$

We compute $X^{\mathrm{T}} X$, and solve the linear system (8.23) to obtain

$$\boldsymbol{\theta}_{\mathrm{LS}} = \begin{bmatrix} \log_{10} k_1 \\ v_a \\ v_b \end{bmatrix} = \begin{bmatrix} -2.6032 \\ 1.0224 \\ 0.9799 \end{bmatrix} \tag{8.33}$$

The estimated rate law for $A + B \rightarrow C$ from those data is then

$$r_{\mathrm{R1}} = k_1 C_A^{v_a} C_B^{v_b} = (0.0025) C_A^{1.0224} C_B^{0.9799} \tag{8.34}$$

For an elementary reaction, we expect $v_a = v_b = 1$, and our fitted values are indeed close to 1. But, is the discrepancy from $v_a = v_b = 1$ due solely to measurement error? To answer this question we need tools for testing hypotheses.

Example. Comparing protein expression data for two bacterial strains

Let us say that we have performed experiments to measure the protein expression rates of wild-type and mutant cell strains.

$$\begin{array}{ll} \text{wild-type} & \text{mutant} \\ \left.\begin{array}{l} 121.9 \\ 113.4 \\ 112.2 \\ 106.1 \end{array}\right\} \begin{array}{l} \text{sample} \\ \text{subset mean} \\ = 113.41 \end{array} & \left.\begin{array}{l} 120.7 \\ 119.5 \\ 116.5 \\ 124.0 \end{array}\right\} \begin{array}{l} \text{sample} \\ \text{subset mean} \\ = 120.16 \end{array} \end{array} \tag{8.35}$$

Is the difference between strains significant in a statistical sense or just an artifact of measurement noise?

To answer this question, we fit the data to the linear model

$$y = \theta_1 + \theta_2 x + \varepsilon \tag{8.36}$$

with the predictor variable

$$x = \begin{cases} 0, & \text{wild-type} \\ 1, & \text{mutant} \end{cases} \tag{8.37}$$

If we estimate probable bounds for the value of θ_2,

$$\theta_{2,\mathrm{lo}} \leq \theta_2 \leq \theta_{2,\mathrm{hi}} \tag{8.38}$$

and find that $\theta_{2,\,\mathrm{lo}} > 0$, then the data suggest that the difference in means between the two subsets is *statistically significant*; i.e., it is probably not due solely to random error.

The least-squares estimates for this model are computed easily. The design matrix X, the vector of measured responses y, and $X^{\mathrm{T}}y$ are

$$
X = \begin{bmatrix} 1 & 0 \\ 1 & 0 \\ 1 & 0 \\ 1 & 0 \\ 1 & 1 \\ 1 & 1 \\ 1 & 1 \\ 1 & 1 \end{bmatrix} \qquad y = \begin{bmatrix} 121.9 \\ 113.4 \\ 112.2 \\ 106.1 \\ 120.7 \\ 119.5 \\ 116.5 \\ 124.0 \end{bmatrix} \qquad X^{\mathrm{T}}y = \begin{bmatrix} 934.3 \\ 480.7 \end{bmatrix} \tag{8.39}
$$

Thus, for $n = 4$ data points for each strain,

$$
X^{\mathrm{T}}X = \begin{bmatrix} (2n) & n \\ n & n \end{bmatrix} = \begin{bmatrix} 8 & 4 \\ 4 & 4 \end{bmatrix}
$$

$$
(X^{\mathrm{T}}X)^{-1} = \begin{bmatrix} n^{-1} & -n^{-1} \\ -n^{-1} & (2n^{-1}) \end{bmatrix} = \begin{bmatrix} 0.25 & -0.25 \\ -0.25 & 0.5 \end{bmatrix} \tag{8.40}
$$

and the least-squares estimate is

$$
\begin{bmatrix} \theta_{\mathrm{LS},\,1} \\ \theta_{\mathrm{LS},\,2} \end{bmatrix} = (X^{\mathrm{T}}X)^{-1}[X^{\mathrm{T}}y] = \begin{bmatrix} 0.25 & -0.25 \\ 0.25 & 0.5 \end{bmatrix}\begin{bmatrix} 934.3 \\ 480.7 \end{bmatrix} = \begin{bmatrix} 113.4 \\ 6.7750 \end{bmatrix} \tag{8.41}
$$

$\theta_{\mathrm{LS},\,1}$ is the sample mean for the wild-type strain, and $\theta_{\mathrm{LS},\,2}$ is the value of the sample mean of the mutant strain minus that of the wild-type strain, $120.16 - 113.41 = 6.75$. But, is it still within the realm of plausibility that the "true" value of θ_2 is zero?

The Bayesian view of statistical inference

In practice, it is insufficient merely to identify the values of the parameters that minimize the sum of squared errors. We need also to consider the accuracy of our estimates. Here, we address this topic with Bayesian statistics, which describes how our uncertainty in the parameter values is changed by doing the experiments. Let us consider single-response regression of a model

$$
y^{[k]} = f\left(x^{[k]}; \theta\right) + \varepsilon^{[k]} \tag{8.42}
$$

We have a particular set of N measured responses $y \in \mathfrak{R}^N$, and wish to estimate the unknown parameter vector $\theta \in \mathfrak{R}^P$ and the statistical properties of the random error $\varepsilon \in \mathfrak{R}$. While the error may not be truly stochastic, we assume that it has the properties of a random variable, since presumably we have no practical way of predicting the error value in any single experiment.

As statistics is based upon probability theory, it is helpful to cite again two possible means of defining probabilities. One way to think about probability – the *frequentist* approach – is based upon the relative occurrences of events in many repetitive trials. Let us say that the probability of observing an event E in an independent random trial is $p(E)$. The frequentist way of defining the value of the probability of observing E, $0 \leq p(E) \leq 1$, is to say that if we perform a large number T of such trials, with the observed number of occurrences of E being N_E, then $p(E) \approx N_E / T$.

We can also define probabilities as statements of belief (de Finetti, 1970). I say that the probability of observing E during a random trial is $p(E)$ if I have no reason to prefer one of the following two bets over the other:

event E is observed in a particular trial;

or

a perfectly uniform random number generator in $[0, 1]$ returns a value u that is less than $p(E)$.

It is then necessary for me, as the holder of this belief system, to ensure that the probability values that I assign satisfy the appropriate conditions, e.g. are nonnegative, sum or integrate to 1, follow all laws of conditional and joint probabilities, etc.

Bayesian statistics is based upon manipulation of the probability $p(\boldsymbol{\theta}|\boldsymbol{y})$ that the model has a parameter vector $\boldsymbol{\theta}$, given a set of measured response data $\boldsymbol{y}$. While the Bayesian approach dates to the work of Thomas Bayes in the mid-1700s, it was slow to gain acceptance because, by treating $\boldsymbol{\theta}$ as a random vector, it violates the philosophical principle of Laplacian determinism, stating that nature is deterministic and predictable. Such criticisms were muted by the interpretation of probabilities as statements of belief, leading to a resurgence in the Bayesian approach.

But by resorting to the use of a belief system to define $p(\boldsymbol{\theta}|\boldsymbol{y})$, we introduce the issue of subjectivity, as it is possible that two different analysts will hold different belief systems, and thus arrive at different conclusions from the same data. The complaints of the frequentist school about the subjectivity of the Bayesian approach persist to this day, but can be countered by showing that implementations of the Bayesian paradigm exist such that two analysts, given the same data and working independently, reach (nearly) the same conclusions. These implementations are not quite objective, but they are highly reproducible from one analyst to another.

The development of tools to ensure the use of such "objective" belief systems, and computational methods such as Monte Carlo simulation have brought the Bayesian approach to the fore in recent years in areas such as parameter estimation, statistical learning, and statistical decision theory. Here, we focus our attention primarily upon parameter estimation, first restricting our discussion to single-response data.

Bayes' theorem

Bayesian analysis is based upon Bayes' theorem, itself simply an axiom of probability theory. It is not the theorem that is controversial; it is its application to statistics. Thus, it is best first to understand the theorem, before considering how it is applied to statistical inference.

Let us consider two random events E_1 and E_2 that are *not* mutually exclusive (i.e. none, one, or both may occur). Let the probability that E_1 occurs be $P(E_1)$ and the probability that E_2 occurs be $P(E_2)$. The joint probability that both occur, $P(E_1 \cap E_2)$, is related to the conditional probability that E_1 occurs if E_2 occurs, $P(E_1|E_2)$, by

$$P(E_1 \cap E_2) = P(E_1|E_2)P(E_2) \tag{8.43}$$

Similarly, if $P(E_2|E_1)$ is the conditional probability that E_2 occurs if E_1 occurs, we may write the joint probability as

$$P(E_1 \cap E_2) = P(E_2|E_1)P(E_1) \tag{8.44}$$

Bayes' theorem simply states that these two expressions for the joint probability must be equal,

$$P(E_1 \cap E_2) = P(E_1|E_2)P(E_2) = P(E_2|E_1)P(E_1) \tag{8.45}$$

and thus

$$P(E_2|E_1) = \frac{P(E_1|E_2)P(E_2)}{P(E_1)} \tag{8.46}$$

So how do we get from this axiom of probability theory to a framework for making inferences from data?

Bayesian view of single-response regression

Let us consider the single-response regression problem, where the response value in experiment k of a set of N experiments equals the value determined by the "true" model plus some random error,

$$y^{[k]} = f\left(x^{[k]}; \theta^{(\text{true})}\right) + \varepsilon^{[k]} \tag{8.47}$$

$\varepsilon^{[k]}$ is a random error whose stochastic properties are unknown. Here, we have used $\theta^{(\text{true})}$ to denote that the random error is added to the model predictions with the "true" values of the parameters (which are, of course, unknown to us). The predicted response value in each experiment, as a function of θ, is

$$\hat{y}^{[k]}(\theta) = f\left(x^{[k]}; \theta\right) \tag{8.48}$$

If we were to do the same set of experiments again, we would get each time a new vector of measurement errors

$$\varepsilon = \begin{bmatrix} \varepsilon_1 & \varepsilon_2 & \cdots & \varepsilon_N \end{bmatrix}^{\text{T}} \tag{8.49}$$

The general difficulty that we face is we do not really know much about the properties of the random error, but the accuracy of the model parameter estimates is largely determined by these errors. Therefore, to obtain a tractable approach to analysis, we make a number of assumptions about the nature of the error. We will need to check these assumptions later for consistency with the data, especially if we use our analysis to test hypotheses.

If ε is truly a measurement error and not a deficiency of the model, we expect that if we do the set of measurements over and over again many times, the expectations (averages) of each $\varepsilon^{[k]}$ will be zero:

$$E\left(\varepsilon^{[k]}\right) = 0 \tag{8.50}$$

We also assume that the error value in each experiment is independent of the values in the other experiments, i.e., they are uncorrelated, so that the covariances among errors are

$$\text{cov}\left(\varepsilon^{[k]}, \varepsilon^{[j]}\right) \equiv \text{E}\left\{\left[\varepsilon^{[k]} - \text{E}\left(\varepsilon^{[k]}\right)\right]\left[\varepsilon^{[j]} - \text{E}\left(\varepsilon^{[k]}\right)\right]\right\} = \delta_{kj}\text{var}\left(\varepsilon^{[k]}\right) \tag{8.51}$$

That is, the covariance matrix of ε is diagonal:

$$\text{cov}(\varepsilon) = \begin{bmatrix} \sigma_1^2 & & & \\ & \sigma_2^2 & & \\ & & \ddots & \\ & & & \sigma_N^2 \end{bmatrix} \qquad \sigma_k^2 = \text{var}\left(\varepsilon^{[k]}\right) \tag{8.52}$$

Finally, we assume that all variances are equal,

$$\sigma_k^2 = \sigma^2 \quad k = 1, 2, \ldots, N \tag{8.53}$$

Combined, these are known as the *Gauss–Markov conditions*:

$$E\left(\varepsilon^{[k]}\right) = 0 \quad \text{cov}\left(\varepsilon^{[k]}, \varepsilon^{[j]}\right) = \delta_{kj}\sigma^2 \tag{8.54}$$

We make one additional assumption: that the error is distributed according to a Gaussian (normal) distribution. This distribution is to be expected when the error in the measured response is the net result of many independent sources of error being added together.

The probability of observing a particular value of the error in the kth experiment between ε and $\varepsilon + d\varepsilon$ is then

$$p\left(\varepsilon^{[k]}|\sigma\right) = \frac{1}{\sigma\sqrt{2\pi}} \exp\left[-\frac{\left(\varepsilon^{[k]}\right)^2}{2\sigma^2}\right] \tag{8.55}$$

We now make a subtle, though important, shift in our point of view about the "true" value of $\boldsymbol{\theta}$. Previously, we have considered the random error to be added to the model prediction with the "true" parameter vector $\boldsymbol{\theta}^{(\text{true})}$ such that $\varepsilon^{[k]} = y^{[k]} - \hat{y}^{[k]}(\boldsymbol{\theta}^{(\text{true})})$. But, as we do not know $\boldsymbol{\theta}^{(\text{true})}$, we cannot relate $\varepsilon^{[k]}$ to $y^{[k]}$. Therefore, let us drop all reference to the "true" value completely, and assume that the relation $\varepsilon^{[k]} = y^{[k]} - \hat{y}^{[k]}(\boldsymbol{\theta})$ holds for all trial values of $\boldsymbol{\theta}$. That is, for any trial $\boldsymbol{\theta}$ and σ, we propose that the probability of observing a response $y^{[k]}$, given $\boldsymbol{\theta}$ and σ, is

$$p\left(y^{[k]}|\boldsymbol{\theta}, \sigma\right) = p\left(\varepsilon^{[k]} = y^{[k]} - \hat{y}^{[k]}(\boldsymbol{\theta})|\sigma\right) = \frac{1}{\sigma\sqrt{2\pi}} \exp\left\{-\frac{\left[y^{[k]} - \hat{y}^{[k]}(\boldsymbol{\theta})\right]^2}{2\sigma^2}\right\} \tag{8.56}$$

As we assume that the errors in each experiment are independent, the joint probability of observing the complete vector of responses $\boldsymbol{y}$ is

$$p(\boldsymbol{y}|\boldsymbol{\theta}, \sigma) = \prod_{k=1}^{N} p\left(y^{[k]}|\boldsymbol{\theta}, \sigma\right)$$

$$= \left(\frac{1}{\sqrt{2\pi}}\right)^N \sigma^{-N} \exp\left[-\frac{1}{2\sigma^2}\sum_{k=1}^{N}\left[y^{[k]} - \hat{y}^{[k]}(\boldsymbol{\theta})\right]^2\right] \tag{8.57}$$

We simplify our notation by defining the *sum of squared errors* as

$$S(\theta) = \sum_{k=1}^{N} \left[y^{[k]} - \hat{y}^{[k]}(\theta) \right]^2 = \sum_{k=1}^{N} \left[y^{[k]} - f\left(x^{[k]}; \theta\right) \right]^2 \tag{8.58}$$

We then write the probability of observing a particular response vector y, given specified values of θ and σ, which follows from the Gauss–Markov conditions and the assumption of normally-distributed errors, as

$$p(y|\theta, \sigma) = \left(\frac{1}{\sqrt{2\pi}} \right)^{N} \sigma^{-N} \exp\left[-\frac{1}{2\sigma^2} S(\theta) \right] \tag{8.59}$$

Now, in our particular application, we know y, but do not know, and wish to estimate, θ and σ. Thus, we turn to Bayes' theorem to invert the order of the arguments, from the $p(y|\theta, \sigma)$ that we propose in (8.59), to the $p(\theta, \sigma|y)$ that describes our uncertainty in θ and σ, given the known value of y.

Bayes' theorem states that

$$p(y, \theta, \sigma) = p(y|\theta, \sigma)p(\theta, \sigma) = p(\theta, \sigma|y)p(y) \tag{8.60}$$

and thus

$$p(\theta, \sigma|y) = \frac{p(y|\theta, \sigma)p(\theta, \sigma)}{p(y)} \tag{8.61}$$

We have proposed a functional form and interpretation for $p(y|\theta, \sigma)$.

What do the probability densities $p(\theta, \sigma)$, $p(\theta, \sigma|y)$, and $p(y)$ represent? We say that $p(\theta, \sigma)$ represents our belief system about the values of θ and σ *before* we do the experiments, and so is called the *prior probability distribution, prior density*, or simply the *prior*. In our analysis, we must propose some functional form for this probability, and we discuss the selection of the prior in further detail below. If we start from a state of near-complete ignorance, the prior should be a very diffuse function, whose density is spread out over a large region of (θ, σ) space.

Our belief system about the values of θ and σ *after* we do the experiments is represented by $p(\theta, \sigma|y)$, and so is called the *posterior probability distribution, posterior density*, or the *posterior*. This is the function that we wish to determine. Ideally, we would like the posterior to be sharply peaked about a localized region in (θ, σ) space, so that there is little uncertainty in the values of these parameters and little dependence upon the prior.

The interpretation of $p(y)$ seems problematic, until we recognize that upon integrating in θ and σ, $p(y)$ is independent of θ and σ as $p(\theta, \sigma|y)$ must normalize to 1:

$$\int d\theta \int d\sigma\, p(\theta, \sigma|y) = 1 = \frac{\int d\theta \int d\sigma\, p(y|\theta, \sigma)p(\theta, \sigma)}{p(y)} \tag{8.62}$$

$$p(y) = \int d\theta \int d\sigma\, p(y|\theta, \sigma)p(\theta, \sigma)$$

Thus, Bayes' theorem simply states that the posterior density is proportional to the product of $p(y|\theta, \sigma)$ and the prior $p(\theta, \sigma)$:

$$p(\theta, \sigma|y) \propto p(y|\theta, \sigma)p(\theta, \sigma) \tag{8.63}$$

Note that $p(y|\theta, \sigma)$ is the probability of observing a particular response y, given specified values of θ and σ. But, at the time of analysis, it is y that is known and θ and σ that are unknown. It is common practice to switch the order of the arguments and so define the *likelihood function* for θ and σ, given specified y,

$$l(\theta, \sigma|y) \equiv p(y|\theta, \sigma) = \left(\frac{1}{\sqrt{2\pi}}\right)^N \sigma^{-N} \exp\left[-\frac{1}{2\sigma^2}S(\theta)\right] \tag{8.64}$$

While we switch the order of the arguments, the meaning of this quantity remains unchanged – it is the probability that the measured response is y, given specified values of θ and σ.

The posterior density is then written as

$$p(\theta, \sigma|y) \propto l(\theta, \sigma|y)p(\theta, \sigma) \tag{8.65}$$

In the Bayesian approach, we take as estimates θ_M, σ_M the values that maximize the posterior density $p(\theta, \sigma|y)$. The frequentist rule is to take the value θ_{MLE} that maximizes the likelihood function $l(\theta, \sigma|y)$. If the prior $p(\theta, \sigma)$ is not uniform in θ, the *Bayesian most probable estimate* θ_M and the *maximum likelihood estimate* θ_{MLE} disagree.

Some general considerations about the selection of a prior

Clearly, the choice of the prior is crucial, as it influences our analysis. It is this subjective nature of the Bayesian approach that is the cause of controversy, since in statistics we would like to think that different people who look at the data will come to the same conclusions. We hope that the data will be sufficiently informative that the likelihood function is sharply peaked around specific values of θ and σ; i.e., that the inference problem is *data-dominated*. In this case, the estimates are rather insensitive to the choice of prior, as long as it is non-zero near the peak of the likelihood function. When this is not the case, the prior influences the results of the analysis.

In some problems, the explicit dependence upon a prior is quite useful. If we know a priori that certain regions of the parameter space are inadmissible (e.g. certain parameters must be nonnegative), then the prior can be set to zero in those regions to exclude them from consideration.

Using an explicit prior also allows us to blend learning from different sets of data in a seamless manner. Let us say that we measure a response vector $y^{(1)}$ in some set of experiments, and compute the posterior density

$$p(\theta|y^{(1)}) \propto l^{(1)}(\theta|y^{(1)})p(\theta) \tag{8.66}$$

We later perform a new set of experiments on the same system and measure a response vector $y^{(2)}$. When we perform the analysis on these new data, an apparent choice of prior for the second set of experiments is the posterior density from the first. The posterior density after having conducted both sets of experiments is

$$p\big(\boldsymbol{\theta}|\boldsymbol{y}^{(1)}, \boldsymbol{y}^{(2)}\big) \propto l^{(2)}\big(\boldsymbol{\theta}|\boldsymbol{y}^{(2)}\big)p\big(\boldsymbol{\theta}|\boldsymbol{y}^{(1)}\big) \tag{8.67}$$

Bayes' theorem provides a framework for learning from new experiences.

Finally, in some instances we may wish to use a prior based upon past experience with similar systems. While such priors violate the ideal of objectivity in statistical analysis, they can provide better estimates if the subjective prior is chosen well (i.e., the "expert" choosing the prior knows what he is talking about). Several approaches to constructing priors to agree with preexisting conditions, data, or experience are discussed in Robert, (2001).

Here, we consider a method for generating priors that is "objective" in the sense that as long as different analysts agree to use this approach, they will independently come to (nearly) the same statistical conclusions given only the predictors, response data, model, and likelihood function. The basic technique, discussed below, is to identify a data-translation, or symmetry, property that the likelihood function has, and choose the prior so that the posterior density retains the same property. Such a prior does not give undue emphasis to any particular region of parameter space and thus is said to be *noninformative*. As we later show, for the single-response likelihood (8.64), a noninformative prior is

$$p(\boldsymbol{\theta}, \sigma) = p(\boldsymbol{\theta})p(\sigma) \qquad p(\boldsymbol{\theta}) \sim c \qquad p(\sigma) \propto \sigma^{-1} \tag{8.68}$$

As we use a prior that is uniform in $\boldsymbol{\theta}$, the Bayesian most probable estimate $\boldsymbol{\theta}_M$ agrees with the maximum likelihood estimate $\boldsymbol{\theta}_{\mathrm{MLE}}$.

The noninformative prior (8.68) is improper as it does not satisfy the normalization condition

$$\int\limits_{\Re^P} \int\limits_{\sigma>0} p(\boldsymbol{\theta}, \sigma)d\sigma d\boldsymbol{\theta} = 1 \tag{8.69}$$

and furthermore does not integrate to a finite value. For some aspects of our analysis, this is not a problem as the posterior density is still proper, but when testing hypotheses, it is important that the prior density also be a proper distribution. We suggest here a very simple fix to this problem, by providing the a priori upper and lower bounds,

$$\Omega_\theta = \{\boldsymbol{\theta}|\theta_{j,\mathrm{lo}} \leq \theta_j \leq \theta_{j,\mathrm{hi}}, j = 1, 2, \ldots, N\} \quad \sigma_{\mathrm{lo}} \leq \sigma \leq \sigma_{\mathrm{hi}} \tag{8.70}$$

with indicator functions $I_\theta(\boldsymbol{\theta})$ and $I_\sigma(\sigma)$ that equal 1 for parameters that satisfy these bounds and equal 0 for values outside of the bounds. We then define the proper prior density

$$p(\boldsymbol{\theta}, \sigma) = c_0\sigma^{-1}I_\theta(\boldsymbol{\theta})I_\sigma(\sigma) \quad c_0^{-1} = \left[\int\limits_{\Re^P} I_\theta(\boldsymbol{\theta})d\boldsymbol{\theta}\right]\left[\int\limits_{\sigma>0} \sigma^{-1}I_\sigma(\sigma)d\sigma\right] \tag{8.71}$$

We show below why this prior is used, but for the moment, let us accept it as valid and continue our discussion.

The least-squares method reconsidered

Our proposed prior (8.68) makes use of the *assumption of prior independence,*

$$p(\boldsymbol{\theta}, \sigma) = p(\boldsymbol{\theta})p(\sigma) \tag{8.72}$$

which states that, prior to the experiment, the belief systems about $\boldsymbol{\theta}$ and σ are independent. The posterior density is then

$$p(\boldsymbol{\theta}, \sigma \,|\, \boldsymbol{y}) \propto l(\boldsymbol{\theta}, \sigma \,|\, \boldsymbol{y})p(\boldsymbol{\theta})p(\sigma) \tag{8.73}$$

Assuming the Gauss–Markov conditions hold and that the errors are normally distributed, the likelihood function is

$$l(\boldsymbol{\theta}, \sigma \,|\, \boldsymbol{y}) = \left(\frac{1}{\sqrt{2\pi}}\right)^{N} \sigma^{-N} \exp\left[-\frac{1}{2\sigma^2}S(\boldsymbol{\theta})\right] \tag{8.74}$$

and thus the posterior is

$$p(\boldsymbol{\theta}, \sigma \,|\, \boldsymbol{y}) \propto \left(\frac{1}{\sqrt{2\pi}}\right)^{N} \sigma^{-N} \exp\left[-\frac{1}{2\sigma^2}S(\boldsymbol{\theta})\right] p(\boldsymbol{\theta})p(\sigma) \tag{8.75}$$

If, as we have assumed in (8.68), the prior is uniform in the region of appreciable nonzero likelihood, $p(\boldsymbol{\theta}) \sim c$, then the most probable value of $\boldsymbol{\theta}$, for any value of σ, is that which minimizes $S(\boldsymbol{\theta})$, (8.58). Therefore, the least-squares method is justified statistically, as long as the Gauss–Markov conditions hold and the errors are normally distributed.

It is shown in the supplemental material in the accompanying website that, in the frequentist view, least squares is an unbiased estimator of the true value (i.e., if we repeat the set of experiments many times, the average estimate is the true value) if merely the zero-mean Gauss–Markov condition (8.50) is satisfied.

Numerical treatment of nonlinear least-squares problems

For a linear model $y^{[k]} = x^{[k]} \cdot \boldsymbol{\theta} + \varepsilon^{[k]}$, we obtain the least-squares estimate by solving an algebraic system $[X^{\mathrm{T}}X]\boldsymbol{\theta}_{\mathrm{LS}} = X^{\mathrm{T}}\boldsymbol{y}$. For a nonlinear model

$$y^{[k]} = f\left(x^{[k]}; \boldsymbol{\theta}\right) + \varepsilon^{[k]} \tag{8.76}$$

we must find the least-squares estimate through numerical optimization. For notational convenience, we define the cost function as

$$F_{\mathrm{cost}}(\boldsymbol{\theta}) = \frac{1}{2}S(\boldsymbol{\theta}) = \frac{1}{2}\sum_{k=1}^{N}\left[y^{[k]} - f\left(x^{[k]}; \boldsymbol{\theta}\right)\right]^2 \tag{8.77}$$

The gradient $\boldsymbol{\gamma} = \boldsymbol{\Delta}F$ of this cost function has components

$$\gamma_a(\boldsymbol{\theta}) = \frac{\partial F_{\mathrm{cost}}}{\partial \theta_a} = -\sum_{k=1}^{N}\left[y^{[k]} - f\left(x^{[k]}; \boldsymbol{\theta}\right)\right]\left(\frac{\partial f}{\partial \theta_a}\bigg|_{x^{[k]};\boldsymbol{\theta}}\right) \tag{8.78}$$

We find a local minimum by applying either the nonlinear conjugate gradient method or, as below, a variation of Newton's method. For the latter technique, we use the Hessian

$H(\boldsymbol{\theta}) = \nabla_\theta^2 F_{\text{cost}}$ with the elements

$$H_{ab}(\boldsymbol{\theta}) = \sum_{k=1}^{N} \left(\left.\frac{\partial f}{\partial \theta_a}\right|_{x^{[k]};\theta} \right) \left(\left.\frac{\partial f}{\partial \theta_b}\right|_{x^{[k]};\theta} \right) - \sum_{k=1}^{N} [y^{[k]} - f(x^{[k]}; \boldsymbol{\theta})] \left(\left.\frac{\partial^2 f}{\partial \theta_a \partial \theta_b}\right|_{x^{[k]};\theta} \right) \tag{8.79}$$

We note again that convergence of Newton's method does not require the use of this exact Hessian, but convergence to a minimum *does* require the approximate Hessian to be positive-definite at each iteration. If we define the *linearized design matrix* with the elements

$$X_{ka}(\boldsymbol{\theta}) = \left.\frac{\partial f}{\partial \theta_a}\right|_{x^{[k]};\theta} \tag{8.80}$$

that agrees with our previous definition in the special case of a linear model, the Hessian then has elements

$$H_{ab}(\boldsymbol{\theta}) = \left(X^{\mathrm{T}} X|_\theta\right)_{ab} - \sum_{k=1}^{N} [y^{[k]} - f(x^{[k]}; \boldsymbol{\theta})] \left(\left.\frac{\partial^2 f}{\partial \theta_a \partial \theta_b}\right|_{x^{[k]};\theta} \right) \tag{8.81}$$

If we approximate the Hessian by retaining only the first contribution, we have an approximation that is always at least positive-semidefinite,

$$X^{\mathrm{T}} X|_\theta \approx H(\boldsymbol{\theta}) \tag{8.82}$$

The gradient components also are expressed simply in terms of $X|_\theta$,

$$\gamma_a(\boldsymbol{\theta}) = -\sum_{k=1}^{N} [y^{[k]} - f(x^{[k]}; \boldsymbol{\theta})](X_{ka}|_\theta) \tag{8.83}$$

As in the linear least-squares method, it is possible that (8.82) may have eigenvalues near or equal to zero that make the Newton update system ill-conditioned. This may be corrected by adding a small positive scalar along the diagonal. Newton iteration using this modification is the most common approach to nonlinear least squares, and is known as the *Levenberg–Marquardt method*. Starting from an initial guess $\theta^{[0]}$, the Newton update at iteration m is

$$\boldsymbol{\theta}^{[m+1]} = \boldsymbol{\theta}^{[m]} + \alpha^{[m]} \boldsymbol{p}^{[m]} \quad [X^{\mathrm{T}} X|_{\theta^{[m]}} + \tau^{[m]} I]\boldsymbol{p}^{[m]} = -\gamma(\boldsymbol{\theta}^{[m]}) \tag{8.84}$$

$\alpha^{[m]}$ is obtained from a weak line search. For a linear model, this method converges after a single iteration. When the approximate Hessian is (nearly) singular, the trust-region Newton method is preferred over the line-search algorithm shown here.

Selecting a prior for single-response data

We now return to the question of proposing a prior, based first upon the assumption of prior independence, $p(\boldsymbol{\theta}, \sigma) = p(\boldsymbol{\theta})p(\sigma)$, such that

$$p(\boldsymbol{\theta}, \sigma | \boldsymbol{y}) \propto l(\boldsymbol{\theta}, \sigma | \boldsymbol{y})p(\boldsymbol{\theta})p(\sigma) \tag{8.85}$$

Using the likelihood function that follows from the Gauss–Markov conditions and the assumption of normally-distributed errors, we have

$$p(\boldsymbol{\theta}, \sigma | \mathbf{y}) \propto \sigma^{-N} \exp\left[-\frac{1}{2\sigma^2} S(\boldsymbol{\theta})\right] p(\boldsymbol{\theta}) p(\sigma) \tag{8.86}$$

How do we propose priors $p(\boldsymbol{\theta})$ and $p(\sigma)$?

Let $\boldsymbol{\theta}_{\mathrm{MLE}} = \boldsymbol{\theta}_{\mathrm{LS}}$ be the *maximum likelihood estimate* of $\boldsymbol{\theta}$, i.e., that maximizing the likelihood by minimizing $S(\boldsymbol{\theta})$. Let us now write $S(\boldsymbol{\theta})$ as

$$S(\boldsymbol{\theta}) = [S(\boldsymbol{\theta}) - S(\boldsymbol{\theta}_{\mathrm{LS}})] + S(\boldsymbol{\theta}_{\mathrm{LS}}) \tag{8.87}$$

so that the posterior becomes

$$p(\boldsymbol{\theta}, \sigma | \mathbf{y}) \propto \sigma^{-N} \exp\left\{-\frac{[S(\boldsymbol{\theta}) - S(\boldsymbol{\theta}_{\mathrm{LS}})]}{2\sigma^2}\right\} \exp\left\{-\frac{S(\boldsymbol{\theta}_{\mathrm{LS}})}{2\sigma^2}\right\} p(\boldsymbol{\theta}) p(\sigma) \tag{8.88}$$

We now define the *sample variance s^2*:

$$s^2 = \frac{1}{\nu} S(\boldsymbol{\theta}_{\mathrm{LS}}) \quad \nu = N - \dim(\boldsymbol{\theta}) \tag{8.89}$$

In the supplemental material in the accompanying website, we show that if the Gauss–Markov conditions (8.54) hold, s^2 provides an unbiased estimate for σ^2. That is, if we redo the set of experiments many times and average the computed s^2 in each, $E[s^2] = \sigma^2$.

Using the definition of the sample variance, the posterior becomes

$$p(\boldsymbol{\theta}, \sigma | \mathbf{y}) \propto \sigma^{-N} \exp\left\{-\frac{1}{2\sigma^2}[S(\boldsymbol{\theta}) - S(\boldsymbol{\theta}_{\mathrm{LS}})]\right\} \exp\left\{-\frac{\nu s^2}{2\sigma^2}\right\} p(\boldsymbol{\theta}) p(\sigma) \tag{8.90}$$

We now define a likelihood function for σ given s,

$$l(\sigma | s) \propto \sigma^{-N} \exp\left\{-\frac{\nu s^2}{2\sigma^2}\right\} \tag{8.91}$$

and a "conditional likelihood" function for $\boldsymbol{\theta}$ given $\mathbf{y}$ and σ,

$$l(\boldsymbol{\theta} | \mathbf{y}, \sigma) \propto \exp\left\{-\frac{1}{2\sigma^2}[S(\boldsymbol{\theta}) - S(\boldsymbol{\theta}_{\mathrm{LS}})]\right\} \tag{8.92}$$

The posterior density then partitions into two contributions,

$$p(\boldsymbol{\theta}, \sigma | \mathbf{y}) \propto [l(\boldsymbol{\theta} | \mathbf{y}, \sigma) p(\boldsymbol{\theta})] \times [l(\sigma | s) p(\sigma)] \tag{8.93}$$

We now consider each contribution independently to search for priors that seem most satisfying, in terms of being reproducible by different analysts.

Noninformative prior for θ

We begin first with the "conditional posterior" for $\boldsymbol{\theta}$, treating σ as known,

$$p(\boldsymbol{\theta} | \mathbf{y}, \sigma) \propto l(\boldsymbol{\theta} | \mathbf{y}, \sigma) p(\boldsymbol{\theta}) = \exp\left\{-\frac{1}{2\sigma^2}[S(\boldsymbol{\theta}) - S(\boldsymbol{\theta}_{\mathrm{LS}})]\right\} p(\boldsymbol{\theta}) \tag{8.94}$$

Although the posterior density above appears to be of a simple functional form, for a general nonlinear model, $S(\theta)$ by (8.58) may depend in a very complicated manner upon θ. Therefore, we approximate $S(\theta)$ in the local neighborhood of θ_{LS} by a quadratic expansion (for a linear model, this expansion is exact),

$$S(\theta) - S(\theta_{LS}) \approx \tfrac{1}{2}(\theta - \theta_{LS})^{\mathrm{T}}[\nabla^2 S(\theta_{LS})](\theta - \theta_{LS}) \tag{8.95}$$

$\nabla^2 S(\theta_{LS})$ is the Hessian of $S(\theta)$, evaluated at θ_{LS}. Following our numerical treatment of nonlinear least squares, we define the matrix $H(\theta_{LS})$ as

$$H(\theta_{LS}) = \nabla^2 \left[\tfrac{1}{2} S(\theta_{LS})\right] \tag{8.96}$$

with the elements

$$H_{ab}(\theta_{LS}) = (X^{\mathrm{T}} X|_{\theta_{LS}})_{ab} - \sum_{k=1}^{N} \left[y^{[k]} - f\left(x^{[k]}; \theta_{LS}\right)\right] \left(\left.\frac{\partial^2 f}{\partial \theta_a \partial \theta_b}\right|_{x^{[k]}; \theta_{LS}}\right) \tag{8.97}$$

The linearized design matrix is again given by (8.80). It is consistent with a quadratic expansion of $S(\theta)$ to neglect the second contribution to $H_{ab}(\theta_{LS})$, so that

$$X^{\mathrm{T}} X|_{\theta_{LS}} \approx H(\theta_{LS}) \tag{8.98}$$

Again, for a linear model, this approximation is exact. The quadratic expansion for $S(\theta)$, using $\tfrac{1}{2}\nabla^2 S(\theta_{LS}) = H(\theta_{LS}) \approx X^{\mathrm{T}} X|_{\theta_{LS}}$, is then

$$S(\theta) - S(\theta_{LS}) \approx (\theta - \theta_{LS})^{\mathrm{T}}[X^{\mathrm{T}} X|_{\theta_{LS}}](\theta - \theta_{LS}) \tag{8.99}$$

Thus, an approximate "conditional posterior" for θ given y and σ is

$$\pi(\theta|y, \sigma) \propto \exp\left\{-\frac{1}{2\sigma^2}(\theta - \theta_{LS})^{\mathrm{T}}[X^{\mathrm{T}} X|_{\theta_{LS}}](\theta - \theta_{LS})\right\} p(\theta) \tag{8.100}$$

The approximate "conditional likelihood function"

$$l_{\mathrm{approx}}(\theta|y, \sigma) \propto \exp\left\{-\frac{1}{2\sigma^2}(\theta - \theta_{LS})^{\mathrm{T}}\left[X^{\mathrm{T}} X\Big|_{\theta_{LS}}\right](\theta - \theta_{LS})\right\} \tag{8.101}$$

is of the form of a multivariate normal distribution

$$p(x|\mu, \Sigma) \propto \exp\left[-\frac{1}{2}(x - \mu)^{\mathrm{T}}\Sigma^{-1}(x - \mu)\right] \tag{8.102}$$

with

$$\mu = \theta_{LS} \qquad \Sigma = \sigma^2(X^{\mathrm{T}} X|_{\theta_{LS}})^{-1} \tag{8.103}$$

For a linear model, the design matrix X is completely specified at the time that we choose the predictor values in each experiment, and thus so is $X^{\mathrm{T}} X$. From one trial set of experiments to the next, the only quantity that varies significantly is θ_{LS}, and this quantity merely sets the center of the distribution, but does not affect its shape. Therefore, the likelihood function (8.101) is said to be *translated by the data*, or *data-translated*. For a linear model, where X does not vary, this data-translation property is exact. For a nonlinear model, this data-translation property is only approximate.

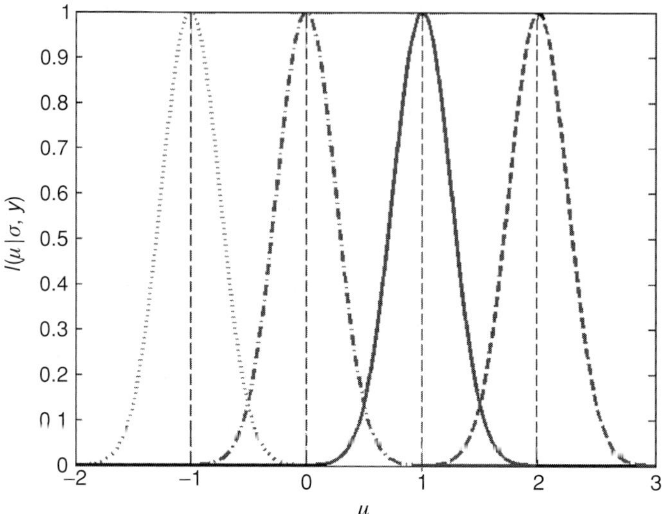

Figure 8.2 Data-translation of conditional likelihood function with a standard deviation of 0.25 and four data sets with sample means of -1, 0, 1, 2. For each data set, the location of the distribution changes, but not the shape.

Data-translation becomes clearer if we consider the simple problem

$$y^{[k]} = \theta + \varepsilon^{[k]} \tag{8.104}$$

After N measurements, $X^{\mathrm{T}}X = N$, and the conditional likelihood is

$$l(\theta|y,\sigma) \propto \exp\left\{-\frac{N}{2\sigma^2}(\theta - \bar{y})^2\right\} \quad \bar{y} = \theta_{\mathrm{LS}} = \frac{1}{N}\sum_{k=1}^{N} y^{[k]} \tag{8.105}$$

Thus of all the data in the response vector y, the only value that affects the shape of this conditional likelihood function is the sample mean $\bar{y}$. Data obtained from different sets of N measurements yield likelihood functions that have the same shape, but are centered at different locations (Figure 8.2).

The conditional posterior density is

$$p(\theta|y,\sigma) \propto \exp\left\{-\frac{N}{2\sigma^2}(\theta - \bar{y})^2\right\} p(\theta) \tag{8.106}$$

If we choose the prior to be uniform in the parameter θ that is data-translated, the posterior density will also be data-translated. The concept of data-translation is important to the generation of priors. Here, the prior is said to be *noninformative about* θ, because the data-translation property for θ of the likelihood function is retained by the posterior density; i.e., the prior does not favor any particular region of θ-space. We posit that by choosing the prior to be noninformative, we try to be as impartial as possible about the value of θ without trying to "spin" the data. We identify a translation property that the likelihood function possesses, and then choose the prior so that we retain this property in the posterior.

A noninformative prior provides a reference standard that other researchers can reproduce and accept as not reflecting the analyst's personal bias. It is not quite true to say that the noninformative prior is objective, as it must be made proper, and this can be done in many different ways. Also, if likelihood function has more than one symmetry property available, it is then a subjective choice of which one to use to generate the noninformative prior. Still, noninformative priors are perhaps "least subjective" or "most reproducible," and several results obtained using (8.68) agree with those of frequentist statistics.

For the single-response model, using a quadratic expansion for $S(\boldsymbol{\theta})$, the likelihood function (8.101) is data-translated by the least-squares estimate $\boldsymbol{\theta}_{\text{LS}}$. Thus the noninformative prior is uniform in this same parameter $\boldsymbol{\theta}$, $p(\boldsymbol{\theta}) \sim c$, at least in the region of appreciable nonzero likelihood. Again, a uniform prior in $\boldsymbol{\theta}$ is justified on the grounds of being noninformative. The Bayesian most probable estimate $\boldsymbol{\theta}_{\text{M}}$ therefore agrees with the maximum likelihood and least squares estimates,

$$\boldsymbol{\theta}_{\text{M}} = \boldsymbol{\theta}_{\text{MLE}} = \boldsymbol{\theta}_{\text{LS}} \tag{8.107}$$

Non informative prior for σ

We next consider the problem of estimating the error standard deviation, using the posterior density for σ,

$$p(\sigma|s) \propto l(\sigma|s)p(\sigma) = \sigma^{-N} \exp\left\{-\frac{\nu s^2}{2\sigma^2}\right\} p(\sigma) \tag{8.108}$$

We see that the likelihood $l(\sigma|s)$ is *not* data-translated in σ. Even though the likelihood $l(\sigma|s)$ is not simply translated by the data, there still may exist a one-to-one transformation $\eta = \eta(\sigma)$ such that $l(\eta|s)$ *is* simply translated by the data. That is, the shape of $l(\eta|s)$ does not change from one data set to the next, but the location of its maximum does. Once we find such a transformation, we should choose our prior to be noninformative in $\eta(\sigma)$; i.e., such that the posterior density in the transformed coordinate, $p(\eta|s)$, is also data-translated. This is the case if the prior is locally uniform in $\eta = \eta(\sigma)$, i.e., $p(\eta) \sim c$, in the region of appreciable nonzero likelihood.

While there exists a technique, due to Jeffreys, for generating noninformative priors directly from the likelihood function, here we simply propose a transformation $\eta(\sigma) = \ln \sigma$, and then show that its likelihood is data-translated. For a discussion of automated procedures for constructing noninformative priors, consult a general text on Bayesian statistics such as Box & Tiao (1973).

Writing $\sigma = e^{\eta}$, we have the transformed likelihood

$$l(\eta|s) \propto (e^{\eta})^{-N} \exp\left\{-\frac{\nu s^2}{2(e^{\eta})^2}\right\} = \exp\left\{-N\eta - \frac{\nu s^2}{2e^{2\eta}}\right\} \tag{8.109}$$

Substituting $1 = s^N s^{-N}$, we write

$$l(\eta|s) \propto s^N s^{-N} \exp\left\{-N\eta - \frac{\nu s^2}{2e^{2\eta}}\right\} = s^{-N} \exp\left\{N(\ln s - \eta) - \frac{\nu s^2}{2e^{2\eta}}\right\} \tag{8.110}$$

Using $s^2 = e^{2\ln s}$, we have

$$l(\eta|s) \propto s^{-N} \exp\left\{ N(\ln s - \eta) - \frac{\nu}{2} \exp[2(\ln s - \eta)] \right\} \tag{8.111}$$

Adding the proper normalization, we have finally

$$l(\eta|s) = \frac{\exp\left\{ N(\ln s - \eta) - \dfrac{\nu}{2} \exp[2(\ln s - \eta)] \right\}}{\displaystyle\int_{\eta_{\min}}^{\eta_{\max}} \exp\left\{ N(\ln s - \eta) - \dfrac{\nu}{2} \exp[2(\ln s - \eta)] \right\} d\eta} \tag{8.112}$$

Thus, $l(\eta|s)$ *is* data-translated. We should choose our noninformative prior to be uniform in $\eta(\sigma) = \ln \sigma$,

$$p(\eta) \sim c \Rightarrow p(\sigma) \propto \left| \frac{d\eta}{d\sigma} \right| = \left| \frac{d}{d\sigma} \ln \sigma \right| = \sigma^{-1} \tag{8.113}$$

where we have used $p(\sigma)d\sigma = p(\eta)d\eta = p(\eta)\,|d\eta/d\sigma|\,d\sigma$.

The posterior density for single-response data

Putting together the results of the two previous sections, we use for single-response data the prior density

$$p(\boldsymbol{\theta}, \sigma) = p(\boldsymbol{\theta})p(\sigma) \quad p(\boldsymbol{\theta}) \sim c \quad p(\sigma) \propto \sigma^{-1} \tag{8.114}$$

such that the posterior density is

$$p(\boldsymbol{\theta}, \sigma|\mathbf{y}) \propto \sigma^{-(N+1)} \exp\left\{ -\frac{1}{2\sigma^2} S(\boldsymbol{\theta}) \right\} \tag{8.115}$$

Because $S(\boldsymbol{\theta})$ may depend in a complicated manner upon $\boldsymbol{\theta}$ for a nonlinear model, use of (8.115) often requires numerical computation. We discuss such techniques later, but first consider analytical calculations based on the approximate posterior density

$$\pi(\boldsymbol{\theta}, \sigma|\mathbf{y}) \propto \sigma^{-(N+1)} \exp\left\{ -\frac{1}{2\sigma^2}(\boldsymbol{\theta} - \boldsymbol{\theta}_{\mathrm{M}})^{\mathrm{T}}[X^{\mathrm{T}}X|_{\boldsymbol{\theta}_{\mathrm{M}}}](\boldsymbol{\theta} - \boldsymbol{\theta}_{\mathrm{M}}) \right\} \exp\left\{ -\frac{\nu s^2}{2\sigma^2} \right\} \tag{8.116}$$

that is obtained from a quadratic expansion of $S(\boldsymbol{\theta})$ about $\boldsymbol{\theta}_{\mathrm{M}} = \boldsymbol{\theta}_{\mathrm{LS}}$. In general, $S(\boldsymbol{\theta})$ varies more rapidly than its quadratic approximation, and so the posterior density $p(\boldsymbol{\theta}, \sigma|\mathbf{y})$ decays to zero as one moves away from $\boldsymbol{\theta}_{\mathrm{M}}$ more rapidly than the approximate posterior $\pi(\boldsymbol{\theta}, \sigma|\mathbf{y})$. It has been traditional to use $\pi(\boldsymbol{\theta}, \sigma|\mathbf{y})$ when generating confidence intervals; however, with the Markov chain Monte Carlo (MCMC) techniques discussed later, this approximation need not be made.

We again note that for a linear model the quadratic expansion for $S(\boldsymbol{\theta})$ is exact, and $\pi(\boldsymbol{\theta}, \sigma|\mathbf{y}) = p(\boldsymbol{\theta}, \sigma|\mathbf{y})$. Thus, the confidence intervals that we compute analytically from $\pi(\boldsymbol{\theta}, \sigma|\mathbf{y})$ are exact for a linear model.

Confidence intervals from the approximate posterior density

The approximate posterior density $\pi(\boldsymbol{\theta}, \sigma|\boldsymbol{y})$ for single-response data (8.116), describes the joint uncertainty in both $\boldsymbol{\theta}$ and σ. Of more interest is the *marginal posterior density for* $\boldsymbol{\theta}$,

$$\pi(\boldsymbol{\theta}|\boldsymbol{y}) = \int_0^\infty \pi(\boldsymbol{\theta}, \sigma|\boldsymbol{y})d\sigma \tag{8.117}$$

This is the posterior for $\boldsymbol{\theta}$, without regard to the exact value of σ. As we discard σ by "integrating it out," it is called a *nuisance parameter*.

For the approximate posterior (8.116), the marginal posterior density can be calculated analytically,

$$\pi(\boldsymbol{\theta}|\boldsymbol{y}) = \int_0^\infty \pi(\boldsymbol{\theta}, \sigma|\boldsymbol{y})d\sigma \propto \left[1 + \frac{1}{\nu s^2}(\boldsymbol{\theta} - \boldsymbol{\theta}_{\mathrm{M}})^{\mathrm{T}}[X^{\mathrm{T}}X|_{\boldsymbol{\theta}_{\mathrm{M}}}](\boldsymbol{\theta} - \boldsymbol{\theta}_{\mathrm{M}})\right]^{-N/2} \tag{8.118}$$

Note again that for a linear model, this marginal posterior is exact.

Confidence interval for the mean of a population and the t-distribution

We now form confidence intervals for the model parameters and the predicted responses using this approximate marginal distribution. First, it is best to consider the simple model (8.104), $y^{[k]} = \theta + \varepsilon^{[k]}$. After N measurements, the posterior density is

$$p(\theta, \sigma|\boldsymbol{y}) \propto \sigma^{-(N+1)} \exp\left\{-\frac{1}{2\sigma^2}[\nu s^2 + N(\theta - \bar{y})^2]\right\} \tag{8.119}$$

where the sample mean and sample variance are

$$\bar{y} = N^{-1}\sum_{k=1}^N y^{[k]} \quad s^2 = \frac{1}{\nu}\sum_{k=1}^N \left(y^{[k]} - \bar{y}\right)^2 \quad \nu = N - 1 \tag{8.120}$$

The marginal posterior density for θ is

$$p(\theta|\boldsymbol{y}) = \int_0^\infty p(\theta, \sigma|\boldsymbol{y})d\sigma \propto \left[1 + \frac{N}{\nu s^2}(\theta - \bar{y})^2\right]^{-(\nu+1)/2} \tag{8.121}$$

Defining the *t-statistic*,

$$t \equiv \frac{\bar{y} - \theta}{(s/\sqrt{N})} \tag{8.122}$$

whose distribution satisfies $p(t|\nu)dt = p(\theta|\boldsymbol{y})d\theta$, we have

$$p(t|\nu) \propto \left[1 + \frac{t^2}{\nu}\right]^{-(\nu+1)/2} \tag{8.123}$$

This is the famous *t-distribution* of Student, a pseudonym for W. S. Gosset (Gosset, 1908). As the number of degrees of freedom ν approaches infinity, the *t*-distribution reduces to a

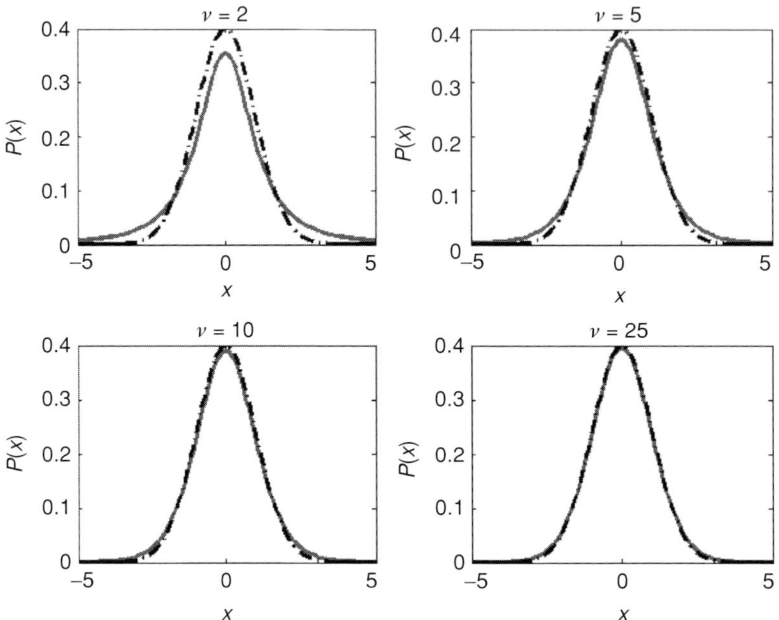

Figure 8.3 Student t-distribution vs. $N(\mu = 0, \sigma^2 = 1)$ for $v = 2, 5, 10, 25$. The t-distribution are the solid lines and the normal distribution are the dashed lines.

Gaussian (normal) distribution of mean zero and variance 1:

$$\lim_{v \to \infty} p(t|v) = \frac{1}{\sqrt{2\pi}} e^{-t^2/2} \tag{8.124}$$

For finite values of v, the t-distribution is somewhat broader than $N(0, 1)$, to account for the extra uncertainty in the estimate of θ due to the uncertainty in σ. plot_t_distribution.m plots the t-distribution and compares it to the normal distribution for various values of v (Figure 8.3).

How do we form a confidence interval for the population mean from the marginal posterior density? Let us say that we know the value of σ so that the conditional posterior is

$$p(\theta|\mathbf{y}, \sigma) \propto \exp\left[-\frac{N(\theta - \bar{y})^2}{2\sigma^2}\right] \tag{8.125}$$

We define next a scaled variable $Z = (\theta - \bar{y})/(\sigma N^{-1/2})$ that is normally distributed with a mean of 0 and a standard deviation of 1,

$$p(Z) = \frac{1}{\sqrt{2\pi}} e^{-Z^2/2} \qquad \int_{-\infty}^{+\infty} p(Z)dZ = 1 \tag{8.126}$$

We then specify $Z_{\alpha/2} > 0$ as the value of Z for which

$$\int_{-Z_{\alpha/2}}^{Z_{\alpha/2}} p(Z)dZ = 1 - \alpha \tag{8.127}$$

There is a $100 \times (1 - \alpha)\%$ chance that

$$|Z| = \frac{|\theta - \bar{y}|}{\sigma N^{-1/2}} \leq Z_{\alpha/2} \tag{8.128}$$

so that the $100 \times (1 - \alpha)\%$ confidence interval for the parameter is

$$\theta = \bar{y} \pm Z_{\alpha/2} \sigma N^{-1/2} \tag{8.129}$$

Typically, we form 95% confidence intervals, with $\alpha = 0.05$, and 99% confidence intervals with $\alpha = 0.01$. Using the **MATLAB Statistics toolkit**, we compute the value of $Z_{\alpha/2}$ with the command

Z_alpha_2 = norminv(1-alpha/2,0,1);

For $\alpha = 0.05$, $Z_{\alpha/2} = 1.9600$ and for $\alpha = 0.01$, $Z_{\alpha/2} = 2.5758$.

Here, we use the frequentist term *confidence interval*, as it is standard usage, but more accurately, in the Bayesian perspective, we should call this a *credible interval*, meaning that the parameter has a probability $(1 - \alpha)$ of lying in this interval. A frequentist confidence interval, by contrast, states that if we repeat the same set of experiments many times, we obtain an estimate of the parameter in this interval $(1 - \alpha)$th of the time. The Bayesian viewpoint is more direct.

Now, for our problem, we do not know the exact value of σ, so instead of forming the confidence interval from the unit normal distribution $N(0, 1)$, we form it using the t-distribution (8.123) which is somewhat broader due to the extra uncertainty in the value of σ. We define $T_{\nu,\alpha/2}$ as the value for which

$$\int_{-T_{\nu,\alpha/2}}^{T_{\nu,\alpha/2}} p(t|\nu)dt = 1 - \alpha \tag{8.130}$$

There is a $100 \times (1 - \alpha)\%$ chance that the t-statistic is in the range

$$|t| = \left| \frac{\bar{y} - \theta}{(s/\sqrt{N})} \right| \leq T_{\nu,\alpha/2} \tag{8.131}$$

The $100 \times (1 - \alpha)\%$ confidence interval on θ is then

$$\theta = \bar{y} \pm T_{\nu,\alpha/2} s N^{-1/2} \tag{8.132}$$

With the **MATLAB Statistics toolkit**, we compute $T_{\nu,\alpha/2}$ by the command

T_val = tinv(1-alpha/2,nu);

Confidence intervals for model parameters

We now use the t-distribution to form confidence intervals from the approximate posterior density $\pi(\theta, \sigma | y)$, (8.116), and its approximate marginal posterior density $\pi(\theta|y)$, (8.118). Once again, let us assume that we know the exact value of σ. The conditional posterior for θ is then

$$\pi(\theta|y, \sigma) \propto \exp\left\{ -\frac{1}{2\sigma^2}(\theta - \theta_{\mathrm{M}})^{\mathrm{T}}[X^{\mathrm{T}}X|_{\theta_{\mathrm{M}}}](\theta - \theta_{\mathrm{M}}) \right\} \tag{8.133}$$

This is of the form of a multivariate Gaussian distribution (8.102) with a mean $\boldsymbol{\theta}_M$ and a covariance matrix

$$\Sigma = \sigma^2 [X^{\mathrm{T}} X|_{\boldsymbol{\theta}_M}]^{-1} = \mathrm{cov}(\boldsymbol{\theta}) \tag{8.134}$$

We wish to obtain a confidence interval for each parameter θ_j. The rule for covariances, with a constant vector c and a random vector x,

$$\mathrm{var}\,(c \cdot x) = c \cdot [\mathrm{cov}(x)]c \tag{8.135}$$

yields for $c = e^{[j]}$ the variance of θ_j,

$$\mathrm{var}(\theta_j) = e^{[j]} \cdot [\mathrm{cov}(\boldsymbol{\theta})]e^{[j]} = \sigma^2 e^{[j]} \cdot \left[X^{\mathrm{T}} X\big|_{\boldsymbol{\theta}_M}\right]^{-1} e^{[j]} = \sigma^2 \left[X^{\mathrm{T}} X\big|_{\boldsymbol{\theta}_M}\right]^{-1}_{jj} \tag{8.136}$$

where $\left[X^{\mathrm{T}} X|_{\boldsymbol{\theta}_M}\right]^{-1}_{jj}$ is the jth diagonal element of $(X^{\mathrm{T}} X)^{-1}$. Thus, the $100 \times (1-\alpha)\%$ confidence interval on θ_j, assuming exact knowledge of σ, is

$$\theta_j = \theta_{M,j} \pm Z_{\alpha/2}\sigma \left\{\left[X^{\mathrm{T}} X\big|_{\boldsymbol{\theta}_M}\right]^{-1}_{jj}\right\}^{1/2} \tag{8.137}$$

As the marginal posterior density (8.118) is a multivariate t-distribution, the confidence interval on the model parameter, accounting for the extra uncertainty in σ, is

$$\theta_j = \theta_{M,j} \pm T_{v,\alpha/2}s \left\{\left[X^{\mathrm{T}} X\big|_{\boldsymbol{\theta}_M}\right]^{-1}_{jj}\right\}^{1/2} \tag{8.138}$$

where $v = N - \dim(\boldsymbol{\theta})$.

Confidence intervals on the model predictions

How much does uncertainty in $\boldsymbol{\theta}$ affect the model predictions? The predictions with $\boldsymbol{\theta}_M$ are

$$\hat{y}^{[k]}(\boldsymbol{\theta}_M) = f(x^{[k]}; \boldsymbol{\theta}_M) \tag{8.139}$$

In keeping with our use of a quadratic approximation for $S(\boldsymbol{\theta})$, we expand the predictions in the vicinity of $\boldsymbol{\theta}_M$:

$$\hat{y}^{[k]}(\boldsymbol{\theta}) - \hat{y}^{[k]}(\boldsymbol{\theta}_M) \approx \sum_{a=1}^{P} \frac{\partial f}{\partial \theta_a}\bigg|_{x^{[k]};\boldsymbol{\theta}_M} (\theta_a - \theta_{M,a}) \tag{8.140}$$

Using the definition (8.80) of the linearized design matrix, we have

$$\hat{y}^{[k]}(\boldsymbol{\theta}) - \hat{y}^{[k]}(\boldsymbol{\theta}_M) \approx \sum_{a=1}^{P} X_{ka}|_{\boldsymbol{\theta}_M}(\theta_a - \theta_{M,a}) = [X|_{\boldsymbol{\theta}_M}(\boldsymbol{\theta} - \boldsymbol{\theta}_M)]_k \tag{8.141}$$

Thus, the expansion of the vector of predicted responses is

$$\hat{\boldsymbol{y}}(\boldsymbol{\theta}) - \hat{\boldsymbol{y}}(\boldsymbol{\theta}_M) \approx X|_{\boldsymbol{\theta}_M}(\boldsymbol{\theta} - \boldsymbol{\theta}_M) \tag{8.142}$$

Treating $X|_{\boldsymbol{\theta}_M}$ as approximately constant (as it is for a linear model), from (8.142) and

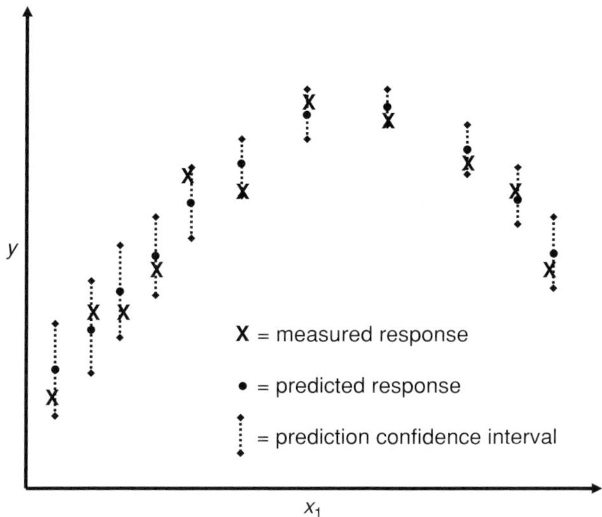

Figure 8.4 Plot of measured vs. predicted responses (with confidence intervals) to check for model adequacy with respect to variation in predictor values. Here, the model appears to agree with the data.

(8.134), the covariance matrix of $\hat{\boldsymbol{y}}(\boldsymbol{\theta})$ is

$$\mathrm{cov}[\hat{\boldsymbol{y}}(\boldsymbol{\theta})] = X|_{\boldsymbol{\theta}_M}\mathrm{cov}(\boldsymbol{\theta})\big(X|_{\boldsymbol{\theta}_M}\big)^{\mathrm{T}} = \sigma^2 X|_{\boldsymbol{\theta}_M}\big(X^{\mathrm{T}}X|_{\boldsymbol{\theta}_M}\big)^{-1}\big(X|_{\boldsymbol{\theta}_M}\big)^{\mathrm{T}} \tag{8.143}$$

The $100 \times (1 - \alpha)\%$ confidence interval for the predicted response in experiment k is then

$$\hat{y}^{[k]}(\boldsymbol{\theta}) = \hat{y}^{[k]}(\boldsymbol{\theta}_M) \pm s\,T_{v,\alpha/2}\left\{\left[\,X|_{\boldsymbol{\theta}_M}\big(X^{\mathrm{T}}X|_{\boldsymbol{\theta}_M}\big)^{-1}\big(X|_{\boldsymbol{\theta}_M}\big)^{\mathrm{T}}\right]_{kk}\right\}^{1/2} \tag{8.144}$$

It is common to plot, as in Figure 8.4, the model predictions and their confidence intervals vs. the model estimates to gauge the reliability of the fitted model. Such plots allow one to identify *outliers*, i.e., points whose residual errors seem to be larger than the others. For an outlier, the measured response lies far outside of the confidence interval of the prediction. Such points may be due to "extra" amounts of error for that particular data point. For example, the investigator may have transposed the digits of a measurement when recording them in a laboratory notebook. As points with such "extra" error corrupt the analysis, they should be removed from the data set. Of course, we should not discard such points based on our subjective expectation of the outcome, as outliers that are actually valid, but have large residuals due to model error, have been known to change the course of scientific history. If possible, redo any experiments that appear to be outliers. If the excessive error appears to be reproducible, it may be due to model inadequacy.

Least-squares fitting and confidence interval generation in MATLAB

The MATLAB statistics toolkit contains several functions that perform least-squares parameter fits and generate confidence intervals for linear and nonlinear models from single-response data.

Linear least-squares calculations in MATLAB

For linear models, we use the function **regress**,

[theta, theta_CI, residuals, res_CI, Stats] = regress(y, X, alpha);

y is the vector $y \in \Re^N$ of measured responses. X is the $N \times P$ design matrix of predictor variables. alpha sets the width of the confidence intervals ($\alpha = 0.05$ for 95% CIs). theta is the vector $\boldsymbol{\theta}_{\mathrm{M}} \in \Re^P$ of fitted parameters. theta_CI contains lower and upper bounds on each parameter. residuals contains the values in each experiment of the residual errors $e^{[k]} = y^{[k]} - \hat{y}^{[k]}(\boldsymbol{\theta}_{\mathrm{M}})$ and res_CI contains lower and upper bounds on these errors. Stats contains various goodness-of-fit measures.

We demonstrate the use of **regress** for the data set comparing the protein expression levels of wild-type and mutant bacterial strains, (8.35). The following MATLAB code performs the least-squares calculation:

y = [121.9; 113.4; 112.2; 106.1; 120.7; 119.5; 116.5; 124.0];
X = [1 0; 1 0; 1 0; 1 0; 1 1; 1 1; 1 1; 1 1];
alpha = 0.05;
[theta, theta_CI, residuals, res_CI, Stats] = regress(y, X, alpha);

The results of these calculations are

theta =
   113.4000
   6.7750
theta_CI =
   107.1646 119.6354
   −2.0432 15.5932

Thus, we have the 95% confidence intervals,

$$\theta_1 = 113.40 \pm 6.33 \quad \theta_2 = 6.78 \pm 8.82 \tag{8.145}$$

As the confidence interval for θ_2 includes $\theta_2 = 0$, the difference in protein expression levels between the two strains is not statistically significant.

Nonlinear least-squares calculations in MATLAB

For nonlinear models, the MATLAB statistics toolkit contains functions to compute the least-squares estimates and generate confidence intervals using the approximate posterior density. First, the least-squares estimate is found using the function **nlinfit**,

[theta, residuals, Jac] = nlinfit(X_pred, y, fun, theta_0);

X_pred is a matrix that contains in each row the set of predictors for the corresponding experiment. y is the vector of measured response values. fun_name is the name of a function,

```
y_hat = feval(fun_name, theta, X_pred);
```

that returns the vector of predicted responses y_hat for input values of the parameters theta and the predictor values in X_pred. theta_0 is an initial guess of the parameter vector. theta is the least-squares estimate of the parameter vector. residuals contains the residual errors, and Jac contains the Jacobian of the model functions (i.e. the linearized design matrix).

For the data of Table 8.1, we fit the model $r_{R1} = k_1 c_A^{\nu_a} c_B^{\nu_b}$ using the code

```
X_pred = [0.1 0.1; 0.2 0.1; 0.1 0.2; 0.2 0.2; 0.05 0.2; 0.2 0.05];
y = [0.0246e-3; 0.0483e-3; 0.0501e-3; 0.1003e-3; 0.0239e-3; 0.0262e-3];
theta_0 = [0.01 1.0 1.0]; % initial guess
fun_name = 'calc_yhat_kinetic_ex1';
[theta, residuals, Jac] = nlinfit(X_pred, y, fun_name, theta_0);
```

where we supply the following routine to compute the predicted responses,

```
function y_hat = calc_yhat_kinetic_ex1(theta, X_pred);
N = size(X_pred,1); % # of experiments
% extract predictors
conc_A = X_pred(:,1); % [A] in each experiment
conc_B = X_pred(:,2); % [B] in each experiment
% extract parameters
k_1 = theta(1); % rate constant
nu_a = theta(2); % exponent for [A]
nu_b = theta(3); % exponent for [B]
% make predictions
y_hat = k_1.*(conc_A.^nu_a).*(conc_B.^nu_b);
return;
```

From the output of **nlinfit**, 95% confidence interval half-widths are generated (using a quadratic expansion of $S(\boldsymbol{\theta})$) for the model parameters and predictions by the commands:

```
theta_CI = nlparci(theta, residuals, Jac);
[y_hat, y_hat_HW] = nlpredci(fun_name,X_pred,theta, residuals,Jac);
```

theta_CI contains the confidence interval bounds on the parameters, y_hat is the vector of model predictions, and y_hat_HW contains the confidence interval half-widths of the predictions. Through additional optional output arguments, **nlpredci** can also return joint confidence intervals and half-widths for other values of α than $\alpha = 0.05$. The results of these calculations are

```
theta =
   0.0025
   1.0001
   1.0001
theta_CI =
```

$$
\begin{array}{cc}
0.0019 & 0.0031 \\
0.9057 & 1.0946 \\
0.9057 & 1.0946
\end{array}
$$

Thus, the fitted rate law is

$$r_{\mathrm{R1}} = (0.0025)c_{\mathrm{A}}^{1.0001}\, c_{\mathrm{B}}^{1.0001} \tag{8.146}$$

with the approximate 95% confidence intervals

$$0.0019 \le k_1 \le 0.0031 \quad 0.9057 \le \nu_a \le 1.0946 \quad 0.9057 \le \nu_b \le 1.0946 \tag{8.147}$$

Therefore, the data do appear to be consistent with an elementary reaction. Below, we test this hypothesis more rigorously using MCMC simulation with the exact posterior.

Example. Fitting the rate constant from the full curve of concentration of C vs. time for a batch reactor experiment

We also can estimate the rate law parameters from the data of Table 8.3. Here, since we consider single-response regression techniques, we use only the data for $c_{\mathrm{C}}(t)$, and assume an elementary reaction $\nu_a = \nu_b = 1$. The following code performs the single-response regression of the $c_{\mathrm{C}}(t)$ data,

```
time_hr = [0.5; 1; 1.5; 2; 3; 4; 5; 6; 7; 8; 9; 10];
cC = [0.001; 0.0357; 0.0501; 0.0512; 0.0682; 0.0747; ...
        0.0809; 0.0818; 0.0858; 0.0863; 0.0872; 0.0928];
time_s = time_hr*3600; % convert hr to sec
X_pred = time_s; % set predictor matrix
y = cC; % set vector of measured resposnes
k1_0 = 0.0025; % initial guess of k1
% call nlinfit to compute fitted parameter
fun_name = 'calc_yhat_reaction_ex_dynamic';
[k1, residuals, Jac] = nlinfit(X_pred, y, fun_name, k1_0);
k1_cI = nlparci(k1, residuals, Jac);
```

We supply a routine that returns the vector of the model predictions by solving the IVP with the trial value of the rate constant:

```
function y_hat = calc_yhat_reaction_ex_dynamic(k1,X_pred);
t_span = X_pred'; % set times at which to report concentrations
x_0 = [0.1; 0.1; 0]; % set initial condition
[t_traj,x_traj] = ode45(@ batch_kinetics_dynamics_ex, ...
        t_span, x_0, [ ], k1);
y_hat = x_traj(:,3); % extract predictions from trajectory
return;
```

and we supply as well a routine that computes the time derivatives of each of the state variables for the batch kinetic model:

```
function x_dot = batch_kinetics_dynamics_ex(t,x,k1);
cA = x(1); cB = x(2);
r1 = k1*cA*cB;
x_dot = zeros(3,1);
x_dot(1) = -r1; x_dot(2) = −r1; x_dot(3) = r1;
return;
```

We obtain from these calculations an estimate $k_1 = 0.0025$ and a 95% confidence interval $0.0022 \leq k_1 \leq 0.0027$.

MCMC techniques in Bayesian analysis

The confidence intervals derived above are based upon a quadratic expansion for $S(\boldsymbol{\theta})$ about $\boldsymbol{\theta}_M$ that is only approximate for a nonlinear model. The exact single-response posterior without this approximation is

$$p(\boldsymbol{\theta}, \sigma | \mathbf{y}) \propto \sigma^{-(N+1)} \exp\left[-\frac{1}{2\sigma^2} S(\boldsymbol{\theta})\right] \tag{8.148}$$

While analytical manipulation of this formula is difficult, *MCMC simulation* is a powerful tool to obtain posterior expectations of the form

$$E[\mathbf{g}|\mathbf{y}] = \int_{\mathfrak{R}^P} \int_0^\infty \mathbf{g}(\boldsymbol{\theta}, \sigma) p(\boldsymbol{\theta}, \sigma | \mathbf{y}) d\sigma d\boldsymbol{\theta} \tag{8.149}$$

Many statistical questions can be posed in this form. For example, let us say that we wish to compute the probability that some hypothesis H_Ω is true. Let Ω be the region in $(\boldsymbol{\theta}, \sigma)$ space in which the hypothesis H_Ω is true, and outside of Ω, the hypothesis is false. Let $I_\Omega(\boldsymbol{\theta}, \sigma)$ be the indicator function

$$I_\Omega(\boldsymbol{\theta}, \sigma) = \begin{cases} 1, & \text{if } (\boldsymbol{\theta}, \sigma) \in \Omega \\ 0, & \text{if } (\boldsymbol{\theta}, \sigma) \notin \Omega \end{cases} \tag{8.150}$$

For example, if we wish to test the hypothesis that $\theta_{lo} \leq \theta_j \leq \theta_{hi}$, we use the indicator function

$$I_\Omega(\boldsymbol{\theta}, \sigma) = \begin{cases} 1, & \text{if } \theta_{lo} \leq \theta_j \leq \theta_{hi} \\ 0, & \text{if } \theta_j < \theta_{lo} \text{ or } \theta_j > \theta_{hi} \end{cases} \tag{8.151}$$

The probability that the hypothesis is true then takes the form of a posterior expectation of the indicator function:

$$p(H_\Omega | \mathbf{y}) = E[I_\Omega | \mathbf{y}] = \int_{\mathfrak{R}^P} \int_0^\infty I_\Omega(\boldsymbol{\theta}, \sigma) p(\boldsymbol{\theta}, \sigma | \mathbf{y}) d\sigma d\boldsymbol{\theta} \tag{8.152}$$

When the dimension of $(\boldsymbol{\theta}, \sigma)$ space is small, it may be possible to compute (8.152) by quadrature, but MCMC simulation is generally more efficient.

Such calculations also arise when we wish to use knowledge gained from analysis of the data to choose an optimal course of action, and form the basis of *statistical decision theory*. We want to consider which of a set $\{a_j\}$ of actions to take, given a *loss function* $L(a_j|\boldsymbol{\theta})$ that measures how "bad" is choosing action a_j, for a particular value of $\boldsymbol{\theta}$. For some values of $\boldsymbol{\theta}$, a_j will be a good action to take (low $L(a_j|\boldsymbol{\theta})$) and for other values of $\boldsymbol{\theta}$, a_j will be a bad action to take (high $L(a_j|\boldsymbol{\theta})$). Given $p(\boldsymbol{\theta}, \sigma|\boldsymbol{y})$, we select the action that has the lowest *posterior expected loss*

$$
L(a_j|\boldsymbol{y}) = \int\limits_{\Re^P} \int\limits_0^\infty L(a_j|\boldsymbol{\theta})p(\boldsymbol{\theta}, \sigma|\boldsymbol{y})d\sigma d\boldsymbol{\theta} = E[L(a_j|\boldsymbol{\theta})|\boldsymbol{y}]
\tag{8.153}
$$

MCMC computation of posterior predictions

To compute an expectation $E[\boldsymbol{g}|\boldsymbol{y}]$ using MCMC simulation, we rewrite (8.149) to integrate over all values of $\sigma \in \Re$ as

$$
E[\boldsymbol{g}|\boldsymbol{y}] = \int\limits_{\Re^P} \int\limits_{-\infty}^{+\infty} I_{\sigma>0}(\sigma)\boldsymbol{g}(\boldsymbol{\theta}, \sigma)p(\boldsymbol{\theta}, \sigma|\boldsymbol{y})d\sigma d\boldsymbol{\theta}
\tag{8.154}
$$

where we introduce the indicator function

$$
I_{\sigma>0}(\sigma) = \begin{cases} 1, & \sigma > 0 \\ 0, & \sigma \leq 0 \end{cases}
\tag{8.155}
$$

We then define the sampling density

$$
\pi_s(\boldsymbol{\theta}, \sigma|\boldsymbol{y}) = I_{\sigma>0}(\sigma)p(\boldsymbol{\theta}, \sigma|\boldsymbol{y})
\tag{8.156}
$$

such that

$$
E[\boldsymbol{g}|\boldsymbol{y}] = \int\limits_{\Re^P} \int\limits_{-\infty}^{+\infty} \boldsymbol{g}(\boldsymbol{\theta}, \sigma)\pi_s(\boldsymbol{\theta}, \sigma|\boldsymbol{y})d\sigma d\boldsymbol{\theta}
\tag{8.157}
$$

We use MCMC simulation to generate a sequence $(\boldsymbol{\theta}^{[m]}, \sigma^{[m]})$ that is distributed according to $\pi_s(\boldsymbol{\theta}, \sigma|\boldsymbol{y})$, such that for a large number N_s of samples, the expectation is approximately

$$
E[\boldsymbol{g}|\boldsymbol{y}] \approx \frac{1}{N_s} \sum_{m=1}^{N_s} \boldsymbol{g}(\boldsymbol{\theta}^{[m]}, \sigma^{[m]})
\tag{8.158}
$$

The error of this approximation is normally distributed with a standard deviation proportional to $\sqrt{N_s}$ if the samples are independent. To generate this sequence, we use the Metropolis algorithm, known in statistics as *Metropolis–Hastings sampling*. While other MC algorithms (e.g. Gibbs sampling) may yield superior performance, here we use only the Metropolis algorithm, which we have encountered already in Chapter 7. From the current state $(\boldsymbol{\theta}^{[m]}, \sigma^{[m]})$, we propose a move $(\boldsymbol{\theta}^{[m]}, \sigma^{[m]}) \rightarrow (\boldsymbol{\theta}^{(\text{new})}, \sigma^{(\text{new})})$ and then decide whether to accept this new state as the next member $(\boldsymbol{\theta}^{[m+1]}, \sigma^{[m+1]})$ of the sequence. We generate a new trial state by displacing at random either θ or σ. For some specified fraction of θ moves $f_\theta \in (0, 1)$, if a random number u, drawn from a uniform distribution on $[0, 1]$, is less that or equal to f_θ, we propose a random displacement of θ:

$$\theta_j^{(\text{new})} = \theta_j^{[m]} + \Delta_\theta M_j \gamma_j$$

(8.159)

$\{\gamma_j\}$ are drawn at random from $N(\mu = 0,\ \sigma = 1)$

where $M_j = \min\{\langle|\theta_j|\rangle, \sqrt{eps}\}$, $\langle|\theta_j|\rangle$ being the running average of the magnitude of θ_j. Else, we propose a random displacement of σ:

$$\sigma^{(\text{new})} = \sigma^{[m]} + \Delta_\sigma \gamma'$$
$$\gamma' \text{ is drawn at random from } N(\mu = 0, \sigma = 1)$$

(8.160)

Δ_θ and Δ_σ are relative step sizes that are adjusted throughout the simulation to meet a target fraction of moves that are accepted (e.g. 0.3).

The probability of accepting the proposed move is

$$\alpha([m] \rightarrow (\text{new})) = \min\left\{1, \frac{\pi_s\left(\theta^{(\text{new})}, \sigma^{(\text{new})}\big|y\right)}{\pi_s\left(\theta^{[m]}, \sigma^{[m]}\big|y\right)}\right\}$$

(8.161)

To decide whether to accept the move or not, we generate a random number u' drawn from a uniform distribution on $[0, 1]$. If $u' \le \alpha([m] \rightarrow (\text{new}))$, we accept the move and $(\theta^{[m+1]}, \sigma^{[m+1]}) = (\theta^{(\text{new})}, \sigma^{(\text{new})})$. Else, we reject the move and reuse the old value $(\theta^{[m+1]}, \sigma^{[m+1]}) = (\theta^{[m]}, \sigma^{[m]})$. It is common to run the MCMC simulation for some number of steps to "equilibrate" the sequence before beginning the sampling stage to compute the expectation.

The routine Bayes_MCMC_pred_SR.m implements this method to compute the posterior prediction of a function $g(\theta, \sigma)$ using single-response data, and is invoked by the command

g_pred = Bayes_MCMC_pred_SR(X_pred, y, fun_yhat, fun_g, . . .
theta_0, sigma_0, MCOPTS, Param);

X_pred is a matrix that contains in each row the predictor values for the corresponding experiment. y is the vector of measured responses. fun_yhat is a user-supplied function that returns the vector of predicted responses:

y_hat = feval(fun_yhat, theta, X_pred);

fun_g is a user-supplied function that returns the value of $g(\theta, \sigma)$:

g = feval(fun_g, theta, sigma, Param);

Param is an optional structure of parameters that can be passed through Bayes_MCMC_pred_SR to fun_g. theta_0 and sigma_0 are the initial values of θ and σ that initiate the Markov chain.

MCOPTS is a structure that controls the operation of the MCMC simulation and contains the following fields (and default values):

MCOPTS.N_equil, the number of equilibration MC iterations to be performed before sampling begins (1 000);

MCOPTS.N_samples, the number of samples taken to compute the prediction (10,000);

MCOPTS.frac_theta, the fraction of Monte Carlo moves that displace θ; the remainder displace σ (0.5);

MCOPTS.delta_theta, the initial relative size of the random Gaussian displacement in each θ_j during proposed move (0.1);

MCOPTS.delta_sigma, the initial size of the random Gaussian displacement of $\sigma (\min\{0.1|\sigma^{[0]}|, \sqrt{eps}\})$;

MCOPTS.accept_tar, the target fraction of moves that are accepted (0.3). The sizes of the proposed MC moves are adjusted dynamically to achieve this target.

Example. Protein expression data for bacterial strains

We consider again the protein expression data (8.35), where we have fitted the linear model

$$y = \theta_1 + \theta_2 x + \varepsilon \quad x = \begin{cases} 0, & \text{wild-type} \\ 1, & \text{mutant} \end{cases} \tag{8.162}$$

and have obtained the least-squares estimates

$$\theta_{\text{LS},1} = 113.4 \quad \theta_{\text{LS},2} = 6.7750 \tag{8.163}$$

We now compute the probability that $\theta_2 \geq 5$; that is, that the protein expression of the mutant strain is five units above that of the wild-type strain. We use MCMC simulation to compute the expectation of the indicator

$$I_{\theta_2 \geq 5} = \begin{cases} 1, & \theta_2 \geq 5 \\ 0, & \theta_2 < 5 \end{cases} \tag{8.164}$$

We perform this calculation with the code

```
X_pred = [1 0; 1 0; 1 0; 1 0; 1 1; 1 1; 1 1; 1 1];
y = [121.9; 113.4; 112.2; 106.1; 120.7; 119.5; 116.5; 124.0];
theta_0 = [113.4; 6.7750]; sigma_0 = 5;
MCOPTS.N_equil = 2000; MCOPTS.N_samples = 50000;
fun_yhat = 'calc_yhat_linear_model';
fun_g = 'calc_g_1Dinterval';
Param.j = 2; Param.val_lo = 5; Param.val_hi = 1e6;
g_pred = Bayes_MCMC_pred_SR(X_pred, y, fun_yhat, fun_g, ...
    theta_0, sigma_0, MCOPTS, Param),
```

where we have supplied the routines,

```
% calc_yhat_linear_model.m
function y_hat = calc_yhat_linear_model(theta,X);
y_hat = X*theta;
return;
```

and

```
% calc_g_1Dinterval.m
function g = calc_g_1Dinterval(theta,sigma,Param);
if((theta(Param.j) >= Param.val_lo) & ...
      (theta(Param.j) <= Param.val_hi))
   g = 1;
else
   g = 0;
end
return;
```

The output from this program is

g_pred = 0.7233

Thus, we compute a $\sim 72.3\%$ chance that the mutant strain has a protein expression level that is greater than or equal to five units above that of the wild-type strain.

MCMC computation of marginal posterior densities

Monte Carlo simulation can also be used to generate marginal posterior densities using indicator functions fitted to histograms – a technique known as *kernel marginalization*. We estimate the marginal posterior $p(\theta_j|y)$ by partitioning the θ_j interval $[\theta_{j,\mathrm{lo}}, \theta_{j,\mathrm{hi}}]$ into N_{bins} bins of width $\Delta_{j,\mathrm{bin}} = (\theta_{j,\mathrm{hi}} - \theta_{j,\mathrm{lo}})/N_{\mathrm{bins}}$. We measure the relative number of visits to each bin through use of the indicator functions

$$g_{j,k}(\boldsymbol{\theta}, \sigma) = \begin{cases} 1, & [\theta_{j,\mathrm{lo}} + (k-1)\Delta_{j,\mathrm{bin}}] \le \theta_j < [\theta_{j,\mathrm{lo}} + k\Delta_{j,\mathrm{bin}}] \\ 0, & \mathrm{otherwise} \end{cases} \tag{8.165}$$

The marginal density at the bin mid-point $\overline{\theta}_{j,k} = \theta_{j,\mathrm{lo}} + (k - \frac{1}{2})\Delta_{j,\mathrm{bin}}$ is then approximately

$$p(\overline{\theta}_{j,k}|y) \approx \frac{E[g_{j,k}|y]}{\Delta_{j,\mathrm{bin}}} \tag{8.166}$$

For a discussion of alternative marginalization methods, consult Chen *et al.* (2000). calc_g_1Dmarginal.m uses the kernel method to construct 1-D marginal densities concurrently for multiple parameters, and is used by

```
function [bin_1Dc, bin_1Dp, frac_above, frac_below] = ...
   Bayes_MCMC_1Dmarginal_SR( X_pred, y, ...
      fun_yhat, j_plot_1D, val_lo, val_hi, ...
      N_bins, theta_0, sigma_0, MCOPTS);
```

X_pred, y, fun_yhat, theta_0, sigma_0, and MCOPTS take the same definitions as in Bayes_MCMC_pred_SR. j_plot_1D contains the numbers of the parameters whose marginal densities are desired. val_lo and val_hi are vectors that set the lower and upper limits of the histogram for each parameter. N_bins is the number of bins in each histogram (we use

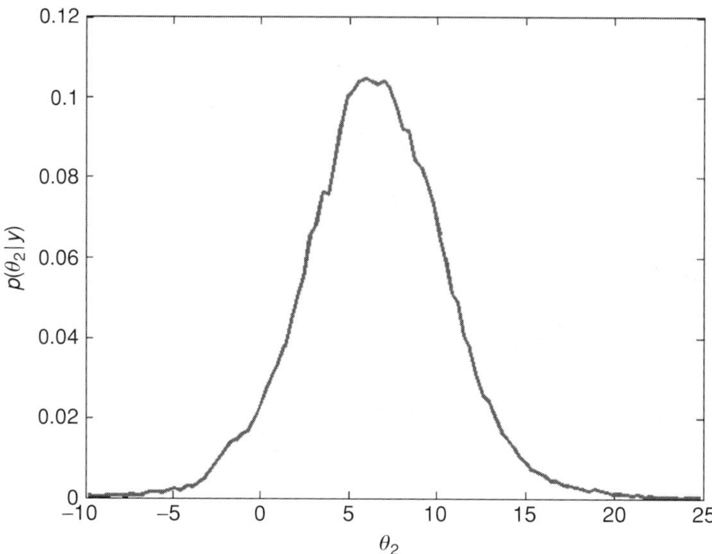

Figure 8.5 Marginal 1-D posterior density for the difference in protein expression levels between the two strains (total probability $= 0.99848$).

the same number for each histogram). bin_1Dc and bin_1Dp contain the histogram representations of the marginal densities. bin_1Dc(m,k) contains the value of $\overline{\theta}_{j\_\text{plot}(m),k}$ and bin_1Dp(m,k) contains the value of $p(\overline{\theta}_{j\_\text{plot}(m),k}|\boldsymbol{y})$. frac_above and frac_below give the fractions of the MC samples that fall outside of the histogram ranges for each parameter.

For the protein expression data, we generate the marginal posterior densities for θ_1 in $90 \leq \theta_1 \leq 130$ and θ_2 in $-10 \leq \theta_2 \leq 25$ by

```
X_pred = [1 0; 1 0; 1 0; 1 0; 1 1; 1 1; 1 1; 1 1];
y = [121.9; 113.4; 112.2; 106.1; 120.7; 119.5; 116.5; 124.0];
theta_0 = [113.4; 6.7750]; sigma_0 = 5;
MCOPTS.N_equil = 20000;
MCOPTS.N_samples = 1000000;
fun_yhat = 'calc_yhat_linear_model';
val_lo = [90; -10]; val_hi = [130; 25];
N_bins = 100; j_plot_1D = [1; 2];
[bin_1Dc, bin_1Dp, frac_above, frac_below] = ...
    Bayes_MCMC_1Dmarginal_SR( X_pred, y, ...
        fun_yhat, j_plot_1D, val_lo, val_hi, ...
        N_bins, theta_0, sigma_0, MCOPTS);
```

The 1-D marginal distribution of θ_2, the difference in expression levels between the two strains, is shown in Figure 8.5.

A similar approach is used by Bayes_MCMC_2Dmarginal_SR.m to compute the 2-D marginal posterior density $p(\theta_i, \theta_j|\boldsymbol{y})$. For the protein expression data, $p(\theta_1, \theta_2|\boldsymbol{y})$ is computed by typing the following commands, after executing the code above:

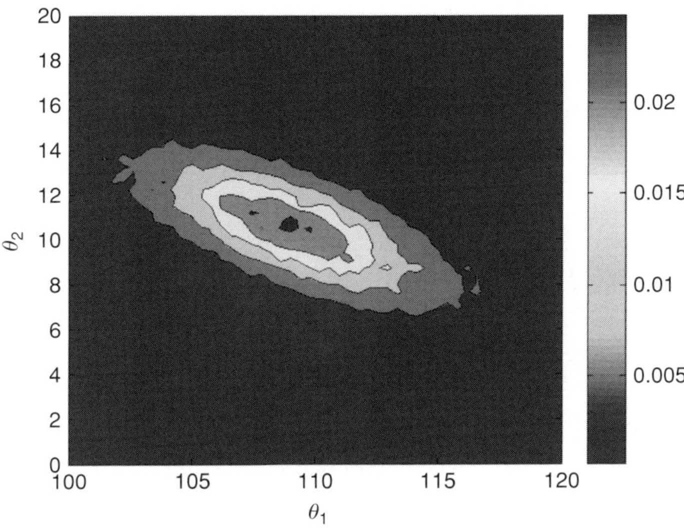

Figure 8.6 Marginal 2-D posterior density for the protein expression data, computed from MCMC simulation.

```
i_plot_2D = 1; j_plot_2D = 2;
[bin_2Dic, bin_2Djc, bin_2Dp] = ...
    Bayes_MCMC_2Dmarginal_SR( X_pred, y, ...
        fun_yhat, i_plot_2D, j_plot_2D, val_lo, val_hi, ...
        N_bins, theta_0, sigma_0, MCOPTS);
```

X_pred, y, fun_yhat, val_lo, val_hi, N_bins, theta_0, sigma_0, and MCOPTS retain the same definitions as when computing 1-D marginal distributions. i_plot_2D and j_plot_2D are the parameters whose 2-D marginal density $p(\theta_{i\_plot}, \theta_{j\_plot}|y)$ is desired. bin_2Dic(m) and bin_2Djc(n) contain the values of $\overline{\theta}_{i\_plot,m}$ and $\overline{\theta}_{j\_plot,n}$ respectively. bin_2Dp(m,n) contains the computed value of $p(\overline{\theta}_{i\_plot,m}, \overline{\theta}_{j\_plot,n}|y)$. The routine generates contour, contourf, and surf plots of $p(\theta_{i\_plot}, \theta_{j\_plot}|y)$. The contourf plot for the protein expression data is shown in Figure 8.6.

Computing highest probability density (HPD) regions from marginal posterior distributions

From marginal posterior densities, we can identify the regions of HPD that contain a specified fraction of the total marginal posterior density. This allows us to compute credible regions without the need of a quadratic expansion of $S(\theta)$. Let $p(\psi|y)$ be a marginal posterior density, where $\psi = \psi(\theta, \sigma)$ and let $Q = \dim(\psi)$ be small enough that $p(\psi|y)$ can be computed feasibly using the histogram technique. Let Ψ_{p_c} be the set of all $\psi \in \Re^Q$, where $p(\psi|y)$ exceeds some contour value p_c:

$$\Psi_{p_c} = \{\psi \in \Re^Q | p(\psi|y) \geq p_c\} \tag{8.167}$$

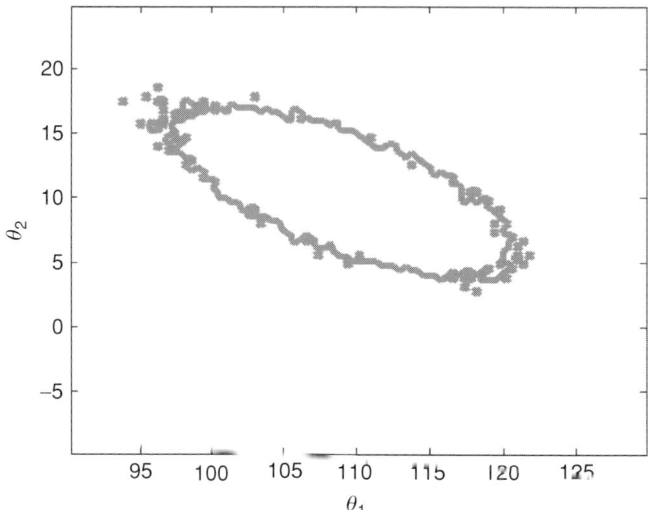

Figure 8.7 The 95% HPD region for (θ_1, θ_2) for the protein expression data is enclosed within the boundary curve plotted above.

Such a set is called a Highest Probability Density (HPD) *region*. We wish to find the HPD region that contains some fraction $(1 - \alpha)$ of the total marginal posterior density; i.e., we identify the contour value $p_{c,\alpha}$ such that

$$\int_{\Psi_{p_{c,\alpha}}} p(\psi|y)d\psi = 1 - \alpha \qquad (8.168)$$

This approach enables us to compute *Bayesian HPD credible regions* for any form of the posterior density, without quadratic approximation of $S(\theta)$. Once we have computed the marginal posterior density on a regular grid, it is fairly straightforward to compute the α for various p_c through quadrature of the grid values. From $p_c = p_c(\alpha)$, we identify the particular p_c that yields the desired αHPD, and then approximate the HPD region as the union of bins whose probabilities exceed this p_c.

From the results of Bayes_MCMC_1Dmarginal.m, HPD regions are identified by the routine Bayes_1D_HPD_SR.m,

alpha = 0.05;
[HPD_lo, HPD_hi] = Bayes_1D_HPD_SR(bin_1Dc, bin_1Dp, . . .
**   j_plot_1D, alpha);**

We compute an approximate 2-D HPD in (θ_1, θ_2) space using the routine Bayes_2D_HPD_SR.m, using as input the results from Bayes_MCMC_2Dmarginal.m:

HPD_2D = Bayes_2D_HPD_SR(bin_2Dic, bin_2Djc, bin_2Dp, . . .
**   i_plot_2D, j_plot_2D, alpha);**

The 95% HPD for (θ_1, θ_2) for the protein expression data is shown in Figure 8.7. The 1-D HPDs are $107.8 \leq \theta_1 \leq 119.0$ and $-1.78 \leq \theta_2 \leq 14.0$, see (8.145).

Example. Batch reactor chemical reaction data

We next use the MCMC approach to study again the hypothesis that the reaction $A + B \rightarrow C$ is elementary, given the data in Table 8.1. We compute the 95% HPD for the 2-D marginal posterior density $p(\theta_2, \theta_3 | y)$. If this HPD region contains the point $(\theta_2 = 1, \theta_3 = 1)$, we cannot support the conclusion that the hypothesis is false (i.e., the reaction is not elementary). We compute this marginal density using MCMC simulation by

```
% input predictor and response data
X_pred = [0.1 0.1; 0.2 0.1; 0.1 0.2; 0.2 0.2; 0.05 0.2; 0.2 0.05];
y = [0.0246e-3; 0.0483e-3; 0.0501e-3; 0.1003e-3; 0.0239e-3; 0.0262e-3];
% provide name of routine that returns predicted responses
fun_yhat = 'calc_yhat_kinetic_ex1';
% provide initial guess of parameters
theta_0 = [0.0025 1.0 1.0];
% compute sample variance with initial guess, and use its
% square root as initial guess for sigma
y_hat = feval(fun_yhat,theta_0,X_pred);
RSS_0 = dot(y-y_hat,y-y_hat);
sample_var_0 = RSS_0/(length(y)-length(theta_0));
sigma_0 = sqrt(sample_var_0);
% select the parameters whose 2-D marginal density
% is desired
i_plot_2D = 2; j_plot_2D = 3;
% set the histogram properties
val_lo = [0.8; 0.8]; val_hi = [1.2; 1.2]; N_bins = 50;
% perform the MCMC simulation to compute the
% 2-D marginal posterior density
MCOPTS.N_equil = 50000;  % # of equilibration iterations
MCOPTS.N_samples = 25000000; % # of samples
[bin_2Dic, bin_2Djc, bin_2Dp] = ...
    Bayes_MCMC_2Dmarginal_SR( X_pred, y, ...
        fun_yhat, i_plot_2D, j_plot_2D, val_lo, val_hi, ...
        N_bins, theta_0, sigma_0, MCOPTS);
% generate from the results the 95% HPD region
alpha = 0.05;
HPD_2D = Bayes_2D_HPD_SR(bin_2Dic, bin_2Djc, bin_2Dp, ...
    i_plot_2D, j_plot_2D, alpha);
```

The computed boundary of the 95% HPD is shown in Figure 8.8. As the HPD contains $(\theta_2 = 1, \theta_3 = 1)$, the data are consistent with an elementary reaction. The code above calls the same routine to predict the responses, calc_yhat_kinetic_ex1.m, as used previously in the least squares fitting of the parameters.

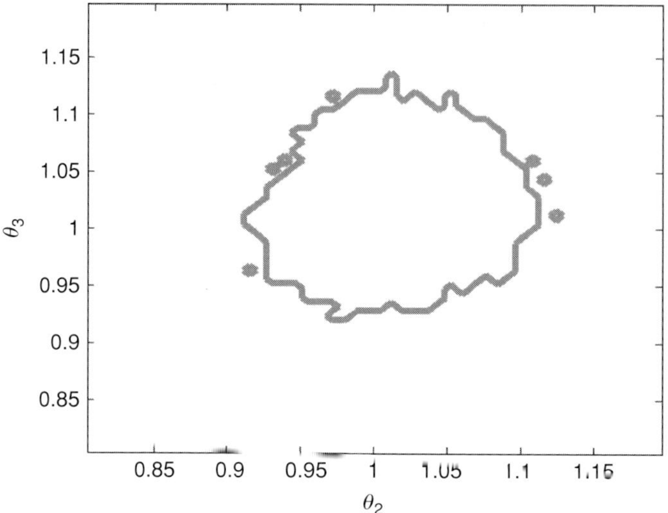

Figure 8.8 Boundary of the 95% HPD for (θ_2, θ_3) for the 2-D marginal posterior density of the rate law exponents from the batch reactor data set.

Applying eigenvalue analysis to experimental design

Above, we have generated confidence intervals from the covariance matrix $\mathrm{cov}(\boldsymbol{\theta}) = \sigma^2 (X^T X)^{-1}$. While the value of σ^2 may be determined by fluctuations in experimental conditions that are beyond our control, we *can* control, through our choice of experimental design, the design matrix X. We would like to design our experiments so that they provide enough information to estimate the parameters to sufficient accuracy. We now consider the application of eigenvalue analysis of $X^T X$ to experimental design.

We diagonalize the positive-semidefinite matrix $X^T X = V \Lambda V^T$, where $\Lambda = \mathrm{diag}(\lambda_1, \lambda_2, \ldots, \lambda_P)$, and $P = \dim(\boldsymbol{\theta})$. The matrix V is an orthogonal $P \times P$ matrix, $V^{-1} = V^T$, whose columns contain the normalized eigenvectors of $X^T X$. The covariance matrix of $\boldsymbol{\theta}$ is

$$\mathrm{cov}(\boldsymbol{\theta}) = \sigma^2 [V \Lambda V^T]^{-1} = \sigma^2 V^{T(-1)} \Lambda^{-1} V^{-1} = \sigma^2 V \Lambda^{-1} V^T \tag{8.169}$$

where $\Lambda^{-1} = \mathrm{diag}(\lambda_1^{-1}, \lambda_2^{-1}, \ldots, \lambda_P^{-1})$. We see that the small eigenvalues of $X^T X$, with the corresponding largest diagonal elements in Λ^{-1}, contribute the most to the uncertainty.

When is an eigenvalue λ_j satisfying $(X^T X) v^{[j]} = \lambda_j v^{[j]}$ small? Writing

$$\lambda_j v^{[j]} = (X^T X) v^{[j]} = X^T \begin{bmatrix} - & x^{[1]} & - \\ & \vdots & \\ - & x^{[N]} & - \end{bmatrix} \begin{bmatrix} | \\ v^{[j]} \\ | \end{bmatrix} = X^T \begin{bmatrix} x^{[1]} \cdot v^{[j]} \\ x^{[2]} \cdot v^{[j]} \\ \vdots \\ x^{[N]} \cdot v^{[j]} \end{bmatrix} \tag{8.170}$$

we see that we obtain a small eigenvalue whenever no row in the design matrix has a significant dot product with the corresponding eigenvector.

As an example, consider fitting the model

$$y = \theta_1 + \theta_2\, x_1 + \theta_3\, x_2 + \varepsilon \tag{8.171}$$

to a data set with the design matrix

$$X = \begin{bmatrix} 1 & 1 & 1 \\ 1 & 2 & 2 \\ 1 & 3 & 3 \\ 1 & 4 & 4 \end{bmatrix} \qquad X^{\mathrm{T}}X = \begin{bmatrix} 4 & 10 & 10 \\ 10 & 30 & 30 \\ 10 & 30 & 30 \end{bmatrix} \tag{8.172}$$

We see that $X^{\mathrm{T}}X$ is singular, as a result of the lack of any experiments with $x_1 \neq x_2$. But, let us say that this deficiency of the data set was not so immediately obvious. We could still diagnose the situation using the eigenvector decomposition (8.27),

$$V = \begin{bmatrix} 0.0000 & 0.9728 & 0.2317 \\ 0.7071 & -0.1639 & 0.6979 \\ -0.7071 & -0.1639 & 0.6879 \end{bmatrix} \qquad \Lambda = \begin{bmatrix} 0.0 & & \\ & 0.6312 & \\ & & 63.3688 \end{bmatrix} \tag{8.173}$$

The eigenvector corresponding to the zero eigenvalue is of the form $[o + c - c]$, suggesting that we need to add an experiment that varies x_1 and x_2 in opposite directions. Therefore, we add an experiment whose predictor variables equal those in the second experiment plus $[0 \ \ 1 \ \ -1]$,

$$[1 \quad 2 \quad 2] + [0 \quad 1 \quad -1] = [1 \quad 3 \quad 1] \tag{8.174}$$

so that the span of the row vectors in the new design matrix contains the eigenvector for the zero eigenvalue. For the new design, we have

$$X = \begin{bmatrix} 1 & 1 & 1 \\ 1 & 2 & 2 \\ 1 & 3 & 3 \\ 1 & 4 & 4 \\ 1 & 3 & 1 \end{bmatrix} \qquad X^{\mathrm{T}}X = \begin{bmatrix} 5 & 13 & 11 \\ 13 & 39 & 33 \\ 11 & 33 & 31 \end{bmatrix}$$

$$\Lambda = \begin{bmatrix} 0.5868 & & \\ & 1.8796 & \\ & & 72.5338 \end{bmatrix} \tag{8.175}$$

The zero eigenvalue has been removed, and we now are able to estimate all parameters to finite accuracy.

This analysis is useful for designing a set of experiments to yield the desired accuracy. Given an a priori estimate of σ^2, we estimate the corresponding width of the confidence intervals. If this accuracy is insufficient, we add more experiments, until the expected accuracy is deemed sufficient.

For a nonlinear model, we also must provide a ballpark estimate of $\boldsymbol{\theta}$, where we evaluate the linearized design matrix. We then apply the eigenvalue analysis above, but use the linearized design matrix. In cases where it may be very costly to come back later to perform additional experiments, we may wish to try multiple estimates of $\boldsymbol{\theta}$, repeat the eigenvalue analysis for each linearized design matrix, and accept only a design that appears to provide sufficient accuracy for all plausible values of $\boldsymbol{\theta}$.

Example. Determining the number of additional experiments necessary for the protein expression data

We consider once again the data for the protein expression levels of wild-type and mutant bacterial strains (8.35) with $X^{\mathrm{T}}X$ and its inverse again given by (8.40). For a specified σ, the standard deviation of θ_2 is

$$\mathrm{std}(\theta_2) = \sigma\sqrt{(X^{\mathrm{T}}X)_{22}^{-1}} = \sigma\sqrt{2n^{-1}} \tag{8.176}$$

The expected width of the confidence interval in this parameter is then

$$\left|\theta_2 - \theta_{M,2}\right| \approx Z_{\alpha/2}\sigma\sqrt{2n^{-1}} \tag{8.177}$$

Or, to account roughly for the extra uncertainty in σ, we could use

$$\left|\theta_2 - \theta_{M,2}\right| \approx T_{n-2,\alpha/2}\,s\sqrt{2n^{-1}} \tag{8.178}$$

We can use (8.178) with $n = 4 + m$ and the s-value from the existing data to estimate the number m of additional experiments necessary to reduce the uncertainty in θ_2 to a desired level.

Here, our emphasis has been upon experimental design; however, eigenvalue analysis and SVD of the design matrix can also be used to extract at least partial results when $X^{\mathrm{T}}X$ is singular. This subject is discussed in further detail in the supplemental material in the accompanying website.

Bayesian multiresponse regression

Previously, we have considered only the analysis of single-response data. Here, we discuss multiresponse regression, focusing primarily upon the extension of the least-squares method to the case of multiple, perhaps correlated, responses in each experiment.

Again, we perform a number N of experiments, where in the kth experiment, we have a known set of M predictor variables, $\boldsymbol{x}^{[k]} \in \Re^M$, and we observe the L responses $\boldsymbol{y}^{[k]} \in \Re^L$. We wish to estimate the values of P unknown parameters $\boldsymbol{\theta} \in \Re^P$, in a model whose predicted responses for each experiment form a vector $\boldsymbol{f}(\boldsymbol{x}^{[k]}; \boldsymbol{\theta}) \in \Re^L$. We assume that the measured responses are equal to the model predictions plus a random error vector,

$$\boldsymbol{y}^{[k]} = \boldsymbol{f}(\boldsymbol{x}^{[k]}; \boldsymbol{\theta}) + \boldsymbol{\varepsilon}^{[k]} \tag{8.179}$$

$\boldsymbol{\varepsilon}^{[k]} \in \Re^L$ is assumed to be independent of the other $\boldsymbol{\varepsilon}^{[l \neq k]}$, but we allow that the components of $\boldsymbol{\varepsilon}^{[k]}$ *may* be correlated. The $L \times L$ covariance matrix (unknown) of each error vector is

$\Sigma = \mathrm{cov}(\varepsilon)$. From the measured responses, we form the $N \times L$ response data matrix

$$
Y = \begin{bmatrix} - & y^{[1]} & - \\ - & y^{[2]} & - \\ & \vdots & \\ - & y^{[N]} & - \end{bmatrix}
\tag{8.180}
$$

We assume that the errors in each experiment are drawn independently from a multivariate normal distribution with zero mean,

$$
p(\varepsilon|\Sigma) = \frac{1}{(2\pi)^{L/2}|\Sigma|^{1/2}} \exp\left[-\frac{1}{2}\varepsilon^{\mathrm{T}}\Sigma^{-1}\varepsilon\right]
\tag{8.181}
$$

Thus, the probability of observing a response $y^{[k]}$ in experiment k is

$$
p\big(y^{[k]}|\theta, \Sigma\big) = (2\pi)^{-L/2}|\Sigma|^{-1/2} \exp\left\{-\frac{1}{2}\big[y^{[k]} - f(x^{[k]}; \theta)\big]^{\mathrm{T}}\Sigma^{-1}\big[y^{[k]} - f(x^{[k]}; \theta)\big]\right\}
\tag{8.182}
$$

Assuming independent errors in each experiment, the likelihood function for the multi-response data is

$$
l(\theta, \Sigma|Y) = p(Y|\theta, \Sigma) = \prod_{k=1}^{N} p\big(y^{[k]}|\theta, \Sigma\big)
\tag{8.183}
$$

Using the rule $e^a e^b = e^{a+b}$, we have

$$
\begin{aligned}
l(\theta, \Sigma|Y) = (2\pi)^{-NL/2}|\Sigma|^{-N/2} \\
\times \exp\left\{-\tfrac{1}{2}\textstyle\sum_{k=1}^{N} \big[y^{[k]} - f(x^{[k]}; \theta)\big]^{\mathrm{T}}\Sigma^{-1}\big[y^{[k]} - f(x^{[k]}; \theta)\big]\right\}
\end{aligned}
\tag{8.184}
$$

We next define the $L \times L$ positive-definite matrix $S(\theta)$ with the elements

$$
S_{ab}(\theta) = \sum_{k=1}^{N} \big[y_{\mathrm{a}}^{[k]} - f_{\mathrm{a}}(x^{[k]}; \theta)\big]\big[y_{\mathrm{b}}^{[k]} - f_{\mathrm{b}}(x^{[k]}; \theta)\big]
\tag{8.185}
$$

and write the likelihood function as

$$
l(\theta, \Sigma|Y) = (2\pi)^{-NL/2}|\Sigma|^{-N/2} \exp\left\{-\tfrac{1}{2}\mathrm{tr}[\Sigma^{-1}S(\theta)]\right\}
\tag{8.186}
$$

For this $l(\theta, \Sigma|Y)$, the noninformative prior is (Box & Tiao, 1973)

$$
p(\theta, \Sigma) = p(\theta)p(\Sigma) \qquad p(\theta) \sim c \qquad p(\Sigma) \propto |\Sigma|^{-(L+1)/2}
\tag{8.187}
$$

Therefore, the posterior density $p(\theta, \Sigma|Y) \propto l(\theta, \Sigma|Y)p(\theta)p(\Sigma)$ is

$$
p(\theta, \Sigma|Y) \propto (2\pi)^{-NL/2}|\Sigma|^{-(N+L+1)/2} \exp\left\{-\frac{1}{2}\mathrm{tr}[\Sigma^{-1}S(\theta)]\right\}
\tag{8.188}
$$

For this posterior density, of the form of a *Wishart distribution*, the marginal posterior density for θ can be computed analytically:

$$
p(\theta|Y) = \int_{\Sigma > 0} p(\theta, \Sigma|Y)d\Sigma \propto |S(\theta)|^{-N/2}
\tag{8.189}
$$

Reduction of the multiresponse posterior density to the previous result for single-response data

We now show that this marginal posterior, $p(\theta|Y) \propto |S(\theta)|^{-N/2}$, agrees with our previous result (8.118) for single-response data, by considering a linear model. For $L = 1$, $\hat{y}^{[k]}(\theta) = (X\theta)_k$,

$$
\begin{aligned}
S(\theta) &= \sum_{k=1}^{N} \left[y^{[k]} - (X\theta)_k\right]^2 = \sum_{k=1}^{N} \left\{\left[y^{[k]} - (X\theta_{\mathrm{LS}})_k\right] + \left[X(\theta_{\mathrm{LS}} - \theta)\right]_k\right\}^2 \\
&= \sum_{k=1}^{N} \left[y^{[k]} - (X\theta_{\mathrm{LS}})_k\right]^2 + 2\sum_{k=1}^{N} \left[y^{[k]} - (X\theta_{\mathrm{LS}})_k\right]\left[X(\theta_{\mathrm{LS}} - \theta)\right]_k \\
&\quad + \sum_{k=1}^{N} \left[X(\theta_{\mathrm{LS}} - \theta)\right]_k^2
\end{aligned}
\tag{8.190}
$$

Writing the sum in the middle term of the right-hand side as

$$
\begin{aligned}
\sum_{k=1}^{N} (y - X\theta_{\mathrm{LS}})_k &\left[\sum_j X_{kj}(\theta_{\mathrm{LS}} - \theta)_j\right] \\
&= \sum_j (\theta_{\mathrm{LS}} - \theta)_j \sum_{k=1}^{N} (X^{\mathrm{T}})_{jk}(y - X\theta_{\mathrm{LS}})_k \\
&= \sum_j (\theta_{\mathrm{LS}} - \theta)_j [X^{\mathrm{T}}(y - X\theta_{\mathrm{LS}})]_j = (\theta_{\mathrm{LS}} - \theta) \cdot [X^{\mathrm{T}}(y - X\theta_{\mathrm{LS}})]
\end{aligned}
\tag{8.191}
$$

we see that this term is zero, as $X^{\mathrm{T}}X\theta_{\mathrm{LS}} = X^{\mathrm{T}}y$. Therefore,

$$
S(\theta) = \sum_{k=1}^{N} \left[y^{[k]} - (X\theta_{\mathrm{LS}})_k\right]^2 + \sum_{k=1}^{N} \left[X(\theta_{\mathrm{LS}} - \theta)\right]_k^2
\tag{8.192}
$$

Using our previous definition (8.89) of the sample variance,

$$
vs^2 = S(\theta_{\mathrm{LS}}) = \sum_{k=1}^{N} \left[y^{[k]} - (X\theta_{\mathrm{LS}})_k\right]^2
\tag{8.193}
$$

and

$$
\sum_{k=1}^{N} \left[X(\theta_{\mathrm{LS}} - \theta)\right]_k^2 = (\theta - \theta_{\mathrm{LS}})^{\mathrm{T}} X^{\mathrm{T}} X (\theta - \theta_{\mathrm{LS}})
\tag{8.194}
$$

we have

$$
S(\theta) = vs^2 + (\theta - \theta_{\mathrm{LS}})^{\mathrm{T}} X^{\mathrm{T}} X (\theta - \theta_{\mathrm{LS}})
\tag{8.195}
$$

Thus, the marginal posterior for θ, $p(\theta|Y) \propto |S(\theta)|^{-N/2}$, becomes

$$
p(\theta|Y) \propto \left[1 + \frac{1}{vs^2}(\theta - \theta_{\mathrm{LS}})^{\mathrm{T}} X^{T} X (\theta - \theta_{\mathrm{LS}})\right]^{-N/2}
\tag{8.196}
$$

This is the same multivariate t-distribution as (8.118).

Numerically computing the parameter estimate

From the marginal posterior density $p(\boldsymbol{\theta}|Y) \propto |S(\boldsymbol{\theta})|^{-N/2}$, we identify the most probable estimate $\boldsymbol{\theta}_{\mathrm{M}}$ as that minimizing $|S(\boldsymbol{\theta})|$. We compute $|S(\boldsymbol{\theta})|$ by performing a LU decomposition,

$$P(\boldsymbol{\theta})S(\boldsymbol{\theta}) = L(\boldsymbol{\theta})U(\boldsymbol{\theta}) \tag{8.197}$$

As $|S(\boldsymbol{\theta})| > 0$ for the positive-definite matrix $S(\boldsymbol{\theta})$, we have

$$|S(\boldsymbol{\theta})| = \prod_{l=1}^{L} |U_{ll}(\boldsymbol{\theta})| \tag{8.198}$$

Here, we have used the fact that $|L| = 1$ and $|P| = \pm 1$. As $f(x)$ and $\ln[f(x)]$ have the same minima, we find $\boldsymbol{\theta}_{\mathrm{M}}$ by minimizing the cost function

$$F_{\mathrm{cost}}(\boldsymbol{\theta}) = \ln|S(\boldsymbol{\theta})| = \sum_{l=1}^{L} \ln|U_{ll}(\boldsymbol{\theta})| \tag{8.199}$$

which varies more slowly with $\boldsymbol{\theta}$ than does $|S(\boldsymbol{\theta})|$.

Because the gradient vector of this cost function is difficult to determine, we use the nonlinear simplex method that only requires us to compute the cost function value for each trial $\boldsymbol{\theta}$. This technique returns only a local minimum; therefore, we augment it by simulated annealing to search for a global minimum, as is done by the routine sim_anneal_MR.m:

[theta, det_S, det_S_0] = sim_anneal_MR(. . .
**   X_pred, Y, fun_yhat, theta_0, N_iter, . . .**
**   freq_quench, freq_reset, T_0);**

X_pred is the $N \times M$ matrix that contains in each row the predictors for the corresponding experiment. Y is an $N \times L$ matrix containing the responses for each experiment in the corresponding row. fun_yhat is the name of a user-supplied routine that returns the predicted response data matrix for input values of theta (the parameters) and X_pred:

Y_hat = feval(fun_yhat, theta, X_pred);

theta_0 contains the initial guesses of the parameters. N_iter is the total number of iterations during the simulated annealing run. freq_quench is a number in [0, 1] that sets the frequency with which the simulation is "quenched" at random times during the simulation. Here, "quenching" refers to running the nonlinear simplex minimizer **fminsearch** to find a local minimum of the cost function. freq_reset specifies the frequency with which the state is reset, at random times, to the best one (lowest cost function) found to date. T_0 is the initial annealing temperature. If this argument is not included, a default value of $10|F_{\mathrm{cost}}(\boldsymbol{\theta}^{[0]})|$ is used.

The routine returns the theta value with the lowest cost function found during the simulation. det_S is the value of $|S(\boldsymbol{\theta})|$ for this optimal parameter value (at least a local minimum). det_S_0 is the value of $|S(\boldsymbol{\theta}^{[0]})|$.

Use of this routine requires the MATLAB optimization toolkit, for access to **fminsearch**, when freq_quench is nonzero. If **fminsearch** is unavailable, the routine may be modified to use the conjugate gradient minimizer routine provided in Chapter 5.

Example. Fitting the rate constant from multiresponse dynamic batch reactor data

Here, we apply this routine to estimate the value of the rate constant k_1 of the reaction, assumed elementary, $A + B \rightarrow C$, from the multiresponse data of Table 8.3. The following code employs sim_anneal_MR.m to fit the rate constant:

```
% input predictors and response data
time_hr = [0.5; 1; 1.5; 2; 3; 4; 5; 6; 7; 8; 9; 10];
time_s = time_hr*3600; % convert hr to sec
X_pred = time_s; % set predictor matrix
cA = [0.0985; 0.0637; 0.0500; 0.0462; 0.0363; 0.0248;  ...
        0.0171; 0.0168; 0.0131; 0.0150; 0.0140; 0.0134];
cB = [0.0995; 0.0651; 0.0596; 0.0453; 0.0384; 0.0247;  ...
        0.0174; 0.0203; 0.0136; 0.0121; 0.0142; 0.0134];
cC = [0.001; 0.0357; 0.0501; 0.0512; 0.0682; 0.0747;  ...
        0.0809; 0.0818; 0.0858; 0.0863; 0.0872; 0.0928];
N = length(time_hr); L = 3;
Y = zeros(N,L); % set response data matrix
Y(:,1) = cA; Y(:,2) = cB; Y(:,3) = cC;
% set parameters for simulated annealing run
N_iter = 1000;
k1_0 = 0.001;
fun_yhat = 'calc_Yhat_reaction_ex_dynamic_MR';
freq_quench = 1e-2; freq_reset = 1e-2;
% perform the simulated annealing run
[k1, det_S, det_S_0] = ...
    sim_anneal_MR(X_pred, Y, fun_yhat, ...
    k1_0, N_iter, freq_quench, freq_reset),
```

The results of this calculation are

```
k1 = 0.0024
det_S = 5.7277e-013
det_S_0 = 2.3605e-011
```

Here, we have supplied the following routine to predict the response data matrix as a function of θ:

```
% calc_Yhat_reaction_ex_dynamic_MR.m
function Y_hat = calc_Yhat_reaction_ex_dynamic_MR(k1, X_pred);
t_span = X_pred'; % set times at which to report concentrations
```

```
x_0 = [0.1; 0.1; 0]; % set initial condition
% perform dynamic simulation
[t_traj,x_traj] = ode45(@batch_kinetics_dynamics_ex, ...
    t_span, x_0, [], k1);
% extract response predictions
Y_hat = x_traj;
return;
```

batch_kinetics_dynamics_ex.m is the same routine as used previously in the least-squares fitting of k_1 from the single-response data set c_C vs. t.

MCMC simulation with the multiresponse marginal posterior density

With the multiresponse marginal posterior density $p(\boldsymbol{\theta}|Y) \propto |S(\boldsymbol{\theta})|^{-N/2}$, we use MCMC simulation to estimate posterior predictions of the form

$$E[\boldsymbol{g}|Y] = \int_{\mathfrak{R}^P} \boldsymbol{g}(\boldsymbol{\theta})p(\boldsymbol{\theta}|Y)d\boldsymbol{\theta} \tag{8.200}$$

We generate, with the Metropolis algorithm, a sequence of states $\{\boldsymbol{\theta}^{[m]}\}$ drawn from $p(\boldsymbol{\theta}|Y)$, and approximate the expectation by

$$E[\boldsymbol{g}|Y] \approx \frac{1}{N_s} \sum_{m=1}^{N_s} \boldsymbol{g}(\boldsymbol{\theta}^{[m]}) \tag{8.201}$$

Such calculations are performed by the routine

```
g_pred = Bayes_MCMC_pred_MR(X_pred, Y, ...
    fun_yhat, fun_g, theta_0, MCOPTS, Param);
```

X_pred, Y, fun_yhat, and theta_0 take their previous definitions. fun_g is the name of a function that computes $\boldsymbol{g}(\boldsymbol{\theta})$:

```
g = feval(fun_g, theta, Param);
```

Param is an optional structure that can be passed to fun_g.

MCOPTS controls the performance of the MCMC simulation and has the following fields (and default values):

MCOPTS.N_equil: the number of equilibration MC iterations to be performed before sampling begins (1000).

MCOPTS.N_samples: the number of MC iterations run during sampling. The larger this number, the more accurate the prediction (10 000).

MCOPTS.delta_theta: the size of the displacements in each parameter during an MC move relative to average magnitudes of each parameter (0.1). This value is changed to meet the target fraction of accepted moves.

MCOPTS.accept_tar: target fraction of proposed MC moves that are accepted (0.3).

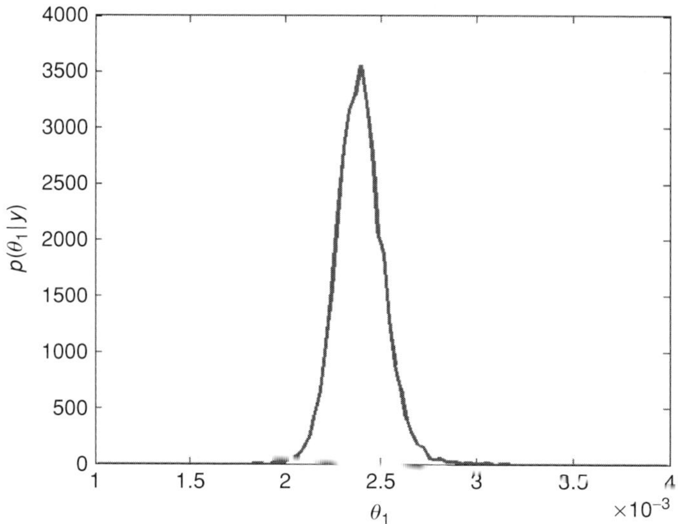

Figure 8.9 Marginal 1-D density for k_1 computed from multiresponse data of a chemical reaction using MCMC simulation.

This MCMC routine is used in turn by other routines that compute marginal posterior densities and generate HPD regions. 1-D marginal densities and their corresponding HPD regions are generated by the routines

```
[bin_1Dc, bin_1Dp, frac_above, frac_below] = . . .
    Bayes_MCMC_1Dmarginal_MR( X_pred, Y, . . .
    fun_yhat, j_plot_1D, val_lo, val_hi, . . .
    N_bins, theta_0, MCOPTS);
```

and

```
[HPD_lo, HPD_hi] = Bayes_1D_HPD_MR( . . .
    bin_1Dc, bin_1Dp, j_plot_1D, alpha);
```

The definitions of the arguments in these routines are the same as for the corresponding single-response routines.

For the multiresponse data of Table 8.3 for the chemical reaction $A + B \rightarrow C$, the following code computes the 1-D marginal posterior density for k_1 (Figure 8.9) and the corresponding 95% HPD.

```
k1_0 = k1;
j_plot_1D = 1;
val_lo = 0.001; val_hi = 0.004;
N_bins = 100;
MCOPTS.N_equil = 1000;
MCOPTS.N_samples = 25000;
[bin_1Dc, bin_1Dp, frac_above, frac_below] = . . .
```

```
Bayes_MCMC_1Dmarginal_MR( X_pred, Y, . . .
fun_yhat, j_plot_1D, val_lo, val_hi, . . .
N_bins, k1_0, MCOPTS);
% compute 95% HPD
alpha = 0.05;
[HPD_lo,HPD_hi] = Bayes_1D_HPD_MR( . . .
   bin_1Dc, bin_1Dp, j_plot_1D, alpha),
```

About the most probable estimate of $k_1 = 0.0024$, this analysis of the multiresponse data of Table 8.3 yields a 95% HPD region for k_1 of

$$0.0022 \leq k_1 \leq 0.0026 \tag{8.202}$$

2-D marginal posterior densities and HPD regions are computed from MCMC simulation for multiresponse data using the routines

```
[bin_2Dic, bin_2Djc, bin_2Dp] = . . .
   Bayes_MCMC_2Dmarginal_MR( X_pred, Y, . . .
      fun_yhat, i_plot_2D, j_plot_2D, val_lo, val_hi, . . .
      N_bins, theta_0, MCOPTS);
```

and

```
HPD_2D = Bayes_2D_HPD_MR(bin_2Dic, bin_2Djc, bin_2Dp, . . .
   i_plot_2D, j_plot_2D, alpha);
```

Again, the arguments in these routines carry the same definitions as the corresponding ones in the routines for single-response data.

Analysis of composite data sets

Above, we have assumed that we measure the same set of responses in each experiment; however, often we estimate parameters from composite data sets that mix different types of data. Here, we treat composite data sets using the sequential learning aspects of Bayesian analysis; hence, the routines take the suffix _MRSL for multiresponse sequential-learning. Let us say that we have some sets of response data $Y^{(1)}, Y^{(2)}, \ldots, Y^{(D)}$ that provide information about the same set of parameters $\boldsymbol{\theta} \in \mathfrak{R}^P$ but that are dimensioned differently and have different error properties. We want to compute the marginal posterior density, taking into account the information provided by all sets of data,

$$p\big(\boldsymbol{\theta} \big| Y^{(1)}, Y^{(2)}, \ldots, Y^{(D)}\big) \tag{8.203}$$

For each data set, we propose a model that predicts the responses $\hat{Y}^{(j)}(\boldsymbol{\theta})$, and compute the $L^{(j)} \times L^{(j)}$ "sum of squared errors" matrix

$$S^{(j)}(\boldsymbol{\theta}) = \big[Y^{(j)} - \hat{Y}^{(j)}(\boldsymbol{\theta})\big]^{\mathrm{T}}\big[Y^{(j)} - \hat{Y}^{(j)}(\boldsymbol{\theta})\big] \tag{8.204}$$

We analyze each data set in turn, using for the first data set a noninformative prior. The marginal posterior after the first data set is

$$p\left(\theta \middle| Y^{\langle 1 \rangle}\right) \propto \left|S^{\langle 1 \rangle}(\theta)\right|^{-N^{\langle 1 \rangle}/2} \tag{8.205}$$

$N^{\langle 1 \rangle}$ is the number of experiments in the first data set.

We then use $p(\theta|Y^{\langle 1 \rangle})$ as the prior for the second set of data, in lieu of the noninformative prior $p(\theta) \sim c$. Thus, after analyzing the two data sets, Bayes' theorem yields the marginal posterior density

$$p\left(\theta \middle| Y^{\langle 1 \rangle}, Y^{\langle 2 \rangle}\right) \propto \left|S^{\langle 1 \rangle}(\theta)\right|^{-N^{\langle 1 \rangle}/2} \left|S^{\langle 2 \rangle}(\theta)\right|^{-N^{\langle 2 \rangle}/2} \tag{8.206}$$

Continuing this process, the marginal posterior density with all D sets is

$$p\left(\theta \middle| Y^{\langle 1 \rangle}, Y^{\langle 2 \rangle}, \dots, Y^{\langle D \rangle}\right) \propto \prod_{d=1}^{D} \left|S^{\langle d \rangle}(\theta)\right|^{-N^{\langle d \rangle}/2} \tag{8.207}$$

Therefore, the most probable θ-value, obtained by considering all sets of data, minimizes the cost function

$$F_{\text{cost}}(\theta) = -\ln\left[p\left(\theta \middle| Y^{\langle 1 \rangle}, Y^{\langle 2 \rangle}, \dots, Y^{\langle D \rangle}\right)\right] = \tfrac{1}{2} \sum_{d=1}^{D} N^{\langle d \rangle} \ln\left|S^{\langle d \rangle}(\theta)\right| \tag{8.208}$$

Let θ_M be the most probable estimate that minimizes (8.208). We then can test hypotheses, compute marginal 1-D or 2-D densities, and generate HPD regions, from MCMC simulation with the marginal posterior density (rescaled to reduce round-off error),

$$p\left(\theta \middle| Y^{\langle 1 \rangle}, Y^{\langle 2 \rangle}, \dots, Y^{\langle D \rangle}\right) \propto \prod_{d=1}^{D} \left\{ \frac{\left|S^{\langle d \rangle}(\theta)\right|}{\left|S^{\langle d \rangle}(\theta_M)\right|} \right\}^{-N^{\langle d \rangle}/2} \tag{8.209}$$

Use of the routines that perform these calculations is demonstrated below for the example of fitting the rate constant k_1 for the reaction $A + B \rightarrow C$ from the combined batch reactor data of Table 8.1 and Table 8.3.

Example. Numerical analysis of composite data sets, applied to the problem of estimating the rate constant of a reaction from multiple reactor data sets

We employ a structure MRSLData to store the predictor and response values for each data set with the following fields:

MRSLData.num_sets: the number of data sets in the composite set of data;
MRSLData.P: the number of parameters to be fitted;
MRSLData.M: a vector containing for each data set the number of predictors;
MRSLData.L: a vector containing for each data set the number of responses;
MRSLData.N: a vector containing for each data set the number of experiments;
MRSLData.X_pred_j: for each data set j, the $N \times M$ matrix of predictors;
MRSLData.Y_j: for each data set j, the $N \times L$ matrix of measured responses;
MRSLData.fun_yhat_j: for each data set j, the name of a routine that predicts the response
    values,

```
Y_hat = feval(fun_yhat_j, theta, X_pred_j);
```

The following code sets the MRSLData structure for a composite data set combining the results of Table 8.1 and Table 8.3:

```
% specify the number of data sets
MRSLData.num_sets = 2;
% specify the number of parameters to be fitted
MRSLData.P = 1;
% allocate memory for dimensioning parameters
MRSLData.N = zeros(MRSLData.num_sets,1);
MRSLData.L = zeros(MRSLData.num_sets,1);
MRSLData.M = zeros(MRSLData.num_sets,1);
% -- DATA SET # 1 -- TABLE 1
% For the first data set (the contents of
% Table 1), input the predictor matrix.
MRSLData.X_pred_1 = [0.1 0.1; 0.2 0.1; 0.1 0.2; ...
     0.2 0.2; 0.05 0.2; 0.2 0.05];
% input the response data matrix
MRSLData.Y_1 = [0.0246e-3; 0.0483e-3; 0.0501e-3; ...
     0.1003e-3; 0.0239e-3; 0.0262e-3];
% specify the name of the routine that predicts
% the responses.
MRSLData.fun_yhat_1 = 'calc_yhat_kinetic_ex_table1';
% set the dimension parameters for the data set
MRSLData.N(1) = size(MRSLData.Y_1,1);
MRSLData.L(1) = size(MRSLData.Y_1,2);
MRSLData.M(1) = size(MRSLData.X_pred_1,2);
% -- DATA SET # 2 -- TABLE 3
% For the second data set (the contents of Table
% 3), input the predictor matrix
time_hr = [0.5; 1; 1.5; 2; 3; 4; 5; 6; 7; 8; 9; 10];
time_s = time_hr*3600; % convert hr to sec
MRSLData.X_pred_2 = time_s; % set predictor matrix
% input the measured response data matrix
cA = [0.0985; 0.0637; 0.0500; 0.0462; 0.0363; 0.0248; ...
       0.0171; 0.0168; 0.0131; 0.0150; 0.0140; 0.0134];
cB = [0.0995; 0.0651; 0.0596; 0.0453; 0.0384; 0.0247; ...
       0.0174; 0.0203; 0.0136; 0.0121; 0.0142; 0.0134];
cC = [0.001; 0.0357; 0.0501; 0.0512; 0.0682; 0.0747; ...
       0.0809; 0.0818; 0.0858; 0.0863; 0.0872; 0.0928];
MRSLData.N(2)= length(cA);
MRSLData.L(2) = 3;
MRSLData.M(2) = size(MRSLData.X_pred_2,2);
MRSLData.Y_2 = zeros(MRSLData.N(2),MRSLData.L(2));
MRSLData.Y_2(:,1) = cA;
```

```
MRSLData.Y_2(:,2) = cB;
MRSLData.Y_2(:,3) = cC;
% specify name of routine that predicts the response
% data matrix for this set of experiments
MRSLData.fun_yhat_2 = 'calc_Yhat_reaction_ex_dynamic_MR';
```

calc_Yhat_reaction_ex_dynamic_MR.m is the same routine as used previously in fitting the rate constant using the data of Table 8.3 alone. The following routine predicts the results in the experiments of Table 8.1 for a trial value of the rate constant:

```
% calc_yhat_kinetic_ex_table1.m
function y_hat = calc_yhat_kinetic_ex_table1(k_1, X_pred);
conc_A = X_pred(:,1); % [A] in each experiment
conc_B = X_pred(:,2); % [B] in each experiment
y_hat = k_1.*conc_A.*conc_B;
return;
```

The cost function and marginal posterior density for the composite data set are computed by the routine

```
[F_cost, det_S, posterior_density] = ...
    calc_MRSL_posterior(theta, MRSLData, det_S_ref);
```

For the input value of theta, and the composite data set MRSLData, F_cost is the cost function value, det_S is a vector of the $|S^{(d)}(\theta)|$ values in each data set. det_S_ref is an optional vector of values $|S^{(d)}(\theta^{(\text{ref})})|$ used to rescale the posterior density, as above for $\theta^{(\text{ref})} = \theta_M$. If this argument is not included, default values of 1 are used.

The following code computes the cost function and $|S^{(d)}(\theta)|$ values for an initial rate constant guess of $k_1^{[0]} = 0.001$:

```
theta_0 = 0.001;
[F_cost_0, det_S_0] = calc_MRSL_posterior(theta_0, MRSLData),
```

The results are

```
F_cost_0 = −203.6059
det_S_0 = 1.0e-008 *
    0.6012
    0.0024
```

From an initial guess of the parameters, simulated annealing with quenching is used to search for the global minimum (the returned value is at least a local minimum). This calculation is performed by the routine

```
[theta, F_cost, det_S] = sim_anneal_MRSL( ...
        MRSLData, theta_0, N_iter, ...
        freq_quench, freq_reset, T_0);
```

The arguments N_iter, freq_quench, freq_reset, and T_0 have the same definitions as previously. If T_0 is not included in the list of arguments, a default value of $10|F_{cost}(\theta^{[0]})|$ is used.

The following code fits the rate constant to the composite data of Table 8.1 and Table 8.3:

```
N_iter = 1000; freq_quench = 0.01; freq_reset = 0.01;
[theta, F_cost, det_S] = sim_anneal_MRSL( ...
        MRSLData, theta_0, N_iter, ...
        freq_quench, freq_reset),
```

The results are

```
theta = 0.0025
F_cost = −246.3639
det_S = 1.0e-011*
   0.5624
   0.0620
```

Thus, we have an estimate $k_1 = 0.0025$ from the composite data.

From MCMC simulation, we can compute posterior expectations and 1-D and 2-D marginal densities with the routines

```
g_pred = Bayes_MCMC_pred_MRSL( ...
        MRSLData, det_S_ref, ...
        fun_g, theta_0, MCOPTS, Param);
[bin_1Dc, bin_1Dp, frac_above, frac_below] = ...
        Bayes_MCMC_1Dmarginal_MRSL( ...
        MRSLData, det_S_ref, ...
        j_plot_1D, val_lo, val_hi, ...
        N_bins, theta_0, MCOPTS);
[bin_2Dic, bin_2Djc, bin_2Dp] = ...
        Bayes_MCMC_2Dmarginal_MRSL( ...
        MRSLData, det_S_ref, ...
        i_plot_2D, j_plot_2D, val_lo, val_hi, ...
        N_bins, theta_0, MCOPTS);
```

The results from these routines can be used with Bayes_1D_HPD_MR.m and Bayes_2D_HPD_MR.m to compute HPD regions.

The following code computes the marginal density for the rate constant from the composite data and the corresponding 95% HPD region:

```
det_S_ref = det_S; % use most probable state as reference.
j_plot_1D = 1;
val_lo = 0.002; val_hi = 0.003;
N_bins = 500;
MCOPTS.N_equil = 1000; MCOPTS.N_samples = 25000;
[bin_1Dc, bin_1Dp, frac_above, frac_below] = ...
```

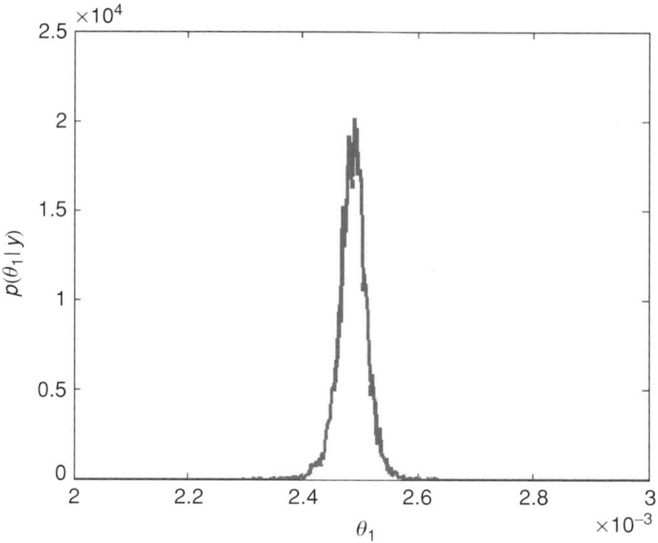

Figure 8.10 Marginal posterior density for the rate constant computed from the composite data set.

```
Bayes_MCMC_1Dmarginal_MRSL( . . .
    MRSLData, det_S_ref, . . .
    j_plot_1D, val_lo, val_hi, . . .
    N_bins, theta, MCOPTS);
% generate the 95% HPD for k1
alpha = 0.05;
[HPD_lo, HPD_hi] = Bayes_1D_HPD_MR(bin_1Dc, bin_1Dp, . . .
    j_plot_1D, alpha);
HPD_HW = 0.5*(HPD_hi-HPD_lo),
```

These calculations return the parameter estimate and the half-width of the 96% confidence interval:

$theta = 0.0025$
$HPD\_HW = 4.8e\text{-}005$

As expected, the confidence interval using both data sets is tighter than those computed using either one individually. The 1-D marginal posterior density is shown in Figure 8.10. Compare the breadth of this marginal density to that obtained from the data of Table 8.3 alone, Figure 8.9.

Bayesian testing and model criticism

Above, we have said that we cannot state with confidence that θ does *not* take any particular value that lies with a HPD credible region. While this argument is common, it falls outside of the structure of statistical decision theory. We present here the Bayesian approach to evaluating the relative merits of alternative hypotheses or models, *Bayesian testing*.

Null hypothesis testing

We begin by considering, after taking the data Y, which of two hypotheses is more probably true. Let the *null hypothesis H_0* be that $\theta \in \Omega_0 \subset \Re^P$, and let the *alternative hypothesis H_1* be that $\theta \in \Omega_1 \subset \Re^P$. Often, Ω_1 is the complement set to Ω_0 so that H_1 is the hypothesis that $\theta \notin \Omega_0$. The prior and posterior probabilities of each hypothesis being true are

$$P(H_j) = \int_{\Omega_j} p(\theta)d\theta \qquad P(H_j|Y) = \int_{\Omega_j} p(\theta|Y)d\theta \qquad (8.210)$$

where the marginalized priors and posteriors are $p(\theta)$ and $p(\theta|Y)$ respectively. For $P(H_j)$ to be defined, we again require $p(\theta)$ to be proper. We would like to control, in some sense, the effect of the prior to see what the data are telling us about which hypothesis is more likely to be true. Therefore, to compare the relative amounts by which the data increase or decrease our beliefs in the hypotheses, we compute the *Bayes factor*

$$B_{01}(Y) = \frac{P(H_0|Y)/P(H_0)}{P(H_1|Y)/P(H_1)} \qquad (8.211)$$

If $B_{01}(Y) \gg 1$, H_0 is favored by the data. If $B_{01}(Y) \ll 1$, H_1 is favored. The exact value of $B_{01}(Y)$ is quite sensitive to the choice of prior, and in particular, we cannot use improper priors here. Let us say that we have used the simple fix (8.71) of setting $p(\theta)$ to zero outside of Ω_θ and to a uniform value inside Ω_θ. Then, $P(H_j)$ is merely the ratio of the volume of Ω_j to that of Ω_θ, and is thus highly sensitive to the (subjective) choice of Ω_θ. For an improper prior with Ω_j of finite volume, $P(H_j)$ is zero and the Bayes factor is not defined. Robert (2001) discusses alternative means to make proper a prior using training sets to fit a prior and then applying Bayesian analysis to the remaining data; however, such methods fall somewhat outside of the Bayesian paradigm. For now, we retain the use of (8.71), yet note that the boundaries of Ω_θ should be selected with some care. Clearly, we should attach significance to the value of the Bayes factor only if it is far greater or far less than 1. Jeffreys (1961) suggests the following interpretation:

$$\begin{aligned}
&0 < \log_{10}[B_{01}(Y)] \le 0.5 &&\text{evidence against } H_0 \text{ is } \quad \textit{poor} \\
&0.5 < \log_{10}[B_{01}(Y)] \le 1 &&\text{evidence against } H_0 \text{ is } \quad \textit{substantial} \\
&1 < \log_{10}[B_{01}(Y)] \le 2 &&\text{evidence against } H_0 \text{ is } \quad \textit{strong} \\
&2 < \log_{10}[B_{01}(Y)] &&\text{evidence against } H_0 \text{ is } \quad \textit{decisive}
\end{aligned} \qquad (8.212)$$

It is common in frequentist statistics to consider a *point-null hypothesis* that θ takes exactly a specific value θ'. When θ is continuous, such a hypothesis has zero probability of being true and is ill posed in the Bayesian framework. We can approximate such a point-null hypothesis as $H_0: \{\theta | \|\theta - \theta'\| \le \varepsilon\}$, as long as the prior is proper.

Model criticism and selection

Similar techniques are used to judge which of several models best fits the data. Let us propose several models $\alpha = 1, 2, \ldots$ of which one is believed to be the "true" model. Let M_α be the event that model α is the true one, and let Y be the event that we observe the particular data set Y. Then, Bayes' theorem applied to the joint probability $P(M_\alpha \cap Y)$ yields the posterior probability that model α is the true one, given the data Y:

$$P(M_\alpha|Y) = \frac{P(Y|M_\alpha)P(M_\alpha)}{\sum_{\alpha'} P(Y|M_{\alpha'})P(M_{\alpha'})} \tag{8.213}$$

To remove the dependence on the priors $P(M_{\alpha'})$, we compare models α and β by computing the Bayes factor

$$B_{\alpha\beta}(Y) = \frac{P(M_\alpha|Y)/P(M_\alpha)}{P(M_\beta|Y)/P(M_\beta)} = \frac{P(Y|M_\alpha)}{P(Y|M_\beta)} \tag{8.214}$$

These probabilities $P(Y|M_\alpha)$ are the priors of the data set, under model α:

$$P(Y|M_\alpha) = \int\limits_{\Re^{P^{\langle\alpha\rangle}}} \int\limits_{\Sigma^{\langle\alpha\rangle}>0} p^{\langle\alpha\rangle}\left(Y|\theta^{\langle\alpha\rangle}, \Sigma^{\langle\alpha\rangle}\right) p^{\langle\alpha\rangle}\left(\theta^{\langle\alpha\rangle}, \Sigma^{\langle\alpha\rangle}\right) d\Sigma^{\langle\alpha\rangle} d\theta^{\langle\alpha\rangle} \tag{8.215}$$

$\theta^{\langle\alpha\rangle} \in \Re^{P^{\langle\alpha\rangle}}$ is the vector of parameters for model α and $\Sigma^{\langle\alpha\rangle}$ is the covariance matrix of the random error. The prior is $p^{\langle\alpha\rangle}(\theta^{\langle\alpha\rangle}, \Sigma^{\langle\alpha\rangle})$ and the likelihood is $p^{\langle\alpha\rangle}(Y|\theta^{\langle\alpha\rangle}, \Sigma^{\langle\alpha\rangle})$.

The integrals (8.215) can be computed by MCMC simulation. We generate a sequence $(\theta^{\langle\alpha\rangle,m}, \Sigma^{\langle\alpha\rangle,m})$ at random from a sampling density $\pi_S(\theta^{\langle\alpha\rangle}, \Sigma^{\langle\alpha\rangle})$. For a number N_s of samples, we approximate $P(Y|M_\alpha)$ as

$$P(Y|M_\alpha) \approx \frac{\sum_{m=1}^{N_s} p^{\langle\alpha\rangle}\left(Y|\theta^{\langle\alpha\rangle,m}, \Sigma^{\langle\alpha\rangle,m}\right)\dfrac{p^{\langle\alpha\rangle}\left(\theta^{\langle\alpha\rangle,m}, \Sigma^{\langle\alpha\rangle,m}\right)}{\pi_S\left(\theta^{\langle\alpha\rangle,m}, \Sigma^{\langle\alpha\rangle,m}\right)}}{\sum_{m=1}^{N_s} \dfrac{p^{\langle\alpha\rangle}\left(\theta^{\langle\alpha\rangle,m}, \Sigma^{\langle\alpha\rangle,m}\right)}{\pi_S\left(\theta^{\langle\alpha\rangle,m}, \Sigma^{\langle\alpha\rangle,m}\right)}} \tag{8.216}$$

A common choice of π_S is the integrand of (8.215). This approach is quite general, but the sampling is carried out more easily for single-response data, for which $\Sigma^{\langle\alpha\rangle} = \sigma_{\langle\alpha\rangle}$. In addition to the importance-sampling method described here, a number of other MCMC techniques tailored to the computation of Bayes factors are available, see Chen *et al.* (2000).

Schwartz's Bayesian information criterion (BIC)

Here we provide an approximation of the Bayes factor for single-response data that does not require MCMC simulation. Let the sum of squared errors for α model be

$$S^{\langle\alpha\rangle}\left(\theta^{\langle\alpha\rangle}\right) = \sum_{k=1}^{N} \left[y_k^{\langle\alpha\rangle}\left(\theta^{\langle\alpha\rangle}\right) - y_k\right]^2 \tag{8.217}$$

where the model prediction for experiment k is $y_k^{\langle\alpha\rangle}(\theta^{\langle\alpha\rangle})$ and the observed responses are stored in the vector $y \in \Re^N$. The likelihood function is

$$p^{\langle\alpha\rangle}\left(y|\theta^{\langle\alpha\rangle}, \sigma^{\langle\alpha\rangle}\right) = (2\pi)^{-N/2}\left(\sigma^{\langle\alpha\rangle}\right)^{-N} \exp\left\{-\frac{S^{\langle\alpha\rangle}\left(\theta^{\langle\alpha\rangle}\right)}{2\left(\sigma^{\langle\alpha\rangle}\right)^2}\right\} \tag{8.218}$$

and the integral (8.215) is

$$p^{\langle\alpha\rangle}(y|M_\alpha) = \int_0^\infty d\sigma^{\langle\alpha\rangle} \int_{\Re^{P^{\langle\alpha\rangle}}} p^{\langle\alpha\rangle}\left(y|\theta^{\langle\alpha\rangle}, \sigma^{\langle\alpha\rangle}\right) p^{\langle\alpha\rangle}\left(\theta^{\langle\alpha\rangle}, \sigma^{\langle\alpha\rangle}\right) d\theta^{\langle\alpha\rangle} \tag{8.219}$$

Let $\theta_{\mathrm{M}}^{\langle\alpha\rangle}$ be the least-squares estimate minimizing $S^{\langle\alpha\rangle}(\theta^{\langle\alpha\rangle})$, and let $\sigma_{\mathrm{MLE}}^{\langle\alpha\rangle}$ be the value maximizing the likelihood (8.218):

$$\left(\sigma_{\mathrm{MLE}}^{\langle\alpha\rangle}\right)^2 = \frac{1}{N} S^{\langle\alpha\rangle}\left(\theta_{\mathrm{M}}^{\langle\alpha\rangle}\right) \tag{8.220}$$

For general forms of the prior $p^{\langle\alpha\rangle}(\theta^{\langle\alpha\rangle}, (\sigma^{\langle\alpha\rangle}))$, we expect the integral over all $\sigma^{\langle\alpha\rangle}$ to be dominated by values in the vicinity of $\sigma_{\mathrm{MLE}}^{\langle\alpha\rangle}$, so that

$$p^{\langle\alpha\rangle}(y|M_\alpha) = \Delta\sigma^{\langle\alpha\rangle} \int_{\Re^{P^{\langle\alpha\rangle}}} p^{\langle\alpha\rangle}\left(y|\theta^{\langle\alpha\rangle}, \sigma_{\mathrm{MLE}}^{\langle\alpha\rangle}\right) p^{\langle\alpha\rangle}\left(\theta^{\langle\alpha\rangle}, \sigma_{\mathrm{MLE}}^{\langle\alpha\rangle}\right) d\theta^{\langle\alpha\rangle} \tag{8.221}$$

where $\Delta\sigma^{\langle\alpha\rangle}$ is a measure of the breadth of the $\sigma^{\langle\alpha\rangle}$ distribution contributing to (8.219). Let us use the prior (8.71) that is zero for $\theta^{\langle\alpha\rangle} \notin \Omega_{\theta^{\langle\alpha\rangle}}$ and/or $\sigma^{\langle\alpha\rangle} \notin \Omega_{\sigma^{\langle\alpha\rangle}}$,

$$p^{\langle\alpha\rangle}\left(\theta^{\langle\alpha\rangle}, \sigma^{\langle\alpha\rangle}\right) = c_0^{\langle\alpha\rangle}\left(\sigma^{\langle\alpha\rangle}\right)^{-1} I_{\theta^{\langle\alpha\rangle}}\left(\theta^{\langle\alpha\rangle}\right) I_{\sigma_{\langle\alpha\rangle}}\left(\sigma^{\langle\alpha\rangle}\right) \tag{8.222}$$

Upon substitution of (8.222), (8.221) becomes

$$p^{\langle\alpha\rangle}(y|M_\alpha) = \Delta_{\sigma^{\langle\alpha\rangle}} c_0^{\langle\alpha\rangle} (2\pi)^{-N/2}\left(\sigma^{\langle\alpha\rangle}\right)^{-(N+1)} Q^{\langle\alpha\rangle} \tag{8.223}$$

where

$$Q^{\langle\alpha\rangle} = \int_{\Re^{P^{\langle\alpha\rangle}}} I_{\theta^{\langle\alpha\rangle}}\left(\theta^{\langle\alpha\rangle}\right) \exp\left\{-\frac{S^{\langle\alpha\rangle}\left(\theta^{\langle\alpha\rangle}\right)}{2\left(\sigma_{\mathrm{MLE}}^{\langle\alpha\rangle}\right)^2}\right\} d\theta^{\langle\alpha\rangle} \tag{8.224}$$

Now, the value of $S^{\langle\alpha\rangle}(\theta^{\langle\alpha\rangle})$ scales roughly linearly with the number N of observations, and is larger when $\sigma_{\mathrm{MLE}}^{\langle\alpha\rangle}$ is larger. To correct for these effects and gain a better measure of the systematic departure of the model predictions from the data, let us define

$$h^{\langle\alpha\rangle}\left(\theta^{\langle\alpha\rangle}\right) = \frac{S^{\langle\alpha\rangle}\left(\theta^{\langle\alpha\rangle}\right)}{N\left(\sigma_{\mathrm{MLE}}^{\langle\alpha\rangle}\right)^2} \tag{8.225}$$

Then, (8.224) becomes

$$Q^{\langle\alpha\rangle} = \int_{\Re^{P^{\langle\alpha\rangle}}} I_{\theta^{\langle\alpha\rangle}}\left(\theta^{\langle\alpha\rangle}\right) \exp\left\{-\frac{N}{2} h^{\langle\alpha\rangle}\left(\theta^{\langle\alpha\rangle}\right)\right\} d\theta^{\langle\alpha\rangle} \tag{8.226}$$

$h^{\langle\alpha\rangle}(\theta^{\langle\alpha\rangle})$ has a minimum at the same $\theta_{\mathrm{M}}^{\langle\alpha\rangle}$ as $S^{\langle\alpha\rangle}(\theta^{\langle\alpha\rangle})$; therefore, let us use the quadratic (Laplace) expansion,

$$h^{\langle\alpha\rangle}\left(\theta^{\langle\alpha\rangle}\right) \approx h^{\langle\alpha\rangle}\left(\theta_{\mathrm{M}}^{\langle\alpha\rangle}\right) + \left(\theta^{\langle\alpha\rangle} - \theta_{\mathrm{M}}^{\langle\alpha\rangle}\right)^{\mathrm{T}} H^{\langle\alpha\rangle}\left(\theta^{\langle\alpha\rangle} - \theta_{\mathrm{M}}^{\langle\alpha\rangle}\right) \tag{8.227}$$

where $H^{\langle\alpha\rangle} > 0$ is the Hessian of $h^{\langle\alpha\rangle}(\boldsymbol{\theta}^{\langle\alpha\rangle})$ at $\boldsymbol{\theta}_{\mathrm{M}}^{\langle\alpha\rangle}$. Substituting this quadratic expansion into (8.226) and assuming $I_{\boldsymbol{\theta}^{\langle\alpha\rangle}}(\boldsymbol{\theta}^{\langle\alpha\rangle}) = 1$ everywhere where $\exp\{-\frac{1}{2}Nh^{\langle\alpha\rangle}(\boldsymbol{\theta}^{\langle\alpha\rangle})\}$ is significantly nonzero yields

$$Q^{\langle\alpha\rangle} = \exp\left\{-\tfrac{1}{2}Nh^{\langle\alpha\rangle}\left(\boldsymbol{\theta}_{\mathrm{M}}^{\langle\alpha\rangle}\right)\right\} T^{\langle\alpha\rangle} \tag{8.228}$$

where

$$T^{\langle\alpha\rangle} = \int\limits_{\mathfrak{R}^{P^{\langle\alpha\rangle}}} \exp\left\{-\frac{1}{2}\left(\boldsymbol{\theta}^{\langle\alpha\rangle} - \boldsymbol{\theta}_{\mathrm{M}}^{\langle\alpha\rangle}\right)^{\mathrm{T}}\left(NH^{\langle\alpha\rangle}\right)\left(\boldsymbol{\theta}^{\langle\alpha\rangle} - \boldsymbol{\theta}_{\mathrm{M}}^{\langle\alpha\rangle}\right)\right\} d\boldsymbol{\theta}^{\langle\alpha\rangle} \tag{8.229}$$

Using the *Gaussian integral*

$$\int\limits_{\mathfrak{R}^n} \exp\left[-\frac{1}{2}z^{\mathrm{T}}Az + \boldsymbol{b}^{\mathrm{T}}z\right] dz = \frac{(2\pi)^{P/2}}{|A|^{1/2}} \exp\left[\frac{1}{2}\boldsymbol{b}^{\mathrm{T}}A^{-1}\boldsymbol{b}\right] \tag{8.230}$$

we obtain

$$T^{\langle\alpha\rangle} = \frac{(2\pi)^{P^{\langle\alpha\rangle}/2}}{|NH^{\langle\alpha\rangle}|^{1/2}} = \left(\frac{2\pi}{N}\right)^{P^{\langle\alpha\rangle}/2}|H^{\langle\alpha\rangle}|^{-1/2} \tag{8.231}$$

Thus, from (8.228) and (8.223), we have

$$p^{\langle\alpha\rangle}(\boldsymbol{y}|M_\alpha) = C^{\langle\alpha\rangle} \exp\left\{-\frac{N}{2}h^{\langle\alpha\rangle}\left(\boldsymbol{\theta}_{\mathrm{M}}^{\langle\alpha\rangle}\right)\right\}\left(\frac{2\pi}{N}\right)^{P^{\langle\alpha\rangle}/2} \tag{8.232}$$

where

$$C^{\langle\alpha\rangle} = \frac{\Delta\sigma^{\langle\alpha\rangle}c_0^{\langle\alpha\rangle}}{(2\pi)^{N/2}(\sigma^{\langle\alpha\rangle})^{(N+1)}\left|H^{\langle\alpha\rangle}\right|^{1/2}} \tag{8.233}$$

We have then an approximation for the Bayes factor (8.214):

$$B_{\alpha\beta}(\boldsymbol{y}) = \frac{p(\boldsymbol{y}|M_\alpha)}{p(\boldsymbol{y}|M_\beta)} \approx \frac{C^{\langle\alpha\rangle} \exp\left\{-\dfrac{S^{\langle\alpha\rangle}\left(\boldsymbol{\theta}_{\mathrm{M}}^{\langle\alpha\rangle}\right)}{2\left(\sigma_{\mathrm{MLE}}^{\langle\alpha\rangle}\right)^2}\right\}\left(\dfrac{2\pi}{N}\right)^{P^{\langle\alpha\rangle}/2}}{C^{\langle\beta\rangle} \exp\left\{-\dfrac{S^{\langle\beta\rangle}\left(\boldsymbol{\theta}_{\mathrm{M}}^{\langle\beta\rangle}\right)}{2\left(\sigma_{\mathrm{MLE}}^{\langle\beta\rangle}\right)^2}\right\}\left(\dfrac{2\pi}{N}\right)^{P^{\langle\beta\rangle}/2}} \tag{8.234}$$

where we have used (8.225). Taking the natural logarithm of (8.234), we have

$$\ln[B_{\alpha\beta}(\boldsymbol{y})] \approx \ln C^{\langle\alpha\rangle} - \frac{S^{\langle\alpha\rangle}\left(\boldsymbol{\theta}_{\mathrm{M}}^{\langle\alpha\rangle}\right)}{2\left(\sigma_{\mathrm{MLE}}^{\langle\alpha\rangle}\right)^2} + \left(\frac{P^{\langle\alpha\rangle}}{2}\right)\ln\left(\frac{2\pi}{N}\right)$$
$$- \left[\ln C^{\langle\beta\rangle} - \frac{S^{\langle\beta\rangle}\left(\boldsymbol{\theta}_{\mathrm{M}}^{\langle\beta\rangle}\right)}{2\left(\sigma_{\mathrm{MLE}}^{\langle\beta\rangle}\right)^2} + \left(\frac{P^{\langle\beta\rangle}}{2}\right)\ln\left(\frac{2\pi}{N}\right)\right] \tag{8.235}$$

If we assume that $C^{\langle\alpha\rangle}$ and $C^{\langle\beta\rangle}$ are of the same order of magnitude, and define Schwartz's Bayesian Information Criterion (BIC),

$$\mathrm{BIC}^{\langle\alpha\rangle}(\boldsymbol{y}) = -\left[\frac{S^{\langle\alpha\rangle}\left(\boldsymbol{\theta}_{\mathrm{M}}^{\langle\alpha\rangle}\right)}{2\left(\sigma_{\mathrm{MLE}}^{\langle\alpha\rangle}\right)^2} + \left(\frac{P^{\langle\alpha\rangle}}{2}\right)\ln\left(\frac{N}{2\pi}\right)\right] \tag{8.236}$$

then

$$\ln[B_{\alpha\beta}(\mathbf{y})] \approx \mathrm{BIC}^{\langle\alpha\rangle}(\mathbf{y}) - \mathrm{BIC}^{\langle\beta\rangle}(\mathbf{y}) \tag{8.237}$$

This approximate result tells us that we should choose the model with the largest value of BIC. The second term in the square brackets in (8.236) adds to the weighted sum of squared errors an additional penalty per parameter, encouraging the use of models with smaller numbers of adjustable parameters. Clearly, this is only an approximation of the Bayes factor, and should be applied only when N is large. For more exact analysis, the Bayes factor should be computed by MCMC evaluation of the integrals (8.219).

Further reading

This chapter has merely introduced the subject of Bayesian statistics, a field that is far broader than the subject of estimating parameters from data with normally-distributed errors discussed here. For further reading, comprehensive graduate-level overviews of Bayesian statistics are provided by Robert (2001) and Leonard & Hsu (2001). A text suitable for undergraduates is Bolstad (2004). Among specialized texts, Box & Tiao (1973) treats in further depth the problem of parameter estimation; however, it does not discuss advanced MCMC techniques. For more on Bayesian Monte Carlo methods, consult Chen *et al.* (2000). For a more philosophical, conceptual treatment of Bayesian statistics see Bernardo & Smith (2000).

MATLAB summary

In this chapter we have addressed the Bayesian approach to estimating parameters from data that are assumed to have normally-distributed errors. The MATLAB statistics toolkit offers several functions for parameter estimation, whose results agree with both the Bayesian approach taken here and the traditional frequentist formalism. **regress** fits a linear model to single-response data and returns confidence intervals on the model parameters and predicted responses. **nlinfit** fits the parameters of a nonlinear model to single-response data, with confidence intervals on the parameters and predictions returned by **nlparci** and **nlpredci** respectively. These routines use a quadratic expansion of the sum of squared errors and thus, in general, give confidence intervals that are too wide. When the MATLAB statistics toolkit is unavailable, linear models may be treated explicitly quite easily, and for nonlinear models the MCMC routines presented here may be used (and give more accurate results without using quadratic expansions of $S(\boldsymbol{\theta})$).

For single-response data, Bayes_MCMC_pred_SR.m computes the expectation of a vector $g(\boldsymbol{\theta}, \sigma)$. This routine is used in turn by Bayes_MCMC_1Dmarginal_SR.m and Bayes_MCMC_2Dmarginal_SR.m to compute 1-D and 2-D marginal densities. The outputs of these routines are used to compute 1-D and 2-D HPD regions by Bayes_1D_HPD_SR.m and Bayes_2D_HPD_SR.m respectively.

Table 8.4 Measured values of Nu for values of Re, Pr in the laminar flow regime of forced convection through a packed bed

Nu	Re	Pr
1.9676	1	0.73
0.8986	0.1	0.73
0.4261	0.01	0.73
2.5098	1	1.5
1.1521	0.1	1.5
0.5520	0.01	1.5

The most probable parameter vector is computed from multiresponse data using sim_anneal_MR.m. The resulting marginal posterior density on θ is used to compute expectations of $g(\theta)$ by Bayes_MCMC_pred_MR.m. Marginal densities and HPD regions are computed by similar routines to those above, with _MR substituted for _SR.

Parameters in a nonlinear model are fit to a composite data set of single- and/or multiresponse data by sim_anneal_MSRL.m. Bayes_MCMC_pred_MSRL.m computes posterior estimates, and is used by Bayes_MCMC_1Dmarginal_MRSL.m and Bayes_MCMC_2Dmarginal_MRSL.m to compute 1-D and 2-D marginal posterior densities. The output from these routines can be used with the _MR HPD algorithms to generate HPD regions.

Problems

8.A.1. We are studying a system in which a fluid flows slowly through a packed bed of solid pellets, and are interested in the transfer of heat between the solid pellets and the fluid. We expect the heat transfer coefficient h to have the dependence

$$h = h(v_f, D, \rho, \mu, k, \hat{C}_p) \tag{8.238}$$

where v_f is the superficial velocity of the fluid, D is the pellet diameter, ρ is the fluid density, μ is the fluid viscosity, k is the fluid thermal conductivity, and $\hat{C}_p$ is the specific heat of the fluid. Through dimensionless analysis, we write this dependence as

$$Nu = Nu(Re, Pr) \tag{8.239}$$

where the Nusselt, Prandtl, and Reynolds numbers are

$$Nu = \frac{hD}{k} \quad Pr = \frac{\mu \hat{C}_p}{k} \quad Re = \frac{\rho v_f D}{\mu} \tag{8.240}$$

We have taken the data of Table 8.4 in the laminar flow regime. We propose the model

$$Nu = \alpha_0 (Re)^{\alpha_1} (Pr)^{\alpha_2} \tag{8.241}$$

Table 8.5 Measured substrate concentrations vs. time in a batch bioreactor

time (min)	$[S]_0 = 0.5\,M$	$[S]_0 = 0.75\,M$	$[S]_0 = 1\,M$	$[S]_0 = 1.5\,M$	$[S]_0 = 2\,M$
10	0.4288	0.6735	0.9299	1.4175	1.9265
20	0.3554	0.6048	0.8504	1.3615	1.8773
30	0.2701	0.5089	0.7767	1.2832	1.8091
45	0.1827	0.3977	0.6652	1.1763	1.7191
60	0.1210	0.2893	0.5448	1.0999	1.6278
90	0.0196	0.1064	0.3030	0.8661	1.4420

To obtain a linear model, we take the base-10 logarithm,

$$\log_{10} Nu = \log_{10} \alpha_0 + \alpha_1 \log_{10} Re + \alpha_2 \log_{10} Pr \tag{8.242}$$

Set up the system of linear algebraic equations that is solved to obtain the least-squares parameter estimate. Then, solve this system by Gaussian elimination. Provide 95% confidence intervals for each of the model parameters. Do all calculations by hand and show complete results.

8.A.2. Repeat problem 8.A.1, but now use **regress**.

8.A.3. Fit again the heat transfer data of problem 8.A.1, but now do not transform the data to make the model linear. Compute the fitted parameter estimate and the 95% confidence intervals using the nontransformed nonlinear model.

8.A.4. Compute more accurate 95% confidence intervals for the data of problem 8.A.1, without taking a quadratic expansion of $S(\boldsymbol{\theta})$, through MCMC simulation.

8.B.1. We are studying the enzymatic conversion of a substrate S to a product P. For several batch reactor kinetic experiments at the same temperature and pH, but at different initial substrate concentrations $[S]_0$, we have measured $[S]$ vs. time (Table 8.5). The reactor has a fluid volume V_R of 0.1 l and contains a mass m_E of 10 mg of enzyme. A balance on the substrate yields the governing ODE

$$\frac{d}{dt}[S] = -\left(\frac{m_E}{\alpha_c V_R}\right) \hat{r}_{S \to P} \tag{8.243}$$

where $\alpha_c = 10^6\ \mu\text{mol/mol}$ is a conversion factor and $\hat{r}_{S \to P}$ is the rate of substrate conversion in micromoles per minute per gram of enzyme. We propose a Michelis–Menten model with substrate inhibition,

$$\hat{r}_{S \to P} = \frac{V_m[S]}{K_m + [S] + K_{si}^{-1}[S]^2} \tag{8.244}$$

From these data, find the most probable parameter values and use MCMC simulation to generate 1-D 95% HPD credible regions for each one.

8.B.2. In problem 8.B.1, we fit the data only to measurements of the substrate conversion as a function of time, but let us say that we also measure the product concentrations (Table 8.6).

Table 8.6 Measured product concentrations vs. time in a batch bioreactor

time (min)	$[S]_0 = 0.5$ M	$[S]_0 = 0.75$ M	$[S]_0 = 1$ M	$[S]_0 = 1.5$ M	$[S]_0 = 2$ M
10	0.0712	0.0695	0.0653	0.0767	0.0656
20	0.1476	0.1576	0.1531	0.1461	0.1123
30	0.2265	0.2382	0.2274	0.2202	0.1912
45	0.3120	0.3423	0.3450	0.3116	0.2821
60	0.3770	0.4671	0.4599	0.4022	0.3682
90	0.4722	0.6402	0.6833	0.6379	0.5502

Table 8.7 Additional rate data for enzymatic substrate conversion reaction

[S] (M)	rate of substrate conversion (μmol/(min mg$_E$))
0.1	36.512
0.25	63.224
0.5	76.881
0.75	77.607
1.0	74.444
1.5	65.001
2.0	58.391

From stoichiometry, we expect the instantaneous substrate and product concentrations to be related by

$$[S]_0 - [S] = [P] \tag{8.245}$$

Due to random measurement errors, (8.245) may not be satisfied exactly, but still it is likely that concurrent measurements of the two concentrations are highly correlated. Using the data from Table 8.5 and Table 8.6, find the best fit of the parameters and generate 1-D 95% HPD credible regions for each.

8.B.3. Using the bioreactor data of Table 8.5 and Table 8.6, compute the probability that the reaction rate (8.244) at a substrate concentration of 1 M is between 70 and 80 μmol/(min mg$_E$).

8.B.4. You have available additional rate data (Table 8.7) for this substrate conversion reaction at the same conditions, but from experiments that were conducted in a different apparatus and so may have a different level of accuracy than those of Table 8.5 and Table 8.6. Using all available data, find the most probable values of the parameters and generate 1-D 95% HPD credible regions for each parameter.

8.C.1. Using all available data on the enzymatic kinetics of the substrate conversion reaction, generate the probability distribution for the value of the reaction rate at a concentration of 1 M.

8.C.2. From the substrate conversion rate data, we would like to design a CSTR bioreactor by choosing the inlet substrate conversion and inlet volumetric flow rate, within the limits

$$0\,M \le [S]_0 \le 2\,M \quad 0\,\frac{l}{\min} \le v \le 1\,\frac{l}{\min} \tag{8.246}$$

that maximize the conversion rate in the bioreactor. If the uncertainty in the parameter values is not too great, we get a reasonable design by simply using the most probable parameter values. However, this approach completely neglects the uncertainty in the parameters themselves. Using statistical decision theory, compute the optimal inlet substrate concentration and inlet flow rate, taking into account the effect of uncertainty.

8.C.3. Using the single-response data of Table 8.5 for the enzymatic substrate conversion reaction, compute the Bayes factor for comparing the following models: (1) Michaelis–Menten kinetics with substrate inhibition and (2) Michelis–Menten kinetics with no substrate inhibition. Do the data strongly suggest that substrate inhibition is significant for this system?

9 Fourier analysis

Fourier analysis treats the representation of periodic functions as linear combinations of sine and cosine basis functions. In chemical engineering, Fourier analysis is applied to study time-dependent signals in spectroscopy and to analyze the spatial structure of materials from scattering experiments. Here, the basic foundation of Fourier analysis is presented, with an emphasis upon implementation in MATLAB.

Fourier series and transforms in one dimension

We begin our discussion of Fourier analysis by considering the representation of a *periodic function* $f(t)$ with a *period* of $2P$, $f(t + 2P) = f(t)$. If $f(t)$ has a finite number of local extrema and a finite number of times $t_j \in [0, 2P]$ at which it is discontinuous, *Dirichlet's theorem* states that it may be represented as the *Fourier series*

$$\tilde{f}(t) = \frac{1}{2}a_0 + \sum_{m=1}^{\infty} \left[a_m \cos\left(\frac{m\pi t}{P}\right) + b_m \sin\left(\frac{m\pi t}{P}\right) \right] \tag{9.1}$$

such that at all t' where $f(t)$ is continuous, $\tilde{f}(t') = f(t')$, and at all points t_j where $f(t)$ is discontinuous, $\tilde{f}(t_j)$ is the average of the right-and left-hand limits:

$$\tilde{f}(t_j) = \frac{f_+(t_j) + f_-(t_j)}{2} \quad f_+(t_j) = \lim_{\varepsilon \to 0} f(t_j + \varepsilon) \quad f_-(t_j) = \lim_{\varepsilon \to 0} f(t_j - \varepsilon) \tag{9.2}$$

a_0, $\{a_1, a_2, \ldots\}$, and $\{b_1, b_2, \ldots\}$ are calculated using the orthogonality properties of sine and cosine functions. First, to compute a_0, we integrate $f(t)$ over the period $[0, 2P]$, and do the same for $\tilde{f}(t)$:

$$\int_0^{2P} f(t)dt = \int_0^{2P} \tilde{f}(t)dt$$

$$= \int_0^{2P} \frac{a_0}{2}dt + \sum_{m=1}^{\infty} \int_0^{2P} \left[a_m \cos\left(\frac{m\pi t}{P}\right) + b_m \sin\left(\frac{m\pi t}{P}\right) \right] dt \tag{9.3}$$

As the cosine and sine terms integrate to 0,

$$a_0 = \frac{1}{P} \int_0^{2P} f(t)\, dt \tag{9.4}$$

Next, we compute a_n, $n = 1, 2, 3, \ldots$ by multiplying both $f(t)$ and $\tilde{f}(t)$ by $\cos(n\pi t/P)$ and integrating over $[0, 2P]$:

$$\int_0^{2P} f(t) \cos\left(\frac{n\pi t}{P}\right) dt = \int_0^{2P} \tilde{f}(t) \cos\left(\frac{n\pi t}{P}\right) dt \tag{9.5}$$

Using the orthogonality properties

$$\int_0^{2P} \cos\left(\frac{m\pi t}{P}\right) \cos\left(\frac{n\pi t}{P}\right) dt = \int_0^{2P} \sin\left(\frac{m\pi t}{P}\right) \sin\left(\frac{n\pi t}{P}\right) dt = P\delta_{mn}$$

$$\int_0^{2P} \sin\left(\frac{m\pi t}{P}\right) \cos\left(\frac{n\pi t}{P}\right) dt = 0 \tag{9.6}$$

we obtain

$$a_n = \frac{1}{P} \int_0^{2P} f(t) \cos\left(\frac{n\pi t}{P}\right) dt \qquad n = 1, 2, 3, \ldots \tag{9.7}$$

A similar procedure, but multiplying by $\sin(n\pi t/P)$ instead, yields

$$b_n = \frac{1}{P} \int_0^{2P} f(t) \sin\left(\frac{n\pi t}{P}\right) dt \qquad n = 1, 2, 3, \ldots \tag{9.8}$$

The summations above are over an infinite number of terms, but if we truncate the series to some finite order N, we obtain an approximate Fourier representation of the function:

$$f(t) \approx \frac{1}{2}a_0 + \sum_{m=1}^{N} \left[a_m \cos\left(\frac{m\pi t}{P}\right) + b_m \sin\left(\frac{m\pi t}{P}\right) \right] \tag{9.9}$$

To compute the $2N + 1$ coefficients of this expansion, we might consider using numerical quadrature for the necessary integrals; however, as N increases, so does the required number of quadrature points, as the sine and cosine basis functions vary more rapidly with increasing m. Thus, the amount of work necessary to obtain an approximate Fourier representation in this manner scales as N^2. Below, we consider an alternative method that requires only $N \log_2 N \ll N^2$ operations.

Gibbs oscillations

Convergence of the Fourier representation to the true function $f(t)$ with increasing N can be quite slow, particularly when the function is discontinuous or varies rapidly over a small interval. As an example, consider the square pulse function

$$f(t) = \begin{cases} 1, & \pi/2 \leq t \leq 3\pi/2 \qquad P = \pi \\ 0, & t < \pi/2 \text{ or } t > 3\pi/2 \quad f(t + 2\pi) = f(t) \end{cases} \tag{9.10}$$

This function, and its approximate Fourier representations for $N = 10, 20$, are shown in Figure 9.1. The Fourier series representations exhibit artificial *Gibbs oscillations* that are not found in the true square pulse function.

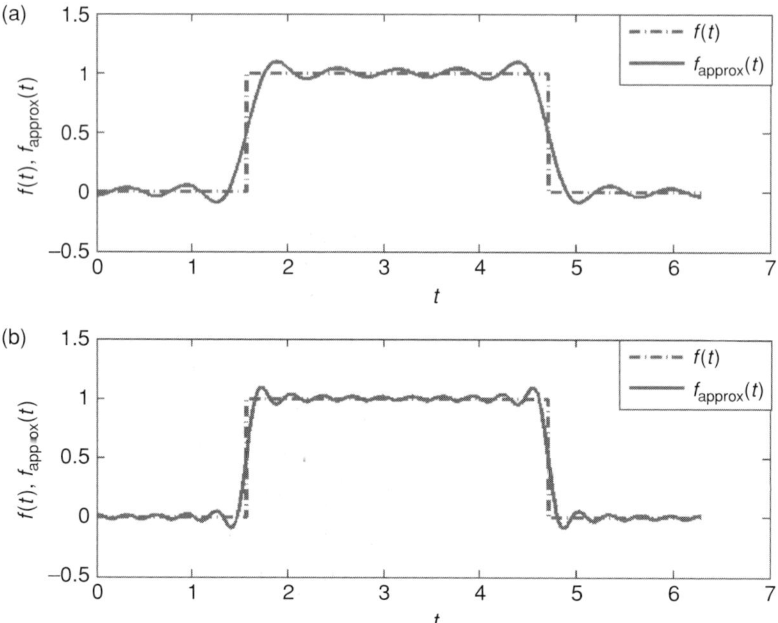

Figure 9.1 Approximate Fourier representations of a square pulse showing Gibbs oscillations for: (a) $N = 10$ and (b) $N = 20$.

Exponential form of the Fourier series

In practice, it is more convenient to write the Fourier series in terms of complex exponential functions, using *Euler's formula*

$$e^{i\theta} = \cos\theta + i\sin\theta \tag{9.11}$$

from which we obtain

$$\cos\theta = \frac{1}{2}[e^{i\theta} + e^{-i\theta}] \qquad \sin\theta = \frac{1}{2i}[e^{i\theta} - e^{-i\theta}] \tag{9.12}$$

Substituting (9.12) for the cosine and sine terms in (9.1), we obtain

$$\tilde{f}(t) = \frac{1}{2}a_0 + \sum_{m=1}^{\infty}\left(\frac{a_m}{2} + \frac{b_m}{2i}\right)e^{im\pi t/P} + \sum_{m=1}^{\infty}\left(\frac{a_m}{2} - \frac{b_m}{2i}\right)e^{-im\pi t/P} \tag{9.13}$$

Collecting terms, we have the *exponential-form Fourier series*

$$\tilde{f}(t) = \sum_{m=-\infty}^{\infty} c_m\, e^{im\pi t/P} \tag{9.14}$$

with the coefficients (complex-valued)

$$c_0 = \frac{1}{2}a_0 \qquad c_{m>0} = \frac{a_m}{2} + \frac{b_m}{2i} \qquad c_{m<0} = \frac{a_m}{2} - \frac{b_m}{2i} \tag{9.15}$$

These coefficients may be computed directly from $f(t)$ through use of the orthogonality property

$$\int_{-P}^{+P} e^{-im\pi t/P} e^{in\pi t/P} dt = 2P\delta_{mn} \tag{9.16}$$

to obtain

$$c_n = \frac{1}{2P} \int_{-P}^{+P} e^{-in\pi t/P} f(t)dt \tag{9.17}$$

The Fourier transform

The Fourier series representation assumes a known period $2P$; however, often we wish to analyze a function whose periodicity is unknown. From the exponential form of the Fourier series, we obtain the Fourier transform by taking the limit $P \to \infty$. We begin by writing the Fourier series as

$$\tilde{f}(t) = \sum_{m=-\infty}^{\infty} c_m e^{im\pi t/P} = \sum_{m=-\infty}^{\infty} \left[\frac{1}{2P} \int_{-P}^{+P} e^{-im\pi t'/P} f(t')dt' \right] e^{im\pi t/P} \tag{9.18}$$

We now define

$$\omega_m = \frac{m\pi}{P} \qquad \Delta\omega = \omega_{m+1} - \omega_m = \frac{\pi}{P} \tag{9.19}$$

such that

$$\frac{1}{2P} = \frac{1}{2\pi} \left(\frac{\pi}{P} \right) = \frac{1}{2\pi} \Delta\omega$$

and thus

$$\tilde{f}(t) = \sum_{m=-\infty}^{\infty} \frac{1}{2\pi} \left[\int_{-P}^{+P} e^{-i\omega_m t'} f(t')dt' \right] e^{i\omega_m t} \Delta\omega \tag{9.20}$$

In the limit $P \to \infty$, $\Delta\omega \to 0$, and the summation becomes an integral,

$$\sum_{m=-\infty}^{\infty} F(\omega_m)\Delta\omega \to \int_{-\infty}^{+\infty} F(\omega)d\omega \tag{9.21}$$

Thus, assuming that $\tilde{f}(t) = f(t)$, we obtain in the limit $P \to \infty$, the relation

$$f(t) = \frac{1}{2\pi} \int_{-\infty}^{+\infty} \left[\int_{-\infty}^{+\infty} e^{-i\omega t'} f(t')dt' \right] e^{i\omega t} d\omega \tag{9.22}$$

This relation is satisfied by the *Fourier transform pair*

$$F(\omega) = \frac{1}{\sqrt{2\pi}} \int_{-\infty}^{+\infty} f(t)e^{-i\omega t} dt \qquad f(t) = \frac{1}{\sqrt{2\pi}} \int_{-\infty}^{+\infty} F(\omega)e^{i\omega t} d\omega \tag{9.23}$$

One can propose alternative definitions that also satisfy (9.22), such as

$$F(\omega) = \int_{-\infty}^{+\infty} f(t)e^{i\omega t}\,dt \qquad f(t) = \frac{1}{2\pi}\int_{-\infty}^{+\infty} F(\omega)e^{-i\omega t}\,d\omega \tag{9.24}$$

As long as one is consistent, the convention used for the Fourier transform is arbitrary. Here, we use the definition consistent with MATLAB, (9.23). Note that as $F(0) = \left(1/\sqrt{2\pi}\right)\int_{-\infty}^{+\infty} f(t)dt$, we assume that $f(t)$ integrates to zero.

The discrete Fourier transform

We next consider computing the Fourier transform from the N sampled data $\{f_1, f_2, \ldots, f_N\}$, $f_k = f(t_k)$, taken at the N uniform times $t_k = (k-1)\Delta t$. If we have samples taken at nonuniform times, we use interpolation to generate the values of $f(t)$ at time values that *are* uniformly-spaced before computing the Fourier transform. We have no data outside of the time period $[0, (N-1)\Delta t]$, so let us assume that $f(t)$ is periodic outside of this range. As generally $f_1 \neq f_N$, the period is

$$2P = N(\Delta t) \tag{9.25}$$

We wish to compute the Fourier transform

$$F(\omega) = \frac{1}{\sqrt{2\pi}}\int_{-\infty}^{+\infty} f(t)e^{-i\omega t}\,dt \tag{9.26}$$

but as we have only N pieces of information $\{f_1, f_2, \ldots, f_N\}$, we can determine $F(\omega)$ independently at only N different frequencies ω_n. What are the frequencies for which we compute $F(\omega)$? First, if $f(t+2P) = f(t)$, we can represent $f(t)$ as the Fourier series

$$f(t) = \sum_{m=-\infty}^{\infty} c_m e^{i\omega_m t} \qquad \omega_m = \frac{m\pi t}{P} \tag{9.27}$$

Thus, $F(\omega)$ takes the form

$$F(\omega) = A \sum_{m=-\infty}^{\infty} F(\omega_m)\delta(\omega - \omega_m) \tag{9.28}$$

as is shown by substitution into the inverse Fourier transform:

$$f(t) = \frac{1}{\sqrt{2\pi}}\int_{-\infty}^{+\infty} F(\omega)e^{i\omega t}\,d\omega = \frac{1}{\sqrt{2\pi}}\int_{-\infty}^{+\infty}\left[A\sum_{m=-\infty}^{\infty} F(\omega_m)\delta(\omega - \omega_m)\right]e^{i\omega t}\,d\omega$$

$$f(t) = \sum_{m=-\infty}^{\infty}\left(\frac{A}{\sqrt{2\pi}}\right)F(\omega_m)\int_{-\infty}^{+\infty}\delta(\omega - \omega_m)e^{i\omega t}\,d\omega = \sum_{m=-\infty}^{\infty}\left[\left(\frac{A}{\sqrt{2\pi}}\right)F(\omega_m)\right]e^{i\omega_m t}$$

$$\tag{9.29}$$

Thus, for a function with $f(t+2P) = f(t)$, the Fourier transform is nonzero only at the discrete frequencies

$$\omega_m = m(\Delta\omega) \qquad \Delta\omega = \frac{m\pi}{P} \qquad m = 0, \pm1, \pm2, \ldots \tag{9.30}$$

While there are an infinite number of such frequencies, we have available only N data $\{f_1, f_2, \ldots, f_N\}$. Let us choose the $\{\omega_n\}$ to be the lowest ones satisfying (9.30),

$$\omega_m = m(\Delta\omega) \qquad \Delta\omega = \frac{\pi}{P} \qquad m = 0, \pm1, \pm2, \ldots, \pm\left(\frac{N}{2} - 1\right), +\frac{N}{2} \tag{9.31}$$

The frequency resolution $\Delta\omega$ is determined by the length of the sampling period $2P$. The highest frequency that we resolve from the sampled data is

$$\omega_{\max} = \left(\frac{N}{2}\right)\Delta\omega = \frac{N\pi}{2P} = \frac{\pi}{(2P/N)} = \frac{\pi}{\Delta t} \tag{9.32}$$

To resolve higher frequencies, we need to sample $f(t)$ more often (decrease Δt). If $F(\omega) = 0$ for all $|\omega| > \omega_{\max}$, $f(t)$ is said to be *bandwidth-limited*, and the sampled data are sufficient to characterize $F(\omega)$. Otherwise, if $F(\omega) \neq 0$ for some $|\omega| > \omega_{\max}$, there exist high-frequency components of $f(t)$ that are sampled incorrectly and that can corrupt the values of $f(\omega_m)$. This is known as *aliasing*, a subject that we discuss later in more detail.
To obtain the normalization constant A in (9.28), we match

$$\int_{-\infty}^{+\infty} F(\omega)G(\omega)d\omega = A \sum_{m=-\infty}^{\infty} F(\omega_m) \int_{-\infty}^{+\infty} \delta(\omega - \omega_m)G(\omega)d\omega$$

$$= A \sum_{m=-\infty}^{\infty} F(\omega_m)G(\omega_m) \tag{9.33}$$

to the quadrature formula

$$\int_{-\infty}^{+\infty} F(\omega)G(\omega)d\omega \approx \sum_{m=-\infty}^{\infty} F(\omega_m)G(\omega_m)(\Delta\omega) \tag{9.34}$$

to yield

$$A = \Delta\omega = \frac{\pi}{P} \tag{9.35}$$

From (9.17) and (9.29), we have

$$\frac{(\Delta\omega)}{\sqrt{2\pi}} F(\omega_m) = c_m = \frac{1}{2P} \int_0^{2P} f(t)e^{-i\omega_m t}dt \tag{9.36}$$

Using

$$\frac{(\sqrt{2\pi})}{2P} \frac{1}{(\Delta\omega)} = \frac{(\sqrt{2\pi})}{2P} \frac{P}{\pi} = \frac{1}{\sqrt{2\pi}}$$

the Fourier transform values are

$$F(\omega_m) = \frac{1}{\sqrt{2\pi}} \int_0^{2P} f(t)e^{-i\omega_m t}\,dt \tag{9.37}$$

Applying quadrature, we write (9.37) as

$$F(\omega_m) \approx \frac{1}{\sqrt{2\pi}} \sum_{k=1}^{N} f(t_k) e^{-i\omega_m t_k}(\Delta t) = \frac{(\Delta t)}{\sqrt{2\pi}} \sum_{k=1}^{N} f_k e^{-im(\Delta\omega)(k-1)(\Delta t)} \tag{9.38}$$

We next show that the $F(\omega_m)$ values have ω-periodicity,

$$F(\omega_m + l2\omega_{\max}) = F(\omega_m) \qquad l = 0, \pm 1, \pm 2, \ldots \tag{9.39}$$

Using (9.31) and (9.32),

$$\omega_m + l2\omega_{\max} = m(\Delta\omega) + l2\left(\frac{N}{2}\right)(\Delta\omega) = [m + lN](\Delta\omega) \tag{9.40}$$

We next use (9.38) and $e^{a+b} = e^a e^b$ to obtain

$$F(\omega_m + l2\omega_{\max}) = \frac{(\Delta t)}{\sqrt{2\pi}} \sum_{k=1}^{N} f_k e^{-im(\Delta\omega)(k-1)(\Delta t)} e^{-ilN(\Delta\omega)(k-1)(\Delta t)} \tag{9.41}$$

Now, as

$$(\Delta\omega)(\Delta t) = \left(\frac{\pi}{P}\right)\left(\frac{2P}{N}\right) = \frac{2\pi}{N}$$

we have

$$e^{-ilN(\Delta\omega)(k-1)(\Delta t)} = e^{-ilN(k-1)(2\pi/N)} = e^{-i(l2\pi)(k-1)} \tag{9.42}$$

Using $e^{ab} = (e^a)^b$, this becomes

$$e^{-i(l2\pi)(k-1)} = \left[e^{-i(l2\pi)}\right]^{(k-1)} = [\cos(-l2\pi) + i\sin(-l2\pi)]^{(k-1)} = 1 \tag{9.43}$$

Therefore, we have ω-periodicity,

$$F(\omega_m + l2\omega_{\max}) = \frac{(\Delta t)}{\sqrt{2\pi}} \sum_{k=1}^{N} f_k e^{-im(\Delta\omega)(k-1)(\Delta t)} = F(\omega_m) \tag{9.44}$$

Rather than evaluating $F(\omega)$ at the frequencies $\omega_m = m(\Delta\omega)$ in $(-\omega_{\max}, \omega_{\max})$, we thus could instead evaluate $F(\omega)$ at $\omega_n = (n-1)(\Delta\omega)$ in $[0, 2\omega_{\max})$, and then obtain the values at "negative" frequencies using $F(\omega_n - 2\omega_{\max}) = F(\omega_n)$. We then have

$$F(\omega_n) \approx \frac{(\Delta t)}{\sqrt{2\pi}} \sum_{k=1}^{N} f_k e^{-i(n-1)(\Delta\omega)(k-1)(\Delta t)} \qquad \omega_n = \frac{(n-1)\pi}{P} \tag{9.45}$$

Again using $e^{ab} = (e^a)^b$,

$$F(\omega_n) \approx \frac{(\Delta t)}{\sqrt{2\pi}} \sum_{k=1}^{N} f_k \left[e^{-i(\Delta\omega)(\Delta t)}\right]^{(n-1)(k-1)} \tag{9.46}$$

As $(\Delta\omega)(\Delta t) = 2\pi/N$, we have $e^{-i(\Delta\omega)(\Delta t)} = e^{-i2\pi/N} \equiv W$ and thus

$$F(\omega_n) \approx \frac{(\Delta t)}{\sqrt{2\pi}} \sum_{k=1}^{N} f_k W^{(n-1)(k-1)} \tag{9.47}$$

From these results, we define the *discrete Fourier transform*

$$F_n = \sum_{k=1}^{N} f_k W^{(n-1)(k-1)} \qquad W = e^{-i2\pi/N} \qquad n = 1, 2, \ldots, N \tag{9.48}$$

such that

$$F(\omega_n) \approx \frac{(\Delta t)}{\sqrt{2\pi}} F_n \qquad \omega_n = \frac{(n-1)\pi}{P} \tag{9.49}$$

Similarly, we approximate the inverse Fourier transform

$$f(t) = \frac{1}{\sqrt{2\pi}} \int_{-\infty}^{+\infty} F(\omega)e^{i\omega t}\, d\omega \rightarrow \frac{1}{\sqrt{2\pi}} \int_{0}^{2\omega_{\max}} F(\omega)e^{i\omega t}\, d\omega \tag{9.50}$$

by

$$f(t) \approx \frac{1}{\sqrt{2\pi}} \sum_{n=1}^{N} F(\omega_n)e^{i\omega_n t}(\Delta\omega) \tag{9.51}$$

Substituting for $F(\omega_n)$ and ω_n, we have at $t_k = (k-1)\Delta t$

$$f(t_k) \approx \left[\frac{(\Delta\omega)}{\sqrt{2\pi}}\right]\left[\frac{(\Delta t)}{\sqrt{2\pi}}\right] \sum_{n=1}^{N} F_n e^{i(n-1)(\Delta\omega)(k-1)(\Delta t)} \tag{9.52}$$

Using again $e^{ab} = (e^a)^b$ and $(\Delta\omega)(\Delta t) = 2\pi/N$,

$$f(t_k) \approx \frac{1}{N} \sum_{n=1}^{N} F_n \left[e^{i(\Delta\omega)(\Delta t)}\right]^{(n-1)(k-1)} = \frac{1}{N} \sum_{n=1}^{N} F_n [W^*]^{(n-1)(k-1)} \tag{9.53}$$

where $W^* = [e^{-i2\pi/N}]^* = e^{i2\pi/N} = e^{i(\Delta\omega)(\Delta t)}$. These relations define the *discrete Fourier transform pair*

$$F_n = \sum_{k=1}^{N} f_k W^{(n-1)(k-1)} \qquad f_k = \frac{1}{N} \sum_{n=1}^{N} F_n [W^*]^{(n-1)(k-1)}$$
$$W = e^{-i2\pi/N} \qquad n = 1, 2, \ldots, N \tag{9.54}$$

that is related to the original, continuous Fourier transform pair by

$$F(\omega_n) \approx \frac{(\Delta t)}{\sqrt{2\pi}} F_n \qquad \omega_n = \frac{(n-1)\pi}{P} \qquad f(t_k) = f_k \qquad t_k = (k-1)\Delta t \tag{9.55}$$

The fast Fourier transform (FFT)

Computing the discrete Fourier transform directly from (9.48) requires a number of operations that scales as N^2. Here, we present an alternative decimation procedure that scales only as $N\log_2 N \ll N^2$. Let us break the summation over k into contributions from the even and odd terms,

$$F_n = \sum_{\substack{k=1 \\ k\ \text{even}}}^{N} f_k W^{(n-1)(k-1)} + \sum_{\substack{k=1 \\ k\ \text{odd}}}^{N} f_k W^{(n-1)(k-1)} \tag{9.56}$$

Let us extract the even and odd f_k as

$$f_l^{(e)} = f_{k=2l} \qquad f_l^{(o)} = f_{k=2l-1} \qquad l = 1, 2, \ldots, N/2 \tag{9.57}$$

such that (9.56) becomes

$$F_n = \sum_{l=1}^{N/2} f_l^{(e)} W^{(n-1)(2l-1)} + \sum_{l=1}^{N/2} f_l^{(o)} W^{(n-1)(2l-2)} \tag{9.58}$$

Using the relations for the even and odd contributions respectively,

$$W^{(n-1)(2l-1)} = W^{(n-1)} (W^2)^{(n-1)(l-1)} \qquad W^{(n-1)(2l-2)} = (W^2)^{(n-1)(l-1)} \tag{9.59}$$

where $W^2 = [e^{-i(\Delta\omega)(\Delta t)}]^2 = e^{-i(\Delta\omega)(2\Delta t)}$, (9.58) becomes

$$F_n = W^{(n-1)} \sum_{l=1}^{N/2} f_l^{(e)}(W^2)^{(n-1)(l-1)} + \sum_{l=1}^{N/2} f_l^{(o)}(W^2)^{(n-1)(l-1)} \tag{9.60}$$

Comparing (9.60) to (9.48), we see that the two summations contributing to F_n themselves are discrete Fourier transforms, but each over only half of the data sampled at an interval $2\Delta t$. If $N = 2^\kappa$, then after this partition, we have two sums over $2^{\kappa-1}$ terms. Performing a similar partition on each of these, we have four sums, each over only one fourth of the data. If we recursively perform this partition $\kappa = \log_2 N$ times, we have "sums" of only one term. The *FFT* algorithm applies this decimation approach to obtain the values of all F_n after only $N\log_2 N \ll N^2$ operations. The computation savings offered by FFT are so significant that many common applications of Fourier analysis would be prohibitively expensive if FFT were not available.

While the decimation procedure may still be applied if N is not a power of 2, the additional book-keeping is expensive. Therefore, it is common in this case to "pad" the data set with zeros until we obtain a total of $2^\kappa > N$ points. The only change in the Fourier transform due to this padding is that the frequencies are now

$$\omega_n = (n-1)\Delta\omega \qquad \Delta\omega = \frac{\pi}{P+Q} \qquad Q = \frac{(2^\kappa - N)(\Delta t)}{2} \tag{9.61}$$

$2Q$ is the time duration of the "padding" of the signal $f(t)$ with zeros.

Special properties of the Fourier transform of a real function and the power spectrum

For a real function with $f^*(t) = f(t)$, we have the property

$$[F(-\omega)]^* = \frac{1}{\sqrt{2\pi}} \int_{-\infty}^{+\infty} f^*(t)e^{+i(-\omega)t}\, dt = \frac{1}{\sqrt{2\pi}} \int_{-\infty}^{+\infty} f(t)e^{-i\omega t}\, dt = F(\omega) \tag{9.62}$$

This holds as well for the Fourier transform computed from sampled data:

$$[F(-\omega_n)]^* = F(\omega_n) \tag{9.63}$$

Using $F(\omega_m + 12\omega_{\max}) = F(\omega_m)$, the discrete Fourier transform (9.48) satisfies

$$F_{N-m}^* = F_m \tag{9.64}$$

Thus, we can extract all of the independent information of the Fourier transform of a sampled real signal $f(t)$ from only one half of the discrete Fourier transform values, say the first half.

Even for a real signal $f(t)$, the Fourier transform $F(\omega)$ is complex; therefore, it is common practice to compute the real-valued, nonnegative *power spectrum* $|F(\omega)|^2$ vs. ω. The quantity $|F(\omega)|^2 \geq 0$ is a measure of the contribution to $f(t)$ from dynamics with a frequency ω. The symmetry $F^*_{N-m} = F_m$ leads to the property $|F(2\omega_{\max} - \omega_n)|^2 = |F(\omega_n)|^2$, and thus all independent information in the power spectrum is contained in the lower-frequency half $0 \leq \omega \leq \omega_{\max}$.

1-D Fourier transforms in MATLAB

The discrete Fourier transform and its inverse (9.54) are computed in MATLAB using **fft** and **ifft** respectively. The following code demonstrates their use for the signal $f(t) = \sin(t) + 2\cos(2t)$, which has two peaks in the power spectrum at $\omega = 1$ and $\omega = 2$. The signal is sampled at uniform times over the period $[0, 12\pi]$, such that $P = l\pi$ and $\Delta t = 2P/N$, where $N = 2^k$ for some integer e.

```
% generate sampled f(t) data
N = 2^8; P = 10*pi; dt = (2*P)/N;
t_val = linspace(0,2*P-dt,N);
f_val = sin(t_val) + 2*cos(2*t_val);
% compute discrete FT
F_DFT = fft(f_val);
% generate sampled frequencies
omega_max = pi/dt; d_omega = pi/P;
omega_val = linspace(0,2*omega_max-d_omega,N);
% compute continuous FT and power spectrum
F_FT = dt/sqrt(2*pi)*F_DFT; F_PS = abs(F_FT);
% plot signal and power spectrum
figure; subplot(2,1,1); plot(t_val,f_val);
xlabel('t'); ylabel('f(t)');
title('f(t) and |F(\omega)|');
subplot(2,1,2); plot(omega_val,F_PS);
xlabel('\omega'); ylabel('|F (\omega)|');
```

Plots of the time signal $f(t)$ and $|F(\omega)|$ are shown in Figure 9.2. There appear two peaks at $\omega = 1$ and $\omega = 2$ with a 2:1 relative intensity. These peaks are repeated again at high frequencies due to the ω-symmetry for a real signal, $F^*_{N-m} = F_m$. Thus, we obtain all the useful information from the low-frequency half $\omega \in [0, \omega_{\max}]$. From the discrete Fourier transform, the time signal can be reconstituted from the inverse discrete FT function **ifft**,

```
f_2 = real(ifft(F_DFT));
```

The **real()** operation is done to remove any near-zero imaginary contributions that are introduced due to numerical error.

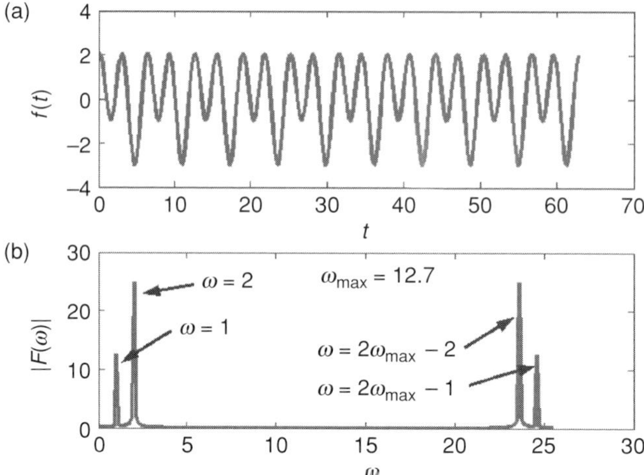

Figure 9.2 (a) The sampled time signal $f(t)$ and (b) the computed power spectrum from **fft**, $f(t) = \sin(t) + 2\cos(2t)$.

Aliasing

In the example above, the sampling interval Δt was sufficiently small to observe all contributing frequencies ω to $F(\omega)$. Let us now consider what happens to the discrete FT when $f(t)$ is *not* bandwidth-limited. Consider sampling the signal with $N = 2^6$ points during an interval $2P$, $P = 3\pi$. The maximum resolvable frequency is

$$\omega_{\max} = \frac{\pi}{\Delta t} = \frac{N\pi}{2P} = \frac{(2^6)\pi}{(2)(3\pi)} = \frac{2^5}{3} = 10.66\overline{6} \tag{9.65}$$

Let us now sample a signal with an added high-frequency component $\omega_{\text{hi}} > \omega_{\max}$,

$$f(t) = \sin(t) + 2\cos(2t) + \sin(\omega_{\text{hi}}t) \qquad \omega_{\text{hi}} = (1.2)\omega_{\max} \tag{9.66}$$

Using a procedure similar to that above, we obtain the power spectrum shown in Figure 9.3. The peaks at $\omega = 1$ and $\omega = 2$ are sampled correctly; however, the peak at $\omega_{\text{hi}} > \omega_{\max}$ generates through the symmetry $F(\omega_m - 2\omega_{\max}) = F(\omega_m)$ a peak at $2\omega_{\max} - \omega_{\text{hi}} < \omega_{\max}$. The true signal has no such frequency component, but our usual experience would lead us to conclude that $f(t)$ *does* contain a component at $2\omega_{\max} - \omega_{\text{hi}}$ and that the peak at ω_{hi} is the "fictitious" one due to sampling artifacts. Thus, inadequate sampling of high-frequency components (Figure 9.4) can corrupt the low-frequency spectrum.

To guard against aliasing, we could filter the data prior to computing the Fourier transform to remove the frequency components above $\omega_{\max}$. Of course, the best approach is to reduce Δt and thus increase $\omega_{\max}$. Increasing N from 2^6 to 2^8 for the same P increases $\omega_{\max}$ to

$$\omega_{\max} = \frac{N\pi}{2P} = \frac{(2^8)\pi}{(2)(3\pi)} = \frac{2^7}{3} = 42.6\overline{6} \tag{9.67}$$

and thus removes the aliasing in the power spectrum (Figure 9.5).

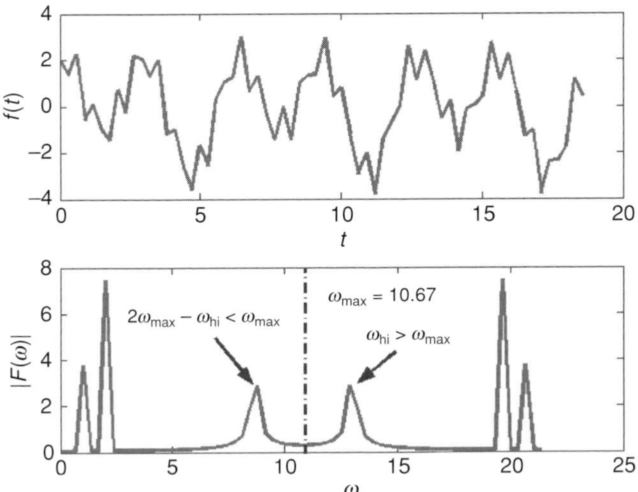

Figure 9.3 Power spectrum of time signal showing aliasing due to incorrectly-sampled high-frequency component.

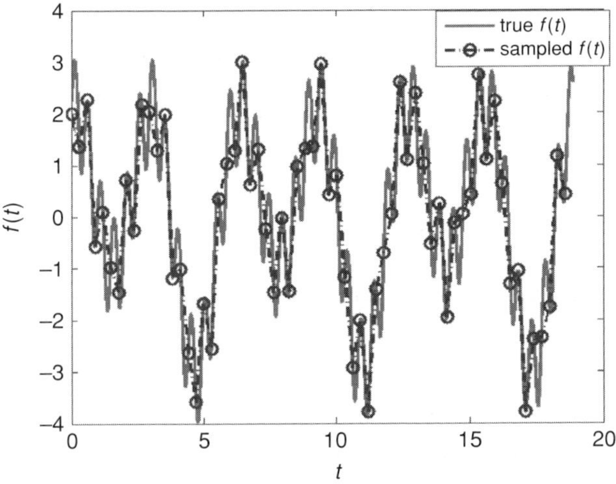

Figure 9.4 Comparison of the "true" $f(t)$ signal to the sampled signal showing the incorrect sampling of the high-frequency component that leads to aliasing.

Convolution and correlation

Convolution of two signals

The *convolution* of two functions $g(t)$ and $f(t)$ is the function of t

$$[g^*f](t) = \frac{1}{\sqrt{2\pi}} \int_{-\infty}^{+\infty} g(\tau)f(t-\tau)d\tau \qquad (9.68)$$

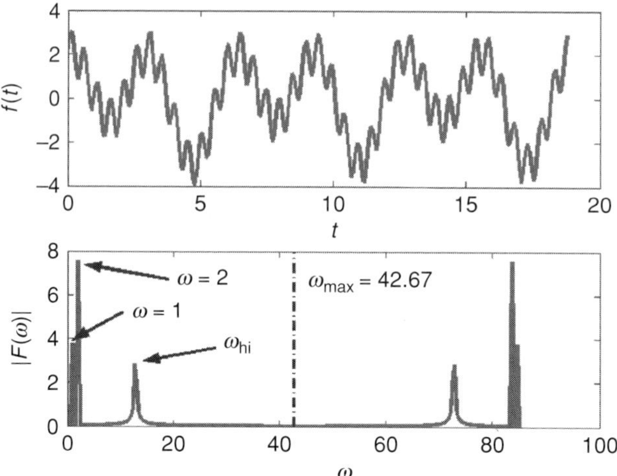

Figure 9.5 Fourier transform of a signal with a high-frequency component shows no aliasing when the sampling interval is sufficiently small.

Such an operation occurs often in the analysis of dynamic systems and signal processing, when one assumes a linear relationship between the time-dependent input to a system $x(t)$ and the time-dependent output $y(t)$:

$$y(t) = \int_{-\infty}^{+\infty} x(\tau)r(t - \tau)d\tau = [x^*r](t) \tag{9.69}$$

For a causal relationship between input and output, $r(t < 0) = 0$.

If we were to apply numerical quadrature directly to the formula above, the required work scales with the number of time values N as N^2. One may use FFT methods to compute the convolution in a much smaller number of operations that scales only as $N\log_2 N \ll N^2$ using the *convolution theorem*:

$$[G^*F](\omega) = G(\omega)F(\omega) \tag{9.70}$$

$G(\omega)$, $F(\omega)$, and $[G^*F](\omega)$ are the Fourier transforms respectively of $g(t)$, $f(t)$, and $[g^*f](t)$.

Proof The Fourier transform of the convolution is

$$[G^*F](\omega) = \frac{1}{\sqrt{2\pi}} \int_{-\infty}^{+\infty} \left\{ \frac{1}{\sqrt{2\pi}} \int_{-\infty}^{+\infty} g(\tau)f(t - \tau)d\tau \right\} e^{-i\omega t}\, dt \tag{9.71}$$

Changing the order of integration,

$$[G^*F](\omega) = \frac{1}{\sqrt{2\pi}} \int_{-\infty}^{+\infty} g(\tau) \left\{ \frac{1}{\sqrt{2\pi}} \int_{-\infty}^{+\infty} f(t - \tau)e^{-i\omega t}\, dt \right\} d\tau \tag{9.72}$$

Introducing the change of variable $t \to s = t - \tau$, $e^{-i\omega t} = e^{-i\omega s} e^{-i\omega \tau}$, $dt = ds$,

$$[G^*F](\omega) = \frac{1}{\sqrt{2\pi}} \int_{-\infty}^{+\infty} g(\tau) \left\{ \frac{1}{\sqrt{2\pi}} \int_{-\infty}^{+\infty} f(s)e^{-i\omega s}\, ds \right\} e^{-i\omega \tau}\, d\tau$$

$$[G^*F](\omega) = \left\{ \frac{1}{\sqrt{2\pi}} \int_{-\infty}^{+\infty} f(s)e^{-i\omega s}\, ds \right\} \left\{ \frac{1}{\sqrt{2\pi}} \int_{-\infty}^{+\infty} g(\tau)e^{-i\omega \tau}\, d\tau \right\}$$

$$[G^*F](\omega) = G(\omega)F(\omega) \tag{9.73}$$

$$QED$$

Correlation of two signals

A similar operation to convolution is the *correlation* of two functions,

$$C_{g,f}(t) = \frac{1}{\sqrt{2\pi}} \int_{-\infty}^{+\infty} g(\tau + t)f(\tau)\, d\tau \tag{9.74}$$

which measures how similar the signal g is to the signal f after a *lag time t*. The *autocorrelation* of a signal g is its correlation to itself, $C_{g,g}(t)$. Like convolution, the correlation of two signals can be computed efficiently by FFT methods. The Fourier transform of the correlation function is

$$C_{g,f}(\omega) = \frac{1}{\sqrt{2\pi}} \int_{-\infty}^{+\infty} C_{g,f}(t)e^{-i\omega t}\, dt \tag{9.75}$$

Substituting for $C_{g,f}(t)$ and switching the integration order, we have

$$C_{g,f}(\omega) = \frac{1}{\sqrt{2\pi}} \int_{-\infty}^{+\infty} \left\{ \frac{1}{\sqrt{2\pi}} \int_{-\infty}^{+\infty} g(\tau + t)f(\tau)e^{-i\omega t}\, dt \right\} d\tau \tag{9.76}$$

Defining $s = \tau + t$, we can write $dt = ds$ in the inner integral, where τ is fixed, and using $e^{-i\omega t} = e^{-i\omega(s-\tau)} = e^{-i\omega s} e^{i\omega \tau}$, we have

$$C_{g,f}(\omega) = \frac{1}{\sqrt{2\pi}} \int_{-\infty}^{+\infty} \left\{ \frac{1}{\sqrt{2\pi}} \int_{-\infty}^{+\infty} g(s)f(\tau)e^{-i\omega s}\, ds \right\} e^{i\omega \tau}\, d\tau \tag{9.77}$$

Thus,

$$C_{g,f}(\omega) = \left\{ \frac{1}{\sqrt{2\pi}} \int_{-\infty}^{+\infty} g(s)e^{-i\omega s}\, ds \right\} \left\{ \frac{1}{\sqrt{2\pi}} \int_{-\infty}^{+\infty} f(\tau)e^{i\omega \tau}\, d\tau \right\} \tag{9.78}$$

and we have the simple result

$$C_{g,f}(\omega) = G(\omega)F(-\omega) \tag{9.79}$$

If f is a real signal, $[F(\omega)]^* = F(-\omega)$, and thus

$$C_{g,f}(\omega) = G(\omega)F(-\omega) = G(\omega)[F(\omega)]^* \qquad \text{for real functions} \tag{9.80}$$

Fourier transforms in multiple dimensions

The *d-dimensional Fourier transform pair* satisfies the relation

$$f(r) = \frac{1}{(2\pi)^d} \int_{\Re^d} \left\{ \int_{\Re^d} f(r')e^{-i(q\cdot r')} \, dr' \right\} e^{i(q\cdot r)} \, dq \tag{9.81}$$

and, in agreement with the result for $d = 1$, (9.23) is defined as

$$F(q) = \frac{1}{(2\pi)^{d/2}} \int_{\Re^d} f(r)e^{-i(q\cdot r)} \, dr \qquad f(r) = \frac{1}{(2\pi)^{d/2}} \int_{\Re^d} F(q)e^{i(q\cdot r)} \, dq \tag{9.82}$$

As in one dimension, alternative definitions of the Fourier transform exist, but as long as one is consistent and satisfies (9.81), the choice is rather arbitrary.

Convolution and correlation

The *convolution* of $f(r)$, $g(r)$, $r \in \Re^d$, is

$$[g^* f](s) = \frac{1}{(2\pi)^{d/2}} \int_{\Re^d} g(r)f(s-r)dr \qquad [G^* F](q) = G(q)F(q) \tag{9.83}$$

The *correlation* of $f(r)$ and $g(r)$ is

$$C_{g,f}(s) = \frac{1}{(2\pi)^{d/2}} \int_{\Re^d} g(r+s)f(r)dr \qquad C_{g,f}(q) = G(q)F(-q) \tag{9.84}$$

When $f(r)$ and $g(r)$ are real, $C_{g,f}(q) = G(q)F(-q) = G(q)[F(q)]^*$.

The discrete d-dimensional Fourier transform

The FFT algorithm is similar in multiple dimensions $d > 2$ to the case $d = 1$. Here, we consider $d = 2$:

$$F(q) = \frac{1}{(2\pi)} \int_{\Re^2} f(r)e^{-i(q\cdot r)} \, dr = \frac{1}{(2\pi)} \int_{-\infty}^{+\infty} \int_{-\infty}^{+\infty} f(x,y)e^{-i(q_x x + q_y y)} \, dx \, dy \tag{9.85}$$

We sample $f(x, y)$ uniformly in $0 \le x < 2P_x$ and $0 \le y < 2P_y$ at

$$\begin{aligned}
x_j &= (j-1)(\Delta x) & \Delta x &= (2P_x)/N_x & j &= 1, 2, \ldots, N_x \\
y_k &= (k-1)(\Delta y) & \Delta y &= (2P_y)/N_y & k &= 1, 2, \ldots, N_y \\
& & N_x &= 2^{K_x} \text{ and } N_y = 2^{K_y}.
\end{aligned} \tag{9.86}$$

We assume periodicity outside of the domain,

$$f(x + l_x 2P_x, y + l_y 2P_y) = f(x, y) \qquad l_{x,y} = 0, \pm 1, \pm 2, \ldots \tag{9.87}$$

and thus determine $F(\boldsymbol{q})$ at $\boldsymbol{q}^{[m,n]}$, where $m = 1, 2, \ldots, N_x$, $n = 1, 2, \ldots, N_y$:

$$\boldsymbol{q}^{[m,n]} = q_x^{[m]} \boldsymbol{e}_x + q_y^{[n]} \boldsymbol{e}_y \qquad q_x^{[m]} = \frac{(m-1)\pi}{P_x} \qquad q_y^{[n]} = \frac{(n-1)\pi}{P_y} \tag{9.88}$$

The maximum resolvable frequencies and the frequency resolutions are

$$q_{j,\max} = \frac{N_j \pi}{2P_j} = \frac{N_j}{2}(\Delta q_j) \qquad \Delta q_j = \frac{\pi}{P_j} \qquad j = x, y \tag{9.89}$$

The computed $F(\boldsymbol{q}^{[m,n]})$ have $\boldsymbol{q}$-periodicity,

$$F(q_x + l_x 2q_{x,\max}, \; q_y + l_y 2q_{y,\max}) = F(q_x, q_y) \qquad l_{x,y} = 0, \pm 1, \pm 2, \ldots \tag{9.90}$$

The $F(\boldsymbol{q}^{[m,n]})$ are obtained by quadrature of (9.85),

$$F(\boldsymbol{q}^{[m,n]}) \approx \frac{(\Delta x)(\Delta y)}{(2\pi)} \sum_{j=1}^{N_x} \sum_{k=1}^{N_y} f(x_j, y_k) e^{-i\left[q_x^{[m]} x_j + q_y^{[n]} y_k\right]} \tag{9.91}$$

In terms of the 2-D discrete Fourier transform values F_{mn}, this becomes

$$F\left(\boldsymbol{q}^{[m,n]}\right) \approx \frac{(\Delta x)(\Delta y)}{(2\pi)} F_{mn} \qquad F_{mn} = \sum_{j=1}^{N_x} \sum_{k=1}^{N_y} f_{jk} e^{-i\left[q_x^{[m]} x_j + q_y^{[n]} y_k\right]} \tag{9.92}$$

Defining $W_x = e^{-i(\Delta q_x)(\Delta x)}$, $W_y = e^{-i(\Delta q_y)(\Delta y)}$, we have

$$F_{mn} = \sum_{j=1}^{N_x} \left\{ \sum_{k=1}^{N_y} f_{jk} \, W_y^{(n-1)(k-1)} \right\} W_x^{(m-1)(j-1)} \tag{9.93}$$

Thus, we obtain the 2-D discrete Fourier transform by first computing the 1-D discrete Fourier transforms over y with x fixed at each x_j,

$$F_k(x_j) = \sum_{k=1}^{N_y} f(x_j, y_k) W_y^{(n-1)(k-1)} \tag{9.94}$$

and then performing a 1-D discrete Fourier transform over x,

$$F_{mn} = \sum_{j=1}^{N_x} F_k(x_j) W_x^{(m-1)(j-1)} \tag{9.95}$$

Thus, multidimensional discrete Fourier transforms can be obtained from recursive application of the 1-D FFT algorithm.

FFT in multiple dimensions in MATLAB

In MATLAB, the multidimensional discrete Fourier transform and its inverse are computed using the definitions above by **fftn** and **ifftn** respectively. A separate routine **fft2** is provided for the 2-D case. The code provided below computes the 2-D power spectrum of the function

$$f(x, y) = \cos(x) + \cos(3x) + 2\sin(x)\cos(2y) + \cos(4y) \tag{9.96}$$

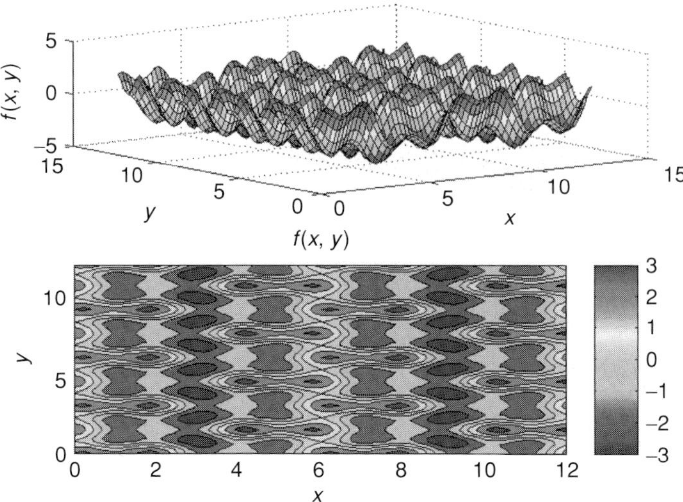

Figure 9.6 Surface and contour plots of a 2-D periodic function $f(x, y)$.

that has four peaks at $(q_x, q_y) = (1, 0), (3, 0), (1, 2)$, and $(0, 4)$ from sampling over the domain $0 \le x < 4\pi$, $0 \le y < 4\pi$,

```
% generate real-space (x,y) grid
P_x = 2*pi; N_x = 2^6; dx = 2*P_x/N_x;
x_val = linspace(0, 2*P_x - dx, N_x);
P_y = 2*pi; N_y = 2^6; dy = 2*P_y/N_y;
y_val = linspace(0, 2*P_y - dy, N_y);
[X,Y] = meshgrid(x_val,y_val);
% generate f(x,y) values
F = zeros(size(X));
F = F + cos(X) + cos(3.*X) + 2*sin(X).*cos(2.*Y) + cos(4.*Y);
% generate the q-space grid
dq_x = pi/P_x; q_x_max = pi/dx;
q_x_val = linspace(0, 2*q_x_max - dq_x, N_x);
dq_y = pi/P_y; q_y_max = pi/dy;
q_y_val = linspace(0, 2*q_y_max - dq_y, N_y);
[QX,QY] = meshgrid(q_x_val, q_y_val);
% compute the 2-D FT and power spectrum
F_FT = (dx*dy/2/pi).*fft2(F); F_PS = abs(F_FT);
```

Plots of $f(x, y)$ and $|F(q_x, q_y)|$ are shown in Figure 9.6 and Figure 9.7.

Scattering theory

This section introduces the theory underpinning scattering experiments, which provide structural information about materials from observing the interaction of a sample with

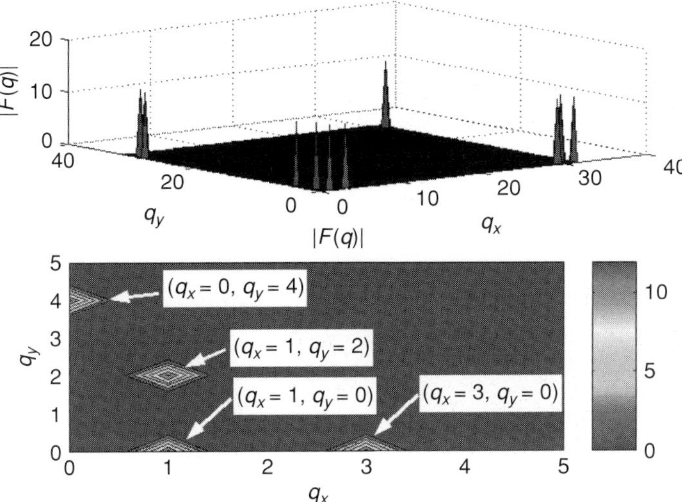

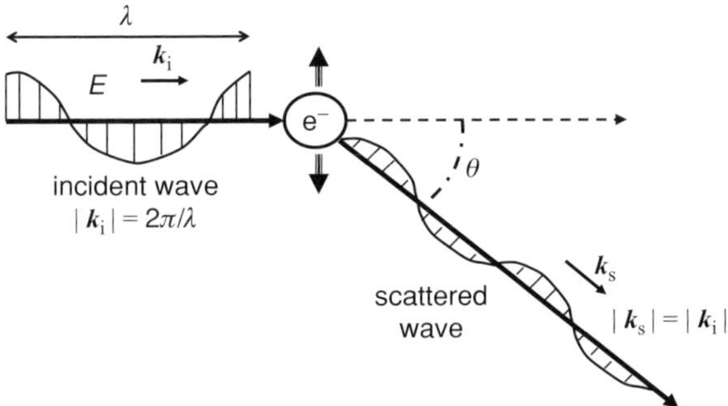

Figure 9.7 Power spectrum of $f(x, y)$ obtained from a 2-D FFT.

Figure 9.8 Generation of scattered light/X-ray radiation from an electron.

a light, X-ray, or neutron beam. In light and X-ray scattering, an incident beam of electromagnetic radiation causes the electrons of the sample to oscillate and emit secondary "scattered" electromagnetic waves, each with the same frequency as the incident beam. The intensity of this scattered radiation is measured as a function of orientation from the incident beam. From the observed interference pattern, the relative spatial positions of the electrons are extracted through Fourier analysis. In neutron scattering, the mechanism for scattering is different, but the mathematical treatment is the same.

Consider a system in which an incident electromagnetic wave interacts with a single electron (Figure 9.8). The wavelength is λ and its speed of propagation is c; thus, the frequency in radians per second is $\omega = 2\pi c/\lambda$. The incident electromagnetic wave imparts a force to the electron and causes it to oscillate at the same frequency ω as the incident radiation. As an accelerating charge emits its own electromagnetic wave, there arise waves of scattered radiation from each electron, each at the same frequency ω as the incident radiation, but propagating in all directions from the electron (although not uniformly – the intensity varies as $1 + \cos^2 \theta$).

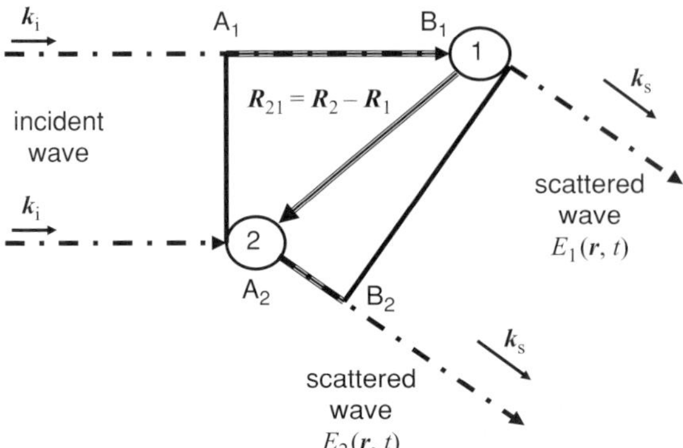

Figure 9.9 Interference diagram for the scattered waves of two electrons.

For the incident beam, the *wave vector* k_i points in the direction of wave propagation and has a magnitude equal to the *wavenumber*

$$|k_i| = \frac{\omega}{c} = \frac{2\pi}{\lambda} \tag{9.97}$$

As the scattered radiation has the same frequency (and thus wavelength) as the incident radiation, $|k_s| = |k_i|$. We define the *scattering vector* q as

$$q = k_s - k_i \tag{9.98}$$

which is related to the angle θ between the incident and scattered waves by

$$|q| = \frac{4\pi \sin(\theta/2)}{\lambda} \tag{9.99}$$

In a scattering experiment, we measure the time-averaged intensity of scattered radiation as a function of $q \neq 0$ with a detector located at a distance $r_D \gg \lambda$ from the sample. q is varied by rotating the sample and/or detector with respect to the incident beam.

This scattered intensity (except at $q = 0$ none of the detected intensity is due to the incident beam) arises from interactions involving a large number of electrons in the sample and in the intervening atmosphere. This latter source is removed by subtracting from the measured intensity the background intensity observed when there is no sample. It is the interference between the scattered waves from each electron in the sample that encodes information about their relative spatial positions.

Let us consider first the interference between the scattered waves coming from only two electrons (Figure 9.9), one at R_1 and the second at R_2. When the distance between the sample and the detector is much greater than the physical extent of the sample, both scattered waves have the same wavevector $k_s = k_i + q$. The interference, dependent upon the relative position of the electrons $R_{21} = R_2 - R_1$, originates from the different path lengths traveled by the two waves from the incident beam source to the detector; specifically the difference between the segment lengths

$$\overline{A_1 B_1} = -\frac{1}{k}(\boldsymbol{k}_i \cdot \boldsymbol{R}_{21}) \qquad \overline{A_2 B_2} = -\frac{1}{k}(\boldsymbol{k}_s \cdot \boldsymbol{R}_{21}) \qquad k = |\boldsymbol{k}_i| = |\boldsymbol{k}_s| \tag{9.100}$$

If $E_s(\theta)$ is the common amplitude of the two scattered waves (we neglect any polarization effects), the scattered electric fields emitted from each electron vary as a function of position and time as

$$E_1(\boldsymbol{r}, t; \boldsymbol{q}) = E_s(\theta)e^{i(\boldsymbol{k}_s \cdot \boldsymbol{r} - \omega t)} \qquad E_2(\boldsymbol{r}, t; \boldsymbol{q}) = E_s(\theta)e^{i(\boldsymbol{k}_s \cdot \boldsymbol{r} - \omega t + \varphi_{21})} \tag{9.101}$$

where φ_{21} is the phase lag due to the path-length difference $\overline{A_2 B_2} - \overline{A_1 B_1}$,

$$\varphi_{21} = \frac{(\overline{A_2 B_2} - \overline{A_1 B_1})}{\lambda}(2\pi) = k(\overline{A_2 B_2} - \overline{A_1 B_1}) \tag{9.102}$$

Substituting for the path lengths,

$$\varphi_{21} = k\left(-\frac{1}{k}\right)[(\boldsymbol{k}_s \cdot \boldsymbol{R}_{21}) - (\boldsymbol{k}_i \cdot \boldsymbol{R}_{21})] = -[(\boldsymbol{k}_s - \boldsymbol{k}_i) \cdot \boldsymbol{R}_{21}] = -(\boldsymbol{q} \cdot \boldsymbol{R}_{21}) \tag{9.103}$$

The total electric field from both scattered waves is

$$E_{\text{tot}}(\boldsymbol{r}, t; \boldsymbol{q}) = E_1(\boldsymbol{r}, t; \boldsymbol{q}) + E_2(\boldsymbol{r}, t; \boldsymbol{q}) = E_s(\theta)e^{i(\boldsymbol{k}_s \cdot \boldsymbol{r} - \omega t)}[1 + e^{i\varphi_{21}}] \tag{9.104}$$

Thus, the net intensity from the two scattered waves at the detector is

$$I_{\text{tot}}(\boldsymbol{r}_{\mathrm{D}}, t; \boldsymbol{q}) = |E_{\text{tot}}(\boldsymbol{r}, t; \boldsymbol{q})|^2 = 2|E_s(\theta)|^2[1 + \cos \varphi_{21}] \tag{9.105}$$

In the decoherent limit $\boldsymbol{q} \to 0$, $\varphi_{21} \to 0$, there is no interference between the waves, and the time-averaged intensity at the detector is

$$[I_{\text{tot}}(\boldsymbol{q})]_{\text{decoh}} = I_{\text{tot}}(\boldsymbol{q} \to 0) = 4|E_s(\theta)|^2 \tag{9.106}$$

Thus, the *structure factor* – the ratio of the observed scattered intensity to that in the decoherent limit – for this two-electron system is

$$S(\boldsymbol{q}) = \frac{\langle I_{\text{tot}}(\boldsymbol{q})\rangle}{\langle I_{\text{tot}}(\boldsymbol{q} \to 0)\rangle} = \frac{1}{2}[1 + \cos \varphi_{21}] = \frac{1}{2}\{1 + \cos[-(\boldsymbol{q} \cdot \boldsymbol{R}_{21})]\} \tag{9.107}$$

From $S(\boldsymbol{q})$, we obtain information about the relative electron position $\boldsymbol{R}_{21}$.

We now extend this analysis to the case with N scattered waves emitted from electrons at the positions $\{\boldsymbol{R}_1, \boldsymbol{R}_2, \ldots, \boldsymbol{R}_N\}$. The phase angle φ_j of the scattered wave at $\boldsymbol{q}$ from electron j is

$$\varphi_j = -(\boldsymbol{q} \cdot \boldsymbol{R}_j) \tag{9.108}$$

and the phase lag between electrons j and m is

$$\varphi_{mj} = \varphi_m - \varphi_j = -(\boldsymbol{q} \cdot \boldsymbol{R}_{mj}) \qquad \boldsymbol{R}_{mj} = \boldsymbol{R}_m - \boldsymbol{R}_j \tag{9.109}$$

Thus, the electric field at the detector arising from all N scattered waves is

$$E_{\text{tot}}(\boldsymbol{r}_{\mathrm{D}}, t; \boldsymbol{q}) = E_s(\theta)e^{i(\boldsymbol{k}_s \cdot \boldsymbol{r} - \omega t)} \sum_{j=1}^{N} e^{i\varphi_j} \tag{9.110}$$

We assume that the sample is so small, or the interaction is so weak, that the scattered waves are not themselves scattered before they strike the detector. The measured intensity is then

$$I_{\text{tot}}(\boldsymbol{r}_{\text{D}}, t; q) = \left| E_{\text{tot}}(\boldsymbol{r}_{\text{D}}, t; q) \right|^2 = |E_s(\theta)|^2 \left[\sum_{j=1}^{N} e^{-i\varphi_j} \right] \left[\sum_{m=1}^{N} e^{i\varphi_m} \right]$$

$$= |E_s(\theta)|^2 \sum_{j=1}^{N} \sum_{m=1}^{N} e^{i(\varphi_m - \varphi_j)} = |E_s(\theta)|^2 \sum_{j=1}^{N} \sum_{m=1}^{N} e^{-i(\boldsymbol{q} \cdot \boldsymbol{R}_{mj})} \tag{9.111}$$

The *structure factor* – the ratio of time-averaged scattered intensity at $\boldsymbol{q} \neq \boldsymbol{0}$ to that in the decoherent limit $\boldsymbol{q} \to \boldsymbol{0}$ – is therefore

$$S(\boldsymbol{q}) = \frac{\langle I_{\text{tot}}(\boldsymbol{q}) \rangle}{\langle I_{\text{tot}}(\boldsymbol{q} \to \boldsymbol{0}) \rangle} = \left\langle \frac{1}{N} \sum_{j=1}^{N} \sum_{m=1}^{N} e^{-i(\boldsymbol{q} \cdot \boldsymbol{R}_{mj})} \right\rangle \tag{9.112}$$

Applying Fourier analysis

We now show that this structure factor is closely related to the Fourier transform of the correlation function of the electron density $\rho(\boldsymbol{r})$. If we know the exact positions $\{\boldsymbol{R}_1, \boldsymbol{R}_2, \ldots, \boldsymbol{R}_N\}$ of each electron (we neglect quantum effects), the density function is merely a sum of Dirac delta functions:

$$\rho(\boldsymbol{r}) = \sum_{j=1}^{N} \delta(\boldsymbol{r} - \boldsymbol{R}_j) \tag{9.113}$$

Let us consider the autocorrelation function of $\rho(\boldsymbol{r})$,

$$C_{\rho,\rho}(\boldsymbol{s}) = \frac{1}{(2\pi)^{d/2}} \int_{\mathfrak{R}^d} \rho(\boldsymbol{r} + \boldsymbol{s})\rho(\boldsymbol{r})d\boldsymbol{r}$$

$$C_{\rho,\rho}(\boldsymbol{q}) = \rho(\boldsymbol{q})\rho(-\boldsymbol{q}) = |\rho(\boldsymbol{q})|^2 = \frac{1}{(2\pi)^{d/2}} \int_{\mathfrak{R}^d} C_{\rho,\rho}(\boldsymbol{s})e^{-i(\boldsymbol{q} \cdot \boldsymbol{s})}d\boldsymbol{s} \tag{9.114}$$

Substituting for $C_{\rho,\rho}(\boldsymbol{s})$ and $\rho(\boldsymbol{r})$ in the expression for $C_{\rho,\rho}(\boldsymbol{q})$,

$$C_{\rho,\rho}(\boldsymbol{q}) = \frac{1}{(2\pi)^d} \int_{\mathfrak{R}^d} \int_{\mathfrak{R}^d} \rho(\boldsymbol{r} + \boldsymbol{s})\rho(\boldsymbol{r})e^{-i(\boldsymbol{q} \cdot \boldsymbol{s})}d\boldsymbol{r}d\boldsymbol{s}$$

$$= \frac{1}{(2\pi)^d} \int_{\mathfrak{R}^d} \int_{\mathfrak{R}^d} \left[\sum_{m=1}^{N} \delta(\boldsymbol{r} + \boldsymbol{s} - \boldsymbol{R}_m) \right] \left[\sum_{j=1}^{N} \delta(\boldsymbol{r} - \boldsymbol{R}_j) \right] e^{-i(\boldsymbol{q} \cdot \boldsymbol{s})}d\boldsymbol{r}d\boldsymbol{s}$$

$$= \frac{1}{(2\pi)^d} \sum_{m=1}^{N} \sum_{j=1}^{N} \left\{ \int_{\mathfrak{R}^d} \int_{\mathfrak{R}^d} \delta(\boldsymbol{r} + \boldsymbol{s} - \boldsymbol{R}_m)\delta(\boldsymbol{r} - \boldsymbol{R}_j)e^{-i(\boldsymbol{q} \cdot \boldsymbol{s})}d\boldsymbol{r}d\boldsymbol{s} \right\} \tag{9.115}$$

From the Dirac delta functions, the integral is nonzero only if

$$\boldsymbol{r} = \boldsymbol{R}_j \quad \boldsymbol{r} + \boldsymbol{s} = \boldsymbol{R}_m \Rightarrow \boldsymbol{s} = \boldsymbol{R}_m - \boldsymbol{R}_j = \boldsymbol{R}_{mj} \tag{9.116}$$

therefore,

$$C_{\rho,\rho}(\boldsymbol{q}) = \frac{1}{(2\pi)^d} \sum_{j=1}^{N} \sum_{m=1}^{N} e^{-i(\boldsymbol{q}\cdot\boldsymbol{R}_{mj})} \tag{9.117}$$

The structure factor thus is related to the time-averaged autocorrelation function of the electron density:

$$\langle C_{\rho,\rho}(\boldsymbol{q})\rangle = \frac{1}{(2\pi)^d} \left\langle \sum_{j=1}^{N} \sum_{m=1}^{N} e^{-i(\boldsymbol{q}\cdot\boldsymbol{R}_{mj})} \right\rangle = \frac{N S(\boldsymbol{q})}{(2\pi)^d} = |\langle \rho(\boldsymbol{q})\rangle|^2 \tag{9.118}$$

The autocorrelation function is obtained from an inverse Fourier transform:

$$\langle C_{\rho,\rho}(\boldsymbol{s})\rangle = \frac{1}{(2\pi)^{d/2}} \int_{\mathfrak{R}^d} \langle \rho(\boldsymbol{r}+\boldsymbol{s})\rho(\boldsymbol{r})\rangle d\boldsymbol{r} = \frac{1}{(2\pi)^{d/2}} \int_{\mathfrak{R}^d} \left[\frac{N S(\boldsymbol{q})}{(2\pi)^d}\right] e^{i(\boldsymbol{q}\cdot\boldsymbol{s})} d\boldsymbol{q} \tag{9.119}$$

Scattering peaks from samples with periodic structure

When $\langle C_{\rho,\rho}(\boldsymbol{s})\rangle$ has a peak at a particular $\boldsymbol{s}'$, it means that when we have an electron at $\boldsymbol{r}'$, we tend to have another electron at $\boldsymbol{r}' + \boldsymbol{s}'$. For example, let us consider the case of a periodic lattice (e.g. a crystal) with *lattice vectors* $\boldsymbol{a}$, $\boldsymbol{b}$, $\boldsymbol{c}$ such that

$$\langle \rho(\boldsymbol{r}+l_1\boldsymbol{a}+l_2\boldsymbol{b}+l_3\boldsymbol{c})\rangle = \langle \rho(\boldsymbol{r})\rangle \qquad l_m = 0, \pm 1, \pm 2, \ldots \tag{9.120}$$

$\langle C_{\rho,\rho}(\boldsymbol{s})\rangle$ then has intense peaks at every $\boldsymbol{s} = l_1\boldsymbol{a}+l_2\boldsymbol{b}+l_3\boldsymbol{c}$, as can be seen by partitioning the sample volume into unit cells $\Omega_\alpha, \alpha = 1, 2, \ldots, N_{\text{cells}}$, each of volume $|\boldsymbol{a} \cdot (\boldsymbol{b} \times \boldsymbol{c})|$:

$$\langle C_{\rho,\rho}(l_1\boldsymbol{a}+l_2\boldsymbol{b}+l_3\boldsymbol{c})\rangle = \frac{1}{(2\pi)^{d/2}} \sum_{\alpha=1}^{N_{\text{cells}}} \int_{\Omega_\alpha} \langle \rho(\boldsymbol{r}+l_1\boldsymbol{a}+l_2\boldsymbol{b}+l_3\boldsymbol{c})\rho(\boldsymbol{r})\rangle d\boldsymbol{r}$$

$$= \frac{N_{\text{cells}}}{(2\pi)^{d/2}} \int_{\Omega_\alpha} \langle \rho(\boldsymbol{r})\rho(\boldsymbol{r})\rangle \, d\boldsymbol{r} \tag{9.121}$$

This periodicity $\langle \rho(\boldsymbol{r}+l_1\boldsymbol{a}+l_2\boldsymbol{b}+l_3\boldsymbol{c})\rangle = \langle \rho(\boldsymbol{r})\rangle$ also yields peaks of $S(\boldsymbol{q})$ and $\langle C_{\rho,\rho}(\boldsymbol{q})\rangle$ in $\boldsymbol{q}$-space. Let us assume that $\langle C_{\rho,\rho}(\boldsymbol{s})\rangle$ consists of infinitely-narrow peaks about each $l_1\boldsymbol{a}+l_2\boldsymbol{b}+l_3\boldsymbol{c}$:

$$\langle C_{\rho,\rho}(\boldsymbol{s})\rangle = n_0 \sum_{l_1,l_2,l_3} \delta[\boldsymbol{s} - (l_1\boldsymbol{a}+l_2\boldsymbol{b}+l_3\boldsymbol{c})] \tag{9.122}$$

In practice, peaks have finite widths, but, assuming this "sharp-peak" limit, the Fourier transform of the autocorrelation function is

$$\langle C_{\rho,\rho}(\boldsymbol{q})\rangle = \frac{1}{(2\pi)^{d/2}} \int_{\mathfrak{R}^d} \left\{ n_0 \sum_{l_1,l_2,l_3} \delta[\boldsymbol{s} - (l_1\boldsymbol{a}+l_2\boldsymbol{b}+l_3\boldsymbol{c})] \right\} e^{-i(\boldsymbol{q}\cdot\boldsymbol{s})} d\boldsymbol{s}$$

$$= \frac{n_0}{(2\pi)^{d/2}} \sum_{l_1,l_2,l_3} e^{-i[\boldsymbol{q}\cdot(l_1\boldsymbol{a}+l_2\boldsymbol{b}+l_3\boldsymbol{c})]} \tag{9.123}$$

Now, if $\boldsymbol{q}' \cdot (l_1\boldsymbol{a} + l_2\boldsymbol{b} + l_3\boldsymbol{c}) = m2\pi$, $m = \pm 1, \pm 2, \ldots$, then

$$\langle C_{\rho,\rho}(\boldsymbol{q}')\rangle = \frac{n_0}{(2\pi)^{d/2}} \sum_{l_1,l_2,l_3} e^{-im2\pi} = \frac{n_0}{(2\pi)^{d/2}} \sum_{l_1,l_2,l_3} (1) = \frac{n_0 N_{\text{cells}}}{(2\pi)^{d/2}} \tag{9.124}$$

Thus, $\langle C_{\rho,\rho}(\boldsymbol{q})\rangle$ has sharp peaks at all $\boldsymbol{q}'$ such that $\boldsymbol{q}' \cdot (l_1\boldsymbol{a} + l_2\boldsymbol{b} + l_3\boldsymbol{c})$ is a multiple of 2π. If we define the *reciprocal lattice vectors* $\boldsymbol{\alpha}, \boldsymbol{\beta}, \boldsymbol{\gamma}$ to satisfy

$$\begin{aligned}
\boldsymbol{\alpha} \cdot \boldsymbol{a} &= 2\pi & \boldsymbol{\alpha} \cdot \boldsymbol{b} &= 0 & \boldsymbol{\alpha} \cdot \boldsymbol{c} &= 0 \\
\boldsymbol{\beta} \cdot \boldsymbol{a} &= 0 & \boldsymbol{\beta} \cdot \boldsymbol{b} &= 2\pi & \boldsymbol{\beta} \cdot \boldsymbol{c} &= 0 \\
\boldsymbol{\gamma} \cdot \boldsymbol{a} &= 0 & \boldsymbol{\gamma} \cdot \boldsymbol{b} &= 0 & \boldsymbol{\gamma} \cdot \boldsymbol{c} &= 2\pi
\end{aligned} \tag{9.125}$$

these peaks occur in $\boldsymbol{q}$-space at

$$\boldsymbol{q}' = n_1\boldsymbol{\alpha} + n_2\boldsymbol{\beta} + n_3\boldsymbol{\gamma} \tag{9.126}$$

since

$$\begin{aligned}
\boldsymbol{q}' \cdot (l_1\boldsymbol{a} + l_2\boldsymbol{b} + l_3\boldsymbol{c}) &= (n_1\boldsymbol{\alpha} + n_2\boldsymbol{\beta} + n_3\boldsymbol{\gamma}) \cdot (l_1\boldsymbol{a} + l_2\boldsymbol{b} + l_3\boldsymbol{c}) \\
&= (n_1 l_1 + n_2 l_2 + n_3 l_3)(2\pi)
\end{aligned} \tag{9.127}$$

Above, we have assumed that the periodicity $\rho(\boldsymbol{r} + l_1\boldsymbol{a} + l_2\boldsymbol{b} + l_3\boldsymbol{c}) = \rho(\boldsymbol{r})$ extends throughout the entire sample; however, often (unless we go to great lengths to anneal the sample or crystallize it slowly from a single nucleation site) the sample contains many different crystal domains, with the crystal lattices in each domain isotropically oriented at random. In such a case, we measure a *powder spectrum*, and obtain not the full structure factor $S(\boldsymbol{q})$ but rather merely an isotropically-averaged structure factor $S(q)$. When $S(q)$ is nonzero (it usually appears as a circular "halo" at some scattering angle θ), this signifies that the material has structural periodicity on a length scale

$$\sigma_{\text{len}} \approx \frac{2\pi}{q} \tag{9.128}$$

Smaller q-values denote structures on larger length scales. Hence, when X-ray scattering (for a Cu-Kα_1 source, λ is 1.54 Å) is used to measure the ångstrom-scale periodicity of crystals, we measure $S(q)$ at large values of q, and thus also at large scattering angles θ (Figure 9.10). By contrast, when we attempt to probe longer nanometer-scale $O(10^{-9}$ m) structure, we must measure the scattering at small angles. Thus, when X-ray scattering is performed to probe the atomic-scale structure of crystals, it is known as WAXS (wide angle X-ray scattering). When it is used to probe the structure on the order of tens of nanometers of materials such as self-assembled block copolymers and micellar emulsions, it is known as SAXS (small angle X-ray scattering). At very small θ, the scattered intensity is buried within the intensity of the incident beam, resulting in an effective upper limit of resolution for SAXS of $O(100$ nm). With special facilities, this technique can be extended to longer length scales, as at the USAXS (ultra small angle X-ray scattering) facility at Argonne National Laboratories.

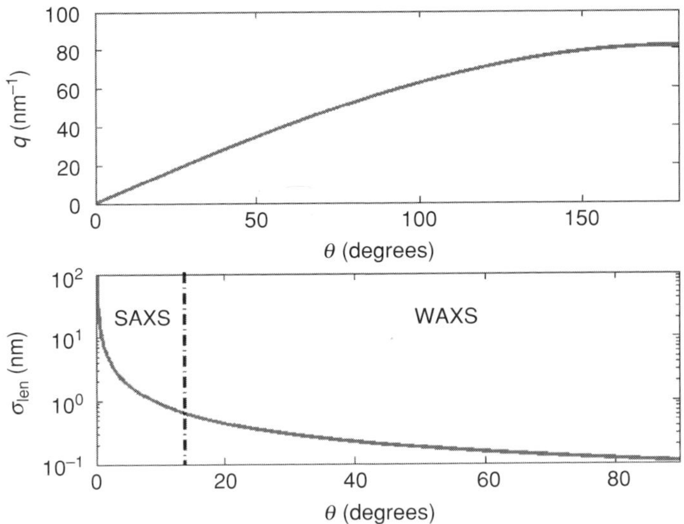

Figure 9.10 Scattering angle dependence for a typical Cu X-ray source. Large (1–10 nm) scale structure is observed at small scattering angles (SAXS) and small (0.1 nm) atomic-scale structure is observed at large angles (WAXS).

MATLAB summary

The discrete Fourier transform and its inverse are implemented as **fft** and **ifft**. In N dimensions, the routines are **fftn** and **ifftn**. In two dimensions, **fft2** and **ifft2** should be used. The examples in this chapter demonstrate the use of these functions. To compute convolutions and correlations, multiply the Fourier transforms appropriately.

Problems

9.A.1. Compute in MATLAB the Fourier transform of $f(t) = \cos(2t) + 3\sin(t) - 0.5\sin(3t)$ from sampled data in $[0, 9]$ at uniform intervals of no greater than 0.1.

9.A.2. Compute in MATLAB the correlation and convolution functions of the signal $f(t)$ from problem 9.A.1 with $g(t) = \cos(2t + 1) - 2\sin(t - 0.5)$.

9.A.3. Assume that you have been given sampled data of the signal $f(t)$ of problem 9.A.1. You identify that it has a component at $\omega = 1$ that you wish to remove. Using Fourier techniques, filter out this component and plot the remaining signal. Compare your result to $\cos(2t) - 0.5\sin(3t)$.

9.B.1. Let $x(t)$ be the position of an object of mass m, connected by a spring to the origin, that experiences a drag force and is acted upon by an external force $F(t)$. The equation of motion is

$$m\frac{d^2x}{dt^2} = -Kx - \zeta\frac{dx}{dt} + F(t) \tag{9.129}$$

Let the Fourier transforms of $x(t)$ and $F(t)$ be $X(\omega)$ and $F(\omega)$ respectively. Relate the two through a convolution, $X(\omega) = R(\omega)F(\omega)$. At what frequency ω_c does $R(\omega)$ become very large, exhibiting resonance? What effect does ζ have on the resonance phenomenon? Hint: Relate first the Fourier transforms of derivatives of $x(t)$ to that of $x(t)$ itself.

9.B.2. For $m = 1$, $K = 1$, and $\zeta = 10^\kappa$, $\kappa = 2, 1, 0, -1, -2, -3$, plot the response $x(t)$ to $F(t) = \cos(\omega t)$ for $\omega/\omega_c = 1 \pm 10^\kappa$, $\kappa = -1, -2, -3$.

9.B.3. Let us say that we wish to measure some signal $x(t)$ by a device whose output $d(t)$ is related to the signal by the convolution $D(\omega) = R(\omega)F(\omega)$. For a device that is only sensitive near ω_{dev},

$$R(\omega) = r_0 \exp\left[-\frac{(|\omega| - \omega_{\text{dev}})^2}{2\sigma^2} \right] \tag{9.130}$$

For $r_0 = 1$, $\omega_{\text{dev}} = 5$, $\sigma = 1$, compute the output signals for input pulsed calibration signals,

$$c(t) = \begin{cases} 1, & 0 \le t \le t_{\text{pulse}} \\ 0, & \text{otherwise} \end{cases} \tag{9.131}$$

with $t_{\text{pulse}} = 0.01, 0.1, 0.5, 1$. Then, let us say that we did not know the device response function but rather only the results of these calibration experiments. For each calibration experiment, estimate $R(\omega)$.

9.C.1. In problem 9.B.3, we estimate $R(\omega)$ separately from the results of each calibration experiment. Propose a method that uses all of the data to generate the best estimate of $R(\omega)$.

References

Abdel-Khalik, S. I., Hassager, O., and Bird, R. B., 1974, *Polym. Eng. Sci.*, **14**, 859–867

Akin, J. E., 1994, *Finite Elements for Analysis and Design*. San Diego: Academic Press

Allen, E. C. and Beers, K. J., 2005, *Polymer*, **46**, 569–573

Arnold, V. I., 1989, *Mathematical Methods of Classical Mechanics*, second edition. New York: Springer

Ascher, U. M. and Petzold, L. R., 1998, *Computer Methods for Ordinary Differential Equations and Differential-Algebraic Equations*. Philadelphia: SIAM

Atkins, P. W. and Friedman, R. S., 1999, *Molecular Quantum Mechanics*, third edition. New York: Oxford University Press

Ballenger, T. F., *et al.*, 1971, *Trans. Soc. Rheol.*, **15**, 195–215

Beers, K. J. and Ray, W. H., 2001, *J. Applied Polym. Sci.*, **79**, 266–274

Bellman, R., 1957, *Dynamic Programming*. Princeton: Princeton University Press

Bernardo, J. M. and Smith, A. F. M., 2000, *Bayesian Theory*. Chichester: John Wiley & Sons Ltd

Bird, R. B., Stewart, W. E., and Lightfoot, E. N., 2002, *Transport Phenomena*, second edition. New York: Wiley

Bolstad, W. M., 2004, *Introduction to Bayesian Statistics*. Hoboken: Wiley

Box, G. E. P. and Tiao, G. C., 1973, *Bayesian Interference in Statistical Analysis*. New York: Wiley

Broyden, C. G., 1965, *Math. Comput.*, **19**, 577–593

Chaikin, P. M. and Lubensky, T. C., 2000, *Principles of Condensed Matter Physics*. Cambridge: Cambridge University Press

Chandler, D., 1987, *Introduction to Modern Statistical Mechanics*. New York: Oxford University Press

Chen, M. H., Shao, Q. M., and Ibrahim, J. G., 2000. *Monte Carlo Methods in Bayesian Computation*. New York: Springer

Cussler, E. L. and Varma, A., 1997, *Diffusion : Mass Transfer in Fluid Systems*. Cambridge: Cambridge University Press

Dean, T., Allen, J., and Aloimonos, Y., 1995, *Artificial Intelligence: Theory and Practice*. Redwood City: Benjamin-Cummings

Deen, W. M., 1998, *Analysis of Transport Phenomena*. New York: Oxford University Press

de Finetti, B., 1970, *Theory of Probability*, new English edition (1990). Chichester: Wiley

Dotson, N. A., Galván, R., Laurence, R. L., and Tirrell, M., 1996, *Polymerization Process Modeling*. New York: VCH

Ferziger, J. H. and Peric, M., 2001, *Computational Methods for Fluid Dynamics*, third edition. Berlin: Springer

Finlayson, B. A., 1992, *Numerical Methods for Problems with Moving Fronts*. Seattle: Ravenna Park

Flory, P. F., 1953, *Principles of Polymer Chemistry*. Ithaca: Cornell University Press

Fogler, H. S., 1999, *Elements of Chemical Reaction Engineering*, third edition. Upper Saddle River: Prentice-Hall

Frenkel D. and Smit B., 2002, *Understanding Molecular Simulation*, second edition. San Diego: Academic Press

Golub, G. H. and van Loan, C. F., 1996, *Matrix Computations*, third edition. Baltimore: Johns Hopkins University Press

Gosset, W. S., 1908, *Biometrika*, **6**, 1–25

Jeffreys, H., 1961, *Theory of Probability*, third edition. Oxford: Oxford University Press

Kennedy, J. and Eberhart, R., 1995, *Proc. IEEE Intl. Conf. on Neural Networks*, Perth, Australia. Piscataway: Institute of Electrical and Electronics Engineers

Kloeden, P. E. and Platen, E., 2000, *Numerical Solution of Stochastic Differential Equations*. Berlin: Springer

Leach, A. R., 2001, *Molecular Modelling: Principles and Applications*, second edition. Harlow: Prentice-Hall

Leonard, T. and Hsu, J. S. J., 2001, *Bayesian Methods*. Cambridge: Cambridge University Press

Macosko, C. W. and Miller, D. R., 1976, *Macromolecules*, **9**, 199–206

Naylor, A. W. and Sell, G. R., 1982, *Linear Operator Theory In Engineering and Science*, second edition. New York: Springer-Verlag

Nocedal, J. and Wright, S. J., 1999, *Numerical Optimization*. New York: Springer

Odian, G., 1991, *Principles of Polymerization*, third edition. New York: Wiley

Oran, E. S. and Boris, J. P., 2001, *Numerical Simulation of Reactive Flow*, second edition. Cambridge: Cambridge University Press

O'Rourke, J., 1993, *Computational Geometry in C*. Cambridge: Cambridge University Press

Öttinger, H. C., 1996, *Stochastic Processes in Polymeric Fluids*. Berlin: Springer-Verlag

Perry and Green, 1984, *Chemical Engineer's Handbook*, sixth edition. New York: McGraw-Hill

Press, W. H., Teukolsky, S. A., Vetterling, W. T., and Flannery, B. P., 1992, *Numerical Recipes in C*, second edition. Cambridge: Cambridge University Press

Quateroni, A., Sacco, R., and Saleri, F., 2000, *Numerical Mathematics*. New York: Springer

Ray, W. H., 1972, *J. Macromol. Sci.-Revs. Macromol. Chem.*, **C8**, 1–56

Reklaitis, G. V., 1983, *Introduction to Mass and Energy Balances*. New York: Wiley

Robert, C., 2001, *The Bayesian Choice*, second edition. New York: Springer

Sontag, E. D., 1990, *Mathematical Control Theory : Deterministic Finite Dimensional Systems*. New York: Springer-Verlag

Stakgold, I., 1979, *Green's Functions and Boundary Value Problems*. New York: Wiley

Stoer, J. and Bulirsch, R., 1993, *Introduction to Numerical Analysis*, second edition. New York: Springer

Stokes, R. J. and Evans, D. F., 1997, *Fundamentals of Interfacial Engineering*. New York: Wiley-VCH

Strang, G., 2003, *Introduction to Linear Algebra*, third edition. Wellesley: Wellesley Cambridge Press

Trottenberg, U., Oosterlee, C. W., and Schuller, A., 2000, *Multigrid*. San Diego: Academic Press

Villadsen, J. and Michelsen, M. L., 1978, *Solution of Differential Equation Models by Polynomial Approximation*. Engleword Cliffs: Prentice-Hall

Wilmott, P., 2000, *Quantitative Finance*. Chichester: Wiley

Yasuda, K., Armstrong, R. C., and Cohen, R. E., 1981, *Rheol. Acta*, **20**, 163–178

Index